# EVOLUTIONARY ECOLOGY

*Fifth Edition*

## Eric R. Pianka
*The University of Texas at Austin*

**HarperCollins***CollegePublishers*

Sponsoring Editor: Susan Penney-McLaughlin
Project Editor: Ellen MacElree
Design Supervisor: Mary Archondes
Cover Design: Mary Archondes
Cover Photo: Caribou and Mt. McKinley, McKinley National Park, south central Alaska.
  Copyright © 1994 by Eric Dragesco, Bruce Coleman, Inc., NYC.
Production Administrator: Jeffrey Taub
Compositor: Publication Services, Inc.
Printer and Binder: R. R. Donnelley & Sons Company
Cover Printer: The Lehigh Press

QH
541
.P5
1994

**Evolutionary Ecology**, Fifth Edition

**Library of Congress Cataloging-in-Publication Data**

Pianka, Eric R.
    Evolutionary ecology / Eric R. Pianka. – 5th ed.
        p.   cm.
    Includes bibliographical references and index.
    ISBN 0-06-501225-9
    1. Ecology.   2. Evolution (Biology)   I. Title.
QH541.P5   1994                                          93-26038
574.5'2–dc20                                             CIP

    94 95 96 9 8 7 6 5 4 3 2

# Contents

# Preface

Language forces us to express ourselves in a one-dimensional stream of words. Nature, however, is multidimensional. This is particularly true of ecology because its subject matter includes many complexly interrelated concepts and phenomena involving several different levels of organization. There is no such thing as an "ideal" outline or a perfect sequential order for presentation of the subject matter treated here. In order to obtain an overview of modern ecology, a student needs to assimilate a great many ideas. Ideally, a reader would know everything in this book even before beginning to read it! Perhaps the only solution is to read it twice. To assist readers following different trains of thought than I have chosen to use, various chapters and sections are cross-referenced.

During the nearly two decades since the appearance of the first edition of this book, evolutionary ecology has blossomed and supplanted other kinds of ecology. A journal goes by this name now. This book, which has become a "citation classic," has been translated into Japanese, Polish, Russian, and Spanish. In the present edition, previous material has been improved, updated, and extensively reorganized. Many new topics have been added dealing with such subjects as speciation, metapopulations, self-deceit, experimental ecology, modern comparative methods, null models, landscape ecology, macroecology, biodiversity, genetic engineering, equilibrium economics, and other aspects of applied ecology. I have made a consistent effort throughout the text to include the most up-to-date research and references. I have also tried to make the figures and art in the text as clear and easy to understand as possible.

*Evolutionary Ecology*, Fifth Edition, is my own eclectic blend and distillation of what I consider to be significant facts, ideas, and principles; they represent the residue remaining after considerable sifting and sorting through many other facts and ideas. My approach is abstract and conceptual, and I strive to provide a reasonably crisp overview of a fairly vast subject matter. This edition of *Evolutionary Ecology* represents an image of a part of my mind, a part that should mirror an external reality common to all living systems anywhere in the cosmos.

Science progresses over decades and centuries, with human knowledge continually expanding and improving. Past genius exerts a strong influence on present-day thinking. But most scientific endeavors are quite pedestrian. Thus, the research projects in which most scientists engage themselves are relatively trivial, constituting mere building blocks for major advances. Such "normal" science is, of course, absolutely essential in that it provides the raw material for progress in understanding. Periodically, an extraordinary event occurs that enables a novel breakthrough. Occasionally, this may be just a serendipitous discovery by a more or less "ordinary" scientist (provided, of course, that someone has the wisdom to appreciate the true significance of the discovery and the creativity to develop it). But more often than not, major new directions are charted by rare individuals with incredible intellectual prowess. Population biology has attracted a few of these people in the past, and ecology today stands poised, awaiting another such genius.

I would like to thank the following individuals for their reviews of the manuscript: Carla M. Delucchi, Augustana College; Dr. Thomas C. Emmel, University of Florida; Brent M. Graves, University of South Dakota; Richard Inouye, Idaho State University; Robert K. Neely, Eastern Michigan University; Thomas M. Niesen, San Francisco State University; Robert W. Paul, St. Mary's College of Maryland; Mark Ridley, Emory University.

Eric R. Pianka

## Chapter 1

# Background

*H*umans seem to delight in animal motifs—thus, we have automobiles, airplanes, and athletic teams named after various animals: cougar, jaguar, lynx, mustang, pinto, ram, eagle, falcon, nighthawk, roadrunner, and the list goes on. Zoos are a popular form of entertainment, particularly for children. Yet, at the same time, many people feel threatened by a free-ranging wild creature, even by a tiny mouse or a harmless snake. Indeed, urbanization is now so complete that, aside from cockroaches and songbirds (and perhaps while on vacation), most of us seldom encounter wild animals.

What is the essential difference between a wild animal versus one in a cage? Clearly, a rattlesnake behind glass does not pose nearly as much physical danger to a human observer as does a wild snake. For the study of many kinds of biological phenomena, there is in fact no difference between a caged specimen, so long as it remains alive, and its wild cousin. The constrained one still has intact cells, molecules, physiological processes, and, to some extent, behavior. But the caged animal, removed from its habitat, is out of context—it has been stripped of its natural history and it no longer interfaces with the environment to which it is adapted and in which it evolved.

The biological discipline of ecology deals with the myriad of ways in which organisms (plants, animals, and other heterotrophs such as bacteria and fungi) interact with, influence, and are in turn influenced by their natural surroundings. Wild plants and animals in their natural communities constitute the subject matter of ecologists. To these scientists, caged organisms might as well be dead for they have no ecology. Ecology is fundamentally different from other

1

sorts of biology in that its perspective is directed *upward* and *outward* from the individual organism to its environment. Other kinds of biology focus on organismal and suborganismal processes and thus involve a reductionistic viewpoint. Ecology has a more holistic perspective.

## SCALING IN BIOLOGY

Biology is a vast discipline ranging from molecules to cells to tissues to whole organisms to kin groups to populations to communities (and clades) to entire ecosystems, or even the entire global biosphere. Across this broad range of scale, factors vary by many orders of magnitude (Figure 1.1). Molecular biology can be done quickly in small spaces, but community ecology requires decades and square kilometers. Biogeographic and historic events take place over many millennia. Continental plates have moved across thousands of kilometers over geological time. Until recently, ecologists have been preoccupied with local phenomena and events occurring on relatively short time scales. However, the emerging new subdisciplines of landscape ecology and macroecology (pp. 375–380) offer promising new regional and global perspectives.

As an example of the effects of scale, consider the movements of organisms across landscapes (Brooks, 1988). At the shortest temporal and spatial scale, individual organisms move during their lifetimes and disperse; over an intermediate scale in space and time, immigration and emigration occur between and among populations (see also section on Metapopulations, pp. 199–200);

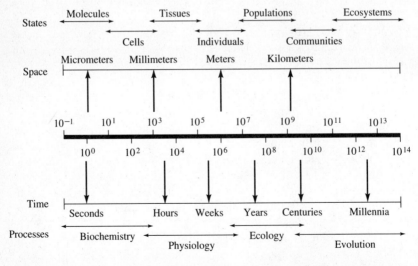

**Figure 1.1** Diagrammatic representation of the time-space scaling of various biological phenomena. Community and ecosystem phenomena occur over longer time spans and more vast areas than suborganismal- and organismal-level processes and entities. [After Anderson (1986) after Osmund et al.]

over a much greater spatial extent and at a much longer time scale, geographical ranges shift in response to vicariant events such as geotectonic movements, leading ultimately to the formation of geographical patterns in species diversity.

## SCIENTIFIC METHODS AND HUMAN KNOWLEDGE

Scientists are motivated by curiosity about their surroundings; they go to great lengths to satisfy their desire to understand natural events and phenomena. All scientists assume that an organized reality exists in nature and that objective principles can be formulated that will adequately reflect this natural order. A fundamental and important way in which biological phenomena can be ordered is by simple and direct enumeration, as in the classification of organisms or biotic communities. Thus, ecologists recognize different ecological systems such as tundra, desert, prairie, savanna, deciduous forest, coniferous forest, and rainforest. Early ecology was primarily descriptive; the originators of the science spent most of their time describing, itemizing, and classifying various ecological elements. This process was absolutely essential before a more process-oriented ecology could be developed.

Founded and firmly based on this older body of descriptive information, modern ecology seeks to develop general theories with predictive powers that can be compared against the real world. Ecologists want to understand and to explain, in general terms, the origin and mechanisms of interactions of organisms with one another and with the nonliving world. To build such general theories of nature, ecologists construct *hypotheses*, hypothetical "models" of reality. All models must make simplifying assumptions—some sacrifice precision for generality, whereas others sacrifice generality for precision. Some models actually sacrifice certain aspects of realism itself! Models have been described as "mere caricatures of nature designed to convey the essence of nature with great economy of detail." No model is "correct" or "true"—any given model merely represents one particular attempt to mimic reality. All models are to some extent incorrect. To be most useful, models are usually designed to generate testable predictions. Most models can therefore be confronted with reality and can be falsified. But not all models are refutable; some conceptual models have proven to be useful in an abstract way even though they do not suggest direct tests. When a model's predictive powers fail, it is either discarded or revised. Models and hypotheses that do not conform adequately to reality are gradually replaced by those that better reflect the real world. The scientific method is thus self-regulating; as time progresses, knowledge expands and is continually refined and improved to reflect external reality better and better. Well-substantiated hypotheses become theories. We are very fortunate indeed to be able to benefit from past genius and research effort; in a few hours of careful reading, you can now learn material that required many lifetimes to acquire.

Due to the multiple meanings of words, verbal models are usually somewhat ambiguous and imprecise and therefore of limited utility. The great complexity of ecological systems necessitates the use of graphical and mathematical models, so much so that ecologists often employ nearly as much mathematics as biology. However, the development of sound ecological principles depends heavily on what might be called "biological intuition," and there is certainly no substitute for a firm foundation in natural history. Models based on erroneous biological assumptions, no matter how elegant and elaborate, can hardly be expected to reflect nature accurately. Hence, a good background for comprehending ecology includes some biology and mathematics as well as a solid basis in general science.

Scientists are often said to be working with *facts*, and human knowledge is supposed to be based on them. But what exactly *is* a "fact"? The dictionary definition is "what has really happened or is the case; truth; reality." Consider, for example, the apparently simple fact that the sun rises each morning. Daily we receive new evidence confirming this fact. We are quite confident that the sun will rise again tomorrow, but there is a remote possibility that someday it may not. Indeed, on a time scale of eternity, we can be certain that someday it will burn itself out and conjecture that eventually it will cease to exist. Appearances to the contrary, even simple facts such as the sunrise are not "clean" of interpretation. Under the now defunct "flat earth–moving sun" hypothesis, the sun's movement across the sky was viewed from the perspective of a fixed non-moving earth at the center of the universe. Indeed, references to sunrise and sunset are based on this interpretation, which is supported by our superficial, commonsense perceptions. But another hypothesis has supplanted the concept of a moving sun; our understanding of cosmic events is greatly enhanced when the sun is viewed as fixed in space at the center of this universe and the earth is interpreted as a rotating globe orbiting around a small star. Perhaps someday another even more powerful model will replace our current working hypothesis "explaining" sunrise.

Observation and experimentation play an important and vital role in science. They are used to test models, to refute inadequate hypotheses, and hence to help formulate improved interpretations of natural phenomena. Certain kinds of natural events cannot be manipulated. Thus, we cannot stop the sun's fusion or earth's rotation to test our current ideas, but each daily observation of sunrise or sunset nevertheless strengthens our confidence in the accepted interpretation of celestial events. Similarly, in many ecological situations (particularly those involving the evolution of adaptations), direct experimentation is often impractical or even impossible, if not illegal or immoral. Some sorts of ecological phenomena do lend themselves to manipulation, but like astronomers, evolutionary ecologists rely heavily on a careful comparative approach (sometimes referred to as a "natural" experiment).

A major goal of science is to understand causality; when we observe that event B always follows event A, we sometimes infer that A causes B. But some other unknown event X could be causing both to occur with a time lag in B. The way scientists test for such spurious correlations is to find a way to cause event A to occur, holding as much else as possible constant in a controlled experiment; then, if B still follows A when X does not occur, we are more confident that A indeed causes B. In the physical sciences and in mechanistic kinds of biology such as molecular biology, causal connections are usually fairly simple and straightforward, allowing construction of mutually exclusive alternative hypotheses. Scientific investigation can then proceed by what has been termed "strong inference" (Platt, 1964), and researchers can choose among logical alternatives. But in population ecology, mutually exclusive alternatives often cannot be formulated. Multiple causality is common and ecologists can seldom eliminate a hypothesis as inadequate under all situations.

## DOMAIN OF ECOLOGY: DEFINITIONS AND GROUNDWORK

*Ecology* has been variously defined as "scientific natural history," "quantitative natural history," "the study of the structure and function of nature," "the sociology and economics of animals," "bionomics," "the study of the distribution and abundance of organisms," and "the study of the interrelationships between organisms and their environments." The last of these definitions is probably the most useful, with "environment" being defined as *the sum total of all physical and biological factors impinging on a particular organismic unit*. For "organismic unit," one can substitute either "individual," "family group," "population," "species," or "community." Thus, we may speak of the environment of an individual or the environment of a population, but to be precise, a particular organismic unit should be understood or specified. The environment of an individual contains fewer elements than the environment of a population, which in turn is a subset of the environment of the species or community.

To avoid the apparent circularity in the preceding definition, ecology might be better defined as *the study of the relationships between organisms and the totality of the physical and biological factors affecting them or influenced by them*. Ecologists therefore begin with the living organism and seek to understand how an organism affects its surroundings and how those surroundings in turn affect the organism.

Environment includes everything from sunlight and rain to soils and other organisms. An organism's environment consists not only of other plants and animals encountered directly (such as foods, trees used for nesting sites, potential predators, and possible competitors), but also of purely physical processes and inorganic substances (such as daily temperature fluctuations and oxygen and carbon dioxide concentrations). Of course, the latter may be affected by other organisms, which then indirectly become a part of the environment of the first organism. Indeed, any remote connection or interaction between two organismic units dictates that each is a part of the other's environment. Pristine natural environments are of particular interest and importance in evolutionary ecology because they constitute the environments to which any particular organism has become adapted over evolutionary time.

Because there are direct or indirect interactions between almost all the organisms in a given area, the biotic component of the environment of most organisms is extremely complex. Coupling this great complexity with a multi-faceted and dynamic physical environment makes ecology an exceedingly vast subject. No other discipline seeks to explain such a wide variety of phenomena at so many different levels. As a consequence, ecology takes in aspects of numerous other fields, including physics, chemistry, mathematics, computer science, geography, climatology, geology, oceanography, economics, sociology, psychology, and anthropology. Ecology is properly classified as a branch of biology, and its students attempt to interweave and correlate the subdisciplines of biology, such as evolution, genetics, systematics, morphology, physiology, ethology (behavior), as well as various taxonomic subdivisions of biology like algology, entomology, ichthyology, herpetology, mammalogy, and ornithology. Sometimes "plant ecology" is distinguished from "animal ecology." As basic as this distinction may be, however, it is most unfortunate; plants and animals inevitably constitute parts of one another's environments, and their ecologies simply cannot be disentangled. Interactions between plants and animals have recently been the subject of considerable thought and field study.

Obviously, no one can master all of such an enormous field, and as a result, there are many different kinds of ecologists with a wide variety of expertise and perspectives on the subject matter of the science. The breadth and difficulty of ecology, combined with its youth and great relevance to urgent human problems, make it a fascinating and exciting field. Major basic discoveries of real importance remain to be made, and the potential for growth and refinement is immense. Young sciences, and especially complex biological ones like ecology, have sometimes been characterized as "soft" sciences, because they are not as precise as the older and better established "hard" sciences such as chemistry and physics. Of course, by its very nature, the subject matter of biology is innately *much more diverse* and complex, not to mention interesting, than that of physics and chemistry. As every science matures, it becomes more and more abstract and its hypotheses are refined, tested, and improved until they eventually attain the status of theories or "laws," as in the familiar laws of chemistry and physics. Ecology at present has few firm laws but many hypotheses, and much work and testing of these hypotheses remains. The single concept closest to deserving the status of "law" in ecology, and one that is shared with all of biology, is *natural selection* (see pp. 9–11; see also pp. 133–136).

Ecology encompasses many complexly interrelated concepts and involves phenomena at several different levels of organization. Indeed, the subject matter of the discipline is inherently multidimensional. But language forces us to express ourselves in a one-dimensional stream of words. How, then, will we unravel this unfinished tapestry that has no beginning and no end? Components of ecological systems can be conveniently considered in order of generally increasing complexity, proceeding from the inorganic to the organic world and then from individuals to populations to communities (Figure 1.2). In this book, the organismic world is treated in much greater detail than the nonorganismic world.

The climate, soils, bacteria, fungi, plants, and animals at any particular place together constitute an *ecosystem*. Ecosystems have both abiotic (nonliv-

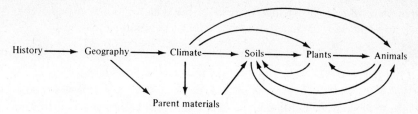

Figure 1.2 Diagrammatic representation of the subject matter of ecology showing the natural sequence proceeding from the inorganic to the organic world. Many other arrows and feedback loops could be added, but those depicted are of major importance. Chapter 4 deals with history; Chapters 3 and 5 treat geography, climate, soils, and plants. Chapter 6 examines how organisms cope with their physical environments. Remaining chapters look at the interactions between and among plants and animals, especially at the population and community levels.

ing) and biotic (living) components. All of the biotic components of an ecosystem, or all the organisms living in it, taken together, comprise an ecological *community*. The abiotic components can be separated into inorganic and organic, whereas the biotic components are usually classified as producers, consumers, and decomposers. Producers, sometimes called autotrophs, are the green plants that trap solar energy and incorporate it into energy-rich chemicals such as glucose. Consumers, or *heterotrophs,* are all the organisms (bacteria, fungi, and animals) that either eat the plants or one another; all heterotrophs thus depend, directly or indirectly, on plants for energy. Several levels of consumers are recognized (primary, secondary, and tertiary), depending on whether they eat plants directly or instead consume other herbivorous or carnivorous animals. Decomposers, which are also heterotrophs, are often bacteria and fungi; they break down plant and animal material into simpler components and thereby return nutrients to the nutrient pool and autotrophs. Decomposers are therefore essential in recycling matter within an ecosystem.

As previously indicated, plants and animals in ecosystems can be considered at several different levels: individuals, family groups, populations, species, and communities (subsequent chapters deal with each organismic level of organization, but most emphasis is given to populations). None of these levels of organization can be adequately understood in isolation because each exerts strong influences upon the others. Every individual is simultaneously a member of a population, a species, and a community, and it must therefore be adapted to cope with each and must be considered in that context. An individual's fitness—its ability to perpetuate itself as measured by its reproductive success—is determined not only by its status within its own population but also by the various interspecific associations of its species and especially by the particular community in which it exists. Similarly, every community is composed of many populations and numerous individuals, which determine many, but by no means all, of its properties. At each level of organization, important new properties emerge that are not properties of the preceding level. Thus, individuals have a fixed genetic makeup and live or die, whereas populations have gene frequencies, birth rates, and death rates. All these population parameters (and others) can change in time as the composition of a population changes in response to a changing environment (evolution).

## THE URGENCY OF BASIC ECOLOGICAL RESEARCH

In one sense, ecology is doubtlessly the oldest and most basic science. Primitive humans simply *had* to have been astute natural historians to survive. As a more rigorous scientific discipline, however, ecology promises to be the shortest-lived science of all time. The word* itself seems to have been first used in Germany in the late 1860s by E. Haeckel, but it did not come into common use until 1895, when it was used in a book title. The first scientific society, the British Ecological Society, was founded in 1913. The earliest textbooks were published in 1927 (Elton) and 1929 (Weaver and Clements). Ecology is thus barely a century old and would seem doomed not to celebrate much of a bicentennial, for by then most natural ecological systems will surely be long gone. Although this realization is saddening, this "doomsday ecologist" nevertheless urges everyone to learn as much ecology as he or she can if for no reason other than simply because very soon we will need all of it that we can possibly get.

Ecology and environment are words frequently encountered in the news and popular media, almost invariably bandied about in conjunction with humans and their environment. As often as not, the terms are misused, especially by politicians and other advertisers. Many people now use "ecology" to refer primarily to applied and human ecology. "Ecologist" has been equated with "rabid environmentalist" (as a result, real ecologists now refer to themselves as population biologists!). The basic science of ecology is *not* synonymous with a study of the effects of people on their own surroundings and on other organisms, but in fact represents a much broader and more fundamental class of subject matter. Some problems facing humans today illustrate what can happen when ecological systems are not used wisely in accordance with sound ecological principles; as such, the content of this book is very pertinent to human ecology. However, throughout the book, emphasis is given to principles of *basic* ecology, particularly as these principles apply to and can be interpreted in terms of the theory of natural selection. Major concepts and principles are stressed more than detail, but references are given at the end of each chapter for those desiring to delve more deeply into particular subjects.

Most people consider the study of biology, particularly ecology, to be a luxury that they can do without. Even medical schools no longer require that premedical students major in biology. The study of basic biology is not a luxury at all, but rather an absolute necessity for living creatures. Despite our anthropocentric attitudes, other life forms are not irrelevant to our own existence. For

---

*Oekologie comes from the Greek *oikos* and *logy*, which translate roughly as "the study of the household."

example, an understanding of basic parasitology is needed to control epidemics in human populations. Similarly, a knowledge of basic principles of community organization and ecosystem function is essential for wise exploitation of both natural and agricultural ecological systems. Beyond such human-oriented arguments, one can argue that other species have a right to exist, too, as proven products of natural selection that have adapted to natural environments over millennia. With human populations burgeoning and pressures on space and other limited resources intensifying, we will need all the biological knowledge that we can get. Ecological understanding will prove to be particularly vital.

There is a great urgency to basic ecological research simply because the worldwide press of humanity is rapidly driving other species to extinction and destroying the very systems that ecologists seek to understand. No natural community remains pristine. Unfortunately, many will disappear without even being adequately *described*, let alone remotely understood. As existing species go extinct and even entire ecosystems disappear, we lose forever the very *opportunity* to study them. Knowledge of their evolutionary history and adaptations vanishes with them: we are thus losing access to biological information itself. Indeed, "destroying species is like tearing pages out of an unread book, written in a language humans hardly know how to read" (Rolston, 1985). Just as ecologists are finally beginning to learn to read this "unread" and rapidly disappearing book of life, they are encountering governmental and public hostility and having a difficult time attracting support. This is simply pitiful. And time is quickly running out.

## NATURAL SELECTION

Darwin's theory of natural selection is a truly fundamental unifying theory of life. A thorough appreciation of the process of selection is essential background to understanding evolutionary ecology. Natural selection comes as close to being a "fact" as anything in biology. Although there is no such thing as "proof" in science (except in mathematics, where all postulates are taken as given), over the past century an enormous body of data has been amassed in support of the theory of natural selection.

**Charles Darwin**

Although natural selection is not a difficult concept, it is frequently misunderstood. Natural selection is not synonymous with evolution. Evolution refers to temporal changes of any kind, whereas natural selection specifies one particular way in which these changes are brought about. Natural selection is the most important agent of evolutionary change simply because it results in conformity between organisms and their environments, or *adaptations*. Other possible mechanisms of evolution besides natural selection include the inheritance of acquired characteristics, gene flow, meiotic drive, and genetic drift. Another frequent misconception is that natural selection occurs mainly through differences between organisms in death rates, or *differential mortality*.

Selection may proceed in a much more subtle and inconspicuous way. Whenever one organism leaves more successful offspring than another, in time its genes will come to dominate the population gene pool. Eventually, the genotype leaving fewer offspring must become extinct in a stable population, unless there are concomitant changes conferring an advantage on it as it becomes rarer. Thus, ultimately, *natural selection operates only by **differential reproductive success***. Differential mortality can be selective but *only* to the degree that it creates differences between individuals in the number of reproductive progeny they produce.

Hence, phrases such as "the struggle for existence" and "survival of the fittest" have had a rather unfortunate consequence. They tend to make people think in terms of a dog-eat-dog world and to consider things such as predation and fighting over food as the prevalent means of selection. All too often, natural selection is couched in terms of differential death rates, with the strongest and fastest individuals considered to have a selective advantage over weaker and slower individuals. But if this were the case, every species would continually gain in strength and speed. Because this is not happening, selection against increased strength and speed (counterselection) must be occurring and must limit the process.

Animals can sometimes be too aggressive for their own good; an extremely aggressive individual may spend so much time and energy chasing other animals that it spends less than average time and energy on mating and reproduction and, as a result, leaves fewer offspring than average. Likewise, an individual can be too submissive and spend too much time and energy running away from other animals.

Differences in survivorship leading to differential mortality can, but need not always, lead to natural selection. A cautious tomcat that seldom crosses noisy streets may live to a ripe old age without leaving as many descendants as another less staid tom killed on a busy road at a much younger age. Unless living longer allows or results in higher reproductive success, long life cannot be favored by natural selection. Similarly, although we might wish otherwise, there is no necessary selective premium on beauty, brains, or brawn, unless such traits are in fact translated into more offspring than average.

Some of the best documented examples of natural selection in action concern pesticide resistance in certain insects and drug resistance in many bacteria. Other examples include industrial melanism in moths and the evolution of myxoma virus and *Oryctolagus* rabbits in Australia (for discussion, see p. 327).

Some words of warning are now appropriate. Overenthusiastic proponents of natural selection have been known to use it to "explain" observed biological phenomena in a somewhat after-the-fact manner. Thus, one might say that an animal "does what it does because that particular behavior increases its fitness." Those who succumb to naively explaining away everything as a result of selection have been aptly labeled "adaptationists." Used in this way, natural selection can be misleading; it is so pervasive and powerful that nearly any observable phenomenon can be interpreted as a result of selection, even though some must not be. There is a real danger of circularity in such arguments. One should always consider alternative explanations for biological phenomena. Historical factors set various sorts of design constraints on organisms that limit directions in which they can evolve. Perfect adaptation is also prevented by continually changing counterevolutionary pressures from other species. Conflicting demands on allocation of an organism's resources may also preclude ideal solutions to environmental exigencies. These kinds of subtleties are considered later.

## LEVELS OF APPROACH IN BIOLOGY

Why do migratory birds fly south in the autumn? A physiologist might tell us that decreasing day length (photoperiod) stimulates hormonal changes that in turn alter bird behavior with an increase in restlessness. Eventually, this "wanderlust" gets the upper hand and the birds head south. In contrast, an evolutionist would most likely explain that, by virtue of reduced winter mortality, those birds that flew south lived longer and therefore left more offspring than their nonmigratory relatives. Over a long period of time, natural selection resulted in intricate patterns of migratory behavior, including the evolution of celestial navigation, by means of differential reproductive success.

The physiologist's answer concerns the *mechanism* by which avian migratory behavior is influenced by *immediate* environmental factors, whereas the evolutionist's response is couched in terms of what might be called the strategy by which individual birds have left the most offspring in response to long-term consistent patterns of environmental change (i.e., high winter mortality). The difference between them is in outlook, between thinking in an "ecological" time scale (the time scale over which individuals live and die) or in an "evolutionary" time scale (geological time). At the physiologist's level of approach to science, the first answer is complete, as is the evolutionist's answer at her or his own level. Mayr (1961) has termed these the "how?" and "why?" approaches to biology. They have also been called the "functional" and "evolutionary" explanations and the "proximate" and "ultimate" factors influencing an event (Baker, 1938). Neither is more correct; a really thorough answer to any question must include both, although often only the first can be examined by direct experiment. Nor are those two ways of looking at biological phenomena mutually exclusive; ecological events can always be profitably considered within an evolutionary framework, and vice versa.

The evolutionary approach to biological questions is relatively new and has resulted in a major revolution in biology during the last half century. Before then, most biologists merely accepted as immutable a broad range of phenomena, such as the fact that sex ratios are often near equality (50:50), without considering why such facts might be so or how they could have evolved. Although we may not fully understand the causes and consequences of many populational phenomena, we can be confident that most have some sort of evolutionary explanation. This is true of a broad spectrum of observations and facts, such as: (1) some genes are dominant, others are recessive; (2) some organisms live longer than others; (3) some organisms produce many more offspring than others; (4) some organisms are common, others are rare; (5) some organisms are generalists, others are specialists; (6) some species are promiscuous, some polygamous, and some monogamous; (7) some species migrate, others do not; and (8) more species coexist in some areas than in others. All these variables are subject to natural selection.

Population biologists are now thinking in an evolutionary time scale, and we have made substantial progress toward a theoretical understanding of why many observed differences occur among organisms. One of the first to recognize the power of rigorous application of the genetical theory of natural selection to population biology (sometimes called "selection thinking") was R. A. Fisher, and his book *The Genetical Theory of Natural Selection* has become a classic. Numerous other biologists have expanded, experimented with, and built on Fisher's groundwork. For example, Lack (1954, 1966, 1968) showed that reproductive rates are subject to natural selection. Others have worked on the evolution of a variety of phenomena, including genetic dominance, mating systems, sex ratio, parental investment, old age, life-history patterns, reproductive tactics, foraging tactics, predator-prey interactions, community structure, and so on. Many of these topics, the subject matter of evolutionary ecology, are taken up in subsequent chapters.

## DEBATES AND PROGRESS IN ECOLOGY

Ecologists are a contentious lot, easily polarized into opposing factions. Early on, there was a separation into the Shelford-Clements community school versus the Gleason individualistic school. During the fifties and sixties, Andrewartha and Birch argued for the importance of density-independent processes in population regulation, whereas Lack and others emphasized the overriding importance of density dependence. In the sixties to seventies, when MacArthur held center stage, resource-partitioning studies blossomed. Group selection fell out of vogue in the sixties, but returned to legitimacy during the seventies.

During the seventies and eighties, an acrimonious debate raged over whether or not competition is important in organizing ecological communities. Anyone who invoked competition was labeled dogmatic. The niche concept was seen by some to be useless, or worse. A "predation" school developed which boldly asserted that predation was more important than competition. Simberloff and Strong and their colleagues at Florida State University ("the Tallahassee turks") argued for the "logical primacy" of null models and urged ecologists to adopt a proper Popperian approach to falsifying testable hypotheses. Experimental ecologists express disdain for descriptive studies such as comparative ecology. Statistically inclined ecologists abhor poorly designed experiments, lack of adequate controls, and pseudoreplication. Some argue about whether or not nature ever reaches an equilibrium, and nonequilibrium perspectives are popular in certain circles.

Theoretical ecologists feel that empiricists do not pay enough attention to their precious theories, whereas empirical ecologists do not find dense mathematical abstract treatments of very much interest. Many population ecologists find community ecology to be lacking in rigor. Healthy debate leads to progress when it forces ecologists to think more rigorously, but deadlock impedes progress. Jargon has sometimes become an end in itself. Vogue has an invisible sway and the pendulum swings back and forth as each generation reacts to the last. There is a tendency to stop citing (and quit reading) someone once he or she dies, in favor of stroking living egos who pass judgment on grant applications and publications. As a result, the older literature is seriously undervalued, and well-meaning scientists are doomed to continually reinvent the wheel.

## SELECTED REFERENCES

### Scaling in Biology

Anderson (1986); Brooks (1988); Brown and Maurer (1987, 1989); Clark (1985); Holling (1992); Kareiva (1987, 1990); Levin (1992); Osmund et al. (1980); Pimm (1991); Ricklefs (1987, 1989); Wiens and Milne (1989).

### Scientific Methods and Human Knowledge

Bender et al. (1984); Colwell and Fuentes (1975); Diamond and Case (1986); Fisher (1935); Hilborn and Stearns (1982); Horn (1979); Hutchinson (1949); Levins (1966); Loehle (1987); Platt (1964); Popper (1959); Quinn and Dunham (1983); Schoener (1983).

### Domain of Ecology: Definitions and Groundwork

Allee (1951); Allee et al. (1949); Andrewartha (1961); Andrewartha and Birch (1954); Billings (1964); Clarke (1954); Collier et al. (1973); Elton (1927, 1958); Greig-Smith (1964); Harper (1967); Hazen (1964, 1970); Kendeigh (1961); Kershaw (1964); Knight (1965); Krebs (1972); MacArthur (1972); MacArthur and Connell (1966); Macfadyen (1963); May (1973, 1976a); Odum (1959, 1963, 1971); Oosting (1958); Pielou (1969); Platt (1964); Ricklefs (1973); Shelford (1963); Smith (1966); Watt (1973); Whittaker (1970); Wilson and Bossert (1971).

### The Urgency of Basic Ecological Research

Ehrlich and Ehrlich (1981); Elton (1927); Lubchenco et al. (1991); Rolston (1985); Weaver and Clements (1929); Wilson (1985).

## Natural Selection

Birch and Ehrlich (1967);  Darwin (1859);  Dobzhansky (1970);  Ehrlich and Holm (1963);  Emlen (1973);  Fisher (1930, 1958a, 1958b);  Ford (1964);  Gould and Lewontin (1979);  Haldane (1932);  Kettlewell (1956, 1958);  Lande and Arnold (1983);  Lewontin (1974);  MacArthur (1962);  Maynard Smith (1958);  Mayr (1959); Mettler and Gregg (1969);  Orians (1962);  Pianka (1976b);  Salthe (1972);  Williams (1966a);  Wilson and Bossert (1971);  Wright (1931).

## Levels of Approach in Biology

Baker (1938);  Fisher (1930);  Lack (1954, 1966, 1968);  MacArthur (1959, 1961); MacArthur and Connell (1966);  Mayr (1961);  Orians (1962);  Pianka (1986b).

## Debates and Progress in Ecology

Connor and Simberloff (1979);  Diamond and Case (1986);  Ehrlich and Birch (1967); Karieva (1989);  Lack (1954);  Roughgarden (1983);  Salt (1983);  Schoener (1982, 1983);  Strong et al. (1984);  Wynne-Edwards (1962).

**Chapter 2**

# Biogeography and Historical Constraints

## SELF-REPLICATING MOLECULAR ASSEMBLAGES

Life began with the first self-replicating molecular assemblage; moreover, natural selection begins to operate as soon as any complex of molecules begins to make replicates of itself. No copying device is perfect, and some variants of the molecular assemblage produced are bound to be better than others in their abilities to survive and replicate themselves under particular environmental conditions. As resources become depleted, competition can occur among various self-replicating units. Furthermore, given enough time, some inferior variants presumably become extinct. Thus, each molecular unit maximizes its own numbers at the expense of other such units. Even as the originally simple self-replicating units become more and more elaborate and eventually attain the complex form of present-day organisms, the same principles of natural selection must remain in effect throughout. Thus, we can make certain statements about life that are entirely independent of the precise mechanism of replication. For example, self-replicating molecular assemblages elsewhere in the cosmos doubtlessly will not obey the laws of Mendelian genetics, yet their basic attributes as living material will not be drastically altered. *Natural selection and competition are inevitable outgrowths of heritable reproduction in a finite environment.* Hence, natural selection exists independently of life on earth. Many of the principles developed here will persist as long as assemblages of molecules replicate themselves anywhere in the cosmos.

Once a self-replicating entity arises, qualitatively new phenomena exist that are not present in an inanimate world. To reproduce, living organisms (or replicating molecular assemblages) must actively gather other materials and energy; that is, they must have some sort of acquisition techniques. Direct and indirect disputes over resources place those units best able to acquire materials and energy (and most successful at transforming them into offspring) at a selective advantage over other such units that are inferior at these processes. Thus, natural selection has the same effect as an efficiency expert, optimizing the allocation of available resources among conflicting demands imposed by foraging, growth, maintenance, and reproduction.

Obviously, the ultimate end point of these processes would be for the one best organismic unit to take over all matter and energy and to exclude all others. This has not occurred for a variety of reasons, as discussed earlier, but especially because of the great variability of the earth's surface, both in time and space. (Humans are making a good stab at it, however!)

## THE GEOLOGICAL PAST

Climatic changes over geological time, called *paleoclimatology*, are of considerable ecological interest because organisms have had to evolve along with such changes. A really thorough ecological study must include consideration of the past history of the area under study. The earth has changed in innumerable ways during the geological past: the planet's orbital motion undergoes several complex celestial cycles measured on time scales of many thousands of years; its poles have shifted and wandered; periods of orogeny (mountain building) by tectonic upheaval of the earth's crust have waxed and waned at different places and times; the continents have "drifted" and moved on its mantle; and the planet itself has alternately warmed and cooled, resulting in periods of extensive glaciation. Sea levels dropped during glacial periods as water accumulated on land as snow and ice (such sea level changes controlled by glacier alterations are called *eustatic*—Figure 2.1).

Although it is difficult to trace past history, a variety of techniques, some of them quite ingenious, have been developed that allow us to deduce many of the changes that the earth has undergone. One way to look into the past is to examine the fossil record. Lake sediments are an ideal source of layered fossils and have often been used to follow the history of an area. Fossilized pollens in a lake's sediments are relatively easily identified; pollens of plants adapted to particular types of climates can be used as indicators of past climates, as well as the types and composition of forests that prevailed near the lake at different times (see Figure 2.2). Of course, such palynological analyses are fraught with difficulties due to both variations between taxa in rates of pollen production

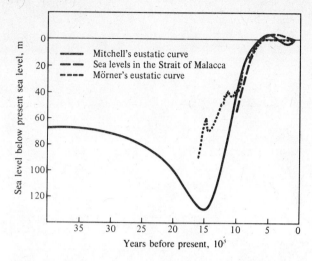

**Figure 2.1** Estimates of sea level changes over the past 35,000 years from three different sources. [After Williamson (1981).]

and differential transport and deposition. Moreover, many deposits may contain mixtures of pollens from several different communities.

A technique known as *carbon dating* allows estimation of the age of fossil plant remains, including pollens and charcoal. Solar radiation converts some atmospheric nitrogen into a radioactive isotope of carbon called *carbon 14* ($^{14}$C). This $^{14}$C is oxidized to carbon dioxide and is taken up by plants in photosynthesis in proportion to its abundance in the air around the plant. All radioactive isotopes emit neutrons and electrons, eventually decaying into nonradioactive isotopes. Half of a quantity of $^{14}$C becomes nonradioactive carbon 12 ($^{12}$C) each 5600 years (this is the half-life of $^{14}$C). When a plant dies, it contains a certain maximal amount of $^{14}$C. Comparison of the relative amounts of $^{14}$C and $^{12}$C in modern-day and fossil plants allows estimation of the age of a fossil. Thus, a fossil plant with half the $^{14}$C content of a modern plant is about 5600 years old, one with one-quarter as much $^{14}$C is 11,200 years old, and so on. The carbon dating method has been checked against ancient Egyptian relics of known age made from plant materials; it accurately estimates their ages, confirming that the rate of production of $^{14}$C and the proportion of $^{14}$C to $^{12}$C have not changed much over the last 5000 years. The technique allows the assignment of fairly accurate ages to all sorts of recently fossilized plant materials. Other methods of radiometric dating, such as uranium-lead, uranium-thorium, and potassium-argon dating, allow older materials to be dated.

A similar technique makes use of the fact that the uptake of two oxygen isotopes, $^{16}$O and $^{18}$O, into carbonates is temperature dependent. Thus, the proportion of these two isotopes in a fossil seashell presumably reflects the temperature of an ancient ocean in which that particular mollusk lived.

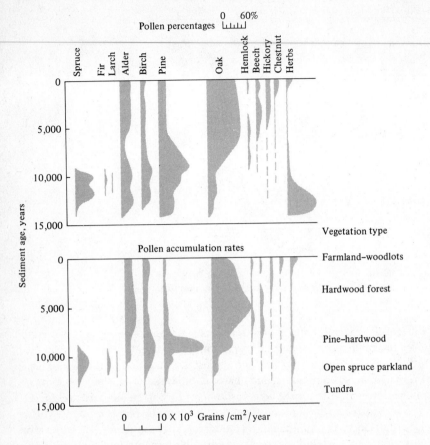

**Figure 2.2** Fossil pollen profiles from dated layers of lake sediments in northeastern United States for the period following the last ice age. Upper plot shows number of pollen grains of each species group as a percentage of the total sample. Lower plot gives estimated rates of deposition of each type of pollen and, at the right, the type of vegetation that probably prevailed in the area. [Adapted from Odum (1971) after M. Davis.]

The geological time scale is summarized in Table 2.1. Several major episodes of extinction stand out in the fossil record (Figure 2.3). These events are so striking that they are used to mark the boundaries between geological periods: (1) 70 percent of marine life went extinct at the Paleozoic-Mesozoic boundary, (2) the last of the dinosaurs went extinct at the Mesozoic-Tertiary transition, and (3) over many different parts of the planet, numerous species of large mammals* died out rather suddenly and dramatically near the end of the Pleistocene ice ages. Long geological periods without major extinctions, followed by abrupt periods of massive extinction, are known as "punctuated

---

*These megafauna include antelope, buffalo, camels, wild oxen, wild pigs, rhinoceros, tapirs, giant sloths, cave bears, as well as mammoths, mastodons, and saber-toothed cats.

TABLE 2.1   **The Geological Time Scale**

| Eras | Periods | Epochs | Years in millions since beginning of period or epoch |
|------|---------|--------|----------------------|
| Cenozoic | Quaternary | Recent | 0.1 |
| | | Pleistocene | 1.6 |
| | Tertiary | Pliocene | 5 |
| | | Miocene | 22 |
| | | Oliogocene | 36 |
| | | Eocene | 55 |
| | | Paleocene | 65 |
| Mesozoic | Cretaceous | | 144 |
| | Jurassic | | 192 |
| | Triassic | | 245 |
| Paleozoic | Permian | | 290 |
| | Carboniferous | | 360 |
| | Devonian | | 408 |
| | Silurian | | 435 |
| | Ordovician | | 485 |
| | Cambrian | | 570 |
| Precambrian | | | 4600 |

equilibria." These widespread extinctions markedly altered almost all ecological communities. Moreover, their synchrony demands attention, since it strongly suggests general underlying causal explanations. Perhaps the most intriguing of those suggested for the extinction of the dinosaurs is the asteroid impact hypothesis (Alvarez et al., 1980), which asserts that extraterrestrial dust from a large comet drastically decreased incident solar radiation and hence primary productivity (this would also have made the planet colder). In support of this hypothesis, a thin fossil layer of iridium, a rare element on the earth but one that is more abundant in meteorites, has been found at widespread localities around the planet dating from this period. Intriguing speculation has also been offered for the Pleistocene mammal extinctions (Martin and Wright, 1967), including the possibility of "overkill" by prehistoric humans (Martin, 1967), although climatological events must certainly have played a major role as well (Guilday, 1967; Graham, 1986). Pleistocene assemblages of small mammals seem to have been more diverse than and fundamentally different from modern ones (Graham, 1986).

South America was virtually completely isolated for most of the past 70 million years—in the absence of many major groups of placentals, a rich fauna of marsupials, including a saber-toothed catlike predator, evolved over this long time period. Edentates and protoungulates were present, and a rodent, a monkey, and a small racoonlike carnivore did eventually reach the isolated

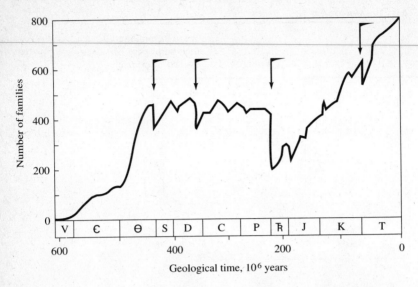

**Figure 2.3** Marine extinctions in the fossil record. V = precambrian, C = cambrian, Θ = ordivician, S = silurian, D = devonian, C = carboniferous, P = permian, R = triassic, J = jurrasic, K = cretaceous, T = tertiary (see Table 2.1). [After Erwin et al. 1987).]

continent. When the Isthmus of Panama was formed about 3 million years ago, numerous different taxa of land and freshwater organisms were exchanged between North America and South America (this process has been termed the "great American biotic interchange"). Although Central America has since come to be dominated by South American elements, the impact of the North American fauna on South America was much more profound than the effect of South American biota on that of North America. Many South American marsupial species went extinct, although a few ancient lineages persisted. South American mammals that have successfully invaded North America include the armadillo, opossum, and porcupine.

Reasons for many of the earth's past changes, such as polar movements and the alternate warming and cooling of the planet, are little known and may well involve solar changes. Piecing together all this diverse and sometimes conflicting evidence on the earth's past history is a most difficult and challenging task and one that occupies many fine minds.

## CLASSICAL BIOGEOGRAPHY

A major goal of ecology is to understand various factors influencing the present distribution and abundance of animals and plants (Andrewartha and Birch, 1954; Krebs, 1972; MacArthur, 1972). Factors affecting abundances and microgeographic distributions (including habitat selection) will be considered later; here, we examine more gross geographic distributions—the spatial distributions of plants and animals over large geographic areas such as major land-masses (continents and islands). The study of the geographical distributions of plants and animals, respectively, are termed *phytogeography* and *zoogeography*. Biogeography encompasses the geography of *all* organisms and involves a search for patterns in the distributions of plants and animals and an attempt to explain how such patterns arose during the geological past. In addition to classifying present distributions, biogeographers seek to interpret and to understand past movements of organisms. Ecology and biogeography are closely related and overlapping disciplines and have profoundly affected one another.

When early naturalists traveled to different parts of the world, they quickly discovered distinctly different assemblages of species. As data were gathered on these patterns, six major biogeographic "realms" or regions were recognized, three of which correspond roughly to the continents of Australia (Australian), North America north of the Mexican escarpment (Nearctic), and South America south of the Mexican escarpment (Neotropical). (The Neotropical region also includes the Antilles.) Africa south of the Sahara is known as the Ethiopian region. Eurasia is divided into two regions, the Palearctic north of the Himalayas (which includes Africa north of the Sahara Desert) and the Oriental south of the Himalayas (India, southern China, Indochina, the Philippines, and Borneo, Java, Sumatra, and other islands of Indonesia east to, and including, the Celebes). Each of the six biogeographic regions (Figure 2.4) is separated from the others by a major barrier to the dispersal of plants and animals, such as a narrow isthmus, high mountains, a desert, an ocean, or an oceanic strait. There is generally a high degree of floral and faunal consistency within regions and a marked shift in higher taxa such as genera and families in going from one region to another. Although biogeographers familiar with different plant and animal groups often disagree on the exact boundaries between regions (Figure 2.5), there is broad agreement on the usefulness of recognizing these six major regions.

High species diversities in the tropics, among other things, have led to the notion that speciation rates in these areas must be extremely high and that such regions often constitute "source areas" for production of new species, many of which then migrate into less hospitable areas, such as the temperate zones. Thus, Darlington (1957, 1959) proposed the "area-climate hypothesis," which states that most dominant animal species have arisen in geographically extensive and climatically favorable areas; he considers the Old World Tropics, which includes the tropical portions of the Ethiopian and Oriental regions, to be the major source area for most vertebrate groups and argues that such dominant forms have migrated centrifugally to other smaller and less favorable areas, including Europe, North and South America, and Australia.

Classical biogeography has produced several so-called biogeographic rules based on recurring patterns of adaptation of organisms. Thus, homeotherms living in cold climates are often larger than those from warmer regions. Such a trend or *cline* can even be demonstrated within some wide-ranging species. This tendency, termed *Bergmann's rule* after its discoverer, has a probable causal basis in that large animals have less surface per unit of body volume

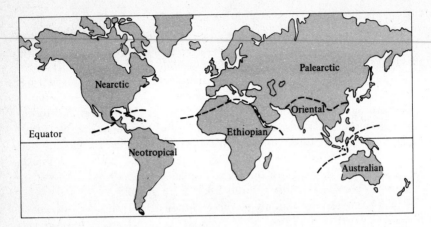

Figure 2.4 The six major biogeographic regions of the world.

than small ones (see pp. 97–99), resulting in more efficient retention of body heat. Many other biogeographic rules have also been proposed, all of which are basically descriptive. *Allen's rule* states that the appendages and/or extremities of homeotherms are either longer or have a larger area in warmer climates; a jackrabbit, for example, has much longer and larger ears than an arctic hare. The presumed functional significance is that large appendages, having a larger relative surface area, are better at heat dissipation than smaller ones. Another

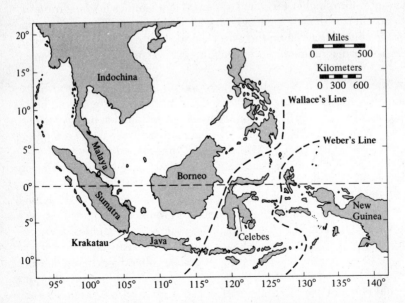

Figure 2.5 Wallace's and Weber's "lines" in southeast Asia, which separate the Oriental from the Australian regions. The position of the volcanic island of Krakatau between Sumatra and Java is indicated at the lower left.

rule (*Gloger's rule*) asserts that animals from hot, dry areas tend to be paler than those from colder, wetter regions. Still another biogeographic rule is that fish from cold waters often have more vertebrae than those from warmer waters. The adaptive significance of many of these biogeographic trends remains obscure, although such geographically variable phenotypic traits are frequently developmentally flexible and respond more or less directly to temperature.

## CONTINENTAL DRIFT

Much of this classical biogeography assumed some permanence in the locations of continents. As a result, interpretations of faunal similarities between them often relied on hypothetical mechanisms of transport from one continent to another, such as "rafting" of organisms across water gaps. Such long-distance dispersal events are exceedingly improbable, although they must occur from time to time (cattle egrets, for example, apparently made a successful trans-Atlantic crossing from Africa to South America without human assistance late in the eighteenth century).

Massive movements of the continents themselves were first suggested by Taylor (1910), but his bold hypothesis was not widely accepted.* Wegener is usually credited with the idea of continental drift. Recent advances in geology indicate that the continents were once joined in a large southern landmass (Pangaea) and have "drifted" apart (Figure 2.6), with a gradual breakup that began in the early Mesozoic era (about 200 million years ago). Geological evidence that the continents have drifted (and are now drifting) is accumulating rapidly (Hallam, 1973; Marvin, 1973; J. T. Wilson, 1973; Skinner, 1986).

Certain types of rocks, particularly basalts, retain a magnetic "memory" of the latitude in which they were solidified. Such paleomagnetic evidence allows mapping of the past position of the North Pole. (An intriguing but as yet still unexplained phenomenon has been discovered: Reversals in magnetic polarity occur in geological time, the last of which has been dated at 700,000 years ago.) Recent rocks from different locations coincide in pinpointing the pole's position, but paleomagnetic records from older rocks from different localities are in discord. These discrepancies strongly suggest that the continents have moved with respect to one another. The continents are formed of light "plates" of siliceous, largely granitic, rocks about 30 km thick, which in turn float on denser mantlelike basaltic blocks. The ocean floors are composed of a relatively thin altered top of the earth's mantle. A mountain range of the sea floor in the mid-Atlantic represents a region of upwelling of the mantle. Under this interpretation, as the upwelling proceeds, sea floors spread and continents

---

*Taylor speculated that our moon was captured by the earth, causing an increase in the speed of rotation, which in turn "pulled" the continents away from the poles and "threw" them toward the equator. He suggested, probably correctly, that the Himalayan Mountains were formed by the collision of two tectonic plates.

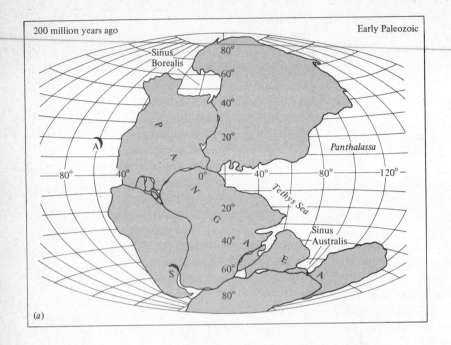

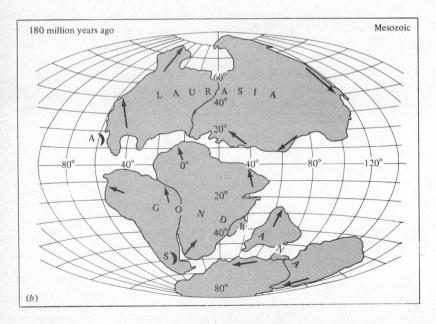

Figure 2.6 Approximate positions of major landmasses at different times in the geological past, showing their probable movements. [Adapted from Dietz and Holden (1970). The Breakup of Pangaea. Copyright ©1970 by Scientific American, Inc. All rights reserved.]

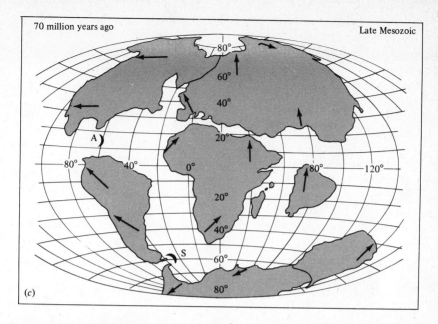

Figure 2.6 (*continued*)

move apart (Figure 2.7). The positions of paleomagnetic anomalies (polarity reversals) in the sea floor allow geologists to calculate the velocity of lateral motion of the ocean floors, which corresponds to comparable estimates for the landmasses. Thus, modern theory holds that, except for the Pacific (which is shrinking), the oceans are growing, with very young ocean floors in midocean and progressively older floors toward the continents. Other evidence, such as the apparent ages of islands and the depths of sediments, nicely corroborates this conclusion.

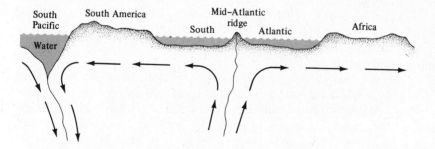

Figure 2.7  Diagrammatic cross-sectional view of the probable movements of the earth's mantle and crust that lead to sea-floor spreading and continental drift. Upwelling of deep mantle materials in the mid-Atlantic is accompanied by a surface movement away from the mid-oceanic ridge. The continents, which float on top of these moving denser materials, are carried along. At the oceanic trench on the far left, these materials sink back down a subduction zone into the mantle, forming a closed system of circulating materials.

Much of classical biogeography is being reinterpreted in light of these new findings. For example, certain very ancient groups of plants, freshwater lungfishes, amphibians, and insects that had spread before the breakup of the continents now occur on several continents, whereas many other more recently evolved groups of plants and animals, such as mammals and birds, are restricted to particular biogeographic regions. These latter, more recent groups follow the regional divisions much more closely than the older groups (Kurtén, 1969).

Imagine the effects of changing climates on plants and animals as continents drift through different latitudes! The Indian subcontinent changed from south temperate to tropical to north temperate and was virtually a "Noah's ark" carrying a flora and fauna (see also Figure 2.16 on p. 38). Australia became arid as it drifted into the horse latitudes; moreover, as this landmass continues to move toward the equator, its climate will gradually become wetter and more tropical.

## ISLAND BIOGEOGRAPHY

Ecosystems are usually difficult to manipulate experimentally; hence, much of modern ecology has had to rely on exploitation of "natural" experiments—situations in which one (or a few) factor(s) affecting a community differs between two (or more) ecosystems. For this reason, ecologists have long been especially interested in islands, which constitute some of the finest of natural ecological experiments. Different islands in an archipelago often contain different combinations of the mainland species, allowing an investigator to observe both ecological and evolutionary responses, such as niche shifts, of various component species to the presence or absence of other species (Figure 2.8). Islands can be exploited as natural ecological experiments in numerous other ways as well. Thus, because islands support fewer predatory species than comparable mainland habitats, they can be used to study the effects of predator exclusion (see also p. 295). Moreover, reduced species densities on islands, such as the land birds of Bermuda (see p. 263 and p. 276), allow partial analysis of the effects of interspecific competition on the ecologies of those species that have populated an island.

Islands of a sort are widespread in the terrestrial landscape, too; a patch of forest separated from a larger stand of trees can be considered a "habitat island." Similarly, isolated lakes and mountain tops (see p. 263) represent "islands." To a nonflying insect, plants in the desert or trees within an open forest may approximate islands in that they are separated from one another by relatively vast open spaces of a different and relatively inhospitable environment. Likewise, cattle droppings scattered about a field are islands to the animals that inhabit them (Mohr, 1943). Hosts are islands to their parasites. Even a teaspoon of water or the body of an insect may be an island to a bacterium.

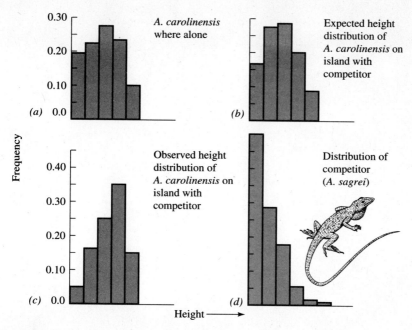

**Figure 2.8** Observed and expected frequency distributions of perch heights of *Anolis* lizards. (*a*) Observed height distribution of *A. carolinensis* where it occurs on an island alone without competitors. (*b*) Expected distribution of perch heights of *A. carolinensis* on another island with different availabilities of various perch heights, assuming no niche shift. (*c*) Observed distribution of *A. carolinensis* on the second island with a competitor [compare with (*b*)]. (*d*) Height distribution of perches of the competing species, *A. sagrei*. [From Schoener (1975a).]

## SPECIES–AREA RELATIONSHIPS

Larger islands generally support more species of plants and animals than smaller ones. In fact, when plotted on a double log scale, the number of species in a given taxon typically increases more or less linearly with island size (Figure 2.9). In most cases, a tenfold increase in area corresponds to an approximate doubling of the number of species. The slope of a linear regression line through such points is designated as that taxon's $z$-value in the particular island system. In a variety of taxa on many different island systems (Table 2.2), $z$-values generally range from about 0.24 to about 0.33. The $z$-value is the exponent in the equation

$$S = CA^z \tag{1}$$

where $S$ is the number of species, $C$ is a constant that varies between taxa and from place to place, and $A$ is the area of the island(s) concerned. Taking logarithms and rearranging, one obtains a linear equation in which $z$ is the slope:

$$log \ S = log \ C + z \ log \ A \tag{2}$$

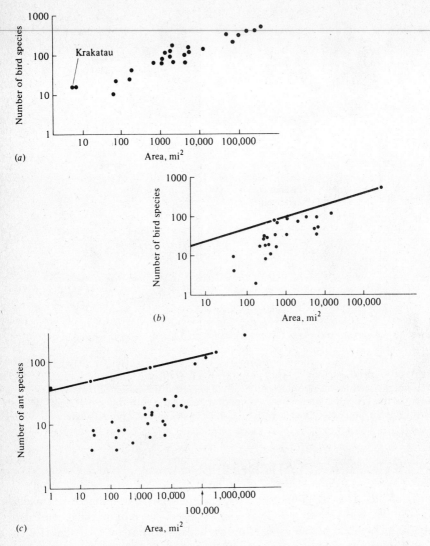

**Figure 2.9** Various species-area relationships. (*a*) Numbers of species of land and freshwater birds on islands of the Sunda group in southeast Asia, with the Philippines and New Guinea. Krakatau is plotted at the extreme left. (*b*) Numbers of species of land and freshwater birds on various often remote islands in the South Pacific, including the Moluccas, Melanesia, Micronesia, Polynesia, and Hawaii. The line is drawn through the two islands (Kei and New Guinea) nearest to source regions to demonstrate the degree of departure of species densities on the more remote islands. (*c*) Number of species of ponerine ants in the faunas of various Mollucan and Melanesian islands. The line represents the number of species with increasing area in subsamples of New Guinea; points represent smaller islands. Note that the islands support fewer species than a comparable sized portion of New Guinea but that the rate of increase of species with area is greater among the islands than it is within New Guinea. [(*a*, *b*) From MacArthur and Wilson (1967). *The Theory of Island Biogeography.* Copyright © 1967 by Princeton University Press. Reprinted by permission of Princeton University Press. (*c*) From Wilson (1961).]

TABLE 2.2    **Estimated z-Values for Various Terrestrial Plants and Animals on Different Island Groups**

| Fauna or flora | Island group | z |
|---|---|---|
| Carabid beetles | West Indies | 0.34 |
| Ponerine ants | Melanesia | 0.30 |
| Amphibians and reptiles | West Indies | 0.301 |
| Breeding land and freshwater birds | West Indies | 0.237 |
| Breeding land and freshwater birds | East Indies | 0.280 |
| Breeding land and freshwater birds | East-Central Pacific | 0.303 |
| Breeding land and freshwater birds | Islands of Gulf of Guinea | 0.489 |
| Land vertebrates | Islands of Lake Michigan | 0.239 |
| Land plants | Galápagos Islands | 0.325 |

*Source:* From MacArthur and Wilson (1967). *The Theory of Island Biogeography.* Reprinted by permission of Princeton University Press. Copyright ©1967 by Princeton University Press.

Large values of $z$ result from topographic diversity and spatial replacement of species, or "islands within islands"; lower values arise with reduced replacement of species in space, as on very homogeneous islands, continents, or subsamples of large islands (see subsequent discussion). Area in itself is probably not the primary factor affecting species density in most situations, but it presumably operates indirectly through increasing the variety of available habitats. However, area can directly affect species densities in some situations.

An area of mainland habitat comparable to and equal in size to an offshore island almost invariably supports more species, especially those at higher trophic levels, than the island does. The number of species in samples of a continental system also increases with the size (area) of the subsample, although not as rapidly as on islands (Figure 2.9). Typically, $z$-values in mainland situations range from about 0.12 to about 0.17. This difference arises because an island is a true "isolate," whereas a similar sized patch of mainland habitat is only a "sample"; rare species can occur in the mainland sample both because of migration from other areas and because areas immediately adjacent to the subsample also support other members of broad-ranging species. A mountain lion requiring a 20-square-kilometer territory would be unlikely to maintain a viable population on a small island of, say, less than 30 to 40 square kilometers, whereas these same cats are able to survive and replace themselves in a similar sized subsample of a larger landmass. This is why islands tend to support fewer species at higher trophic levels than mainland areas do.

## EQUILIBRIUM THEORY OF ISLAND BIOGEOGRAPHY

For many years, islands were considered to be in some sense "impoverished" of species both because of the obvious problems species have in colonizing them and because islands typically support fewer species than a comparable area of mainland habitat. However, the regularity of species-area patterns led MacArthur and Wilson (1963, 1967) to examine the possibility that islands might in fact be supporting as many species as possible [this possibility was first considered by Munroe (1948)].

MacArthur and Wilson reasoned that the rate of immigration of *new* species to an island should decrease as the number of species on that island increases. The immigration rate must drop to zero as the species density of the island reaches the total number of species in the "species pool" available for colonization of the island at which no immigrant can be a new species. (The species pool corresponds to the total number of species in source areas surrounding a particular island system.) MacArthur and Wilson argue that the rate of extinction of species already present on an island should *increase* as the number of species on an island increases; this seems likely because, as more species invade an island, average population size must decrease and both the intensity of interspecific competition and the incidence of competitive exclusion should increase. Moreover, there are more species to go extinct.

When the rate of immigration equals the rate of extinction (Figure 2.10), existing species go extinct at the same rate that new ones invade; thus, species density reaches a dynamic equilibrium. Although species density stays constant, the continual turnover of species means that the actual composition of species on an island can be changing.

MacArthur and Wilson's equilibrium theory is somewhat analogous to the model of the Verhulst-Pearl logistic equation for growth processes within a

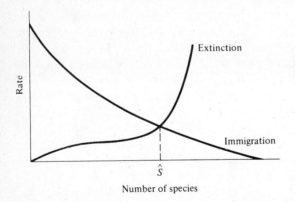

Figure 2.10 Illustration of the equilibrium theory for island species densities, with the immigration rate of new species falling and the rate of extinction of existing species rising as the total number of species on an island increases. At equilibrium, immigration just balances extinction and $\hat{S}$ different species exist on the island. The composition of the island's biota may change as some of the existing species go extinct and are replaced by other, different species.

population (see Chapter 9); thus, number of individuals (density), $N$, is replaced by the number of species (species density), $S$, and density-dependent birth and death rates, $bN$ and $dN$, are replaced by a falling immigration rate, $\lambda$, and a rising rate of extinction, $\mu$, as the species density of an island increases. As a first approximation, we might assume that rates of immigration ($\lambda$) and extinction ($\mu$) vary linearly with species density according to the equations

$$\lambda_S = \lambda_0 - \alpha S \tag{3}$$

$$\mu_S = \beta S \tag{4}$$

where $\lambda_0$ is the rate of immigration with no species present on the island and $\alpha$ and $\beta$ represent rates of change in rates of immigration and extinction, respectively, as species density increases (Figure 2.11). [MacArthur and Wilson (1963, 1967) point out that this assumption of linearity is not as stringent as it might at first seem, because transformations of the ordinate may allow simultaneous straightening of immigration and extinction curves.] At equilibrium, or $\hat{S}$, the rate of immigration must exactly equal the rate of extinction—that is, $\lambda_S$, must equal $\mu_S$. Setting equation (3) equal to (4),

$$\lambda_0 - \alpha\hat{S} = \beta\hat{S} \tag{5}$$

and rearranging, one obtains an expression for the number of species at equilibrium:

$$\hat{S} = \frac{\lambda_0}{\alpha + \beta} \tag{6}$$

Equation (6) is, of course, identical in form to the expression for carrying capacity $K$, in the logistic equation which is $K = r/(x + y)$ [see also equation (9) in Chapter 9, p. 187].

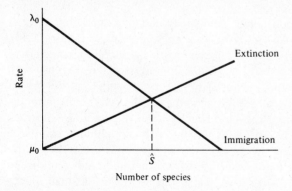

Figure 2.11 Immigration and extinction rates that change linearly with island species density. Equilibrium species density, $\hat{S}$, is a simple function of the slopes and intercepts of the two lines.

At equilibrium, the total rate of immigration of species must equal the total extinction rate. However, because species going extinct will undoubtedly often differ from those that successfully invade an island, the *composition* of an island's biota will be continually changing, even at equilibrium.

As previously developed, $\lambda$'s and $\mu$'s represent total rates of immigration and extinction and thus indicate little about the relative rates *per species* either already present on the island or available in the species pool ($P$). The average rate of immigration per species, $\bar{\lambda}$, and the average rate of extinction per species, $\bar{\mu}$, can be obtained by dividing by, respectively, the number of species not yet on the island ($P - S$) and the number already present on the island ($S$):

$$\bar{\lambda} = \frac{\lambda_S}{P - S} \qquad \text{or} \qquad \lambda_S = \bar{\lambda}(P - S) \tag{7}$$

$$\bar{\mu} = \frac{\mu_S}{S} \qquad \text{or} \qquad \mu_S = \bar{\mu} S \tag{8}$$

Again, at equilibrium, total extinction rate ($\mu_S$) must equal the total rate of immigration ($\lambda_S$); that is, $\lambda_S = \mu_S$, or in terms of the average rates per species (which are the rates with which an ecologist will usually be working):

$$\bar{\lambda}(P - \hat{S}) = \bar{\mu}\hat{S} \tag{9}$$

Solving for the equilibrium number of species, $\hat{S}$, gives

$$\hat{S} = \frac{\bar{\lambda}P}{\bar{\mu} + \bar{\lambda}} \tag{10}$$

Equation (10) demonstrates that $\hat{S}$ increases with increasing $P$ and $\bar{\lambda}$ and decreases with increased $\bar{\mu}$. Notice also that $\bar{\lambda}P$ is $\lambda_0$ [compare equation (10) with equation (6)] and that $\bar{\lambda}$ is identical to $\alpha$ in equation (3), whereas $\bar{\mu}$ is $\beta$ in equation (4).

Because dispersal falls off more or less exponentially with distance (Figure 2.12), MacArthur and Wilson (1963, 1967) reasoned that immigration rates should decrease with increasing distance from source areas (Figure 2.13). Further, they argued that rates of extinction should be largely unaffected by distance from source areas per se but should instead generally increase with decreasing island size because smaller islands support smaller, more tenuous populations (Figure 2.14). (Because they present a smaller "target" for potential invaders, smaller islands might also have slightly lower immigration rates than other equivalent but larger islands. But this change should be minor compared with the expected decline due to the exponential decay in the number of immigrants with distance; see Figure 2.12.) Note also that on islands equidistant from source areas, rates of species turnover should be higher on small islands than on large ones (turnover rate should thus vary inversely with equilibrium species density). Simple islands with little topographic relief and relatively few different habitats should have generally higher extinction rates than more complex and more diverse islands with a greater variety of habitats, because

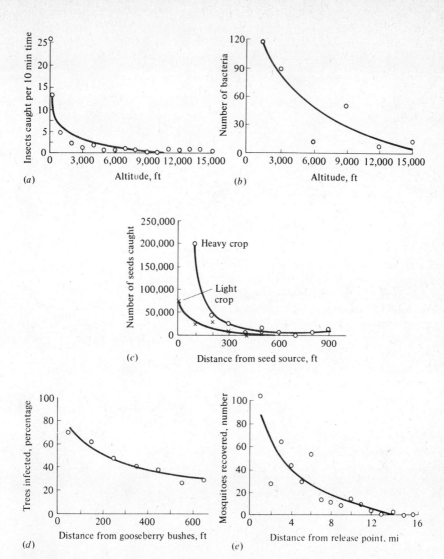

**Figure 2.12** Some actual patterns of dispersal, both vertical and horizontal. The number of organisms decays rapidly at first and then more and more slowly with increasing distance. [From Odum (1959) after Wolfenbarger.]

the latter would provide a greater variety of immigrants with suitable opportunities for successful invasion and persistence on the island. Finally, clumped islands such as archipelagos should have higher rates of immigration than more scattered or isolated islands due to interchanges of plants and animals among islands.

Some predictions of equilibrium theory have now been supported by observations; others, especially those involving turnover rates, have proven to be difficult to test.

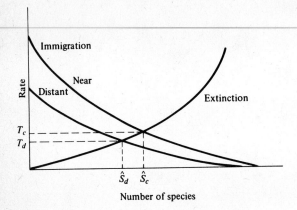

**Figure 2.13** Immigration rates should decrease with increasing distance from source areas so that distant islands should reach equilibrium with fewer species, $\hat{S}_d$, than close-in islands, $\hat{S}_c$, all else being equal. Moreover, turnover rates should also be higher on nearby islands than on comparable but more distant islands ($T_c > T_d$).

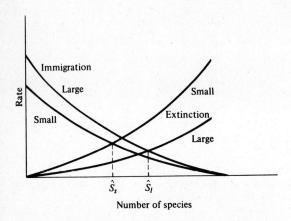

**Figure 2.14** Extinction rates should be little affected by distances from source areas, but they should often vary inversely with island size, complexity, or both. Immigration rates may also be slightly higher on larger islands because they present a larger "target" for potential invaders. Thus, all else being equal, a small island should equilibrate with fewer species, $\hat{S}_s$, than a larger island, $\hat{S}_l$.

# ISLANDS AS ECOLOGICAL EXPERIMENTS: SOME EXAMPLES

## Hawaiian Drosophilidae

The Hawaiian Islands constitute the most isolated archipelago on the earth. They support a truly remarkable natural fauna, unfortunately now seriously disturbed by inane human activities. Among the most interesting of Hawaiian organisms are its famous fruit flies, family Drosophilidae. A diverse group of nearly 800 different species has evolved there, almost certainly from a single common ancestor. These flies are placed in two genera, *Drosophila* and *Scaptomyza*, both of which are also found elsewhere in the world, although in greatly reduced diversity. Hawaiian drosophilids have adopted a wide variety of ecological niches and have diversified into predators, parasites, nectarivores, detritovores, and herbivores. Some are generalists (most larvae eat parts of plants, including leaves, bark, stems, roots, sap, nectar, and fruit), but others are very specialized, frequently to certain specific positions on a single species of host plant. In contrast, one of the generalized species has been found on plants of 21 different families! Most *Scaptomyza* are tiny flies 2 mm or less in length, with leaf-mining larvae. Some of the so-called "picture-winged" *Drosophila* are relatively gigantic flies, measuring up to 7 or 8 mm long (*D. cyrtoloma* has a wingspread of nearly 20 mm!). Detailed genetic work has allowed construction of a phylogeny for 100 species of picture-winged *Drosophila*. The islands themselves have been dated by geological means, as they arise over a "hot spot" in the earth's mantle and then drift off to the northwest (the Hawaiian Islands become progressively older from the southeast to the northwest). Almost all species of picture-winged *Drosophila* are endemic to a single island. The most primitive species are in fact found on the oldest islands, while the most derived species occur on the youngest islands.

Many Hawaiian *Drosophila* have experienced strong sexual selection. Males compete intensely for virgin females, advertising from small territories, sometimes termed "leks" (although these differ from avian leks). Males are very aggressive toward other males and courtship behavior is complex. Reproductive tactics of Hawaiian *Drosophila* are very varied; females of some species deposit large numbers of eggs in clusters, others produce many eggs but lay them singly, while females in still other species lay just a single egg per day. This model system clearly invites much more extensive ecological and evolutionary research.

## Krakatau

In 1883, the small volcanic island of Krakatau, located between Java and Sumatra (see Figure 2.5), erupted repeatedly over a three-month period. All of Krakatau and two adjacent islands were covered with red-hot lava, pumice, and ash to a depth of many meters. The islands were so hot that months afterward falling rain turned to steam on contact. It is most unlikely that any organisms survived. Repopulation from adjacent Sumatra (about 25 kilometers away) and Java proceeded rapidly, and by 1921 the number of resident species of birds was comparable to that expected on a small island of 13 km$^2$ (the size of Krakatau after the eruptions) in the general region (see Figure 2.9a). The total number of bird species did not change much between 1921 and 1933, although the composition of the avifauna did (Table 2.3). This example suggests that mobile organisms such as birds rapidly reach an equilibrium species density. On the other hand, plant species were still being added rapidly in 1934 but had leveled out by 1983 (Figure 2.15; Whittaker et al., 1989).

## The Taxon Cycle

Many island species are thought to progress through a series of evolutionary changes, termed a *taxon cycle*, that eventually may greatly increase their probability of going extinct (Wilson, 1961; MacArthur and Wilson, 1967; Ricklefs and Cox, 1972). Under this hypothesis, early in the taxon cycle a species is widespread and occurs on many islands, is often in the process of invading new islands, and is only slightly (if at all) differentiated into distinct populations on the various islands. It is adapted to marginal, relatively unstable habitats such as riverbanks and forest clearings. Later in the cycle, populations of a species become progressively more and more differentiated on different islands, at first remaining widespread. At this stage, it penetrates more stable habitats, such

TABLE 2.3   **Number of Species of Land and Freshwater Birds on Krakatau and Verlaten During Three Collection Periods and the Number of Species "Lost" Between Intervals**

| | 1908 | | | 1919–1921 | | |
|---|---|---|---|---|---|---|
| | Nonmigrant | Migrant | Total | Nonmigrant | Migrant | Total |
| Krakatau | 13 | 0 | 13 | 27 | 4 | 31 |
| Verlaten | 1 | 0 | 1 | 27 | 2 | 29 |

| | 1932–1934 | | | Number "lost" | |
|---|---|---|---|---|---|
| | Nonmigrant | Migrant | Total | 1908 to 1919–1921 | 1919–1921 to 1932–1934 |
| Krakatau | 27 | 3 | 30 | 2 | 5 |
| Verlaten | 29 | 5 | 34 | 0 | 2 |

*Source:* From MacArthur and Wilson (1967) after Dammerman (1948).

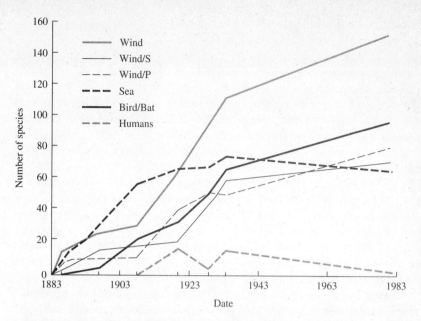

**Figure 2.15** Numbers of plant species recorded on the three islands of the Krakatau group, from 1883 to 1983. Wind-dispersed Spermatophyta (higher plants) and wind-dispersed Pteridophyta (ferns) as well as species dispersed by other means, such as rafting (sea), birds/bats, and humans, are also shown. [From Whittaker et al. (1983).]

as old forests, where it must coexist with larger numbers of native species. Still later, after local extinction on some islands, a differentiated species becomes more restricted and its geographic range is fragmented. Finally, species at the end of the taxon cycle are found only on a single island (i.e., they are *endemic* to one island). Occasionally, some species are able to shift back into marginal, species-poor habitats, thus restarting the cycle. Ricklefs and Cox (1972) argue that the taxon cycle for a particular species is driven by counteradaptations of other members of an island biota against the species concerned. Dispersal ability decreases as species depend less and less on the ability to colonize marginal habitats and more on their capacity to coexist with competing species in stable habitats. Such reduced dispersal and more pronounced local adaptation in turn favors speciation and endemism. Newly arrived colonists are relatively free of a counteradaptive load, allowing them to spread successfully throughout an island system in relatively unstable habitats. On small and remote islands, old populations of endemics (pp. 138–139), such as the Cocos Island finch, may persist.

Although the taxon cycle may not apply to all species, the concept underlying it presumably could operate in mainland faunas as well as on islands. Little attempt has yet been made to interpret the ecology of mainland populations in terms of such counteradaptations, although the concept of a fugitive species is clearly relevant.

# VICARIANCE BIOGEOGRAPHY
# AND PHYLOGENETIC SYSTEMATICS

As a consequence of the emerging awareness of continental drift, a vigorous new branch of biogeography has arisen. Prior attempts at explaining the current geographic distributions of organisms relied heavily on improbable movements and episodes of dispersal. But the recognition that landmasses themselves actually break apart and move (aquatic systems behave analogously) has enabled distributions to be understood in terms of such geographic or "vicariant" events as well as dispersal.

The goal of phylogenetic systematics or "cladistics" is to construct classifications that accurately reflect phylogeny. To accomplish this end, the states of numerous characters are determined and carefully analyzed for all the species that are members of a given monophyletic group. Comparison with appropriate related "outgroups" allows systematists to identify traits that are probably shared derived characteristics versus those likely to be ancestral ones; these can then be used to construct a phyletic tree. Using a closely related "outgroup," the tree can be "rooted" to produce a "cladogram" that shows probable genealogical relationships among members of the group concerned. Past history can thus be recovered, at least to some extent, from current character states. Exploitation of these techniques to their fullest potential obviously requires examination of all related taxa—extinctions of existing species truly become "lost pages" in the unread book of life (Rolston, 1985).

Modern molecular techniques, particularly DNA amplification and sequencing, now allow systematists to reconstruct probable phylogenies of diverse monophyletic groups of organisms. Any phylogeny is no more than a hypothesis, but procedures have been worked out that greatly increase confidence in a particular "resolved" phylogenetic tree. Armed with such a phylogeny showing the probable relationships among members of a monophyletic lineage, one can overlay various character states, such as behavioral and ecological traits, to examine the actual course of evolution and to begin to understand the sequence of changes that led to the adaptive radiation of that group. Modern comparative

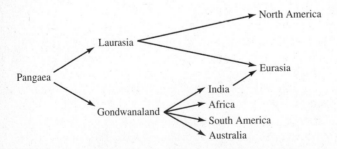

Figure 2.16 Continental movements, including the breakup of Pangaea and Gondwanaland, plus the collision of the Indian plate with southern Asia are depicted in an "area cladogram."

methods have also been developed that allow scientists to reconstruct probable ancestral states from those of surviving descendants (see also section entitled "Phylogeny and the Modern Comparative Method," pp. 301–302). Obviously, we have
made substantial progress in beginning to decipher the vanishing book of life!

Concordances among cladograms for different plant and animal groups
in the same general region suggest that common vicariant events underlie
their phylogenies. Ultimately, such patterns can be related to features of the
geographic history of the area concerned in an "area" cladogram (Figure 2.16).
Indeed, cladistic vicariance biogeographers have actually suggested that certain
physical geographic events must have occurred (for many years, biogeographic
distributions were the strongest evidence for continental drift!).

## SELECTED REFERENCES

### Self-Replicating Molecular Assemblages

Bernal (1967);  Blum (1968);  Calvin (1969);  Ehrlich and Holm (1963);  Fox and
Dose (1972);  Oparin (1957);  Ponnamperuma (1972);  Salthe (1972);  Wald (1964).

### The Geological Past

Alvarez et al. (1980);  Birch and Ehrlich (1967);  Brown (1982);  Dansereau (1957);
Darlington (1957, 1965);  Dietz and Holden (1970);  Erwin et al. (1987);  Graham
(1986);  Guilday (1967);  Hesse et al. (1951);  Imbrie et al. (1984);  Jelgersma (1966);
Martin (1967);  Martin and Klein (1984);  Martin and Mehringer (1965);  Martin and
Wright (1967);  Olsen (1986);  Sawyer (1966);  Stehli and Webb (1985);  Udvardy
(1969);  Wilson (1971, 1973);  Wiseman (1966);  Wright and Frey (1965).

### Classical Biogeography

Andrewartha and Birch (1954);  Cain (1944);  Dansereau (1957);  Darlington (1957,
1959, 1965);  Grinnell (1924);  Hesse et al. (1951);  Krebs (1972);  MacArthur (1959,
1972);  MacArthur and Wilson (1967);  Nelson and Platnick (1981);  Newbigin (1936);
Schall and Pianka (1978);  Terborgh (1971);  Udvardy (1969);  Wallace (1876);  Watts
(1971).

### Continental Drift

Cracraft (1974);  Dietz and Holden (1970);  DuToit (1937);  Hallam (1973);  Kurtén
(1969);  Marvin (1973);  Skinner (1986);  Tarling and Runcorn (1973);  Taylor (1910);
Wegener (1924);  J. T. Wilson (1971, 1973).

### Island Biogeography

Carlquist (1965);  Gilbert (1980);  MacArthur and Wilson (1963, 1967);  Maguire
(1963, 1971);  Mohr (1943);  Williamson (1981);  Wilson (1969);  Wilson and Bossert
(1971).

## Species–Area Relationships

Gleason (1922, 1929); Krebs (1972); MacArthur and Wilson (1967); May (1975a); Odum (1959, 1971); Preston (1948, 1960, 1962a, 1962b); Schoener (1976a).

## Equilibrium Theory of Island Biogeography

Brown (1971); Gilbert (1980); MacArthur and Wilson (1963, 1967); Munroe (1948); Simberloff (1976); Wilson (1969); Wilson and Bossert (1971).

## Islands as Ecological Experiments: Some Examples

Carlquist (1965); Greenslade (1968); MacArthur (1972); MacArthur and Wilson (1967); Williamson (1981).

### Hawaiian Drosophilidae

Beverley and Wilson (1985); Carson (1973, 1983); Carson and Kaneshiro (1976); Kaneshiro and Boake (1987); Simon (1987); Williamson (1981).

### Krakatau

Dammerman (1948); Docters van Leeuwen (1936); MacArthur and Wilson (1967); Whittaker et al. (1989).

### The Taxon Cycle

MacArthur and Wilson (1967); Ricklefs and Cox (1972); Wilson (1961).

## Vicariance Biogeography and Phylogenetic Systematics

Eldredge and Cracraft (1980); Felsenstein (1985, 1988); Garland (1992); Garland et al. (1991, 1992); Harvey and Pagel (1991); Hennig (1966); Nelson and Platnick (1981); Platnick and Nelson (1978); Ridley (1983); Rolston (1985); Rosen (1978); Wiley (1981).

## Chapter
## 3

# Meteorology

*T*he earth supports an enormous variety of organisms. Plants range from microscopic short-lived aquatic phytoplankton to small annual flowering plants to larger perennials to gigantic ancient sequoia trees. Animals, although they never attain quite the massive size of a redwood tree, include forms as diverse as marine zooplankton, jellyfish, sea stars, barnacles, clams, snails, beetles, butterflies, worms, frogs, fish, lizards, sparrows, hawks, bats, elephants, whales, and lions. Different species have evolved and live under different environmental conditions. Some organisms are relatively specialized either in the variety of foods they eat or in the microhabitats they exploit, whereas others are more generalized; some are widespread, occurring in many different habitats, whereas still others have more restricted habitat requirements and geographic ranges. Temporal and spatial variation in the physical conditions for life often make possible or even actually necessitate variety among organisms, both directly and indirectly. Of course, interactions among organisms also contribute substantially to the maintenance of this great diversity of life. Before considering such biological interactions, we first examine briefly the nonliving world, which sets the background for all life and often strongly influences the ecology of any particular organismic unit. A major factor in the physical environment is climate, which in turn is the ultimate determinant of water availability and the thermal environment; moreover, the latter two interact to determine the actual amount of solar energy that can be captured by plants (primary productivity) at any given time and place. Finally, because climate is a major determinant of both soils and vegetation, there is a close correspondence between particular

41

climates and the types of natural biological communities that exist under those climatic conditions. The interface between climate and vegetation is considered in Chapter 4.

Major global and local patterns of climate are described briefly in this chapter. Entire books have been devoted to some of these subjects, and the reader interested in greater detail is referred to the references at the end of the chapter.

## MAJOR DETERMINANTS OF CLIMATE

The elements of climate (sun, wind, and water) are complexly interrelated. Incident solar energy produces thermal patterns that, coupled with the earth's rotation and movements around its sun, generate the prevailing winds and ocean currents. These currents of air and water in turn strongly influence the distribution of precipitation, both in time and space.

The amount of solar energy intercepting a unit area of the earth's surface varies markedly with latitude for two reasons. First, at high latitudes, a beam of light hits the surface at an angle, and its energy is spread out over a large surface area. Second, a beam that intercepts the atmosphere at an angle must penetrate a deeper blanket of air, and hence more solar energy is reflected by particles in the atmosphere and radiated back into space. (Local cloud cover also limits the amount of the sun's energy that reaches the ground.) A familiar result of both these effects is that average annual temperatures tend to decrease with increasing latitude (Table 3.1). The poles are cold and the tropics are generally warm (seasons are discussed later).

Water in the atmosphere is warmed by heat radiating from the earth's surface, and much of this heat is radiated back to the earth again. The result is the so-called *greenhouse effect*, which leads to the retention of heat, keeping the earth relatively warm even at night when there is temporarily no influx of solar energy. Without this effect, the earth's surface would cool to many degrees below zero—like the dark side of the moon's surface. Thus, the atmosphere buffers day-night thermal change.

Hot air rises. The ground and air masses above it receive more solar energy at low latitudes than at higher ones (Figure 3.1). Thus, tropical air masses, especially those near the equator, are warmed relatively more than temperate air masses, and an equatorial zone of rising air is created. These equatorial air masses cool as they rise and eventually move northward and southward high in the atmosphere above the earth's surface (Figure 3.2a). As this cold air moves toward higher latitudes, it sinks slowly at first and then descends rapidly to the surface at the so-called *horse latitudes* of about 30°N and 30°S. At ground level at these latitudes, some air moves toward the equator again and some of it moves toward the poles. (The amount of air in the atmosphere is finite, so air masses leaving one place must always be replaced by air coming from somewhere else; thus, a closed system of circulating air masses is set

**TABLE 3.1  Average Annual Temperature (°C) at Different Latitudes**

| Latitude | Year | January | July | Range |
|----------|------|---------|------|-------|
| 90°N | −22.7 | −41.1 | −1.1 | 40.0 |
| 80°N | −18.3 | −32.2 | 2.0 | 34.2 |
| 70°N | −10.7 | −26.3 | 7.3 | 33.6 |
| 60°N | −1.1 | −16.1 | 14.1 | 30.2 |
| 50°N | 5.8 | −7.1 | 18.1 | 25.2 |
| 40°N | 14.1 | 5.0 | 24.0 | 19.0 |
| 30°N | 20.4 | 14.5 | 27.3 | 12.8 |
| 20°N | 25.3 | 21.8 | 28.0 | 6.2 |
| 10°N | 26.7 | 25.8 | 27.2 | 1.4 |
| Equator | 26.2 | 26.4 | 25.6 | 0.8 |
| 10°S | 25.3 | 26.3 | 23.9 | 2.4 |
| 20°S | 22.9 | 25.4 | 20.0 | 5.4 |
| 30°S | 16.6 | 21.9 | 14.7 | 7.2 |
| 40°S | 11.9 | 15.6 | 9.0 | 6.6 |
| 50°S | 5.8 | 8.1 | 3.4 | 4.7 |
| 60°S | −3.4 | 2.1 | −9.1 | 11.2 |
| 70°S | −13.6 | −3.5 | −23.0 | 19.5 |
| 80°S | −27.0 | −10.8 | −39.5 | 28.7 |
| 90°S | −33.1 | −13.5 | −47.8 | 34.3 |

*Source:* Adapted from Haurwitz and Austin (1944).

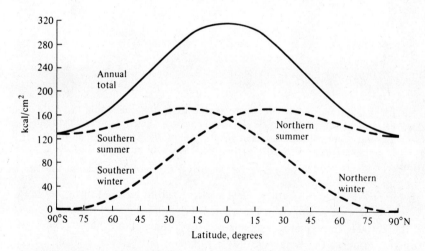

**Figure 3.1** Estimated amount of incoming solar radiation that would intercept the earth's surface in the absence of an atmosphere as a function of latitude. The six-month period from the spring equinox to the fall equinox (see Figure 3.7) is labeled "northern summer" and "southern winter," whereas the six months from the fall equinox to the spring equinox represent the "northern winter" and the "southern summer." [After Haurwitz and Austin (1944).]

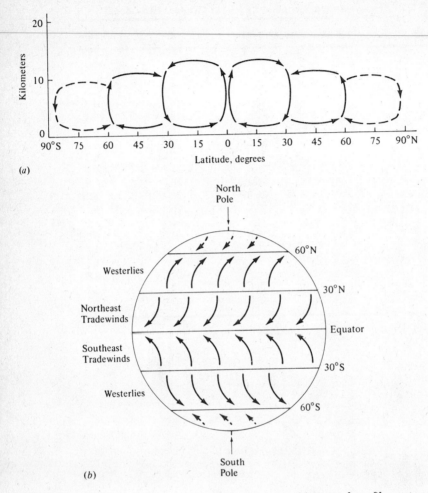

**Figure 3.2** Idealized atmospheric circulation patterns. (*a*) Vertical profile against latitude. (*b*) Prevailing wind currents on the earth's surface. These belts of moving air move north and south with the seasons. Winds high in the atmosphere move in an opposite manner to those at the surface. [After MacArthur and Connell (1966), MacArthur (1972), and others.]

up.) An idealized diagram of the typical vertical and horizontal movements of atmospheric currents is shown in Figure 3.2.

At the surface, the equator is a zone of convergence of air masses, whereas they are diverging at the horse latitudes. Between latitudes 0° and 30°, surface air generally moves toward the equator; between latitudes 30° and 60°, it generally moves away from the equator. As air masses move along the surface, they are slowly warmed and eventually rise again.

Movements of air masses are not strictly north-south as suggested by Figure 3.2*a*; instead, they acquire an east-west component due to the rotation of the earth about its axis (Figure 3.2*b*). The earth rotates from west to east. A person standing on either pole would rotate slowly around and do a full "about face" each 12 hours. (Near the North Pole the ground moves counterclockwise under one's feet, whereas near the South Pole it moves clockwise; other related important differences between the hemispheres are considered in the following paragraphs.) However, someone located near the equator travels much farther during a 24-hour period; indeed, such a person would traverse a distance equal to the earth's circumference, or about 40,000 kilometers, during each rotation' of the globe. Hence, the velocity of a body near the equator is approximately 1600 kilometers/hour (relative to the earth's axis), whereas a body at either pole is, relatively speaking, at a standstill.

Previous considerations, plus the physical law of conservation of momentum, dictate that objects moving north in the Northern Hemisphere must speed up, relative to the earth's surface, and thus veer toward the right. Similarly, objects moving south in the Northern Hemisphere slow down relative to the earth's surface, which means that they also veer to the right. In contrast, moving objects in the Southern Hemisphere always veer to the left; northward-moving objects slow down and southward-moving ones speed up. These forces, known collectively as the *Coriolis force*, act on north-south wind and water currents to give them an east-west component. The Coriolis force is maximal at the poles, where a slight latitudinal displacement is accompanied by a large change in velocity, and minimal at the equator, where a slight latitudinal change has little effect upon the velocity of an object.

Equator-bound surface air between latitudes 0° and 30° slows relative to the surface and veers toward the west in both hemispheres, producing winds from the east ("easterlies"); these constitute the familiar tradewinds, known as the northeast trades between 0° and 30°N and the southeast trades between 0° and 30°S. Between latitudes 30° and 60°, surface air moving toward the poles speeds up (again, relative to the earth's surface) and veers toward the east, producing the familiar prevailing "westerly" winds at these latitudes in both hemispheres (Figure 3.2b).

These wind patterns, coupled with the action of the Coriolis force on water masses moving north to south, drive the world's ocean currents; in the Northern Hemisphere, ocean waters rotate generally clockwise, whereas they rotate counterclockwise in the Southern Hemisphere (Figure 3.3).

During their movement westward along the equator, oceanic waters are warmed by solar irradiation. (These waters also "pile up" on the western sides of the oceanic basins; in Central America, the sea level of the Atlantic is about a meter higher than that of the Pacific.) As this warm equatorial water approaches the eastern side of landmasses, it is diverted northward and/or southward to higher latitudes, carrying equatorial heat toward the poles along the eastern coasts of continents. Cold polar waters flow toward the equator on west coasts (this is the main reason the Pacific Ocean off southern California is cold but at the same latitude the Atlantic Ocean off Georgia is quite warm).

Heat is molecular movement. Compressing a volume of air results in more collisions between molecules and increased molecular movement; hence, compression causes an air mass to heat up. Exactly analogous considerations apply

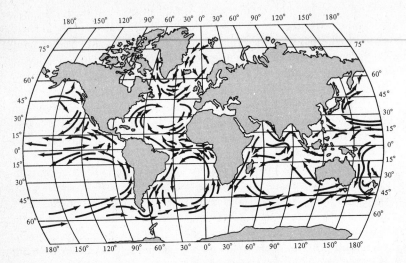

Figure 3.3 Major surface currents of the world's oceans.

in the reverse case. Allowing compressed air to expand decreases the number of molecular collisions and the air mass cools off. As warm air rises, atmospheric pressure decreases, and the air expands and is cooled *adiabatically,* or without change in total heat content. Some of the air's own heat is used in its expansion. As descending cold air is compressed, it is adiabatically warmed in a similar manner.

Warm air can carry more water vapor than cold air. As warm, water-laden equatorial air rises and cools adiabatically, it first becomes saturated with water at its dew point, and then its water vapors condense to form precipitation. As a result, regions of heavy rainfall tend to occur near the equator (Figure 3.4). In contrast, at the horse latitudes, as cold dry air masses descend and warm adiabatically, they take up water; such desiccating effects help to produce the earth's major deserts at these latitudes (see Figures 3.4 and 3.5). Deserts occur mainly on the western sides of continents, where cold offshore waters are associated with a blanket of cold, and therefore dry, air; westerly winds coming in off these oceans do not contain much water to give up to the descending dry air, and precipitation must therefore be scanty. The global pattern of annual precipitation shown in Figure 3.5 conforms in general outline to the expected, but there are many local anomalies and exceptions.

## LOCAL PERTURBATIONS

Major climatic trends are modified locally by a variety of factors, most notably by the size(s) and position(s) of nearby water bodies and landmasses and by topography (especially mountains). Ascending a mountain is, in many ways,

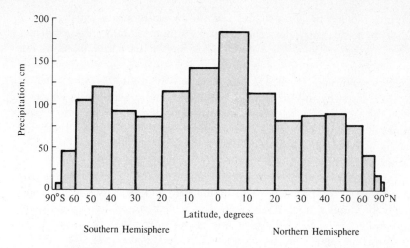

Figure 3.4 Histogram of average annual precipitation versus latitudinal zones. Note the effect of the earth's major deserts in the horse latitudes. [From Haurwitz and Austin (1944).]

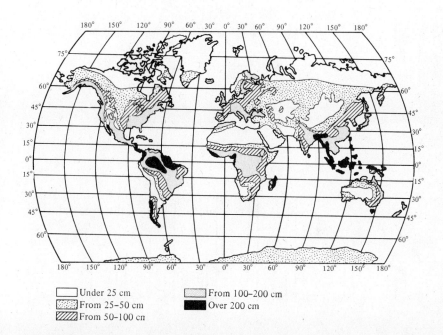

Figure 3.5 Geographic distribution of average annual precipitation. Annual world precipitation modified van der Griten Projection. [After MacArthur and Connell (1966) after Koeppen.]

comparable to moving toward a higher latitude. Mountains are usually cooler and windier than adjacent valleys and generally support communities of plants and animals characteristic of lower elevations at higher latitudes (100 meters of elevation corresponds roughly to 50 kilometers of latitude). In addition to this thermal effect, mountains markedly modify water availability and precipitation patterns. Water rapidly runs off a slope but sits around longer on a flatter place and soaks into the ground. For this reason, precipitation falling on a mountain-side is generally less effective than an equivalent amount falling on a relatively flat valley floor. Precipitation patterns themselves are also directly affected by the presence of mountains. Consider a north-south mountain range receiving westerly winds, such as the Sierra Nevada of the western United States. Air is forced upward as it approaches the mountains, and as it ascends, it cools adiabatically, becomes saturated with water, and releases some of its water content as precipitation on the windward side of the sierra. After going over the ridge, this same air, now cold and dry, descends, and, as it warms adiabatically, it sucks up much of the moisture available on the leeward (downwind) side of the mountains. The so-called *rainshadow effect* (Figure 3.6) is produced, with

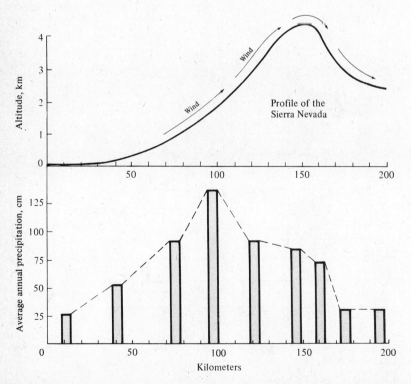

Figure 3.6 Illustration of the "rainshadow" effect of the Sierra Nevada in central California.

windward slopes being relatively wet and leeward slopes being much drier. Desiccating effects of these warm dry air masses extend for many miles beyond the sierra and help to produce the Mojave Desert in southern California and Nevada.

Water has a high specific heat; that is, a considerable amount of heat energy is needed to change its temperature. Conversely, a body of water can give up a relatively large amount of heat without cooling very much. A result of these heat "sink" attributes is that large bodies of water, particularly oceans, effectively reduce temperature changes of nearby landmasses. Thus, coastal "maritime" climates are distinguished from inland "continental" climates, with the former being much milder and less variable. Large lakes, such as the Great Lakes, also decrease thermal changes on adjacent landmasses and produce a more constant local temperature.

## VARIATIONS IN TIME AND SPACE

The seasons are produced by the annual elliptical orbit of the earth around its sun and by the inclination of the planet's axis relative to this orbital plane (Figure 3.7). These orbital movements do not repeat themselves exactly but follow a complex celestial periodicity, known as *Milankovitch cycles*, which are measured on a time scale of many thousands of years. For example, earth's orbit around the sun changes from near-circular to elliptical and back again

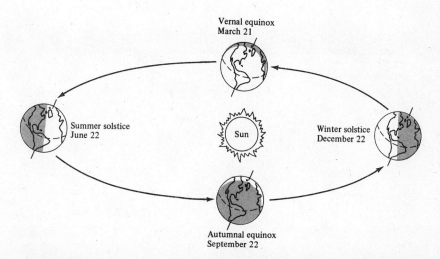

Figure 3.7 Diagram of the earth's annual elliptical orbit around the sun, which produces the seasons. [After MacArthur and Connell (1966).]

over a period of 95,000 years. At present, earth is closest to the sun during winter in the Northern Hemisphere, but halfway through a 22,000-year cycle, some 11,000 years ago, it was farthest from the sun during the northern winter. Such orbital cycles generate long-term climate cycles such as glacial and interglacial (pluvial) periods.

For historical reasons, these movements and patterns have been described from the point of view of the Northern Hemisphere, although by symmetry the same events occur some six months out of phase in the Southern Hemisphere. Twice each year, at the vernal equinox (March 21) and the autumnal equinox (September 22), solar light beams intercept the earth's surface perpendicularly on the equator (i.e., the sun is "directly overhead" at its zenith and equatorial shadows point exactly east to west). At two other times of year, the summer solstice (June 22) and the winter solstice (December 22), the earth's axis is tilted maximally with respect to the sun's rays. Viewed from the Northern Hemisphere, the axis at present inclines approximately 23.5° toward the sun during the summer solstice and 23.5° away from it during the winter solstice (this angle of inclination itself varies cyclically between 22° and 24.5°, with a periodicity of about 41,000 years). At each of the solstices, rays of light hit

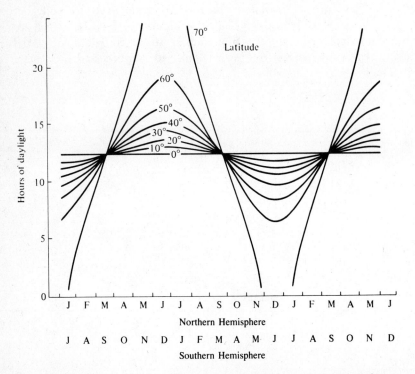

**Figure 3.8** Seasonal changes in photoperiod at different latitudes [After Sadleir (1973).]

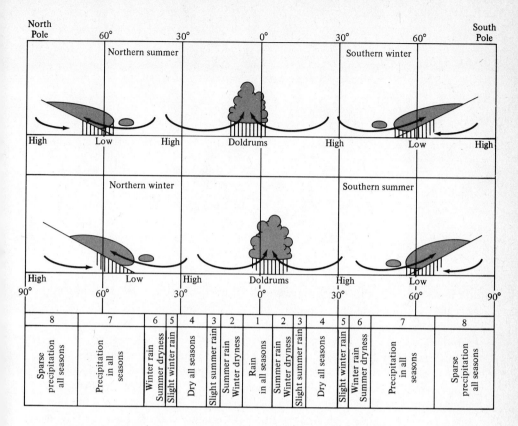

**Figure 3.9** Precipitation zones vary latitudinally in a more-or-less regular fashion, as indicated. [After Haurwitz and Austin (1944).]

the surface perpendicularly (the sun is at its zenith) near the Tropic of Cancer (23°N) and the Tropic of Capricorn (23°S), respectively. At summer solstice, the North Pole is in the middle of its six-month period of sunlight (the "polar summer"). The excess of daylight in summer is exactly balanced by the winter deficit, so the total annual period of daylight is precisely six months at every latitude; the equator has invariate days of exactly 12 hours duration, whereas the poles receive their sunshine all at once over a six-month interval and then face six months of twilight and total darkness. Day length is one of the most dependable indicators of seasonality, and many temperate zone plants and animals rely on photoperiod as an environmental cue (Figure 3.8).

Of course, prevailing winds and ocean currents are not static, as suggested by Figures 3.2 and 3.3; in fact, they vary seasonally with the earth's movement about its sun (Figure 3.9). The latitudinal belt receiving the most solar radiation (the "thermal equator") gradually shifts northward and southward between latitudes 23°S and 23°N; moreover, latitudinal belts of easterlies

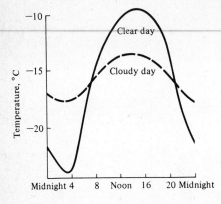

Figure 3.10 Cloud cover reduces both the amplitude and the extremes of the daily march of temperature. [After Haurwitz and Austin (1944).]

and westerlies also move northward and southward with the seasons, producing seasonal weather changes at higher latitudes. Because of the earth's spherical shape, seasonal changes in insolation increase markedly with increasing latitude.

Although temperatures are modified by prevailing winds, topography, altitude, proximity to bodies of water, cloud cover (Figure 3.10), and other factors, annual marches of average daily temperatures at any given place nevertheless closely reflect earth's movement around the sun. Thus, average daily temperatures on the equator change very little seasonally, whereas those at higher latitudes usually fluctuate considerably more (Figure 3.11); moreover, the annual range in temperature is also much greater in the temperate zones (see Table 3.1).

Annual patterns of precipitation also reflect earth's orbital movements (Figure 3.12), although precipitation patterns are perhaps modified locally more

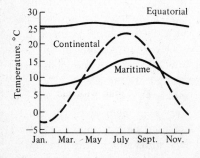

Figure 3.11 Annual marches of average daily temperature at an equatorial locality (Batavia) and two temperate areas, one coastal (Scilly) and one continental (Chicago). [After Haurwitz and Austin (1944).]

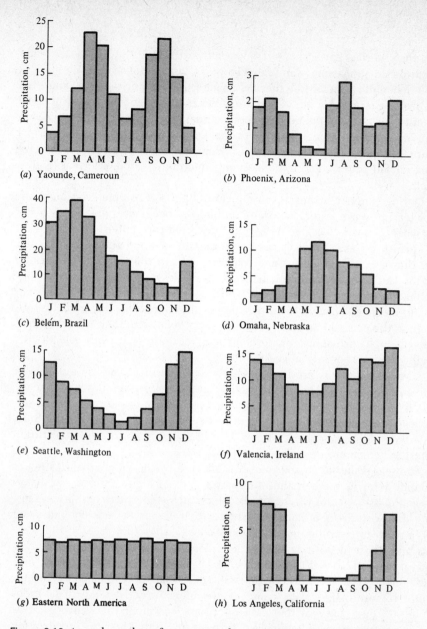

Figure 3.12 Annual marches of average yearly precipitation from eight selected localities (see text). (*a*) Bimodal equatorial rainfall. (*b*) Bimodal annual precipitation pattern of the Sonoran Desert. (*c*) Unimodal annual rainfall pattern of an equatorial area. (*d*) Typical continental summer rainfall regime. (*e*) Coastal area with winter precipitation. (*f*) Typical maritime rainfall regime with more winter precipitation than summer rain. (*g*) The annual march of precipitation in eastern North America is distributed fairly evenly over the year. (*h*) An area with a pronounced summer drought that would support a chaparral vegetation. [After Haurwitz and Austin (1944).]

easily than thermal patterns. At very low latitudes, say, 10°S to 10°N, there are often two rainfall maxima each year (Figure 3.12a), one following each equinox as the region of rising air (the thermal equator) passes over the area. Thus, the sun's passing over these equatorial areas twice each year produces a bimodal annual pattern of precipitation. Bimodal annual precipitation patterns also occur in other regions at higher latitudes, such as the Sonoran Desert in Arizona (Figure 3.12b), but for different reasons. However, not all equatorial areas have such a rainfall pattern, and some have only one rainy season each year (Figure 3.12c).

During summer, air masses in the central parts of continents at high latitudes tend to warm faster than those around the periphery, which are cooled somewhat by nearby oceans. As hot air rises from the center of continents, a "low" pressure area is formed and cooler, but still warm and water-laden, coastal air is drawn in off the oceans and coastlines. When this moist air is warmed over the land, it rises, cools adiabatically, and releases much of its water content on the interiors of continents. Thus, continental, in contrast to coastal, climates are often characterized by summer thunderstorms (Figure 3.12d). In winter, as the central regions of continents cool relative to their water-warmed edges, a "high" pressure area develops and winds reverse, with cold dry air pouring outward toward the coasts.

A weather *front* is produced when a cold (usually polar) air mass and a warm air mass collide. The warmer lighter air is displaced upward by the heavy cold air; as this warm air rises, it is cooled adiabatically, and provided it contains enough water vapor, clouds form and eventually precipitation falls. Such fronts typify the boundaries between the polar easterlies and the midlatitude westerlies, which move north to south with the seasons.

The seasonality of coastal rainfall is affected by the differential heating of land and water in another, almost opposite, way. At high latitudes, say, 40° to 60°, landmasses cool in winter and become much colder than offshore oceans; along the west coasts of continents, water-rich westerly winds coming in off such a relatively warm ocean are rapidly cooled when they meet a cold landmass and hence deposit much of their moisture as cold winter rain and/or snow along the coast (Figure 3.12e). Maritime climates typically have precipitation through the entire year, with somewhat more during the winter months (Figure 3.12f). As a result of prevailing westerly winds at latitudes 30° to 60°, east coasts at these latitudes have continental climates, whereas west coasts have maritime climates. In eastern North America, precipitation is spread fairly evenly over the year (Figure 3.12g). Areas with a long summer drought and winter rains (Figure 3.12h) typically support a vegetation of chaparral.

A convenient means of graphically depicting seasonal climates is the *climograph*, a plot of average monthly temperature against average monthly precipitation (Figure 3.13). Although such graphs do not reflect year to year variability in climate, they do show at a glance the changes in both temperature

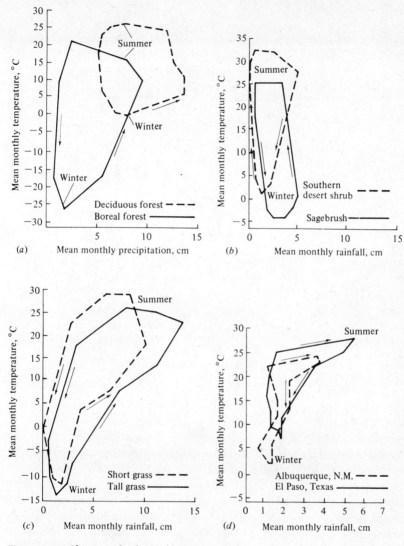

**Figure 3.13** Climographs for eight different areas with different vegetation types. [After Smith (1940).]

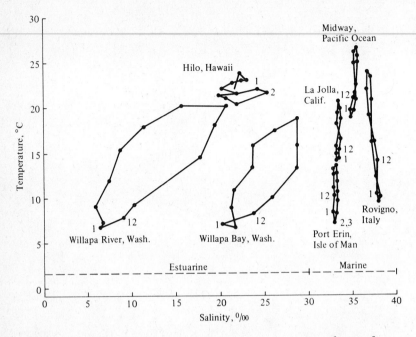

**Figure 3.14** Graphs of average monthly temperature versus salinity of some estuarine and marine waters; seasonal variation is great in the brackish waters, whereas salinity varies little in true oceanic waters. [After Odum (1971) after Hedgepeth.]

and precipitation within an average year, as well as the season(s) during which the precipitation usually falls. (Without actually identifying points by months, however, spring cannot be distinguished from fall since both are seasons of moderate temperatures—hence the arrows in (Figure 3.13.) Exactly analogous plots are often made for any two physical variables, such as temperature versus humidity, or temperature versus salinity in aquatic systems (Figure 3.14). Many other variables of biological importance, such as pH and dissolved phosphorus or nitrogen, can be treated similarly. When coupled with information on the tolerance limits of organisms, climographs and their analogues can be useful in predicting responses of organisms to changes in their physical environments (Figure 3.15).

Very cold climates are usually dry, whereas warmer regions show a wide range of average annual precipitation (Figure 3.16). Although there are an infinite number of different types of climates, attempts have been made to classify them. One scheme (Köppen's) recognizes five major climatic types (as well as many minor ones): tropical rainy, dry, warm temperate rainy, cool snow forest, and polar. Another classification is shown in Figure 3.17.

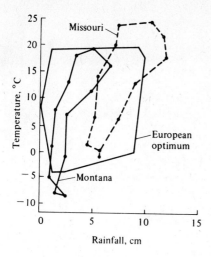

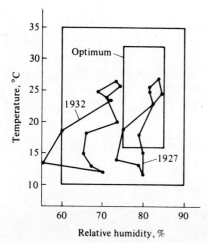

**Figure 3.15** Two plots of temperature against moisture. (*a*) Climographs for an area in Montana where the Hungarian partridge was introduced successfully and a Missouri locality where its introduction failed, compared to the average climatic conditions of its European geographic range. Apparently, Missouri summers are too hot or too wet for these birds. (*b*) Plots of temperature versus relative humidity in 1927 and 1932 in Israel superimposed on optimal (inner rectangle) and favorable (outer rectangle) conditions for the Mediterranean fruit fly. [Note that as drawn these rectangles assume no interaction between temperature and humidity; in actuality, the edges would presumably be rounded due to the principle of allocation (see also Chapter 5).] Damage to fruit crops by these flies was much greater in 1927 than in 1932. [After Odum (1959).]

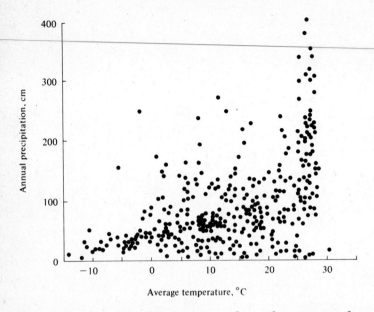

Figure 3.16 Average annual temperature and annual precipitation for many localities scattered more or less evenly over the land area of the earth. [Adapted from Ricklefs (1973) based on data of Clayton. By permission of the Smithsonian Institution Press, from Smithsonian Misc. Col. 79, *World Weather Records*. H. H. Clayton, Smithsonian Institution, Washington, D. C., 1927.]

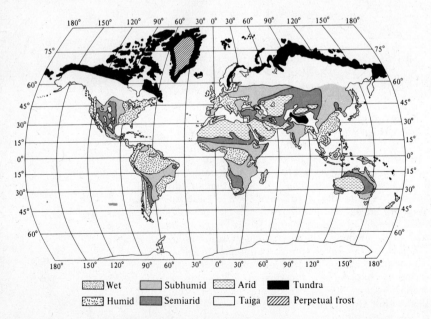

Figure 3.17 Geographic distribution of the principal climates, according to the Thornthwaite classification. [After Blumenstock and Thornthwaite (1941).]

# SELECTED REFERENCES

Blair and Fite (1965);   Blumenstock and Thornthwaite (1941);   Byers (1954);   Chorley and Kennedy (1971);   Collier et al. (1973);   Finch and Trewartha (1949);   Flohn (1969);   Gates (1962);   Haurwitz and Austin (1944);   Lowry (1969);   MacArthur (1972);   MacArthur and Connell (1966);   Taylor (1920);   Thornthwaite (1948);   Trewartha (1943);   U. S. Department of Agriculture (1941).

**Chapter**
**4**

# The Interface Between Climate and Vegetation

$C$limate is the major determinant of vegetation. Plants in turn exert some degree of influence on climate. Both climate and vegetation profoundly affect soil development and the animals that live in an area. Here we examine some ways in which climate and vegetation interact. More emphasis is given to terrestrial ecosystems than to aquatic ones, although some aquatic analogues are briefly noted. Topics presented rather succinctly here are treated in greater detail elsewhere (see references at end of chapter).

## PLANT LIFE FORMS AND BIOMES

Terrestrial plants adapted to a particular climatic regime often have similar morphologies, or plant growth forms. Thus, climbing vines, epiphytes, and broad-leafed species characterize tropical rainforests. Evergreen conifers dominate very cold areas at high latitudes and/or altitudes, whereas small frost-resistant tundra species occupy still higher latitudes and altitudes. Seasonal temperate zone areas with moderate precipitation usually support broad-leafed, decidu-ous trees, whereas tough-leafed (sclerophyllous) evergreen shrubs, or so-called chaparral-type vegetation, occur in regions with winter rains and a pronounced long water deficit during spring, summer, and fall. Chaparral vegetation is found wherever this type of climate prevails, including southern California, Chile, Spain, Italy, southwestern Australia, and the northern and southern tips of Africa (see Figure 4.1), although the actual plant species comprising the flora

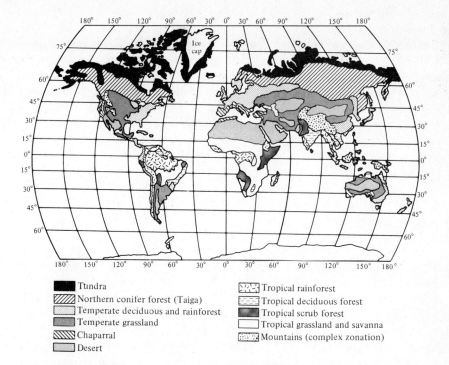

**Figure 4.1** Geographic distribution of major vegetation types. [After MacArthur and Connell (1966) after Odum.]

usually differ. Areas with very predictable and stable climates tend to support fewer different plant life forms than regions with more erratic climates. In general, there is a close correspondence between climate and vegetation (compare Figures 3.5 and 3.17 with Figure 4.1); indeed, climatologists have sometimes used vegetation as the best indicator of climate! Thus, rainforests occur in rainy tropical and rainy warm-temperate climates, forests exist under more moderate mesic climates, savannas and grasslands prevail in semiarid climates, and deserts characterize still drier climates. Of course, topography and soils also play a part in the determination of vegetation types, which are sometimes termed *plant formations*. Such major communities of characteristic plants and animals are known as *biomes*. Classification of natural communities is discussed later in this chapter.

## MICROCLIMATE

Even in the complete absence of vegetation, major climatic forces, or macroclimates, are expressed differently at a very local spatial level, which has resulted in the recognition of so-called *microclimates*. Thus, the surface of the ground undergoes the greatest daily variation in temperature, and daily thermal flux is progressively reduced with both increasing distance above and below ground level (Figure 4.2). During daylight hours the surface intercepts most of the incident solar energy and rapidly heats up, whereas at night this same surface cools more than its surroundings. Such plots of temperature versus height above and below ground are called *thermal profiles*. An analogous type of graph, called a *bathythermograph*, is often made for aquatic ecosystems by plotting temperature against depth (see Figure 4.17 on p. 78).

Daily temperature patterns are also modified by topography even in the absence of vegetation. A slope facing the sun intercepts light beams more perpendicularly than does a slope facing away from the sun; as a result, a south-facing slope in the Northern Hemisphere receives more solar energy than a north-facing slope, and the former heats up faster and gets hotter during the day (Figure 4.3). Moreover, such a south-facing slope is typically drier than a north-facing one because it receives more solar energy and therefore more water is evaporated.

By orienting themselves either parallel to or at right angles to the sun's rays, organisms (and parts of organisms such as leaves) may decrease or increase the total amount of solar energy they actually intercept. Leaves in the brightly illuminated canopy often droop during midday, whereas those in the shaded understory typically present their full surface to incoming beams of solar radiation. Similarly, many desert lizards position themselves on the ground perpendicular to the sun's rays in the early morning when environmental temperatures are low, but during the high temperatures of midday, these same animals reduce their heat load by climbing up off the ground into cooler air temperatures and orienting themselves parallel to the sun's rays by facing into the sun.

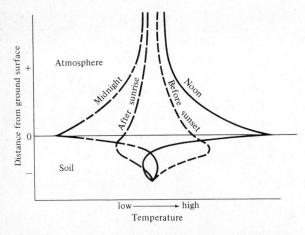

Figure 4.2 Idealized thermal profile showing temperatures at various distances above and below ground at four different times of day. [After Gates (1962).]

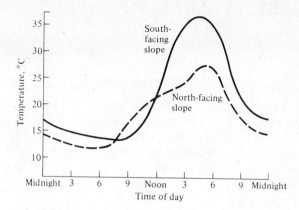

Figure 4.3 Daily marches of temperature on an exposed south-facing slope (solid line) and on a north-facing slope (dashed line) during late summer in the Northern Hemisphere. [After Smith (1966) after van Eck.]

The major effect of a blanket of vegetation is to moderate most daily climatic changes, such as changes in temperature, humidity, and wind. (However, plants generate daily variations in concentrations of oxygen and carbon dioxide through their photosynthetic and respiratory activities.) Thermal profiles at midday in cornfields at various stages of growth are shown in Figure 4.4, demonstrating the marked reduction in ground temperature due especially to shading. In the mature field, air is warmest at about a meter above ground. Similar vegetational effects on microclimates occur in natural communities. For example, a patch of open sand in a desert might have a daily thermal profile somewhat like that shown in Figure 4.2, whereas temperatures in the litter underneath a nearby dense shrub would vary much less with the daily march of temperature.

Humidities are similarly modified by vegetation, with relative humidities within a dense plant being somewhat greater than those of the open air adjacent

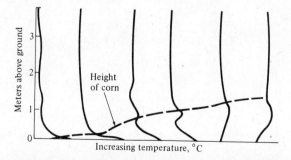

Figure 4.4 Temperature profiles in a growing cornfield at midday, showing the effect of vegetation on thermal microclimate. [After Smith (1966).]

to the plant. An aphid may spend its entire lifetime in the very thin zone (only about a millimeter thick) of high humidity that surrounds the surface of a leaf. Moisture content is more stable, and therefore more dependable, deeper in the soil than it is at the surface, where high temperatures periodically evaporate water to produce a desiccating effect.

Wind velocities are also reduced sharply by vegetation and are usually lowest near the ground (Figures 4.5 and 4.6). Moving currents of air promote rapid exchange of heat and water; hence an organism cools or warms more rapidly in a wind than it does in a stationary air mass at the same temperature. Likewise, winds often carry away moist air and replace it with drier air, thereby promoting evaporation and water loss. The desiccating effects of such dry winds can be extremely important to an organism's water balance.

In aquatic systems, water turbulence parallels wind in many ways, and rooted vegetation around the edges of a pond or stream reduces water turbulence. At a more microscopic level, algae and other organisms that attach themselves to underwater surfaces (so-called *periphyton*) create a thin film of distinctly modified microenvironment in which water turbulence, among other things, is reduced. Localized spatial patches with particular concentrations of hydrogen ions (pH), salinities, dissolved nitrogen and phosphorus, and the like form similar aquatic microhabitats.

By actively or passively selecting such microhabitats, organisms can effectively reduce the overall environmental variation they encounter and enjoy more optimal conditions than they could without microhabitat selection. Innu-

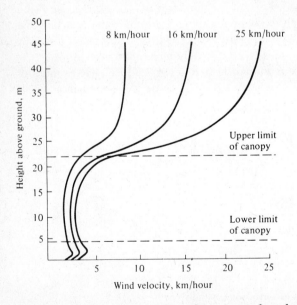

Figure 4.5 Wind velocities within a forest vary relatively little with changes in the wind velocity above the canopy. [After Fons (1940).]

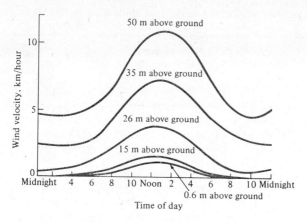

**Figure 4.6** Daily march of average wind velocities during June at various heights inside a coniferous forest in Idaho. [After Smith (1966) after Gisborne.]

merable other microclimatic effects could be cited, but these should serve to illustrate their existence and their significance to plants and animals.

## PRIMARY PRODUCTION AND EVAPOTRANSPIRATION

In terrestrial ecosystems, climate is by far the most important determinant of the amount of solar energy plants are able to capture as chemical energy, or the gross primary productivity (Table 4.1). In warm arid regions, water is a master limiting factor, and in the absence of runoff, primary production is strongly positively correlated with rainfall in a linear fashion (see Figure 5.1 on p. 83). Above about 80 centimeters of precipitation per year, primary production slowly decreases with increasing precipitation and then levels off (asymptotes; Figure 4.7 on p. 67). Notice that some points fall below the curve in this figure, presumably because there is some water loss by runoff and seepage into groundwater supplies.

*Evapotranspiration* refers to the release of water into the atmosphere as water vapor, both by the physical process of evaporation and by the biological processes of transpiration and respiration. The amount of water vapor thus returned to the atmosphere depends strongly on temperature, with greater evapotranspiration at higher temperatures. The *theoretical* temperature-dependent amount of water that *could* be "cooked out" of an ecological system, given its input of solar energy and provided that much water fell on the area, is called its *potential evapotranspiration* (PET). In many ecosystems, water is frequently in short supply, so *actual evapotranspiration* (AET) is somewhat less than potential (clearly, AET can never exceed PET and is equal to PET only in water-saturated habitats). Actual evapotranspiration can be thought of as the reverse of rain, for it is the amount of water that actually goes back into the atmosphere at a given spot.

**TABLE 4.1 Net Primary Productivity and World Net Primary Production for the Major Ecosystems**

| | Area[a] ($10^6$ km$^2$) | Net primary productivity per unit area[b] (dry g/m$^2$/yr) Normal range | Mean | World net primary production[c] ($10^9$ dry tons/yr) |
|---|---|---|---|---|
| Lake and stream | 2 | 100–1500 | 500 | 1.0 |
| Swamp and marsh | 2 | 800–4000 | 2000 | 4.0 |
| Tropical forest | 20 | 1000–5000 | 2000 | 40.0 |
| Temperate forest | 18 | 600–2500 | 1300 | 23.4 |
| Boreal forest | 12 | 400–2000 | 800 | 9.6 |
| Woodland and shrubland | 7 | 200–1200 | 600 | 4.2 |
| Savanna | 15 | 200–2000 | 700 | 10.5 |
| Temperate grassland | 9 | 150–1500 | 500 | 4.5 |
| Tundra and alpine | 8 | 10–400 | 140 | 1.1 |
| Desert scrub | 18 | 10–250 | 70 | 1.3 |
| Extreme desert, rock, and ice | 24 | 0–10 | 3 | 0.07 |
| Agricultural land | 14 | 100–4000 | 650 | 9.1 |
| Total land | 149 | | 730 | 109.0 |
| Open ocean | 332 | 2–400 | 125 | 41.5 |
| Continental shelf | 27 | 200–600 | 350 | 9.5 |
| Attached algae and estuaries | 2 | 500–4000 | 2000 | 4.0 |
| Total ocean | 361 | | 155 | 55.0 |
| Total for Earth | 510 | | 320 | 164.0 |

[a]Square kilometers $\times$ 0.3861 = square miles.

[b]Grams per square meter $\times$ 0.01 = t/ha, $\times$ 0.1 = dz/ha or m centn/ha (metric centers, 100 kg, per hectare, $10^4$ square meters), $\times$ 10 = kg/ha, $\times$ 8.92 = lb/acre.

[c]Metric tons ($10^6$ g) $\times$ 1.1023 = English short tons.

*Source:* Adapted from Whittaker (1970), *Communities and Ecosystems.* Reprinted with permission of Macmillan Publishing Co., Inc. ©Copyright Robert H. Whittaker, 1970.

The potential evapotranspiration for any spot on earth is determined by the same factors that regulate temperature, most notably latitude, altitude, cloud cover, and slope (topography). There is a nearly one-to-one correspondence between PET and temperature, and an annual march of PET can be plotted in centimeters of water. By superimposing the annual march of precipitation on these plots (Figure 4.8 on p. 68), seasonal changes in water availability can be depicted graphically. A water deficit occurs when PET exceeds precipitation; a water surplus exists when the situation is reversed.

During a period of water surplus, some water may be stored by plants and some may accumulate in the soil as soil moisture, depending on runoff and the capacity of soils to hold water; during a later water deficit, such stored water can be used by plants and released back into the atmosphere. Winter rain is generally much less effective than summer rain because of the reduced activity (or complete inactivity) of plants in winter; indeed, two areas with the

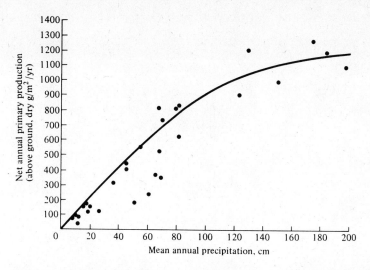

**Figure 4.7** Net primary productivity (above ground) plotted against average annual precipitation. [From Whittaker (1970). Reprinted with permission of Macmillan Publishing Co., Inc., from *Communities and Ecosystems* by Robert H. Whittaker. Copyright © 1970 by Robert H. Whittaker.]

same annual march of temperature and total annual precipitation may differ greatly in the types of plants they support and in their productivity as a result of their seasonal patterns of precipitation. An area receiving about 50 centimeters of precipitation annually supports either a grassland vegetation or chaparral, depending on whether the precipitation falls in summer or winter, respectively.

Net annual primary production above ground is strongly correlated with actual evapotranspiration, or AET (Figure 4.9 on p. 69). This correlation is remarkable in that AET was crudely estimated using only monthly macroclimatic statistics with no allowance for either runoff and water-holding capacities of soils or for groundwater usage.

Rosenzweig (1968) suggested that the reason for the observed correlation is that AET measured simultaneously two of the most important factors limiting primary production on land: water and solar energy. Photosynthesis, that fundamental process on which nearly all life depends for energy, is represented by the chemical equation

$$6CO_2 + 12H_2O \rightarrow C_6H_{12}O_6 + 6O_2 + 6H_2O$$

where $C_6H_{12}O_6$ is the energy-rich glucose molecule. Carbon dioxide concentration in the atmosphere is fairly constant at about 0.03 to 0.04 percent and does not strongly influence the rate of photosynthesis, except under unusual conditions of high water availability and full or nearly full sunlight, when $CO_2$ is limiting (Meyer, Anderson, and Bohning, 1960). Rosenzweig notes that, on a geographical scale, each of the other two requisites for photosynthesis, water and solar energy, is much more variable in its availability and more often lim-

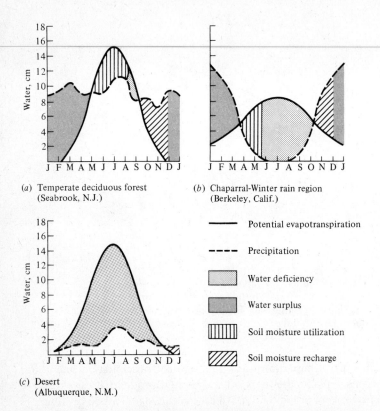

(a) Temperate deciduous forest
(Seabrook, N.J.)

(b) Chaparral-Winter rain region
(Berkeley, Calif.)

Potential evapotranspiration

Precipitation

Water deficiency

Water surplus

Soil moisture utilization

Soil moisture recharge

(c) Desert
(Albuquerque, N.M.)

**Figure 4.8.** Plots of annual march of potential evapotranspiration superimposed on the annual march of precipitation for three ecologically distinct regions showing the water relations of each. (a) Temperate deciduous forest, (b) chaparral with winter rain, (c) desert. [From Odum (1959) after Thornthwaite.]

iting; moreover, AET measures the availability of both. Temperature is often a rate-limiting factor and markedly affects photosynthesis; it, too, is presumably incorporated into an estimated AET value. Primary production may be influenced by nutrient availability as well, but in many terrestrial ecosystems, these effects are often relatively minor. In aquatic ecosystems, nutrient availability is often a major determinant of the rate of photosynthesis. Primary production in aquatic systems of fairly constant temperature (such as the oceans) is often strongly affected by light. Because light intensity diminishes rapidly with depth in lakes and oceans, most primary productivity is concentrated near the surface (Figure 4.10a). However, short wavelengths (blue light) penetrate deeper into water than longer ones, and some benthic (bottom-dwelling) marine "red" algae have evolved unique photosynthetic adaptations to utilize these wavelengths.

Similarly, within forests, light intensity varies markedly with height above ground (Figure 4.10b). Tall trees in the canopy receive the full incident solar radiation, whereas shorter trees and shrubs receive progressively less light. In

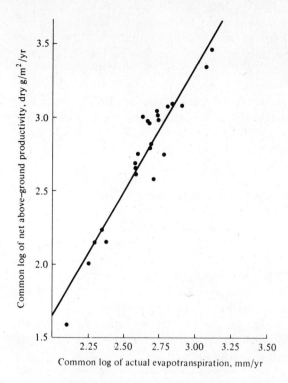

**Figure 4.9** Log-log plot of net primary productivity above ground against estimated actual evapotranspiration for 24 areas, ranging from barren desert to luxurious tropical rainforest. [After Rosenzweig (1968).]

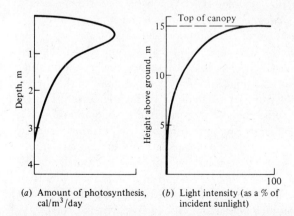

(a) Amount of photosynthesis, cal/m³/day

(b) Light intensity (as a % of incident sunlight)

**Figure 4.10** Vertical profiles of (a) the amount of photosynthesis versus depth in a lake and (b) light intensity versus height above ground in a forest.

really dense forests, less than 1 percent of the incident solar energy impinging on the canopy actually penetrates to the forest floor. Although a tree in the canopy has more solar energy available to it than a fern on the forest floor, the canopy tree must also expend much more energy on vegetative supporting tissues (wood) than the fern. (Understory plants are usually very shade tolerant and able to photosynthesize at very low light intensities. Also, herbs of the forest floor often grow and flower early in the spring before deciduous trees leaf out.) Hence, each plant life form and each growth strategy has its own associated costs and profits.

## SOIL FORMATION AND PRIMARY SUCCESSION

Soils are a key part of the terrestrial ecosystem because many processes critical to the functioning of ecosystems occur in the soil. This is where dead organisms are decomposed and where their nutrients are retained until used by plants and, indirectly, returned to the remainder of the community. Soils are essentially a meeting ground of the inorganic and organic worlds. Many organisms live in the soil; perhaps the most important are the decomposers, which include a rich biota of bacteria and fungi. Also, the vast majority of insects spend at least a part of their life cycle in the soil. Certain soil-dwelling organisms, such as earthworms, often play a major role in breaking down organic particles into smaller pieces, which present a larger surface area for microbial action, thereby facilitating decomposition. Indeed, activities of such soil organisms sometimes constitute a "bottleneck" for the rate of nutrient cycling, and as such they can regulate nutrient availability and turnover rates in the entire community. In aquatic ecosystems, bottom sediments like mud and ooze are closely analogous to the soils of terrestrial systems.

Much of modern soil science, or *pedology*, was anticipated in the late 1800s by the prominent Russian pedologist V. V. Dokuchaev. He devised a theory of soil formation, or *pedogenesis*, based largely on climate, although he also recognized the importance of time, topography, organisms (especially vegetation), and parent materials (the underlying rocks from which the soil is derived). The relative importance of each of these five major soil-forming factors varies from situation to situation. Figure 4.11 shows how markedly the soil changes along the transition from prairie to forest in the midwestern United States, where the only other conspicuous major variable is vegetation type.

Grasses and trees differ substantially both in mineral requirements and in the extent to which the products of their own primary production contribute organic materials to the soil. Typically, natural soils underneath grasslands are considerably deeper and richer in both organic and inorganic nutrients than are the natural soils of forested regions. Jenny (1941) gives many examples of how each of the five major soil-forming factors influences particular soils.

The marked effects some soil types can have on plants are well illustrated by so-called *serpentine soils* (Whittaker, Walker, and Kruckeberg, 1954), which are formed over a parent material of serpentine rock. These soils often occur

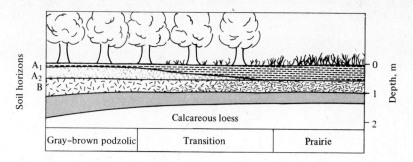

Figure 4.11 Diagram of typical soil changes along the transition from prairie (deep black topsoil) to forest (shallow topsoil) at the edge of the North American great plains. [After Crocker (1952).]

in localized patches surrounded by other soil types; typically, the vegetation changes abruptly from nonserpentine to serpentine soils. Serpentine soils are rich in magnesium, chromium, and nickel, but they contain very little calcium, molybdenum, nitrogen, and phosphorus. They usually support a stunted vegetation and are relatively less productive than adjacent areas with different, richer soils. Indeed, entire floras of specialized plant species have evolved that are tolerant of the conditions of serpentine soils (particularly their low calcium levels). Introduced Mediterranean "weeds" have replaced native Californian coastal grasses and forbs almost everywhere except on serpentine soils, where the native flora still persists.

Soil development from bare rock, or primary succession, is a very slow process that often requires centuries. Rock is fragmented by temperature changes, by the action of windblown particles, and in colder regions by the alternate expansion and contraction of water as it freezes and thaws. Chemical reactions, such as the formation of carbonic acid ($H_2CO_3$) from water and carbon dioxide, may also help to dissolve and break down certain rock types like limestones. Such weathering of rocks releases inorganic nutrients that can be used by plants. Eventually, lichens establish themselves, and, as other plants root and grow, root expansion further breaks up the rock into still smaller fragments. As these plants photosynthesize, they convert inorganic materials into organic matter. Such organic material, mixed with inorganic rock fragments, accumulates, and soil is slowly formed. Early in primary succession, production of new organic material exceeds its consumption and organic matter accumulates; as soil "maturity" is approached, soil eventually ceases to accumulate (see also Figure 4.16 on p. 76).

Whereas most organic material is contributed to soils from above as leaf and litter fall, mineral inorganic components tend to be added from the underlying rocks below. These polarized processes thus generate fairly distinct layers, termed *soil horizons*.

Even though litter fall is high in tropical forests, it does not accumulate to nearly as great an extent as it does in the temperate zones, presumably because decomposition rates are very high in the warm tropics. As a result, tropical soils tend to be poor in nutrients (high rainfall in many tropical areas

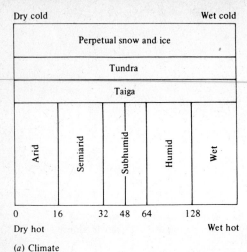

(a) Climate

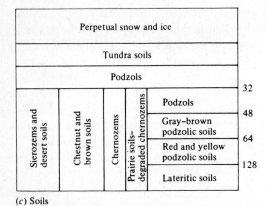

(b) Vegetation

(c) Soils

Figure 4.12 Relationships between temperature and precipitation and (a) climatic types, (b) vegetational formations, and (c) major zonal soil groups. [From Blumenstock and Thornthwaite (1941).]

further depletes these soils by leaching out water-soluble nutrients). For both reasons, tropical areas simply cannot support sustained agriculture nearly as well as can temperate regions (in addition, diverse tropical communities are probably much more fragile than simpler temperate-zone systems).

There are fairly close parallels between concepts of soil development and those of developing ecological communities; pedologists speak of "mature" soils at the steady state, whereas ecologists recognize the "climax" communities that grow on and live in these same soils. These two components of the ecosystem (soils and vegetation) are intricately interrelated and interdependent; each strongly influences the other. Except in forests and rainforests, there is usually a one-to-one correspondence between them (Figure 4.12). (Compare also the geographic distribution of soil types shown in Figure 4.13 with the distribution of vegetation types shown in Figure 4.1.)

Once a mature soil has been formed, a disturbance such as the removal of vegetation by fire or human activities often results in gradual sequential changes in the organisms comprising the community. Such a temporal sequence of communities is termed a *secondary succession*.

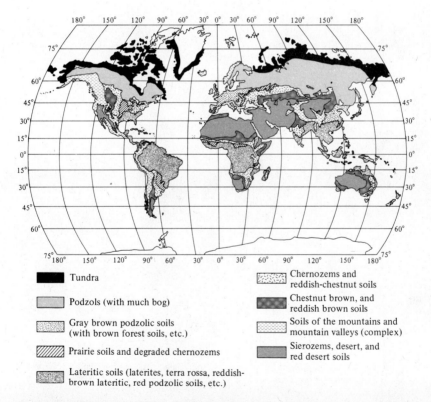

Tundra

Podzols (with much bog)

Gray brown podzolic soils (with brown forest soils, etc.)

Prairie soils and degraded chernozems

Lateritic soils (laterites, terra rossa, reddish-brown lateritic, red podzolic soils, etc.)

Chernozems and reddish-chestnut soils

Chestnut brown, and reddish brown soils

Soils of the mountains and mountain valleys (complex)

Sierozems, desert, and red desert soils

**Figure 4.13** Geographic distribution of the primary soil types. Compare with vegetation map shown in Figure 4.1. [After Blumenstock and Thornthwaite (1941).]

## ECOTONES AND VEGETATIONAL CONTINUA

Communities are seldom discrete entities; in fact, they usually grade into one another in both space and time. A localized "edge community" between two other reasonably distinct communities is termed an *ecotone*. Typically, such ecotonal communities are rich in species because they contain representatives from both parent communities and may also contain species distinctive of the ecotone itself. Often a series of communities grade into one another in an almost continuous fashion (Figure 4.14); such a gradient community is called an *ecocline*. Ecoclines may occur either in space or in time.

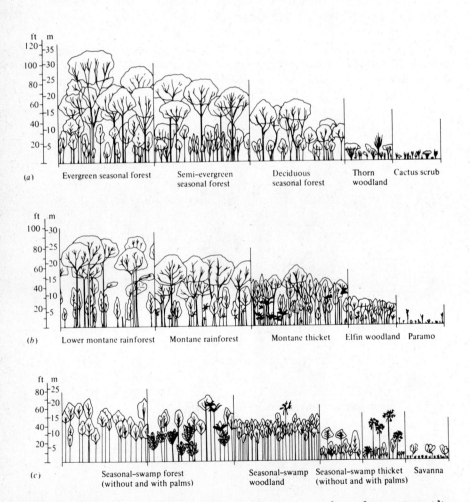

**Figure 4.14** Vegetation profiles along three ecoclines. (*a*) A gradient of increasing aridity from seasonal rainforest to desert. (*b*) An elevation gradient up a tropical mountainside from tropical rainforest to alpine meadow (paramo). (*c*) A humidity gradient from swamp forest to savanna. [From Beard (1955).]

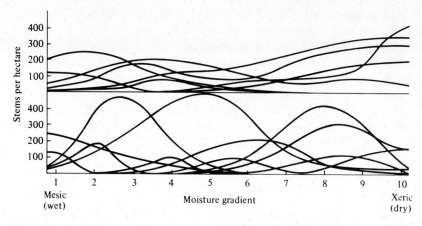

**Figure 4.15** Actual distributions of some populations of plant species along moisture gradients from relatively wet ravines to dry southwest-facing slopes in the Siskiyou mountains of northern California (above) and the Santa Catalina mountains of Arizona (below). [After Whittaker (1967).]

Spatial ecoclines on a more local scale have led to so-called *gradient analysis* (Whittaker, 1967). The abundance and actual distribution of organisms along many environmental gradients have shown that the importances of various species along any given gradient typically form bell-shaped curves reminiscent of tolerance curves (considered in the next chapter). These curves tend to vary independently of one another and often overlap broadly (Figure 4.15), indicating that each species' population has its own particular habitat requirements and width of habitat tolerance—and hence its own zone of maximal importance. Such a continuous replacement of plant species by one another along a habitat gradient is termed a *vegetational continuum* (Figure 4.15).

A temporal ecocline, or a change in community composition in time, both by changes in the relative importance of component populations and by extinction of old species and invasion of new ones, is termed a *succession*. Primary succession, as we have just seen, is the development of communities from bare rock; secondary successions are changes that take place after destruction of the natural vegetation of an area that already has soil present. Secondary succession is often a more or less orderly sequential replacement of early succession species, typically rapidly growing, colonizing species, by other more competitive species that succeed in later stages, usually slow-growing and shade-tolerant species. The final stage in succession is termed the *climax*. Secondary succession is discussed in more detail in Chapter 16 (pp. 359–361).

## CLASSIFICATION OF NATURAL COMMUNITIES

Biotic communities have been classified in various ways. An early attempt to classify communities was that of Merriam (1890), who recognized a number of different "life zones" defined solely in terms of temperature (ignoring precipitation). His somewhat simplistic scheme is no longer used, but his approach did link climate with vegetation in a more or less predictive manner.

Shelford (1913a, 1963) and his students have taken a somewhat different approach to the classification of natural communities that does not attempt to correlate climate with the plants and animals occurring in an area. Rather, they classify different natural communities into a large number of so-called *biomes* and *associations*, relying largely upon the characteristic plant and animal species that compose a particular community. As such, this scheme is descriptive rather than predictive. Such massive descriptions of different communities

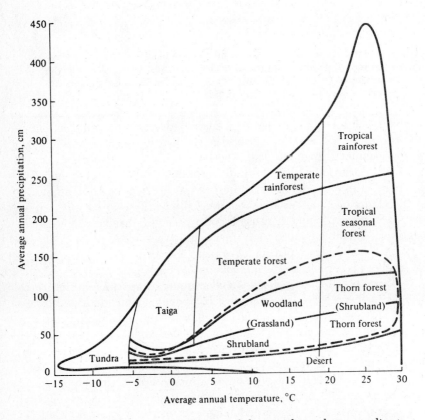

**Figure 4.16** Diagrammatic representation of the correlation between climate, as reflected by average annual temperature and precipitation, and vegetational formation types. Boundaries between types are approximate and are influenced locally by soil type, seasonality of rainfall, and disturbances such as fires. The dashed line encloses a range of climates in which either grasslands or woody plants may constitute the prevailing vegetation of an area, depending on the seasonality of precipitation. Compare this figure with Figure 3.15. [After Whittaker (1970). Reprinted with permission of Macmillan Publishing Co., Inc., from *Communities and Ecosystems* by Robert H. Whittaker. Copyright © 1970 by Robert H. Whittaker.]

(see, e.g., Dice, 1952, and Shelford, 1963) can often be quite useful in that they allow one to become familiar with a particular community with relative ease.

Workers involved in such attempts at classification typically envision communities as discrete entities with relatively little or no intergradation between them; thus, the Shelford school considers biomes to be distinct and real entities in nature rather than artificial and arbitrary human constructs. Another school of ecologists, represented by McIntosh (1967) and Whittaker (1970), takes an opposing view, emphasizing that communities grade gradually into one another and form so-called *continua* or *ecoclines* (Figures 4.14 and 4.15).

In a plot space of average annual precipitation versus average annual temperature (Figure 4.16), various vegetational formations typically occurring under various climatic regimes can be plotted. Macroclimate determines the vegetation of an area—these correlations are not hard and fast, and assume that local vegetation type depends on other factors such as soil types, seasonality of rainfall regime, and frequency of disturbance by fires and floods.

# SOME CONSIDERATIONS OF AQUATIC ECOSYSTEMS

Although the same ecological principles presumably operate in both aquatic and terrestrial ecosystems, there are striking and interesting fundamental differences between these two ecological systems. For example, primary producers on land are sessile and many tend to be large and relatively long-lived (air does not provide much support and woody tissues are needed), whereas, except for kelp, producers in aquatic communities are typically free-floating, microscopic, and very short-lived (the buoyancy of water may make supportive plant tissues unnecessary; a large planktonic plant might be easily broken by water turbulence). Most ecologists study either aquatic or terrestrial systems, but seldom both. Various aquatic subdisciplines of ecology are recognized, such as aquatic ecology and marine ecology. Limnology is the study of freshwater ecosystems (ponds, lakes, and streams); oceanography is concerned with bodies of salt water. Because the preceding part of this chapter and most of the remainder of the book emphasize terrestrial ecosystems, certain salient properties of aquatic ecosystems, especially lakes, are briefly considered in this section. Lakes are particularly appealing subjects for ecological study in that they are self-contained ecosystems, discrete, and largely isolated from other ecosystems. Nutrient flow into and out of a lake can often be estimated with relative ease. The study of lakes is fascinating, and the interested reader is referred to Ruttner (1953), Hutchinson (1957b, 1967), Cole (1975), and Wetzel (1975).

Water has peculiar physical and chemical properties that strongly influence the organisms that live in it. As indicated earlier, water has a high specific heat; moreover, in the solid (frozen) state, its density is less than it is in the liquid state (i.e., ice floats). Water is most dense at 4°C and water at this temperature "sinks." Furthermore, water is nearly a "universal solvent" in that many important substances dissolve into aqueous solution.

A typical, relatively deep lake in the temperate zones undergoes marked and very predictable seasonal changes in temperature. During the warm summer months, its surface waters are heated up, and because warm water is less dense than colder water, a distinct upper layer of warm water, termed the *epilimnion*, is formed (Figure 4.17*a*). (Movement of heat within a lake is due to water currents produced primarily by wind.) Deeper waters, termed the *hypolimnion*, remain relatively cold during summer, often at about 4°C; an intermediate layer of rapid temperature change, termed the *thermocline*, separates the epilimnion from the hypolimnion (Figure 4.17*a*). (A swimmer sometimes experiences these layers of different temperatures when diving into deep water or when in treading water his or her feet drop down into the cold hypolimnion.) A lake with a thermal profile, or bathythermograph, like that

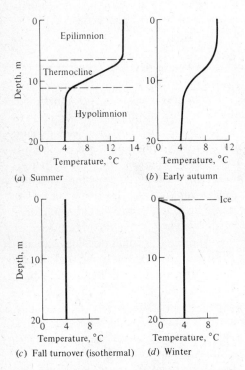

(a) Summer

(b) Early autumn

(c) Fall turnover (isothermal)

(d) Winter

**Figure 4.17** Hypothetical bathythermographs showing seasonal changes typical of a deep temperate zone lake. (a) A stratified lake during summer. (b) In early autumn, upper waters cool. (c) In late autumn or early winter, the lake's waters are all at exactly the same temperature, here 4°C. (The lake is "isothermal.") (d) During the freezing winter months, a layer of surface ice chills the uppermost water.

shown in Figure 4.17*a* is said to be "stratified" because of its layer of warm water over cold water. Typically, there is little mixing of the warm upper layer with the heavier deeper water.

With the decrease in incident solar energy in autumn, surface waters cool and give up their heat to adjacent landmasses and the atmosphere (Figure 4.17*b*). Eventually, the epilimnion cools to the same temperature as the hypolimnion and the lake becomes isothermal (Figure 4.17*c*). This is the time of the "fall turnover." With winter's freezing temperatures, the lake's surface turns to ice and its temperature-versus-depth profile looks something like that in Figure 4.17*d*. Finally, in spring the ice melts and the lake is briefly isothermal once again (it may have a spring turnover) until its surface waters are rapidly warmed, when it again becomes stratified and the annual cycle repeats itself.

Because prevailing winds produce surface water currents, a lake's waters circulate. In stratified lakes, the epilimnion constitutes a more or less closed cell of circulating warm water, whereas the deep cold water scarcely moves or mixes with the warmer water above it. During this period, as dead organisms and particulate organic matter sink into the noncirculating hypolimnion, the lake undergoes what is known as *summer stagnation*. When a lake becomes isothermal, its entire water mass can be circulated and nutrient-rich bottom waters brought to the surface; limnologists thus speak of the spring and the fall "turnover." Meteorological conditions, particularly wind velocity and duration, strongly influence such turnovers; indeed, if there is little wind during the period a lake is isothermal, its waters might not be thoroughly mixed and many nutrients may remain locked in its depths. After a thorough turnover, the entire water mass of a lake is equalized and concentrations of various substances, such as oxygen and carbon dioxide, are similar throughout the lake.

Lakes differ in their nutrient content and degree of productivity and they can be arranged along a continuum ranging from those with low nutrient levels and low productivity (*oligotrophic* lakes) to those with high nutrient content and high productivity (*eutrophic* lakes). Clear, cold, and deep lakes high in the mountains are usually relatively oligotrophic, whereas shallower, warmer, and more turbid lakes such as those in low-lying areas are generally more eutrophic. Oligotrophic lakes typically support game fish such as trout, whereas eutrophic lakes contain "trash" fish such as carp. As they age and fill with sediments, many lakes gradually undergo a natural process of eutrophication, steadily becoming more and more productive. People accelerate this process by enriching lakes with wastes, and many oligotrophic lakes have rapidly become eutrophic under our influence. A good indicator of the degree of eutrophication is the oxygen content of deep water during summer. In a relatively unproductive lake, oxygen content varies little with depth and there is ample oxygen at the bottom of the lake. In contrast, oxygen content diminishes rapidly with depth in productive lakes, and anaerobic processes sometimes characterize their depths during the summer months (an example of a succession). With the autumn turnover, oxygen-rich waters again reach the bottom sediments and aerobic processes become possible. However, once such a lake becomes stratified, the oxygen in

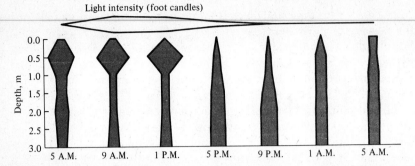

Figure 4.18 Many small freshwater planktonic animals move vertically during the daily cycle of illumination somewhat as shown here (widths of bands represent the density of animals at a given depth at a particular time). [After Hutchinson (1967) after Cowles and Brambel.]

its deep water is quickly used up by benthic organisms (in the dark depths there is little or no photosynthesis to replenish the oxygen).

These seasonal physical changes profoundly influence the community of organisms living in a lake. During the early spring and after the fall turnover, surface waters are rich in dissolved nutrients such as nitrates and phosphates and temperate lakes are very productive, whereas during midsummer, many nutrients are unavailable to phytoplankton in the upper waters and primary production is much reduced.

Organisms within a lake community are usually distributed quite predictably in time and space. Thus, there is typically a regular seasonal progression of planktonic algae, with diatoms most abundant in the winter, changing to desmids and green algae in the spring, and gradually giving way to blue-green algae during summer months. The composition of the zooplankton also varies seasonally. Such temporal heterogeneity may well promote a diverse plankton community by periodically altering competitive abilities of component species, hence facilitating coexistence of many species of plants and animals in the relatively homogeneous planktonic environment (Hutchinson, 1961). Although plankton are moved about by water currents, many are strong enough swimmers to select a particular depth. Such species often actually "migrate" vertically during the day and with the seasons, being found at characteristic depths at any given time (Figure 4.18). Some zooplankters sink into the dark depths during the daylight hours (probably an adaptation to avoid visually hunting predators such as fish), but ascend to the surface waters at night to feed on phytoplankton.

## SELECTED REFERENCES

Allee et al. (1949);   Andrewartha and Birch (1954);   Clapham (1973);   Clarke (1954); Colinvaux (1973);   Collier et al. (1973);   Daubenmire (1947, 1956, 1968);   Gates

(1972); Kendeigh (1961); Knight (1965); Krebs (1972); Lowry (1969); Odum (1959, 1971); Oosting (1958); Ricklefs (1973); Smith (1966); Watt (1973); Weaver and Clements (1938); Whittaker (1970).

## Plant Life Forms and Biomes

Cain (1950); Clapham (1973); Givnish and Vermeij (1976); Horn (1971); Raunkaier (1934); Whittaker (1970).

## Microclimate

Collier et al. (1973); Fons (1940); Gates (1962); Geiger (1966); Gisborne (1941); Lowry (1969); Schmidt-Nielsen (1964); Smith (1966).

## Primary Production and Evapotranspiration

Collier et al. (1973); Gates (1965); Horn (1971); Meyer et al. (1960); Odum (1959, 1971); Rosenzweig (1968); Whittaker (1970); Woodwell and Whittaker (1968).

## Soil Formation and Primary Succession

Black (1965); Burges and Raw (1967); Crocker (1952); Crocker and Major (1955); Doeksen and van der Drift (1963); Eyre (1963); Fried and Broeshart (1967); Jenny (1941); Joffe (1949); Oosting (1958); Richards (1974); Schaller (1968); Waksman (1952); Whittaker et al. (1954).

## Ecotones and Vegetational Continua

Clements (1920, 1949); Horn (1971, 1975a, 1975b, 1976); Kershaw (1964); Loucks (1970); Margalef (1958b); McIntosh (1967); Pickett (1976); Shimwell (1971); Terborgh (1971); Whittaker (1953, 1965, 1967, 1969, 1970, 1972).

## Classification of Natural Communities

Beard (1955); Braun-Blanquet (1932); Clapham (1973); Dice (1952); Gleason and Cronquist (1964); Holdridge (1947, 1959, 1967); McIntosh (1967); Merriam (1890); Shelford (1913a, 1963); Tosi (1964); Whittaker (1962, 1967, 1970).

## Some Considerations of Aquatic Ecosystems

Clapham (1973); Cole (1975); Cowles and Brambel (1936); Ford and Hazen (1972); Frank (1968); Frey (1963); Grice and Hart (1962); Henderson (1913); Hochachka and Somero (1973); Hutchinson (1951, 1957b, 1961, 1967); Mann (1969); National Academy of Science (1969); Perkins (1974); Russell-Hunter (1970); Ruttner (1953); Sverdrup et al. (1942); Watt (1973); Welch (1952); Wetzel (1975); Weyl (1970).

## Chapter 5

# Resource Acquisition and Allocation

## LIMITING FACTORS AND TOLERANCE CURVES

Ecological events and their outcomes, such as growth, reproduction, photosynthesis, primary production, and population size, are often regulated by the availability of one or a few factors or requisites in short supply, whereas other resources and raw materials present in excess may go partially unused. This principle has become known as the *law of the minimum* (Liebig, 1840). For instance, in arid climates, primary production (the amount of solar energy trapped by green plants) is strongly correlated with precipitation (Figure 5.1); here water is a "master limiting factor." Of many different factors that can be limiting, frequently among the most important are various nutrients, water, and temperature.

When considering populations, we often speak of those that are food-limited, predator-limited, or climate-limited. Populations may be limited by other factors as well; for example, density of breeding pairs of blue tits (*Parus caeruleus*) in an English woods was doubled by the addition of many new nesting boxes (Lack, 1954, 1966), providing an indication that nest sites were limiting. However, limiting factors are not always so clear-cut but may usually interact so that a process is limited simultaneously by several factors, with a change in any one of them resulting in a new equilibrium. For instance, both increased food availability and decreased predation pressures might result in a larger population size.

A related concept, developed by Shelford (1913b), is now known as the *law of tolerance*. Too much or too little of anything can be detrimental to an

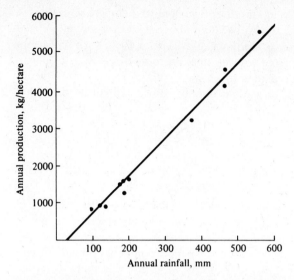

**Figure 5.1.** An example of the strong correlation between annual rainfall and primary production along a precipitation gradient in a desert region of Namibia. [Adapted from Odum (1959) after Walter (1939).]

organism. In the early morning, a desert lizard finds itself in an environment that is largely too cold, whereas later in the day its environment is too hot. The lizard compensates somewhat for this by spending most of its time during the early morning in sunny places, whereas later on most of its activities take place in the shade. Each lizard has a definite optimal range of temperature, with both upper and lower *limits of tolerance*. More precisely, when measures of performance (such as fitness, survivorship, or foraging efficiency) are plotted against important environmental variables, bell-shaped curves usually result (e.g., Figures 5.2 and 5.8).

Organisms can be viewed as simple input-output systems, with foraging or photosynthesis providing an input of materials and energy that are in turn "mapped" into an output consisting of progeny. Fairly extensive bodies of theory now exist both on optimal foraging and reproductive tactics. In optimal foraging theory, the "goal" usually assumed to be maximized is energy uptake per unit time (successful offspring produced during an organism's lifetime would be a more realistic measure of its foraging ability, but fitness is exceedingly difficult to measure). Similarly, among organisms without parental care, reproductive effort has sometimes been estimated by the ratio of calories devoted to eggs or offspring over total female calories at any instant (rates of uptake versus expenditure of calories have unfortunately not yet infiltrated empirical studies of reproductive tactics). To date, empirical studies of resource partitioning have been concerned largely with "input" phenomena such as overlap in and efficiency of resource utilization and have neglected to relate these to "output" aspects. In contrast, empirical studies of reproductive tactics have done the reverse and almost entirely omitted any consideration of foraging. Interactions and constraints between foraging and reproduction have barely begun to be considered. A promising area for future work will be to merge aspects of optimal foraging with optimal reproductive tactics to specify rules by which input is translated into output; optimal reproductive tactics ("output" phenomena) surely must often impose substantial constraints upon "input" possibilities, and vice versa.

## RESOURCE BUDGETS: THE PRINCIPLE OF ALLOCATION

Any organism has a limited amount of resources available to devote to foraging, growth, maintenance, and reproduction. The way in which an organism allocates its time and energy and other resources among various conflicting demands is of fundamental interest because such apportionments provide insight into how the organism copes with and conforms to its environment. Moreover, because any individual has finite resource and energy budgets, its capacity for regulation is necessarily limited. Organisms stressed along any one environmental variable are thus able to tolerate a lesser range of conditions along other environmental variables (see Figure 5.9, p. 97). Various tolerance and performance curves (Figure 5.2) are presumably subject to certain constraints. For example, their breadth (variance) usually cannot be increased without a simultaneous reduction in their height, and vice versa (Levins, 1968). This useful notion of trade-offs, known as the *principle of allocation*, has proven to be quite helpful in interpreting and understanding numerous ecological phenomena.

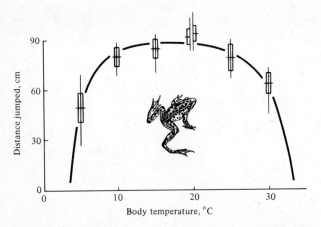

Figure 5.2. Distance jumped by a frog as a function of its body temperature. Notice that performance diminishes at both low and high temperatures. [From Huey and Stevenson (1979).]

As an example of allocation, imagine an animal of a given size and mouth-part anatomy. A certain size of prey item is optimal, whereas other prey are suboptimal because they are either too large or too small for efficient capture and swallowing. Any given animal has its own *utilization curve* that indicates the actual numbers of prey of different sizes taken per unit time under partic-ular environmental conditions. In an idealized, perfectly stable, and infinitely productive environment, a utilization curve might become a spike with no variance, with the organism using only its most optimal prey resource type. In actuality, limited and changing availabilities of resources, in both time and space, result in utilization curves with breadth as well as height. In terms of the principle of allocation, an individual with a generalized diet adapted to eat prey of a wide range of available sizes presumably is not so effective at exploiting prey of intermediate size as another, more specialized, feeder. In other words, a jack-of-many-trades is a master of none. We will consider this subject in more detail later.

## TIME, MATTER, AND ENERGY BUDGETS

Time, matter, and energy budgets vary widely among organisms. For example, some creatures allot more time and energy to reproduction at any instant than do others. Varying time and energy budgeting is a potent means of coping with a changing environment while retaining some degree of adaptation to it. Thus, many male songbirds expend a great deal of energy on territorial defense during the breeding season but little or none at other times of the year. Similarly, in animals with parental care, an increasing amount of energy is spent on growing offspring until some point when progeny begin to become independent of their parents, whereupon the amount of time and energy de-voted to them decreases. Indeed, adult female red squirrels, *Tamiascurus*, at the height of lactation consume an average of 323 kilocalories of food per day compared with an average daily energy consumption of a similar-sized adult male of only about 117 kilocalories (C. Smith, 1968). The time budgets of these squirrels also vary markedly with the seasons. In a bad dry year, many annual plants "go to seed" while still very small, whereas in a good wet year, these plants grow to a much larger size before becoming reproductive; presumably more seeds are produced in good years, but perhaps none (or very few) would be produced in a bad year if individuals attempted to grow to the sizes they reach in good years.

An animal's time and energy budget provides a convenient starting point for clarifying some ways in which foraging influences reproduction, and vice versa. Any animal has only a certain finite period of time available in which to perform all its activities, including foraging and reproduction. This total time budget, which can be considered either on a daily basis or over the animal's life-time, is determined both by the diurnal rhythm of activity and by the animal's ability to "make time" by performing more than one activity at the same time (such as a male lizard sitting on a perch, simultaneously watching for potential prey and predators while monitoring mates and competing males). Provided that a time period is profitable for foraging (expected gains in matter and energy exceed inevitable losses from energetic costs of foraging), any increase in time devoted to foraging clearly will increase an animal's supply of matter and energy. However, necessarily accompanying this increase in matter and energy is a concomitant *decrease* in time available for nonforaging activities such as mating and reproduction. Thus, profits of time spent foraging are measured in matter and energy, while costs take on units of time lost. Conversely, increased time spent on nonforaging activities confers profits in time, while costs take the form of decreased energy availability. Hence, gains in energy correspond to losses in time, while dividends in time require reductions in energy availability. (Of course, risks of foraging and reproduction also need to be considered.)

The preceding arguments suggest that optimal allocation of time and energy ultimately depend on how costs in each currency vary with profits. However, because units of costs and profits in time and energy differ, one would like to be able to convert them into a common currency. Costs and profits in time might be measured empirically in energetic units by estimating the net gain in energy per unit of foraging time. If all potential foraging time is equivalent, profits would vary linearly with costs; under such circumstances, the loss in energy associated with nonforaging activities would be directly proportional to the amount of time devoted to such activity. Optimal budgeting of time and energy into foraging versus nonforaging activities is usually profoundly influenced by various circadian and seasonal rhythms of physical conditions, as well as those of predators and potential prey. Clearly, certain time periods favorable for foraging return greater gains in energy gathered per unit time than other periods. Risks of exposure to both harsh physical conditions and predators must often figure into the optimal amount of time to devote to various activities. Ideally, one would ultimately like to measure both an animal's foraging efficiency and its success at budgeting time and energy by its lifetime reproductive success, which would reflect all such environmental "risks."

Foraging and reproductive activities interact in another important way. Many organisms gather and store materials and energy during time periods that are unfavorable for successful reproduction but then expend these same resources on reproduction at a later, more suitable, time. Lipid storage and utilization systems obviously facilitate such temporal integration of uptake and expenditure of matter and energy. This temporal component greatly complicates the empirical measurement of reproductive effort.

Prey density can strongly affect an animal's time and energy budget. Gibb (1956) watched rock pipits, *Anthus spinoletta*, feeding in the intertidal along the English seacoast during two consecutive winters. The first winter was relatively mild; the birds spent $6\frac{1}{2}$ hours feeding, $1\frac{3}{4}$ hours resting, and $\frac{3}{4}$ hour fighting in defense of their territories (total daylight slightly exceeded 9 hours). The next winter was much harsher and food was considerably scarcer; the birds spent $8\frac{1}{4}$ hours feeding, 39 minutes resting, and only 7 minutes on territorial defense! Apparently, the combination of low food density and extreme cold (endotherms require more energy in colder weather) demanded that over 90 percent of the bird's waking hours be spent feeding and no time remained for frivolities. This example also illustrates that food is less defendable at lower densities, as indicated by reduced time spent on territorial defense. Obviously, food density in the second year was near the lower limit that would allow survival of rock pipits. When prey items are too sparse, encounters may be so infrequent that an individual cannot survive. Gibb (1960) calculated that, to balance their energy budget during the winter in some places, English tits must find an insect on the average once every $2\frac{1}{2}$ seconds during daylight hours.

Time and energy budgets are influenced by a multitude of other ecological factors, including body size, mode of foraging, mode of locomotion, vagility, trophic level, prey size, resource density, environmental heterogeneity, rarefaction, competition, risks of predation, and reproductive tactics.

## LEAF TACTICS

Leaves take on an almost bewildering array of sizes and shapes (Figure 5.3). Some leaves are deciduous, others evergreen; some are simple, others compound; and their actual spatial arrangement on a given plant differs considerably both within and between species. Some leaves are much more costly to produce and maintain than others (elementary economic considerations dictate

Figure 5.3. Leaves have evolved a spectacular variety of shapes and sizes, and yet repeatable patterns do occur.

that any given leaf must pay for itself plus generate a net energetic profit). Presumably, this great diversity of leaf tactics is a result of natural selection maximizing the lifetime reproductive success of individual plants under diverse environmental conditions. Leaf tactics are influenced by many factors that include light, water availability, prevailing winds, and herbivores. When grown in the shade, individuals of many species grow larger, less dissected leaves than when grown in the sun. Similarly, shade-tolerant plants of the understory usually have larger and less lobed leaves than canopy species. Similar types of leaves often evolve independently in different plant lineages subjected to comparable climatic conditions at different geographic localities, especially among trees (Bailey and Sinnot, 1916; Ryder, 1954; Stowe and Brown, 1981). Compound leaves, thought to conserve woody tissue, with small leaflets are found in hot dry regions, whereas those with larger leaflets occur under warm moist conditions. Lowland wet tropical rainforest trees have large evergreen leaves with nonlobed or continuous margins, chaparral plants tend to have small sclerophyllous evergreen leaves, arid regions tend to support leafless stem succulents such as cacti or plants with entire leaf margins (especially among evergreens), plants from cold wet climates often have notched or lobed leaf margins, and so on. Such repeated patterns of leaf size and shape suggest that a general theory of leaf tactics is possible.

Several models for optimal leaf size under differing environmental conditions have been developed. Efficiency of water use (grams of carbon dioxide assimilated per gram of water lost) was the measure of plant performance maximized by Parkhurst and Loucks (1971). A similar model for size and shape of vine leaves was developed by Givnish and Vermeij (1976). Even these relatively simple models predict several observed patterns in leaf size, such as large leaves in warm shady wet places and small leaves in colder areas or warmer and sunnier locales (Figure 5.4).

The evergreen versus deciduous dichotomy can be approached similarly using cost-benefit arguments (Orians and Solbrig, 1977; Miller, 1979). In con-

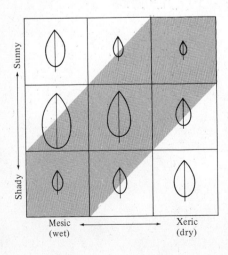

Figure 5.4. Patterns of leaf sizes predicted under differing environmental conditions from wet to dry (mesic to xeric) and shady to sunny. The shaded region represents conditions likely to prevail in nature. [From Givnish and Vermeij (1976). Copyright © 1976 by The University of Chicago Press.]

sidering leaf tactics of desert plants, Orians and Solbrig contrast leaf types along a continuum ranging from the relatively inexpensive deciduous "mesophytic" leaf to the more costly evergreen "xerophytic" leaf. Mesophytic leaves photosynthesize and transpire at a rapid rate and hence require high water availability (low "soil water potential"). In deserts, such plants grow primarily along washes. In contrast, xerophytic leaves cannot photosynthesize as rapidly when abundant water is available, but they can extract water from relatively dry soil. Each plant leaf tactic has an advantage either at different times or in different places, thereby promoting plant life form diversity. During wet periods, plants with mesophytic leaves photosynthesize rapidly, but under drought conditions, they must drop their leaves and become dormant. During such dry periods, however, the slower photosynthesizers with xerophytic leaves are still able to function by virtue of their ability to extract water from dry soils. Of course, all degrees of intermediate leaf tactics exist, each of which may enjoy a competitive advantage under particular conditions of water availability (Figure 5.5). In a predictable environment, net annual profit per unit of leaf surface area determines the winning phenotype. Even a relatively brief wet season could suffice to give mesophytic leaves a higher annual profit (which accounts for the occurrence of these plant life forms in deserts).

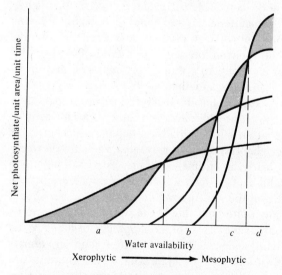

Figure 5.5. Probable relationship between efficiency of photosynthesis and water availability among different types of leaves. The most xerophytic leaf performs best under conditions of low water availability (zone *a*), whereas the most mesophytic leaf does best when soils are wet (zone *d*). Intermediate types are superior under intermediate conditions of water availability. Shaded areas indicate superiority of various leaf types under different conditions of soil moisture availability. [Adapted from Orians and Solbrig (1977). Copyright © 1977 by The University of Chicago Press.]

In an interesting discussion of leaf arrangement and forest structure, Horn (1971, 1975a, 1976) distinguished "monolayers" from "multilayers." Each plant in the multilayer of a forest (usually sunnier places such as the canopy) has leaves scattered throughout its volume at several different levels, whereas monolayer plants have essentially a single blanket or shell of leaves. Plants in the multilayer gain from a geometry that allows some light to pass through to their own leaves at lower levels. Horn points out that lobing facilitates passage of light and that such plants do well in the sun (in the shade, inner leaves may respire more than they photosynthesize). In contrast, the optimal tree design in the shade is a monolayer in which each leaf typically intercepts as much light as possible (leaves are large and seldom lobed). Moreover, slow-growing monolayered plants eventually outcompete fast-growing multilayered plants that persist by regular colonization of newly vacated areas created by continual disturbance (Horn, 1976).

## FORAGING TACTICS AND FEEDING EFFICIENCY

Foraging tactics involve the ways in which animals gather matter and energy. As explained above, matter and energy constitute the profits gained from foraging, in that they are used in growth, maintenance, and reproduction. But foraging has its costs as well; a foraging animal may often expose itself to potential predators, and much of the time spent in foraging is rendered unavailable for other activities, including reproduction. An optimal foraging tactic maximizes the difference between foraging profits and their costs. Presumably, natural selection, acting as an efficiency expert, has often favored such optimal foraging behavior. Consider, for example, prey of different sizes and what might be termed "catchability." How great an effort should a foraging animal make to obtain a prey item of a given catchability and of a particular size (and therefore matter and energy content)? Clearly, an optimal consumer should be willing to expend more energy to find and capture food items that return the most energy per unit of expenditure upon them. Moreover, an optimal forager should take advantage of natural feeding routes and should not waste time and energy looking for prey either in inappropriate places or at inappropriate times. What is optimal in one environment is seldom optimal in another, and an animal's particular anatomy strongly constrains its optimal foraging tactic. There is considerable evidence that animals actually do attempt to maximize their foraging efficiencies, and a substantial body of theory on optimal foraging tactics exists.

Numerous aspects of optimal foraging theory are concisely summarized in an excellent chapter, "The Economics of Consumer Choice," by MacArthur (1972). He makes several preliminary assumptions: (a) Environmental structure is repeatable, with some statistical expectation of finding a particular resource (such as a habitat, microhabitat, and/or prey item). (b) Food items can be arranged in a continuous and unimodal spectrum, such as size distributions of insects (Schoener and Janzen, 1968; Hespenhide, 1971). (This assumption is clearly violated by foods of some animals, such as monophagous insects or herbivores generally, because plant chemical defenses are typically discrete; see pp. 323–324.) (c) Similar animal phenotypes are usually closely equivalent in their harvesting abilities; an intermediate phenotype is thus best able to exploit foods intermediate between those that are optimal for two neighboring phenotypes (see also pp. 281–284). Conversely, similar foods are gathered with similar efficiencies; a lizard with a jaw length that best adapts it to exploit 5-mm-long insects is only slightly less efficient at eating 4- and 6-mm insects. (d) The principle of allocation applies, and no one phenotype can be maximally efficient on all prey types; improving harvesting efficiency on one food type necessitates reducing the efficiency of exploiting other kinds of items. (e) Finally, an individual's economic "goal" is to maximize its total intake of food resources. (Assumptions b, c, and d are not vital to the argument.)

MacArthur (1972) then breaks foraging down into four phases: (1) deciding where to search, (2) searching for palatable food items, (3) upon locating a potential food item, deciding whether or not to pursue it, and (4) pursuit itself, with *possible* capture and eating. Search and pursuit efficiencies for each food type in each habitat are entirely determined by the preceding assumptions about morphology (assumption c) and environmental repeatability (assumption a); moreover, these efficiencies dictate the probabilities associated with the searching and pursuing phases of foraging (2 and 4, respectively). Thus, MacArthur considers only the two decisions: where to forage and what prey items to pursue (phases 1 and 3 of foraging). Clearly, an optimal consumer should forage where its expectation of yield is greatest—an easy decision to make, given knowledge of the previous efficiencies and the structure of its environment (in reality, of course, animals are far from omniscient and must make decisions based on incomplete information). The decision as to which prey items to pursue is also simple. Upon finding a potential prey item, a consumer has only two options: either pursue it or go on searching for a better item and pursue that one instead. Both decisions end in the forager beginning a new search, so the best choice is clearly the one that returns the greatest yield per unit time. Thus, an optimal consumer should opt to pursue an item only when it cannot expect to locate, catch, and eat a better item (i.e., one that returns more energy per unit of time) during the time required to capture and ingest the first prey item.

Many animals, such as foliage-gleaning insectivorous birds, spend much of their foraging time searching for prey but expend relatively little time and energy pursuing, capturing, and eating small sedentary insects that are usually easy to catch and quickly swallowed. In such "searchers," mean search time per item eaten is large compared to average pursuit time per item; hence, the optimal strategy is to eat essentially all palatable insects encountered. Other animals ("pursuers") that expend little energy in finding their prey but a great deal of effort in capturing it (such as, perhaps, a falcon or a lion) should select prey with small average pursuit times (and energetic costs). Thus, pursuers should generally be more selective and more specialized than searchers. Moreover, because a food-dense environment offers a lower average search time per item than does a food-sparse area, an optimal consumer should restrict its diet to only the better types of food items in the former habitat. To date, optimal foraging theory has been developed primarily in terms of the rate at which energy is gathered per unit of time. Limiting materials such as nutrients in short supply and the risks of predation have so far been largely neglected.

Carnivorous animals forage in extremely different ways. In the "sit-and-wait" mode, a predator waits in one place until a moving prey item comes by and then "ambushes" the prey; in the "widely foraging" mode, the predator actively searches out its prey (Pianka, 1966b; Schoener, 1969a, 1969b). The second strategy normally requires a greater energy expenditure than the first. The success of the sit-and-wait tactic usually depends on one or more of three conditions: a fairly high prey density, high prey mobility, and low predator energy requirements. The widely foraging tactic also depends on prey density and mobility and on the predator's energy needs, but here the distribution of prey in space and the predator's searching abilities assume paramount importance. Although these two tactics are endpoints of a continuum of possible foraging strategies (and hence somewhat artificial), foraging techniques actually employed by many organisms are rather strongly polarized. The dichotomy of sit-and-wait versus widely foraging therefore has substantial practical value. Among snakes, for example, racers and cobras forage widely when compared with boas, pythons, and vipers, which are relatively sit-and-wait foragers. Among hawks, accipiters such as Cooper's hawks and goshawks often hunt by ambush using a sit-and-wait strategy, whereas most buteos and many falcons are relatively more widely foraging. Web-building spiders and sessile filter feeders such as barnacles typically forage by sitting and waiting. Many spiders expend considerable amounts of energy and time building their webs rather than moving about in search of prey; those that do not build webs, such as wolf spiders, forage much more widely. Some general correlates of these two modes of foraging are listed in Table 5.1.

Similar considerations can be applied in comparing herbivores with carnivores. Because the density of plant food almost always greatly exceeds the density of animal food, herbivores often expend little energy, relative to carnivores, in finding their prey (to the extent that secondary chemical compounds of plants, such as tannins, and other antiherbivore defenses reduce palatability of plants or parts of plants, the effective supply of plant foods may be greatly reduced). Because cellulose in plants is difficult to digest, however, herbivores must expend considerable energy in extracting nutrient from their plant food. (Most herbivores have a large ratio of gut volume to body volume, harbor intestinal microorganisms that can digest cellulose, and spend much of their time eating or ruminating—envision a cow chewing its cud.) Animal food, composed of readily available proteins, lipids, and carbohydrates, is more readily digested; carnivores can afford to expend considerable effort in searching for their prey because of the large dividends obtained once they find it. As would be expected, the efficiency of conversion of food into an animal's own tissues (*assimilation*) is considerably lower in herbivores than it is in carnivores.

Many carnivores have extremely efficient prey-capturing devices (see Chapter 15); often the size of a prey object markedly influences this efficiency.

TABLE 5.1  **Some General Correlates of Foraging Mode**

|  | Sit-and-wait | Widely foraging |
|---|---|---|
| Prey type | Eat active prey | Eat sedentary and unpredictable (but clumped or large) prey |
| Volume prey captured/day | Low | Generally high, but low in certain species |
| Daily metabolic expense | Low | High |
| Types of predators | Vulnerable primarily to widely foraging predators | Vulnerable to both sit-and-wait and to widely foraging predators |
| Rate of encounters with predators | Probably low | Probably high |
| Morphology | Stocky (short tails) | Streamlined (generally long tails) |
| Probably physiological correlates | Limited endurance (anaerobic) | High endurance capacity (aerobic) |
| Relative clutch mass | High | Low |
| Sensory mode | Visual primarily | Visual or olfactory |
| Learning ability | Limited | Enhanced learning and memory, larger brains |
| Niche breadth | Wide | Narrow |

*Source:* Adapted from Huey and Pianka (1981).

Using simple geometry (Figure 5.6), Holling (1964) estimated the diameter of a prey item that should be optimal for a praying mantid of a particular size. He then offered five hungry mantids prey objects of various sizes and recorded percentages attacked (Figure 5.7). Mantids were noticeably reluctant to attack prey that were either much larger or much smaller than the estimated optimum! Thus, natural selection has resulted in efficient predators both by producing efficient prey-capturing devices and by programming animals so that they are unlikely to attempt to capture decidedly suboptimal items. Larger predators tend to take larger prey than smaller ones (see Figure 12.12, p. 261). It may in fact be better strategy for a large predator to overlook prey below some minimal size threshold and to spend the time that would have been spent in capturing and eating such small items in searching out larger prey (see also above). Similarly, the effort a predator will expend on any given prey item is proportional to the expected return from that item (which usually increases with prey size). Thus, a lizard waiting on a perch will not usually go far for a very small prey item but will often move much greater distances in attempts to obtain larger prey.

Because small prey are generally much more abundant than large prey, most animals encounter and eat many more small prey items than large ones. Small animals that eat small prey items encounter prey of suitable size much

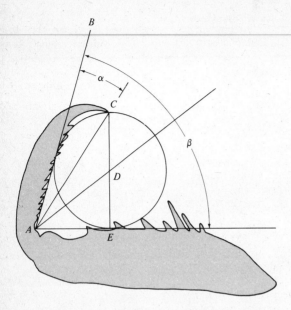

Figure 5.6. Diagram showing how Holling used geometry to calculate an estimated optimal size of a prey item from the anatomy of a mantid's foreleg. Optimal prey diameter, $D$, is simply $T \sin(\beta - \alpha)$, where $T$ is the distance $A$ to $C$. [From Holling (1964).]

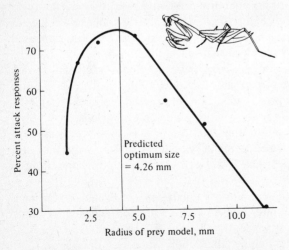

Figure 5.7. The percentage of prey items actually attacked by hungry mantids versus prey size. [From Holling (1964).]

more frequently than do larger animals that rely on larger prey items; as a result, larger animals tend to eat a wider range of prey sizes. Because of such increased food niche breadths of larger animals, size *differences* between predators increase markedly with increasing predator size (MacArthur, 1972).

## PHYSIOLOGICAL ECOLOGY

The concern of environmental physiology, or ecophysiology, is *how* organisms function within, adapt and respond to, and exploit their physical environments. Physiological ecologists are interested primarily in the immediate functional and behavioral mechanisms by which organisms cope with their abiotic environments. Physiological mechanisms clearly must reflect ecological conditions; moreover, mutual constraints between physiology and ecology dictate that both must evolve together in a synergistic fashion.

A fundamental principle of physiology is the notion of *homeostasis*, the maintenance of a relatively stable internal state under a much wider range of external environmental conditions. Homeostasis is achieved not only by physiological means but also by appropriate behavioral responses. An example is temperature regulation in which an organism maintains a fairly constant internal body temperature over a considerably greater range of ambient thermal conditions (homeostasis is never perfect). Homeostatic mechanisms have also evolved that buffer environmental variation in humidity, light intensity, and concentrations of various substances such as hydrogen ions (pH), salts, and so on. By effectively moderating spatial and temporal variation in the physical environment, homeostasis allows organisms to persist and be active within a broad range of environmental conditions, thereby enhancing their fitness. The subject of environmental physiology is vast; some references are given at the end of this chapter.

## PHYSIOLOGICAL OPTIMA AND TOLERANCE CURVES

Physiological processes proceed at different rates under different conditions. Most, such as rate of movement and photosynthesis, are temperature dependent (Figure 5.8). Other processes vary with availability of various materials such as water, carbon dioxide, nitrogen, and hydrogen ions (pH). Curves of performance, known as *tolerance curves* (Shelford, 1913b), are typically bell shaped and unimodal, with their peaks representing optimal conditions for a particular physiological process and their tails reflecting the *limits of tolerance*. Some individuals and species have very narrow peaked tolerance curves; in others these curves are considerably broader. Broad tolerance curves are described with the prefix *eury-* (e.g., eurythermic, euryhaline), whereas *steno-* is used for narrow ones (e.g., stenophagous). An organism's use of environmental resources such as foods and microhabitats can profitably be viewed similarly, and performance can be measured in a wide variety of units such as survivorship, reproductive success, foraging efficiency, and fitness.

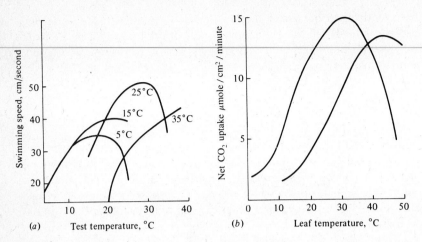

Figure 5.8. Two plots of performance against temperature. (*a*) Goldfish swimming speed versus temperature for fish acclimated to different thermal conditions; in most cases, performance peaks near the temperature to which fish are accustomed. [From Ricklefs (1973) after Fry and Hart (1948).] (*b*) Photosynthetic rate versus leaf temperature in the plant *Atriplex lentiformis* at two different localities. [From Mooney, Bjorkman, and Berry (1975).]

Performance curves can sometimes be altered during the lifetime of an individual as it becomes exposed to different ambient external conditions. Such short-term alteration of physiological optima is known as *acclimation* or acclimatization (Figure 5.8). Within certain design constraints, tolerance curves clearly must change over evolutionary time as natural selection molds them to reflect changing environmental conditions. However, very little is known about the evolution of tolerance; most researchers have merely described the range(s) of conditions tolerated or exploited by particular organisms. Tolerance curves are often taken almost as given and immutable, with little or no consideration of the ecological and evolutionary forces that shape them.

Performance or tolerance is often sensitive to two or more environmental variables. For example, the fitness of a hypothetical organism in various microhabitats might be a function of relative humidity (or vapor pressure deficit), somewhat as shown in Figure 5.9*a*. Assume that fitness varies similarly along a temperature gradient (Figure 5.9*b*). Figure 5.9*c* combines humidity and temperature conditions to show variation in fitness with respect to both variables simultaneously (a third axis, fitness, is implicit in this figure). The range of thermal conditions tolerated is narrower at very low and very high humidities than it is at intermediate and more optimal humidities. Similarly, an organism's tolerance range for relative humidity is narrower at extreme temperatures than it is at more optimal ones. The organism's thermal optimum depends on humidity conditions (and vice versa). Fitness reaches its maximum at intermediate temperatures and humidities. Hence, temperature tolerance and tolerance

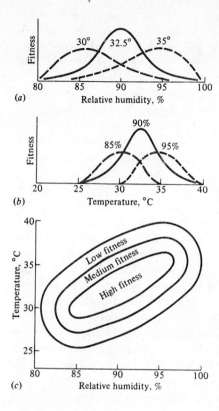

**Figure 5.9.** Hypothetical response curves showing how two variables can interact to determine an organism's fitness. Fitness is reduced at extremes of either temperature or humidity, and the range of humidities tolerated is less at extreme temperatures than it is at intermediate ones.

of relative humidities interact in this example. The concept of a single fixed optimum is in some ways an artifact of considering only one environmental dimension at a time.

## ENERGETICS OF METABOLISM AND MOVEMENT

Some of the food ingested by any animal passes through its gut unused. Such egested material can be as high as 80 to 90 percent of the total intake in some caterpillars (Whittaker, 1975). The food that is actually digested is termed *assimilation*: a fraction of this must be used in respiration to support maintenance metabolism and activity. The remainder is incorporated into the animal concerned as secondary productivity and ultimately can be used either in growth or in reproduction. These relationships are summarized as:

Ingestion = Assimilation + Egestion

Assimilation = Productivity + Respiration

Productivity = Growth + Reproduction

The *total amount of energy* needed per unit time for maintenance increases with increasing body weight (Figure 5.10). However, because small animals have relatively high ratios of body surface to body volume, they generally have much higher metabolic rates and hence have greater energy requirements *per unit of body weight* than larger animals (Figure 5.11). Animals that maintain relatively constant internal body temperatures are known as *homeotherms;* those whose temperatures vary widely from time to time, usually approximating the temperature of their immediate environment, are called *poikilotherms.* These two terms are sometimes confused with a related pair of useful terms. An organism that obtains its heat from its external environment is an *ectotherm;* one that produces most of its own heat internally by means of oxidative metabolism is known as an *endotherm.* All plants and the vast majority of animals are ectothermic—the only continuously endothermic animals are found among birds and mammals (but even some of these become ectothermic at times). Some poikilotherms (some large reptiles and certain large fast-swimming fish such as tuna) are at times at least partially endothermic. Certain ectotherms (many lizards and temperate-zone flying insects) actually regulate their body temperatures fairly precisely during periods of activity by appropriate behavioral means. Thus, ectotherms at times can be homeotherms! An active bumblebee or desert lizard may have a body temperature as high as that of a bird or mammal (the layperson's terms "warm-blooded" and "cold-blooded" can thus be quite misleading and should be abandoned). Because energy is required to maintain a constant internal body temperature, endotherms have considerably higher metabolic rates, as well as higher energy needs (and budgets), than ectotherms of the same body weight. There is a distinct lower limit on body size for endotherms—about the size of a small hummingbird or shrew (2 or 3 grams). Indeed, both hummingbirds and shrews have very high metabolic rates and hence rather precarious energetic relationships; they depend on continual supplies of energy-rich foods. Small hummingbirds would starve to death during cold nights if they did not allow their body temperatures to drop and go into a state of torpor.

Body size, diet, and movements are complexly intertwined with the energetics of metabolism. Energy requirements do not scale linearly with body mass, $m$, but instead as a fractional exponent: $E = cm^{0.67}$, where $c$ is a taxon-specific constant. Large animals require more matter and energy for their maintenance than small ones, and in order to obtain it they usually must range over larger geographical areas than smaller animals with otherwise similar food requirements. Food habits also influence movements and home range size. Because the foods of herbivorous animals that eat the green parts of plants (such as grazers, which eat grasses and ground-level vegetation, and browsers, which eat tree leaves) are usually quite dense, these animals usually do not have very large home ranges. In contrast, carnivores and those herbivores that must search for their foods (such as granivores and frugivores, which eat seeds and fruits, respectively) frequently spend much of their foraging time and energy in search, ranging over considerably larger areas. McNab (1963) termed the first group "croppers" and the latter "hunters." Croppers generally exploit foods that occur in relatively dense supply, whereas hunters typically utilize less dense foods. Hunters are often territorial, but croppers seldom defend territories.

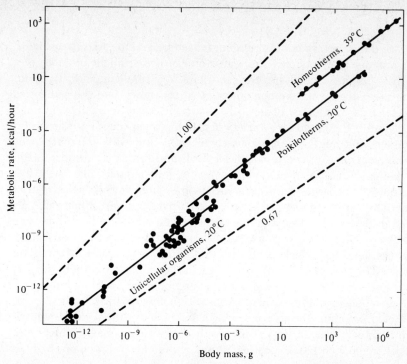

**Figure 5.10.** Metabolic rates of a variety of organisms of different sizes (log-log plot). Total oxygen consumption increases with increasing body size. [From Schmidt-Nielsen (1975).]

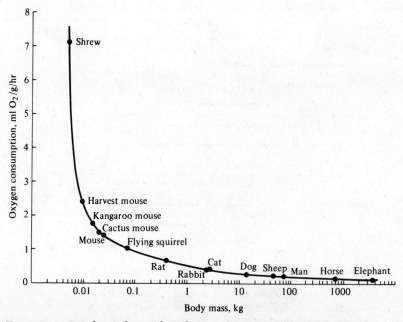

**Figure 5.11.** Semilogarithmic plot of rates of oxygen consumption per unit of body mass for a wide variety of mammals plotted against body mass. [From Schmidt-Nielsen (1975).]

Croppers and hunters are not discrete but in fact grade into each other (Figure 5.12). A browser that eats only the leaves of a rare tree might be more of a hunter than a granivore that eats the seeds of a very common plant. However, such intermediates are uncommon enough that separation into two categories is useful for many purposes. Figure 5.13 shows the correlation between home range size and body weight for a variety of mammalian species, here separated into croppers and hunters. Analogous correlations, but with different slopes and/or intercepts, have been obtained for birds and lizards (Schoener, 1968b; Turner, Jennrich, and Weintraub, 1969). Very mobile animals, like birds, frequently range over larger areas than less mobile animals such as terrestrial mammals and lizards. In areas of low productivity (e.g., deserts), most animals may be forced to range over a greater area to find adequate food than they would in more productive regions. Large home ranges or territories usually result in low densities, which in turn markedly limit possibilities for evolution of sociality. Thus, McNab (1963) points out that complex social behavior has usually evolved only in croppers and among exceptionally mobile hunters.

The metabolic cost of movement varies with both an animal's body size and its mode of locomotion. The cost of moving a unit of body weight some standard distance is actually less in larger animals than in smaller ones (Figure 5.14). Terrestrial locomotion is the most expensive mode of transportation, flight is

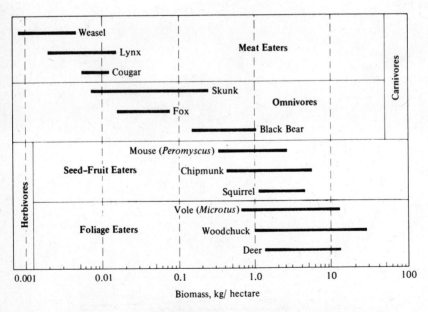

Figure 5.12. Biomass, in kilograms/hectare, of various mammals, arranged according to their food habits. Although mammals as a class vary over five orders of magnitude, the range among those that eat any given type of food is considerably smaller. Meat eaters and omnivores are much less dense than herbivores. [From Odum (1959) after Mohr (1940).]

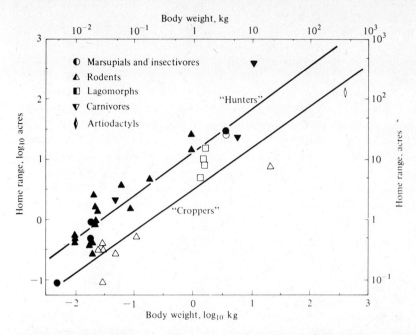

**Figure 5.13.** Log-log plot of average home range area against mean body weight for a variety of mammals, separated into "croppers" (open symbols and lower regression line) and "hunters" (closed symbols and upper line). [After McNab (1963).]

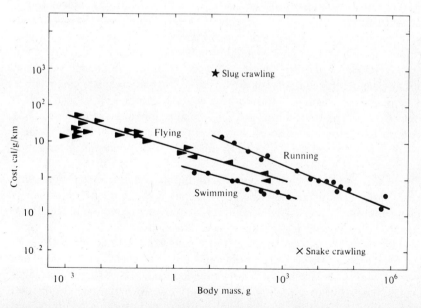

**Figure 5.14.** Comparison of the energetic cost of moving a unit of body weight one kilometer for several different modes of locomotion. [From Schmidt-Nielsen (1975).]

intermediate in cost, and swimming is the most economical means of moving about—provided body shape is fusiform and buoyancy is neutral (Figure 5.14).

Physiologists have documented numerous consistent size-related trends in organs and metabolic properties. For example, among mammals, heart weight is always about 0.6 percent of total body weight, whereas blood volume is almost universally about 5.5 percent of body mass over a great range of body sizes (these organ systems are thus directly proportional to size). Other physiological attributes, such as lung surface in mammals, vary directly with metabolic rate rather than with size. However, some organ systems, such as the kidney and liver, are not scaled directly either to size or metabolic rate (Schmidt-Nielsen, 1975). Such "physiological rules" or design constraints (see also pp. 113–118) apparently dictate available avenues for physiological change, thereby constraining possible ecological adaptations.

## ADAPTATION AND DETERIORATION OF ENVIRONMENT

Organisms are adapted to their environments in that, to survive and reproduce, they must meet their environment's conditions for existence. Evolutionary adaptation can be defined as *conformity between the organism and its environment*. Plants and animals have adapted to their environments both genetically and by means of physiological, behavioral, and/or developmental flexibility. The former includes instinctive behavior and the latter learning. Adaptation has many dimensions in that most organisms must conform simultaneously to numerous different aspects of their environments. Thus, for an organism to adapt, it must cope not only with various aspects of its physical environment, such as temperature and humidity conditions, but also with competitors, predators, and escape tactics of its prey. Conflicting demands of these various environmental components often require that an organism compromise in its adaptations to each. Conformity to any given component takes a certain amount of energy that is then no longer available for other adaptations. The presence of predators, for example, may require that an animal be wary, which in turn is likely to reduce its foraging efficiency and hence its competitive ability.

Organisms can conform to and cope with highly predictable environments relatively easily, even when they change in a regular way, as long as the changes are not too extreme. Adaptation to an unpredictable environment may usually be much more difficult; adapting to extremely erratic environments may even prove impossible. Many organisms have evolved dormant stages that allow them to survive unfavorable periods, both predictable and unpredictable. Annual plants everywhere and brine shrimp in deserts are good examples. Brine shrimp eggs survive for years in the salty crust of dry desert lakes; when a rare desert rain fills one of these lakes, the eggs hatch, the shrimp grow rapidly to adults, and they produce many eggs. Some seeds known to be many centuries old are still viable and have been germinated. Changes in the environment that reduce overall adaptation are collectively termed the *deterioration of environment*; such changes cause directional selection, resulting in accommodation to the new environment.

A simple but elegant model of adaptation and undirected environmental deterioration was developed by Fisher (1930). He reasoned that no organism is "perfectly adapted"—all must fail to conform to their environments in some ways and to differing degrees. However, a hypothetical, perfectly adapted organism can always be imagined (actually this reflects the environment) against which existing organisms may be compared. Fisher's mathematical argument is phrased in terms of an infinite number of "dimensions" for adaptation (only three are used here for ease of illustration). Imagine an adaptational space of three coordinates representing, respectively, the competitive, predatory, and physical environments (Figure 5.15). An ideal "perfectly adapted" organism lies at a particular point (say A) in this space, but any given real organism is at another point (say B), some distance, d, away from the point of perfect adaptation. Changes in the position of A correspond to environmental changes, making the optimally adapted organism different; changes in B represent changes in the organisms concerned, such as mutations. The distance between the two

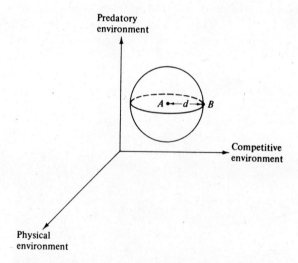

Figure 5.15. Fisher's model of adaptation and deterioration of environment. Point A represents a hypothetical "perfectly adapted" organism; an actual organism (point B) is never perfectly adapted and thus lies at some distance, d, from point A. The surface of the sphere represents all possible points with a level of adaptation equal to the organism under consideration. Very small undirected changes in either organism (point B) or environment (point A) are equally likely to increase or to decrease the level of adaptation, d. Notice the duality of the model (make A → B and B → A).

points, $d$, represents the degree of conformity between the organism and environment, or the level of adaptation. Fisher noted that very small *undirected* changes in either organism or environment have a 50:50 chance of being to the organism's advantage (i.e., reducing the distance between $A$ and $B$). The probability of such improvement is inversely related to the magnitude of the change (Figure 5.16). Very great changes in either organism or environment are always maladaptive because even if they are in the correct direction, they "overshoot" points of closer adaptation. (Of course, it is remotely possible that such major environmental changes or "macromutations" could put an organism into a completely new adaptive realm and thereby improve its overall level of adaptation.) Fisher makes an analogy with focusing a microscope. Very fine changes are as likely as not to improve the focus, but gross changes will almost invariably throw the machine further out of focus. Organisms may be thought of as "tracking" their environments in both ecological and evolutionary time, changing as their environments change; thus, as point $A$ shifts because of daily, seasonal, and long-term environmental fluctuations, point $B$ follows it. Such environmental tracking may be physiological (as in acclimation), behavioral (including learning), and/or genetic (evolutionary), depending on the time scale of environmental change. Reciprocal counterevolutionary responses to other species (prey, competitors, parasites, and predators) constitute examples

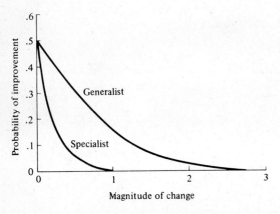

Figure 5.16. The probability of improvement of the level of adaptation (i.e., of reducing $d$) is plotted against the magnitude of an undirected change when the number of dimensions is large. Two hypothetical organisms are shown, one highly adapted such as a specialist with narrow tolerance limits and one less highly adapted such as a generalist with broader tolerance limits, or a greater number of niche dimensions, or both. A random change of a given magnitude is more likely to improve the level of adaptation of the generalist than the specialist. [Partially adapted from Fisher (1958a).]

of such evolutionary tracking,* and have been termed *coevolution* (see also pp. 329–334).

Individual organisms with narrow tolerance limits, such as highly adapted specialists, generally suffer greater losses in fitness due to a unit of environmental deterioration than generalized organisms with more versatile requirements, all else being equal. Thus, more specialized organisms or those with restricted homeostatic abilities cannot tolerate as much environmental change as generalists or organisms with better developed homeostasis (Figure 5.16). Fisher's model applies only to *nondirected* changes in either party of the adaptational complex—such as mutations and perhaps certain climatic fluctuations, or other random events. However, many environmental changes are probably nonrandom. Changes in other associated organisms, especially predators and prey, are invariably directed so as to reduce an organism's degree of conformity to its environment; thus, they inevitably constitute a deterioration of that organism's environment. Directed changes in competitors can either increase or decrease an organism's level of adaptation, depending on whether they involve avoidance of competition or improvements in competitive ability per se. Directed changes in mutualistic systems would usually tend to improve the overall level of adaptation of both parties.

# HEAT BUDGETS AND THERMAL ECOLOGY

When averaged over a long enough period of time, heat gained by an organism must be exactly balanced by heat lost to its environment; otherwise, the plant or animal would either warm up or cool off. Many different pathways of heat gains and heat losses exist (Figure 5.17). The notion of a heat budget is closely related to the concept of an energy budget; balancing a heat budget requires very different adaptations under varying environmental conditions. At different times of day, ambient thermal conditions may change from being too cold to being too warm for a particular organism's optimal performance. Organisms living in hot deserts must avoid overheating by being able to minimize heat loads and to dissipate heat efficiently; in contrast, those that live in colder places such as at high altitudes or in polar regions must avoid overcooling—and they have hence evolved efficient means of heat retention, such as insulation by blubber, feathers, or fur, that reduce the rate of heat exchange with the external environment.

As seen in previous chapters, environmental temperatures fluctuate in characteristic ways at different places over the earth's surface, both daily and seasonally. In the absence of a long-term warming or cooling trend, environmental temperatures at any given spot remain roughly constant when averaged

---

*The phenomenon of evolving as fast as possible just to maintain a given current level of adaptation in the face of a continually deteriorating environment has been called the "Red Queen" hypothesis by Van Valen (1973), after the famous character in Alice in Wonderland who ran as fast as she could just to stay where she was.

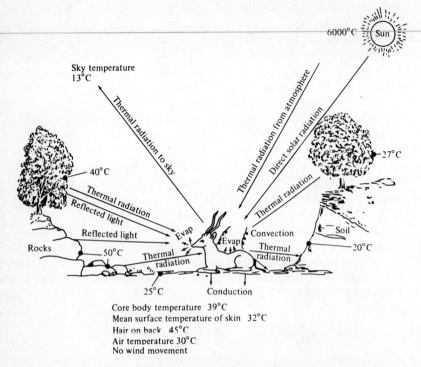

**Figure 5.17.** Diagrammatic representation of various pathways of heat gains and losses between an animal and its environment under moderately warm conditions. Heat exchange between an animal and its environment is roughly proportional to its body surface: Because small animals have a larger surface area relative to their weight than larger ones, the former gain or lose relatively more heat than the latter. [From Bartholomew (1972). Reprinted with permission of Macmillan Publishing Co., Inc., from *Animal Physiology: Principles and Adaptations*, M. S. Gordon (ed.). Copyright © 1972 by Malcolm S. Gordon.]

over an entire annual cycle. Recall that the range in temperature within a year is much greater at high latitudes than it is nearer the equator. An organism could balance its annual heat budget by being entirely passive and simply allowing its temperature to mirror that of its environment. Such a passive thermoregulator is known as a *thermoconformer* (Figures 5.18 and 5.19). Of course, it is also an ectotherm. Another extreme would be to maintain an absolutely constant body temperature by physiological or behavioral means, dissipating (or avoiding) excess bodily heat during warm periods but retaining (or gaining) heat during cooler periods (in endotherms, energy intake is often increased during cold periods and more metabolic heat is produced to offset the increased heat losses).

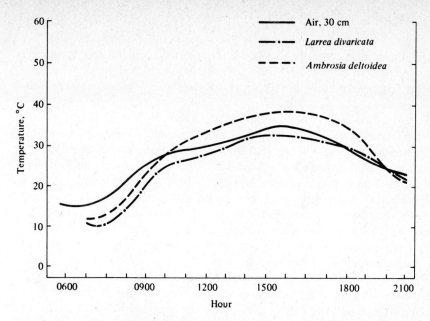

Figure 5.18. Average temperatures of leaves of *Larrea divaricata* and *Ambrosia deltoidea* during three sunny days in June with slight breezes prevailing. [From Patten and Smith (1975).]

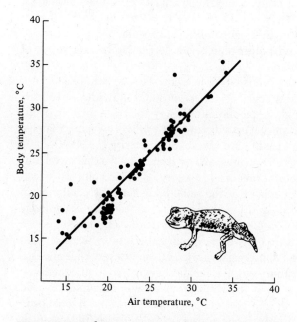

Figure 5.19. Body temperatures of 86 active Australian nocturnal gekkonid lizards (*Nephrurus laevissimus*) plotted against ambient air temperature. The line represents body temperatures equal to air temperatures. Like most nocturnal lizards, these animals are nearly perfect thermoconformers.

107

Organisms that carefully regulate their internal temperatures are called *thermoregulators*, or *homeotherms*. (Recall that both endotherms and ectotherms may regulate their body temperatures.) There is, of course, a continuum between the two extremes of perfect conformity and perfect regulation (see Figure 5.22 on p. 117). Homeostasis, remember, is never perfect. Because regulation clearly has costs and risks as well as profits, an emerging conceptual framework envisions an optimal level of regulation that depends on the precise form of the constraints and interactions among the costs and benefits arising from a particular ecological situation (Huey and Slatkin, 1976). Thermoregulation often involves both physiological and behavioral adjustments; as an example of the latter, consider a typical terrestrial diurnal desert lizard. During the early morning, when ambient temperatures are low, such a lizard locates itself in warmer microclimates of the environmental thermal mosaic (e.g., small depressions in the open or on tree trunks), basking in the sun with its body as perpendicular as possible to the sun's rays and thereby maximizing heat gained. With the daily march of temperature, ambient thermal conditions quickly rise and the lizard seeks cooler shady microhabitats. Individuals of some species retreat into burrows as temperatures rise; others climb up off the ground into cooler air and orient themselves by facing into the sun's rays, thereby reducing heat load. Many lizards change colors and their heat reflectance properties, being dark and heat absorbent at colder times of day but light and heat reflectant at hotter times. Such adjustments allow individual lizards to be active over a longer period of time than they could be if they conformed passively to ambient thermal conditions; presumably, they are also more effective competitors and better able to elude predators as a result of such thermoregulatory behaviors.

Hot, arid regions typically support rich lizard faunas, whereas cooler forested areas have considerably fewer lizard species and individuals. Lizards can enjoy the benefits of a high metabolic rate during relatively brief periods when conditions are appropriate for activity and yet can still become inactive during adverse conditions. By facilitating metabolic inactivity on both a daily and a seasonal basis, poikilothermy thus allows lizards to capitalize on unpredictable food supplies. Ectotherms are low-energy animals; one day's food supply for a small bird will last a lizard of the same body mass for a full month! Most endothermic diurnal birds and mammals must wait out the hot midday period at considerable metabolic cost, whereas lizards can effectively reduce temporal heterogeneity by retreating underground, becoming inactive, and lowering their metabolic rate during harsh periods (some desert rodents estivate when food or water is in short supply). Poikilothermy may well contribute to the apparent relative success of lizards over birds and mammals in arid regions. Temperate zone forests and grasslands are probably simply too shady and too cold for ectothermic lizards to be very successful because these animals depend on basking to reach body temperatures high enough for activity; in contrast, birds and mammals do quite well in such areas partly because of their endothermy.

# WATER ECONOMY IN DESERT ORGANISMS

Because water conservation is a major problem for desert organisms, their physiological and behavioral adaptations for acquisition of water and for economy of its use have been well studied. These interesting adaptations are quite varied. Like energy and heat budgets, water budgets must balance; losses must be replaced by gains. For examples of water acquisition mechanisms, consider rooting strategies. Desert plants usually invest considerably more in root systems than plants from wetter areas; one study showed that perennial shrubs in the Great Basin Desert allocate nearly 90 percent of their biomass to underground tissues (Caldwell and Fernandez, 1975), whereas roots apparently represent a much smaller fraction (only about 10 percent) of the standing-crop biomass of a mesic hardwood forest. The creosote bush *Larrea divaricata* has both a surface root system and an extremely deep tap root that often reaches all the way down to the water table. This long tap root provides *Larrea* with water even during long dry spells when surface soils contain little moisture. Cacti, in contrast, have an extensive but relatively shallow root system and rely on water storage to survive drought. Many such desert plants have tough sclerophyllous xerophytic leaves that do not allow much water to escape (they also photosynthesize at a low rate as a consequence). Mesophytic plants occur in deserts too, but they photosynthesize rapidly and grow only during periods when moisture is relatively available; they drop their leaves and become inactive during droughts. Plants also reduce water losses during the heat of midday by closing their stomata and drooping their leaves (wilting). Many desert plants and animals absorb and use atmospheric or substrate moisture; most can also tolerate extreme desiccation.

Camels do not rely on water storage to survive water deprivation, as is commonly thought, but can lose as much as a quarter of their body weight, primarily as water loss (Schmidt-Nielsen, 1964). Like many desert organisms, camels also conserve water by allowing their temperature to rise during midday. Moreover, camels tolerate greater changes in plasma electrolyte concentrations than less drought-adapted animals. In deserts, small mammals like kangaroo rats survive without drinking by relying on metabolic water derived from the oxidation of their food (here, then, is an interface between energy budgets and water budgets). Most desert rodents are nocturnal and avoid using valuable water for heat regulation by spending hot daytime hours underground in cool burrows with high relative humidity, thereby minimizing losses to evaporation (most desert organisms resort to evaporative cooling mechanisms such as panting only in emergencies). The urine of kangaroo rats is extremely concentrated, and their feces contain little water (Schmidt-Nielsen, 1964). Most other desert animals minimize water losses in excretion, too. Birds and lizards produce solid uric acid wastes rather than urea, thereby requiring little water for excretion. Desert lizards also conserve water by retreating to burrows and lowering their metabolic rate during the heat of the day.

## OTHER LIMITING MATERIALS

Numerous other materials, including calcium, chloride, magnesium, nitrogen, potassium, and sodium, may be in short supply for particular organisms and must therefore be budgeted. Neural mechanisms of animals depend on sodium, potassium, and chloride ions, which are sometimes available in limited quantities. Because many herbivorous mammals obtain little sodium from their plant foods (plants lack nerves, and sodium is not essential to their physiology), these animals must conserve sodium or find supplemental sources at salt licks—indeed, Feeny (1975) suggests that plants may actually withhold sodium as an antiherbivore tactic! Similarly, amino acids are in short supply for many insects, such as in *Heliconius* butterflies, which supplement their diets with protein-rich pollen (Gilbert, 1972).

An organism's nutrient and vitamin requirements are strongly influenced by the evolution of its metabolic pathways; likewise, these same pathways may themselves determine certain of the organism's nutritional needs. To illustrate: Almost all species of vertebrates synthesize their own ascorbic acid, but humans and several other primate species that have been tested cannot; they require a dietary supplement of ascorbic acid, known as vitamin C. Of thousands of other species of mammals, only two—the guinea pig and an Indian fruit-eating bat—are known to have lost the ability to synthesize their own ascorbic acid. A few species of birds must supplement their diets with ascorbic acid, too. Thus, a frog, a lizard, a sparrow, or a rat can make its own ascorbic acid, but we cannot. Why should natural selection favor the loss of the ability to produce a vital material? Pauling (1970) suggests that species of animals that have lost this capacity evolved in environments with ample supplies of ascorbic acid in available foods. It might actually be advantageous to dismantle a biochemical pathway in favor of another once it becomes redundant. Conversely, natural selection should favor evolution of the ability to synthesize any necessary materials that cannot be predictably obtained from available foods where this is possible (clearly organisms cannot synthesize elements; e.g., herbivores cannot make sodium).

## SENSORY CAPACITIES AND ENVIRONMENTAL CUES

Animals vary tremendously in their perceptive abilities. Most (except some cave dwellers and deep-sea forms) use light to perceive their environments. But visual spectra and acuity vary greatly. Some, such as insects, lizards, and birds, have color vision, whereas others (most mammals, except squirrels and primates) do not. Ants, bees, and some birds can detect polarized light and exploit this ability to navigate by the sun's position; pigeons have poorly understood backup systems that enable them to return home remarkably well, even when their vision is severely impaired with opaque contact lenses. Many temperate-zone species of plants and animals rely on changes in day length to anticipate seasonal changes in climatic conditions. This, in turn, requires an accurate "biological clock." (Some organisms may also use barometric pressure to anticipate climatic changes.) Certain snakes, such as pit vipers and boas, have infrared receptors that allow them to locate and capture endothermic prey in total darkness. Most animals can hear, of course, although response to different frequencies varies considerably (some actually perceive ultrasonic sounds). Bats and porpoises emit and exploit sonar signals to navigate by echolocation. Similarly, nocturnal electric fish perceive their immediate environments by means of self-generated electrical fields. Certain bioluminescent organisms, such as "fireflies" (actually beetles), produce their own light for a variety of purposes, including the attraction of mates and prey as well as (possibly) the evasion of predators (in some cases, a mutualistic relationship is formed with a bioluminescent bacteria). Certain deep-sea fish probably use their "headlights" to find prey in the dark depths of the ocean. Pigeons can detect magnetic fields. Although a few animals have only a relatively feeble sense of smell (e.g., birds and humans), most have keen chemoreceptors and olfactory abilities. Certain male moths can detect exceedingly dilute pheromones released by a female a full kilometer upwind, allowing these males to find females at considerable distances. Similarly, dung beetles use a zigzag flight to "home in" on upwind fecal material with remarkable precision.

Various environmental cues clearly provide particular animals with qualitatively and quantitatively different kinds and amounts of information. Certain environmental cues are useful in the context of capturing prey and escaping predators; others may facilitate timing of reproduction to coincide with good conditions for raising young. Some environmental signals are noisier and less reliable than others. Moreover, the ability to process information received from the environment is limited by a finite neural capacity. The principle of allocation and the notion of trade-offs dictate that an individual cannot perceive all environmental cues with high efficiency. If ability to perceive a broad range of environmental stimuli actually requires lowered levels of performance along each perceptual dimension, natural selection should improve perceptual abilities along certain critical dimensions at the expense of other less useful ones. Clearly, echolocation has tremendous utility for a nocturnal bat, whereas vision is relatively much less useful. In contrast, the values of these two senses are reversed for a diurnal arboreal squirrel. Within phylogenetic constraints imposed by its evolutionary history, an animal's sensory capacities can be viewed as a bioassay of the importance of particular perceptual dimensions and environmental cues in that animal's ecology.

## ADAPTIVE SUITES

A basic point of this chapter is that any given organism possesses a unique coadapted complex of physiological, behavioral, and ecological traits whose functions complement one another and enhance that organism's reproductive success. Such a constellation of adaptations has been called an *optimal design* (Rosen, 1967) or an *adaptive suite* (Bartholomew, 1972).

Consider the desert horned lizard *Phrynosoma platyrhinos* (Figure 5.20). Various features of its anatomy, behavior, diet, temporal pattern of activity, thermoregulation, and reproductive tactics can be profitably interrelated and interpreted to provide an integrated view of the ecology of this interesting animal (Pianka and Parker, 1975a). Horned lizards are ant specialists and usually eat essentially nothing else. Ants are small and contain much undigestible chitin, so that large numbers of them must be consumed. Hence, an ant specialist must possess a large stomach for its body size. When expressed as a proportion of total body weight, the stomach of this horned lizard occupies a considerably larger fraction of the animal's overall body mass (about 13 percent) than do the stomachs of all other sympatric desert lizard species, including the herbivorous desert iguana *Dipsosaurus dorsalis* (herbivores typically have lower assimilation rates and larger stomachs than carnivores). Possession of such a large gut necessitates a tanklike body form, reducing speed and decreasing the lizard's ability to escape from predators by flight. As a result, natural selection has favored a spiny body form and cryptic behavior rather than a sleek body and rapid movement to cover (as in most other species of lizards). Risks of predation are likely to be increased during long periods of exposure while foraging in the open. A reluctance to move, even when actually threatened by a potential predator, could well be advantageous; movement might attract attention of predators and negate the advantage of concealing coloration and contour. Such decreased movement doubtless contributes to the observed high

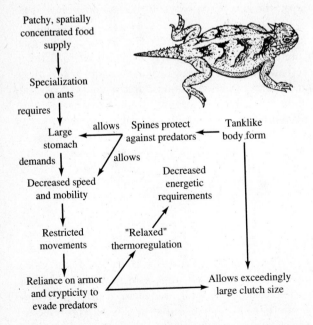

Figure 5.20. Diagrammatic portrayal of factors influencing the ecology and body form of the North American desert horned lizard, *Phrynosoma platyrhinos*.

variance in body temperature of *Phrynosoma platyrhinos*, which is significantly greater than that of all other species of sympatric lizards.

*Phrynosoma platyrhinos* are also active over a longer time interval than any sympatric lizard species. Wide fluctuations in horned lizard body temperatures under natural conditions presumably reflect both the long activity period and perhaps their reduced movements into or out of the sun and shade (most of these lizards are in the open sun when first sighted). More time is thus made available for activities such as feeding. A foraging anteater must spend considerable time feeding. Food specialization on ants is economically feasible only because these insects usually occur in a clumped spatial distribution and hence constitute a concentrated food supply. To make use of this patchy and spatially concentrated, but at the same time not overly nutritious, food supply, *P. platyrhinos* has evolved a unique constellation of adaptations that include a large stomach, spiny body form, an expanded period of activity, and "relaxed" thermoregulation (eurythermy). The high reproductive investment of adult horned lizards is probably also a simple and direct consequence of their robust body form. Lizards that must be able to move rapidly to escape predators, such as racerunners (*Cnemidophorus*), would hardly be expected to weight themselves down with eggs to the same extent as animals like horned lizards that rely almost entirely upon spines and camouflage to avoid their enemies.

Energetics of metabolism of weasels provide another, somewhat more physiological, example of a suite of adaptations (Brown and Lasiewski, 1972). Due to their long thin body shape, weasels have a higher surface-to-volume ratio than mammals with a more standard shape, and as a consequence, they have an increased energy requirement. Presumably, benefits of the elongate body form more than outweigh associated costs; otherwise, natural selection would not have favored evolution of the weasel body shape. Brown and Lasiewski (1972) speculate that a major advantage of the elongate form is the ability to enter burrows of small mammals (weasel prey), which results in increased hunting success and thus allows weasels to balance their energy budgets (Figure 5.21). A further spinoff of the elongate shape is the evolution of a pronounced sexual dimorphism in body size, which allows male and female weasels to exploit prey of different sizes and hence reduces competition between the sexes (related mustelids such as skunks and badgers do not have the marked sexual size dimorphism characteristic of weasels).

# DESIGN CONSTRAINTS

Most biologists are acutely aware that possible evolutionary pathways are somehow constrained by basic body plans. Although natural selection has "invented," developed, and refined a truly amazing variety of adaptations,* selection is

---

*Consider, for example, photosynthesis, the immune response, vision, flight, echolocation, and celestial navigation to name only a few of the more magnificent biotechnological innovations. Feathers remain one of the very best insulators (down-filled sleeping bags).

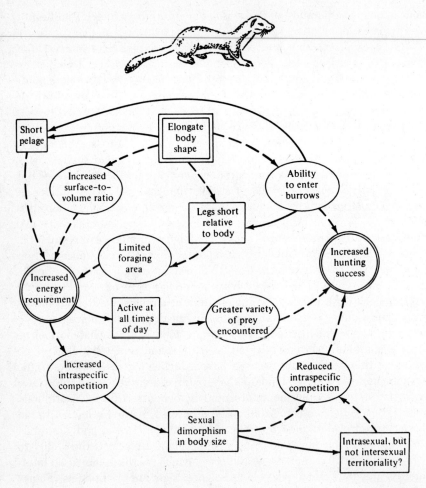

**Figure 5.21.** A schematic representation of factors involved in the evolution of elongate body shape in weasels. Circles indicate primary consequences of evolving a long, thin body configuration; ellipses show secondary consequences; rectangles indicate phenotypic characteristics of weasels affected by evolution of this body shape. Unbroken arrows indicate selective pressures and dashed arrows show causal sequences. Changes proceed in the direction of the arrows as long as selection favors a more elongate shape. [From Brown and Lasiewski (1972). Copyright © 1972 by the Ecological Society of America.]

clearly far from omnipotent. Wheels might be a desirable solution to certain environmental contingencies and yet they have not been evolved. Such "design constraints" are usually elusive and not easily demonstrable.

Students of thermoregulation have often noted an apparent upper thermal limit of about 40°C for most of the earth's eukaryotic creatures (most plants, invertebrates, and vertebrates). This thermal "lid" has frequently been used as evidence for some extremely archaic and inflexible fundamental physiological process (perhaps an enzyme basic to life processes, such as a dehydrogenase, denatures above this limit). The major exceptions are certain heat-tolerant bacteria and blue-green algae, inhabitants of hot springs and oceanic volcanic vents. These prokaryotic organisms may well have arisen before the origin of the heat-sensitive metabolic pathway that seems to limit the eukaryotes.

An example of such a physiological design constraint involves the thermal relationships of vertebrates, spanning classes from reptiles to mammals (Pianka, 1985, 1986a). Detailed consideration of behavioral thermoregulation in lizards enables a fairly accurate prediction of the active body temperatures of mammalian homeotherms. A provocative biological "constant" can thus be identified that suggests a substantial degree of physiological inertia.

An intriguing hypothesis for the evolution of homeothermy was offered by Hamilton (1973), who suggested that homeothermy is a by-product of advantages gained from maintaining maximum body temperatures in the face of such an innate physiological ceiling. Ecologically optimal temperatures need not coincide with physiological optima.

Remember that not all homeotherms are endotherms; many ectotherms have attained a substantial degree of homeothermy by means of behavioral thermoregulation. Typically, such organisms actively select thermally suitable microhabitats, orient their bodies (or parts thereof) to control heat exchange, and/or shuttle between sun and shade as necessary to maintain a more or less constant internal body temperature.

Thermoregulation in lizards is not nearly as simple as it might appear to be at first glance, but rather encompasses a wide diversity of very different thermoregulatory tactics among species ranging from ectothermic poikilothermy to and including ectothermic homeothermy. Even a casual observer quickly notices that various species of desert lizards differ markedly in their times and places of activity. Some are active early in the morning, but other species do not emerge until late morning or midday. Most geckos and pygopodids and some Australian skinks are nocturnal. Certain species are climbers, others subterranean, while still others are strictly surface dwellers. Among the latter, some tend to be found in open areas, whereas others frequent the edges of vegetation. Thermal relations of active lizards vary widely among species and are profoundly influenced by their spatial and temporal patterns of activity. Body temperatures of some diurnal heliothermic species average 38°C or higher, whereas those of nocturnal thigmothermic species are typically in the mid-twenties, closely paralleling ambient air temperatures.

Interesting interspecific differences also occur in the variance in body temperature as well as in the relationship between body temperatures and air temperatures. For example, among North American lizards, two arboreal species (*Urosaurus graciosus* and *Sceloporus magister*) display narrower variances in body temperature than do terrestrial species. Presumably, arboreal habits often facilitate efficient, economic, and rather precise thermoregulation. Climbing lizards have only to shift position slightly to be in the sun or shade or on a warmer or cooler substrate, and normally do not move through a diverse thermal environment. Moreover, arboreal lizards need not expend energy making long runs as do most ground dwellers, and thus climbing species do not raise their body temperatures metabolically to as great an extent as do terrestrial lizards.

Such differences in temporal patterns of activity, the use of space, and body temperature relationships are hardly independent. Rather, they complexly constrain one another, sometimes in intricate and obscure ways. For example, thermal conditions associated with particular microhabitats change in characteristic ways in time; a choice basking site at one time of day becomes an inhospitable hot spot at another time. Perches of arboreal lizards receive full sun early and late in the day when ambient air temperatures tend to be low and basking is therefore desirable, but these same tree trunks are shady and cool during the heat of midday when heat-avoidance behavior becomes necessary. In contrast, the fraction of the ground's surface in the sun is low when shadows are long early and late, but reaches a maximum at midday. Terrestrial heliothermic lizards may thus experience a shortage of suitable basking sites early and late in the day; moreover, during the heat of the day, their movements through relatively extensive patches of open sun can be severely curtailed. Hence, ground-dwelling lizards encounter fundamentally different and more difficult thermal challenges than do climbing species.

Radiation and conduction are the most important means of heat exchange for the majority of diurnal lizards, although the thermal background in which these processes occur is strongly influenced by prevailing air temperatures. Ambient air temperatures are critical to nocturnal lizards as well as to certain very cryptic diurnal species.

In an analysis of the costs and benefits of lizard thermoregulatory strategies, Huey and Slatkin (1976) identified the slope of the regression of body temperature against ambient environmental temperature as a useful indicator (in this case, an inverse measure) of the degree of passiveness in regulation of body temperature. On such a plot of active body temperature versus ambient temperature, a slope of one indicates true poikilothermy or totally passive thermoconformity (a perfect correlation between air temperature and body temperature results), whereas a slope of zero reflects the other extreme of perfect thermoregulation. Lizards span this entire thermoregulation spectrum. Among active diurnal heliothermic species, regressions of body temperature on air temperature are fairly flat (for several species, including some quite small ones, slopes do not differ significantly from zero); among nocturnal species, slopes of similar plots are typically closer to unity. Various other species (nocturnal, diurnal, and crepuscular), particularly Australian ones, are intermediate, filling in this continuum of thermoregulatory tactics.

A straight line can be represented as a single point in the coordinates of slope versus intercept; these two parameters are plotted for linear regressions of body temperatures on air temperatures among some 82 species of lizards in Figure 5.22. Each data point represents the least-squares linear regression of body temperature against air temperature for a given species of desert

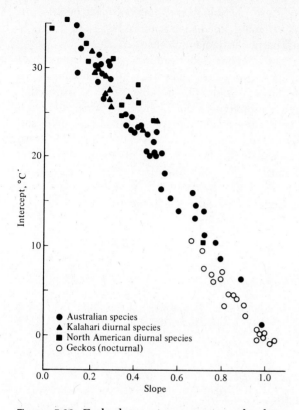

**Figure 5.22.** Each data point represents the least-squares linear regression of body temperature against air temperature for a given species of desert lizard (data given in Pianka, 1986a). Sample sizes are usually substantial (average is 145). The horizontal axis represents the spectrum of thermoregulatory tactics ranging from active thermoregulators (slope of zero) to entirely passive thermoconformity (slope of one). The intriguing "intercept" of the intercepts (38.8°C) approximates the point of intersection of all 82 regression lines and presumably represents an innate design constraint imposed by lizard physiology and metabolism.

lizard. Interestingly enough, these data points fall on yet another, transcendent, straight line. The position of any particular species along this spectrum reflects a great deal about its complex activities in space and time. The line plotted in Figure 5.22 thus offers a potent linear dimension on which various species can be placed in attempts to formulate general schemes of lizard ecology (Pianka, 1985, 1986a). Various other ecological parameters, including reproductive tactics, can be mapped on to this emergent spatial-temporal axis.

The intriguing "intercept" of the intercepts (38.8°C) approximates the point of intersection of all 82 regression lines and presumably represents an innate design constraint imposed by lizard physiology and metabolism. It is presumably not an accident that this value also corresponds more or less to the body temperature of homeotherms, particularly mammals!

Birds, which maintain slightly higher body temperatures than mammals (Hamilton, 1973), descended from another reptilian stock, the archosaurs, represented today by the crocodilians. Would a comparable study of crocodilian thermoregulation yield a higher intercept of the intercepts? (This prediction could be doomed to failure by the mere fact that crocodilians are aquatic and very large—yet they obviously thermoregulate when out of the water.) Although most insects are so small that convective heat exchange prevents them from attaining body temperatures much higher than that of ambient air, some, such as bumblebees and butterflies, do exhibit behavioral thermoregulation. Would a plot for insects show more scatter and a different intercept?

## SELECTED REFERENCES

### Limiting Factors and Tolerance Curves

Ehrlich and Birch (1967); Errington (1956); Hairston et al. (1960); Lack (1954, 1966); Liebig (1840); Murdoch (1966a); Odum (1959, 1963, 1971); Shelford (1913b); Terborgh (1971); Walter (1939).

### Resource Budgets: The Principle of Allocation

Fitzpatrick (1973); Levins (1968); Randolph et al. (1975); Townsend and Calow (1981).

### Time, Matter, and Energy Budgets

Emlen (1966); Gadgil and Bossert (1970); Gibb (1956, 1960); Grodzinski and Gorecki (1967); Hickman (1975); Pianka (1976b, 1981a); C. Smith (1968); Townsend and Calow (1981); Whittaker (1975); Willson (1972a); Wootton (1979); Zeuthen (1953).

### Leaf Tactics

Bailey and Sinnott (1916); Esser (1946a, 1946b); Gentry (1969); Givnish (1979); Givnish and Vermeij (1976); Horn (1971, 1975a, 1975b, 1976); Howland (1962); Janzen (1976); Miller (1979); Mooney et al. (1975); Orians and Solbrig (1977); Parkhurst and Loucks (1971); Ryder (1954); Stowe and Brown (1981); Vogel (1970); Wolf and Hainsworth (1971); Wolf et al. (1972).

### Foraging Tactics and Feeding Efficiency

Charnov (1973, 1976a, 1976b); Charnov et al. (1976); Cody (1968); Emlen (1966, 1968a); Hespenhide (1971); Holling (1964); Huey and Pianka (1981);

Kamil and Sargent (1982);   MacArthur (1959, 1972);   MacArthur and Pianka (1966); Morse (1971);   Orians and Pearson (1979);   Pianka (1966b);   Pulliam (1974);   Rapport (1971);   Royama (1970);   Schoener (1969a, 1969b, 1971);   Schoener and Janzen (1968); Tullock (1970a);   Werner and Hall (1974);   Wolf et al. (1972).

## Physiological Ecology

Bligh (1973);   Cloudsley-Thompson (1971);   Florey (1966);   Folk (1974);   Gates and Schmerl (1975);   Gordon (1972);   Guyton and Horrobin (1974);   Hadley (1975); Hochachka and Somero (1973);   Levitt (1972);   Patten and Smith (1975);   Prosser (1973);   Schmidt-Nielsen (1964, 1975);   Townsend and Calow (1981);   Vernberg (1975);   Vernberg and Vernberg (1974);   Wieser (1973);   Yousef, Horvath, and Bullard (1972).

## Physiological Optima and Tolerance Curves

Brown and Feldmeth (1971);   Huey and Stevenson (1979);   Ruibal and Philibosian (1970);   Schmidt-Nielsen (1975);   Shelford (1913b).

## Energetics of Metabolism and Movement

Bennett and Nagy (1977);   Denny (1980);   McNab (1963);   Mohr (1940);   Nagy (1987);   Pearson (1948);   Schmidt-Nielsen (1972, 1975);   Schoener (1968b);   Tucker (1975);   Turner, Jennrich, and Weintraub (1969);   Whittaker (1975);   Zeuthen (1953).

## Adaptation and Deterioration of Environment

Fisher (1930, 1958a, 1958b);   Henderson (1913);   Maynard Smith (1976);   Ødum (1965);   Van Valen (1973).

## Heat Budgets and Thermal Ecology

Bartholomew (1972);   Bartlett and Gates (1967);   Brown and Feldmeth (1971);   Brown and Lasiewski (1972);   Cowles and Bogert (1944);   Dawson (1975);   Gates (1962); Hamilton (1973);   Huey and Slatkin (1976);   Huey and Stevenson (1979);   Patten and Smith (1975);   Porter and Gates (1969);   Porter et al. (1973);   Ruibal (1961); Ruibal and Philibosian (1970);   Schmidt-Nielsen (1964);   Schmidt-Nielsen and Dawson (1964); Wieser (1973);   Wittow (1970).

## Water Economy in Desert Organisms

Caldwell (1979);   Caldwell and Fernandez (1975);   Cloudsley-Thompson (1971);   Folk (1974);   Gindell (1973);   Hadley (1975);   Main (1976);   Schmidt-Nielsen (1964, 1975).

## Other Limiting Materials

Feeny (1975);   Gilbert (1972);   Pauling (1970).

## Sensory Capacities and Environmental Cues

Griffin (1958); Lloyd (1965, 1971, 1975); Machin and Lissman (1960); Schmidt-Nielsen (1975).

## Adaptive Suites

Bartholomew (1972); Brown and Lasiewski (1972); Frazzetta (1975); Pianka and Parker (1975a); Rosen (1967); Wilbur (1977).

## Design Constraints

Brown and Feldmeth (1971); Duellman and Trueb (1986); Hamilton (1973); Huey and Pianka (1977); Huey and Slatkin (1976); LaBarbera (1983); Liem and Wake (1985); Pianka (1985, 1986a).

# Rules of Inheritance for Life on Earth

## BASIC MENDELIAN GENETICS

Although a background in genetics is certainly not essential for appreciation of many populational and ecological phenomena, it is a useful aid for application to some such phenomena and is required for a full understanding of others. Precise rules of inheritance were actually unknown when Darwin (1859) developed the theory of natural selection, but they were formulated a short time afterward (Mendel, 1865). Darwin accepted the hypothesis of inheritance in vogue at the time: *blending inheritance*. Under the blending inheritance hypothesis, the genetic makeups of both parents are imagined to be blended in their progeny, and all offspring produced by sexual reproduction should be genetically intermediate between their parents; genetic variability is thus lost rapidly unless new variation is continually being produced. (Under blending inheritance and random mating, genetic variability is *halved* each generation.) Darwin was forced to postulate extremely high mutation rates to maintain the genetic variability observed in most organisms (Fisher, 1930) and he was painfully aware of the inadequacy of knowledge on inheritance. Mendel's discovery of *particulate inheritance* represents one of the major empirical breakthroughs in biology.

Mendel performed breeding experiments with different varieties of peas, paying close attention to a single trait at a time. He had two types that bred "true" for yellow and green peas, respectively. When a purebred green pea plant was crossed with a purebred yellow pea plant, all progeny, or individuals of the first filial generation ($F_1$), had yellow peas. However, when these $F_1$ plants were crossed with each other or self-fertilized, about one out of every four offspring in the second filial generation ($F_2$) had green peas. Furthermore, only about one-third of the yellow $F_2$ pea plants bred true; the other two-thirds, when self-fertilized, produced some offspring with green peas. All green pea plants bred true. Mendel proposed a very simple quantitative hypothesis to explain his results and performed many other breeding experiments on a variety of other traits that corroborated and confirmed his interpretations. Subsequent work has strengthened Mendel's hypothesis, although it has also led to certain modifications and improvements.

Mendel postulated that each pea plant had a double dose of the "character" controlling pea color but that only a single dose was transmitted into each of its sexual cells, or *gametes* (pollen and ovules or sperm and eggs). Purebred plants, with identical doses, produced genetically identical single-dosed gametes; the above-mentioned $F_1$ plants, on the other hand, with two different doses, produced equal numbers of the two kinds of gametes, half bearing the character for green and half that for yellow. In addition, Mendel proposed that yellow masked green whenever the two occurred together in double dose; hence, all plants had yellow peas, but when self-fertilized, produced some $F_2$ progeny with green peas. All green pea plants, which had a double dose of green, always bred true.

**Gregor Mendel**

Modern terminology for various aspects of Mendelian inheritance is as follows: (1) the "character" or "dose" controlling a particular trait is termed an *allele*; (2) its position on a chromosome (below) is termed its *locus*; (3) a single dose is the *haploid* condition, designated by $n$, whereas the double-dosed condition, designated by $2n$, is *diploid* (polyploids, such as triploids and tetraploids, are designated with still higher numbers); (4) the set of alternative alleles that may occur at a given locus (there can be only two alleles in a diploid individual, but there may be more than two in any given population) is termed a *gene*; (5) purebred diploid individuals with identical alleles are *homozygotes*, homozygous for the trait concerned; (6) individuals with two different alleles, such as the preceding $F_1$ plants, are *heterozygotes*, heterozygous at that locus; (7) an allele that masks the expression of another allele is said to be *dominant*, whereas the one that is masked is *recessive;* (8) unlinked alleles separate, or *segregate*, from each other in the formation of gametes; (9) whenever heterozygotes or two individuals that are homozygous for different alleles mate, new combinations of alleles arise in the following generation by *reassortment* of the genetic material; (10) observable traits of an individual (e.g., yellow or green in the previous example) are aspects of its *phenotype*, which includes all observable characteristics of an organism; and (11) whether or not an organism breeds true is determined by its *genotype*, which is the sum total of all its genes.

Occasionally, some organisms have pairs of alleles with *incomplete dominance*. In such cases, the phenotype of the heterozygote is intermediate between that of the two homozygotes; that is, phenotype accurately reflects genotype, and vice versa. Presumably, alleles conferring advantages upon their bearers usually evolve dominance over time because such dominance ensures that a maximal number of the organism's progeny and descendants will benefit from possession of that allele. The apparent rarity of incomplete dominance is further evidence that dominance has evolved. Moreover, so-called wild-type alleles—that is, those most prevalent in natural wild populations—are nearly always dominant over other alleles occurring at the same locus. Geneticists have developed numerous theories of the evolution of dominance, but the exact details of the process have not been completely resolved.

Cytological observations of appropriately prepared cell nuclei confirm Mendel's hypothesis beautifully (Figure 6.1). Microscopic examination of such cells reveals elongated dense bodies in cell nuclei; these are the *chromosomes*, which contain the actual genetic material, deoxyribonucleic acid, or DNA. Nuclei of diploid cells, including the zygote (the fertilized ovule or egg) and the somatic (body) cells of most organisms, always contain an even number of chromosomes. (The exact number varies widely from species to species, with as few as two in certain arthropods to hundreds in some plants.) Moreover, pairs of distinctly similar *homologous chromosomes* are always present and often easily detected visually. However, gametes contain only half the number of chromosomes found in diploid cells, and, except in polyploids, none of them is homologous. Thus, haploid cells contain only one full set of different chromosomes and alleles, or one *genome*, whereas all diploid cells contain two. During the reduction division (*meiosis*) in which diploid gonadal cells give rise to haploid gametes, homologous chromosomes separate (Figure 6.1). Later, when male and female gametes fuse to form a diploid *zygote* that will develop into a new diploid organism, homologous chromosomes come together again. Hence, one genome in every diploid organism is of paternal origin while the other is of maternal ancestry. Because each member of a pair of homologous chromosomes separates from its homologue independently of other chromosome pairs, the previous generation's chromosomes are reassorted with each reduction division. Thus, the genetic material is regularly rearranged and mixed up by the dual processes of meiosis and the actual fusion of gametes (this has been termed the *Mendelian lottery*).

Numerous different loci and allelic systems occur on each chromosome. Two different traits controlled by different alleles located on the same chromosome do not segregate truly independently but are statistically associated with or dissociated from one another. This is the phenomenon of *linkage*. Dur-

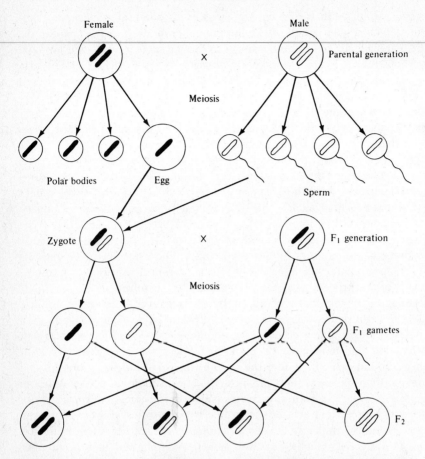

**Figure 6.1.** Diagrammatic representation of the cytological events in cell nuclei showing how the two parental genomes are sorted and recombined in the next generation, or the $F_2$. For simplicity, only one pair of chromosomes is shown and the complex events of the reduction division (meiosis) are omitted.

ing meiosis, homologous chromosomes can effectively exchange portions by means of *crossovers*; this process is referred to as *recombination*. Because the frequency of occurrence of crossovers between two loci is a function of the distance between them on the chromosome, geneticists use crossover frequencies to "map" the effective distance between loci, as well as their positions relative to one another on the chromosome. By means of close linkage, whole blocks of statistically associated alleles can be passed on to progeny as a functionally integrated unit of "coadapted" alleles. Certain kinds of chromosomal rearrangements, such as inversions, may suppress crossovers. Indeed, one of the major advantages of chromosomes is presumably that they enhance the degree to which clusters of genes can occur together.

In many organisms a single pair of chromosomes, termed *sex chromosomes*, determines the sex of their bearer (the remaining chromosomes, which are not involved in sex determination, are *autosomes*). Typically, one homologue of the sex chromosome pair is smaller. In the diploid state, an individual heterozygous for the sex chromosomes is *heterogametic*. In mammals, males are the heterogametic sex with an XY pair of sex chromosomes, whereas females are the homogametic sex with an XX pair. Because male-male matings do not result in progeny, the homozygous genotype YY can never occur. In birds and some other organisms, the female is the heterogametic sex.

Although natural selection actually operates on phenotypes of individuals (i.e., an organism's immediate fitness is determined by its total phenotype), the effectiveness of selection in changing the composition of a population depends on the *heritability* of phenotypic characteristics or the percentage of phenotypic variability attributable to genotype. Traits that are under strong selection usually display low heritability because the genetic component of phenotypic variability has been reduced by selection. Because nongenetic traits are not inherited, differential reproduction by different phenotypes stemming from such nontransmittable traits obviously cannot alter a population's gene pool. Different genotypes may often have fairly similar phenotypes and thus similar fitnesses. Selection may even favor alleles that are "good mixers" and work well with a wide variety of other genes to increase their bearer's fitness in various genetic backgrounds (Mayr, 1959). Conversely, of course, identical genotypes can develop into rather different phenotypes under different environmental conditions.

Genes that act to control the expression of other genes at different loci are called *modifier* or *regulatory genes*, whereas those that code for specific cell products are termed *structural* genes (some genes probably do not fit this dichotomy but may serve in both capacities). Although relatively little is known, geneticists imagine that an intricate hierarchy must exist leading from regulator genes to structural genes to proteins to other nonproteinaceous metabolic products to specific phenotypes. Moreover, complex interactions must occur among regulators as they do among proteins and other metabolites, such as neurotransmitters and hormones.

Humans and the great apes share many genes, including the familiar ABO blood groups. A detailed molecular comparison of human proteins with those of chimpanzees by three different techniques (sequencing, immunological distance, and electrophoresis) revealed nearly identical amino acid sequences in the vast majority of proteins (99% similarity, with concordant results at 44 loci), presumably the products of structural genes (King and Wilson, 1975). Apparently, a relatively few genetic changes in major regulatory genes can have profound phenotypic effects without much difference at the level of proteins. Thus, relatively trivial genetic differences can lead to major phenotypic differences. The genetic similarity between humans and chimps is comparable to that of sibling species in insects and other mammals. Recent DNA hybridization studies have demonstrated 98.4% similarity between humans and chimpanzees (Sibley et al., 1990). Whereas humans are placed in the family Hominidae, chimps are placed with the Great Apes in the family Pongidae. However, phylogenies based on molecular similarities show that humans are embedded within the apes (Figure 6.2). Clearly, it is time to consider reclassification!

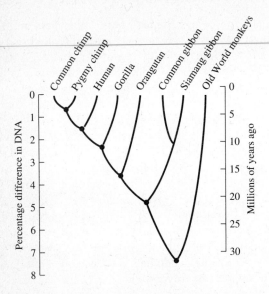

**Figure 6.2.** Phylogeny of primates based on DNA hybridization. [Adapted from Diamond (1991) after Sibley and Ahlquist.]

## NATURE VERSUS NURTURE

A widespread misconception is that any phenotypic trait can always be assigned to either one of two mutually exclusive categories: genetic or environmental. However, this dichotomy is not only oversimplified but can be rather misleading. Because natural selection acts only on heritable phenotypic traits, even environmentally flexible traits must usually have an underlying genetic basis.

For example, when grown on dry plant foods, the Texas grasshoppers *Syrbula* and *Chortophaga* become brown, but when fed on moist grasses, these same insects develop green phenotypes—this classic "environmentally induced" polymorphism is presumably highly adaptive since it produces background color-matching cryptic green grasshoppers when environments are green, but brown ones in brown environments (Otte and Williams, 1972). The capacity for developmental plasticity itself has almost surely evolved in response to the unpredictable environment these grasshoppers must face. If enough were known, much environmentally determined phenotypic variation would presumably have a somewhat comparable basis in natural selection. Thus, truly nongenetic traits are unimportant and uninteresting simply *because they cannot evolve* and hence do not affect fitness. Indeed, for purposes of evolutionary ecology, virtually all traits can be considered as being subject to natural selection (those that are not cannot easily persist and have little if any evolutionary significance).

# SELFISH GENES

Certain alleles do not obey the Mendelian lottery of meiosis and recombination; instead, these "outlaw genes" obtain disproportionate representation in a carrier's gametes at the expense of alternate alleles on homologous chromosomes. An example of such a selfish gene is the "segregation distortion" allele in the fruit fly *Drosophila*. Males heterozygous for this sex-linked trait produce sperm, some 95 percent of which carry the allele (rather than the expected ratio of 50 percent). This process is known as *meiotic drive*. Why don't more genes behave in this manner? Very probably the intense contest for representation in the gametes has itself ended in a stalemate, yielding the traditional Mendelian ratio (as viewed in this way, segregation itself is clearly a product of natural selection).

Advocates of the "selfish gene" hypothesis (Dawkins, 1976, 1982) argue somewhat as follows. Barring mutations, genes are perfect replicators, always making exact copies of themselves. However, phenotypes of individual organisms are transmitted to their offspring only imperfectly, at least in sexually reproducing species. Individuals can thus be viewed as mere "vehicles" for the genes that they carry.

Except for viruses, genes usually do not exist in isolation but must occur together in large clusters. A sort of "packaging problem" arises, with the number of copies left by any given allele dependent upon the particular combination of other genes, or "genetic background," in which the allele concerned actually occurs. (Actually, of course, selection "sees" the phenotypic expression resulting from the interactions among all the alleles in a particular genetic constellation in a given environment.) Also, genes clearly cannot replicate themselves except by means of successful reproduction of the entire organism. Hence selection must *usually** favor genes that work together to enhance an individual's ability to perpetuate them, thus increasing its fitness. A "parliament of genes" acts to govern the phenotype. A mutant gene that prevented its bearer from reproducing would be short-lived indeed! Likewise, mutations that reduce the reproductive success of individuals will *normally** be disfavored by natural selection. Viruses are indeed selfish genes, as their interests need not coincide with the fitness of their hosts unless they cause their premature demise.

# POPULATION GENETICS

Each generation, sexually reproducing organisms mix their genetic materials. Such shared genetic material is called a *gene pool*, and all the organisms contributing to a gene pool are collectively termed a *Mendelian population*. Gene pools have continuity through time, even as individuals are added and removed

---

*The qualifiers "usually" and "normally" in the above sentences are necessary because other copies of the genes concerned, which are identical by descent, occur in the bodies of related individuals (the important phenomenon of kin selection is considered later).

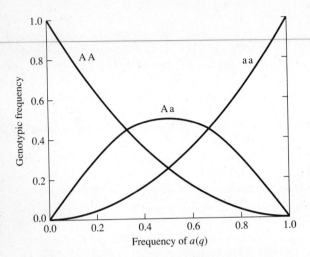

Figure 6.3. Frequencies of the three diploid genotypes for various gene frequencies in a two-allele system in Hardy-Weinberg equilibrium.

by births and deaths. One of the most fundamental concepts of population genetics is the notion of *gene frequency*. An allele's frequency in the haploid gene pool, or its proportional representation, is traditionally represented by the symbol $p$ or $q$. Changes in gene frequencies in a gene pool in time constitute evolution. An individual's ability to contribute its own genes to the gene pool represents that individual's *fitness*.

Equilibrium frequencies of various diploid genotypes that emerge in a given gene pool, given random mating and no evolution, can be calculated from the haploid gene frequencies using the binomial* expansion (also known as the Hardy-Weinberg equation):

$$1 = (p + q)(p + q) = p^2 + 2pq + q^2$$

If $p$ is the frequency of allele A and $q$ the frequency of allele a, the expected equilibrium frequencies of the three diploid genotypes A/A, A/a, and a/a are given by the three terms at the right: $p^2$, $2pq$, and $q^2$, respectively. With random mating and no evolution, gene frequencies remain unchanged from generation to generation. The equilibrium frequency of heterozygotes reaches its maximum of 50 percent when the two alleles are equally represented ($p = q = 0.5$; Figure 6.3).

Population geneticists have relaxed these limiting assumptions and elaborated and extended these equations to model various phenomena such as random genetic drift, gene flow, nonrandom mating, and frequency-dependent selection. Genotypes can also be assigned fixed relative fitnesses; when the

---

*With only two alleles, $q = 1 - p$. For three alleles, the appropriate equation is a trinomial, with $p + q + r = 1$, and so on.

heterozygote is fitter than either homozygote, both alleles are maintained indefinitely, even if one or both of the homozygotes is actually lethal! This widespread and important phenomenon is known as heterozygote advantage or *heterosis*. Such a reduction in fitness due to genetic segregation is termed *genetic load*. If one of the homozygotes enjoys the highest fitness, it is favored by natural selection until that locus becomes "fixed" with an allele frequency of 1.0. Ultimately, the maintenance of genetic variation depends on the precise rules coupling gene frequencies to genotype frequencies.

## MAINTENANCE OF VARIABILITY

The fundamental source of variation between individuals is sexual reproduction; reassortment and recombination of genes in each generation ensure that new genotypes will arise regularly in any population with genetic variability. In most higher organisms, no two individuals are genetically identical (except identical twins and progeny produced asexually). Population biologists are interested in understanding factors that create and maintain genetic variability in natural populations. Numerous genetic mechanisms, including linkage and heterosis, produce genetic variability, both within and between populations.

At the outset, we must distinguish phenotypic from genotypic variation. The phenotypic component of variability is the total observable variability; the genotypic component is that with a genetic basis. It is usually difficult to distinguish genetically induced variation from environmentally induced variation. However, by growing clones of genetically identical individuals (i.e., with the same genotype) under differing environmental conditions, biologists have been able to determine how much interindividual variation is due to the developmental plasticity of a particular genotype in different environments. Pedigree studies show that approximately half the phenotypic variation in height observed in human populations has a genetic basis and the remaining variation is environmentally induced. Because natural selection can act only on heritable traits, many phenotypic variants may have little direct selective value. The degree of developmental flexibility of a given phenotypic trait strongly influences an organism's fitness; such a trait is said to be *canalized* when the same phenotypic character is produced in a wide range of genetic and environmental backgrounds. Presumably, some genes are rather strongly canalized, such as those that produce "wild-type" individuals, whereas others are less determinant, allowing individuals to adapt and regulate via developmental plasticity. Such environmentally induced phenotypic varieties are abundant in plants, but they are less common among animals, probably because mobile organisms can easily select an appropriate environment. Presumably, it is selectively advantageous for certain genetically induced traits to be under tight control, whereas others enhance individual fitness by allowing some flexibility of response to differing environmental influences.

Genotypic and phenotypic variation between individuals, in itself, is probably seldom selected for directly. But it may often arise and be maintained in a number of more or less indirect ways. Especially important are changing environments; in a temporally varying environment, selective pressures vary from time to time and the phenotype of highest fitness is always changing. There is inevitably some lag in response to selection, and organisms adapted to tolerate a wide range of conditions are frequently at an advantage. (Heterozygotes may often be better able to perform under a wider range of conditions than homozygotes.) Indeed, in unpredictably changing environments, reproductive success may usually be maximized by the production of offspring with a broad spectrum of phenotypes (which may well be the major advantage of sexual reproduction).

Similar considerations apply to spatially varying environments because phenotypes best able to exploit various "patches" usually differ. On a broader geographic level, differences from one habitat to the next presumably often result in different selective milieus and therefore in different gene pools adapted to local conditions. Gene flow between and among such divergent populations can result in substantial amounts of genetic variability, even at a single spot.

Competition among members of a population for preferred resources may often confer a relative advantage on variant individuals that are better able to exploit marginal resources; thus, competition within a population can directly favor an increase in its variability. By virtue of such variation between individuals, the population exploits a broader spectrum of resources more effectively and has a larger populational "niche breadth"; the "between-phenotype" component of niche breadth is great (Roughgarden, 1972). Because such increased phenotypic variability between individuals promotes a broader populational niche, this has been termed the *niche-variation hypothesis* (Soulé and Stewart, 1970). Similarly, environments with low availability of resources usually require that individuals exploiting them make use of a wide variety of available resources; in this case, however, because each individual must possess a broad niche, variation between individuals is not great (i.e., the "between-phenotype" component of niche breadth is slight, whereas the "within-phenotype" component is great).

One further way in which variability can be advantageous involves the coevolutionary interactions between individuals belonging to different species, especially interspecific competition and predation. Fisher (1958b) likened such interspecific interactions and coevolution to a giant evolutionary game in which moves alternate with countermoves. He suggested that it may well be more difficult to evolve against an unpredictable and variable polymorphic species than against a better standardized and more predictable monomorphic species. A possible example may be foraging birds developing a "search image" for prey items commonly encountered, often bypassing other less abundant kinds of suitable prey.

## UNITS OF SELECTION

Classical Darwinian natural selection acts only on heritable phenotypic traits of individuals. As discussed earlier, selfish genes are also known to exist. Can selection operate on entire groups of individuals such as families, colonies, populations, species, communities, and ecosystems? To what extent is the individual a natural "unit" of selection? How are conflicts between suborganismal, organismal, and superorganismal levels of selection resolved? These questions are often discussed both by geneticists and by ecologists, but there is no clear consensus as to correct answers.

Many behavioral and ecological attributes can be interpreted as having evolved for the benefit of the group rather than the individual. As an example of such group selection, consider the assertion that "mockingbirds lay fewer eggs during a drought year because competition for limited food supplies would be detrimental to the species." Such statements have a fatal flaw: "cheaters" that laid as many eggs as possible would reap a higher reproductive success than individuals that voluntarily decreased their clutch size for the "benefit of the species." The same phenomenon can be interpreted more plausibly in terms of classical Darwinian selection at the level of the individual. During droughts, parental birds cannot bring as many insects to their nest and therefore cannot feed and fledge as many chicks as they can when food supplies are more ample. Birds can actually leave more surviving offspring to breed in the next generation by laying fewer eggs. Most evolutionary biologists now dismiss the preceding sort of "naive" group selection as untenable.

In the last decade, thinking about group selection has achieved considerably greater sophistication, although it remains speculative. Two distinct types of selection at the level of groups emerge from these mathematical arguments. For "extinction" group selection to oppose natural selection at the level of the individual, isolated selfish subgroups must go extinct faster than selfishness arises within altruistic subgroups and most newly founded isolates must be altruistic. "Graded" group selection requires that distinct subpopulations contribute differentially to reproduction in a bigger population at large. In essence, entire groups must possess differential rates of survivorship or reproduction (i.e., differential fitness).

A major consideration is the extent to which an individual's own best interests are in conflict with those of the group to which it belongs. Ultimately, the frequency of occurrence of socially advantageous behaviors depends largely on the precise form of the trade-offs between group benefit(s) versus individual cost(s). Any individual sacrificing its own reproductive success for the benefit of a group is obviously at a selective disadvantage (within that group) to any other individual not making such a sacrifice. Hence, classical Darwinian selection will always favor individuals that maximize their own reproductive success. Clearly, the course of selection acting within groups cannot be altered by selection operating between groups. Group selection requires very restricted conditions.

Williams (1966a) reemphasized, restated, and expanded the argument against naive group selection, pointing out that classical Darwinian selection at the level of the individual is adequate to explain most putatively "group-selected" attributes of populations and species, such as those suggested by Wynne-Edwards (1962) and Dunbar (1960, 1968, 1972). Williams reminds us that group selection has more conditions and is therefore a more onerous process than classical natural selection; furthermore, he urges that it be invoked only after the simpler explanation has clearly failed. Although group selection is certainly possible, it probably would not actually oppose natural selection at the individual level except under most unusual circumstances. A special form of selection at the level of the individual, *kin selection*, may frequently be the mechanism behind many phenomena interpreted as evidence for group selection. We will return to this issue from time to time in later chapters.

## SELECTED REFERENCES

### Basic Mendelian Genetics

Darlington and Mather (1949);  Darwin (1859);  Ehrlich and Holm (1963);  Fisher (1930);  Ford (1931, 1964);  King and Wilson (1975);  Maynard Smith (1958);  Mayr (1959);  Mendel (1865);  Mettler and Gregg (1969);  Sheppard (1959);  Sibley et al. (1990).

### Nature versus Nurture

Bradshaw (1965);  Clausen et al. (1948);  Greene (1989);  Otte and Williams (1972); Quinn (1987).

### Selfish Genes

Alexander and Borgia (1978);  Dawkins (1976, 1982);  Hamilton (1967);  Leigh (1977); Orgel and Crick (1980).

### Population Genetics

Crow (1986); Crow and Kimura (1970); Falconer (1981); Fisher (1930, 1958a); Ginzburg and Golenberg (1985); Haldane (1932, 1941, 1964); Hedrick (1983); Mayr (1959); Mettler and Gregg (1969); Wright (1931, 1968, 1969, 1977, 1978).

### Maintenance of Variability

Ehrlich and Raven (1969); Fisher (1958b); Mettler and Gregg (1969); Roughgarden (1972); Somero (1969); Soulé and Stewart (1970); Van Valen (1965); Wilson and Bossert (1971).

### Units of Selection

Alexander and Borgia (1978); Boorman and Levitt (1972, 1973); Brown (1966); Cole (1954b); Darlington (1971); Darnell (1970); Dawkins (1976, 1982); Dunbar (1960, 1968, 1972); Emerson (1960); Emlen (1973); Eshel (1972); Fisher (1958a); Gilpin (1975a); Leigh (1977); Levins (1970, 1975); Lewontin (1970); Maynard Smith (1964); Uyenoyama (1979); Van Valen (1971); Wade (1976, 1977, 1978); Wiens (1966); Williams (1966a, 1971); D. S. Wilson (1975, 1980, 1983); E. O. Wilson (1973, 1976); Wright (1931); Wynne-Edwards (1962, 1964, 1965a, 1965b).

## Chapter
## 7

# Evolution, Natural Selection, and Speciation

## AGENTS OF EVOLUTION

Evolution occurs whenever gene frequencies in a population change in time. Individuals do not evolve, but populations do. In addition to natural selection, there are several other agents of evolution, including genetic drift, gene flow, meiotic drive, and mutation. Genetic drift operates by random sampling bias and is confined to relatively small populations. Gene flow occurs by migration movements of plants and animals among and between populations with different gene frequencies. Meiotic drive, or segregation distortion, was briefly considered in Chapter 6. Foreward and backward rates of mutation are seldom the same, and the resulting mutation pressure can cause gene frequencies to change. Of these five different agents of evolution, only natural selection is directed in that it results in conformity between organisms and their environments.

## TYPES OF NATURAL SELECTION

Under *stable* conditions, *intermediates* in a population typically leave more descendants, on the average, than do the extremes. We say that they are more "fit." An individual's "fitness" is measured by the proportion of its genes left in

133

the population gene pool. Selection of this sort, which continually crops the extremes and tends to hold constant the intermediate or average phenotype, is termed *stabilizing selection* (Figure 7.1a). In a stable environment, genetic recombination increases the population's variance each generation, whereas stabilizing selection reduces it to approximately what it was in the previous generation.

However, in a *changing* environment, average individuals (modal phenotypes) may not be the most fit members of the population. Under such a situation, *directional selection* occurs and the population mean shifts toward a new phenotype (Figure 7.1b) that is better adapted to the altered environment. Eventually, of course, unless the environment continues to change, an equilibrium is reached in which the population is readjusted to the new environment, whereupon stabilizing selection resumes.

A third type of selection, *disruptive selection*, takes place when two or more phenotypes with high fitnesses are separated by intermediate phenotypes of lower fitness (Figure 7.1c). This usually occurs in distinctly heterogeneous environments with a discrete number of different "patches." Disruptive selection is one mechanism that produces and maintains polymorphisms, such as the green-brown color polymorphisms of many insects. For instance, some butterflies (commonly called "leaf butterflies") mimic leaves; one population may contain both green and brown animals, with the former matching living leaves and the latter dead ones. Through appropriate behavior and selection of matching resting sites, each color morph enjoys a relatively high fitness; in

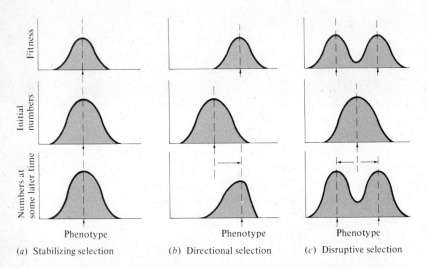

Figure 7.1. Graphic portrayal of the three types of selection. (a) Stabilizing selection, which occurs in constant environments, holds the modal phenotype constant. (b) Directional selection takes place in a changed environment and causes a shift in the modal phenotype. (c) Disruptive selection, with two or more modal phenotypes, occurs in patchy environments with more than one discrete phase.

contrast, a butterfly intermediate between green and brown would presumably match its surroundings less well and thus have a considerably lower fitness.

Another important type of selection is known as *frequency-dependent selection*—this occurs when the fitness of a particular phenotypic trait varies with its frequency in the population. Negative frequency-dependent selection promotes genetic variability. An example would be a bird forming a search image for an abundant prey type, but switching to an alternate prey type when the abundance of the first type is reduced. Such predator switching behaviors favor the rarer prey type, which can then increase in abundance until a flip-flop occurs and the predator resumes eating the first prey type again. Both disruptive selection and frequency-dependent selection can act to help maintain genetic polymorphisms. Other types of selection to be considered in subsequent chapters include *age-specific selection*, *density-dependent selection*, *density-independent selection*, *kin selection*, and *sexual selection*.

## ECOLOGICAL GENETICS

The European land snail, *Cepaea nemoralis*, is polymorphic for shell color, with three phases: brown, pink, and yellow. This polymorphism is genetic, based on a three-allele system, with each of the six diploid genotypes varying in color between the endpoints of brown and yellow. In parts of England, a major predator on these snails is the song thrush *Turdus ericetorum*. These birds break open snail shells on stones ("thrush anvils"). Proportions of shells accumulated at these anvils usually differ from those in the population at large, thereby reflecting the relative intensity of thrush predation on the various snail morphs (Figure 7.2). During April, the predominant background color of the

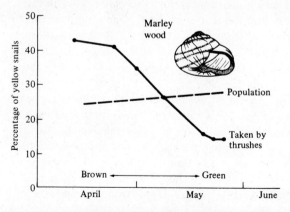

Figure 7.2. Percentages of yellow snails taken by thrushes versus those in the population at large, showing differential predation in time. [From Sheppard (1951).]

woods is brown; conditions become progressively greener during May. The percentage of yellow shells (with the animal inside, yellow shells take on a greenish hue) found at thrush anvils is higher than in the population at large in April, but lower than in the population in late May (Figure 7.2), indicating differential predation by thrushes. Thrushes form search images for common *Cepaea* morphs, resulting in negative frequency-dependent selection (Harvey et al., 1975) favoring rare morphs. Thrush predation may well help to maintain the polymorphism in shell color, since the brown and pink morphs appear to have a fitness advantage early in the season, while yellow snails seem to be at an advantage later.

## ALLOPATRIC SPECIATION AND SYMPATRIC SPECIATION

How do new species arise? How can one set of interbreeding populations break into two? There are two distinct modes of speciation, the process by which new species arise. *Allopatric* or *geographic speciation* occurs in different areas slowly over thousands of years. When the geographic range of a species is broken into two, sets of populations are isolated and gene exchange is prevented, thus allowing such populations to diverge if they are subjected to different selection pressures (this may usually occur when different local conditions require divergent adaptations). Barriers that can cause such events to occur include glaciers, mountain ranges, oceanic straits and isthmuses, as well as habitat changes caused by long-term shifts in climate.

For example, during the pluvial glacial periods in the recent past (about 10,000 years ago), Australia was considerably wetter and an uninterrupted belt of lush mesic forested habitat extended from east to west across the southern third of the continent. Species of birds, lizards, and other animals had geographic ranges coincident with this contiguous belt of habitat. With the retreat of glaciers and the advent of interglacial conditions (also called interpluvials), an extensive arid area developed in the center of the continent and spread to the south coast. Wetter habitats survive today in the southeast and southwest, but these areas are now separated by a vast arid zone that includes the Nullarbor plain. Closely related species pairs of birds and lizards now occur in these southeastern and southwestern refugia.

Due to the lack of gene flow, geographically isolated populations of a species are free to adapt to local conditions. Following an extensive period of experiencing different selective pressures in isolation, the two subsets of what was formerly a single species may have diverged greatly from one another. Whether or not speciation has actually occurred may be put to the test at some later point in time if the two subsets come back together again in sympatry (e.g., this might happen with the advent of another ice age). If the two incipient "species" interbreed and hybridize extensively, the two subsets merge together into one species again in a process known as *introgression*. However, if the two incipient species are different enough that hybrid individuals suffer reduced fitness, natural selection can favor the evolution of reproductive isolating mechanisms (see next section) which prevent introgression from occurring and reinforces the differences between the two new species.

When speciation takes place without geographic isolation, it is termed *sympatric speciation*. This can occur more or less instantly, in several different ways. Many species of plants appear to have arisen by the hybridization of two parental species and subsequent chromosome doubling. This process, known as *allotetraploidy*, occurs as follows: let the two parent species' diploid genomes be represented as AA BB CC and XX YY ZZ, respectively (A, B, C, X, Y, and Z could represent chromosomes). Because the hybrid's genotype is ABCXYZ, chromosomes have no homologues to pair with in meiosis and the hybrid is sterile. However, since the hybrid has attributes of both parental species, it might well be superior to both of them in intermediate habitats (e.g., if one parent species is cold-adapted and the other is hot-adapted, the hybrid could outperform both parental species under warm conditions). Such hybrids can survive and even increase in numbers by clonal or vegetative reproduction. Eventually, endoduplication or a nondisjunction event during meiosis converts the hybrid's genome from ABCXYZ to AA BB CC XX YY ZZ, restoring fertility (chromosomes can now pair, and the hybrid can therefore produce viable gametes). Species that arose through such biparental origin and consequent chromosome doubling are often phenotypically intermediate between their parental species and are often found under intermediate environmental conditions.

Another example of sympatric speciation occurs in tephritid fruit flies (*Rhagoletis*) which oviposit on hawthorn and apple trees. During the past century, these flies have expanded their host plant range to include several other members of the plant family Rosaceae, including cherries, roses, dogwood, and pears. During their tenancy as larvae, individuals learn their host food plant from the plant species on which they find themselves feeding (Prokopy et al., 1982). When a female "accidentally" oviposits on a new (and wrong) species of host plant, her brood effectively becomes an instant new species, effectively isolated from its parent's population (Bush, 1974, 1975).

## REPRODUCTIVE ISOLATING MECHANISMS

Many closely related species that do not interbreed in nature have been hybridized in captivity. For example, all pairs of species of falcons (*Falco*) can produce viable progeny. Mechanisms that prevent interbreeding between species, known as *reproductive isolating mechanisms*, have frequently evolved. Isolating mechanisms can be prezygotic or postzygotic, depending on whether or not fertilization actually occurs. Hybrid sterility is an example of a postzygotic isolating mechanism. Prezygotic behavioral isolating mechanisms can involve courtship behavior, pheromones, vocalizations, and color patterns that promote species recognition and prevent mismatings from occurring between species. For example, female fireflies (actually beetles) respond only to the flashing pattern and flight paths of males belonging to their own species (Lloyd, 1986).

## DARWIN'S FINCHES

The Galápagos Islands are volcanic in origin and are located about a thousand kilometers west of the Ecuadorian coast. An archipelago of relatively small and remote deep water islands, the Galápagos (Figure 7.3) support a remarkable group of birds that nicely illustrate a number of evolutionary and ecological principles. Named Darwin's finches after the first evolutionist to appreciate and study them, these birds dominate the avifauna of these islands. Only 26 species of land birds occurred in the archipelago naturally (i.e., before human introductions), and 13 of these are finches (the islands also support 4 species

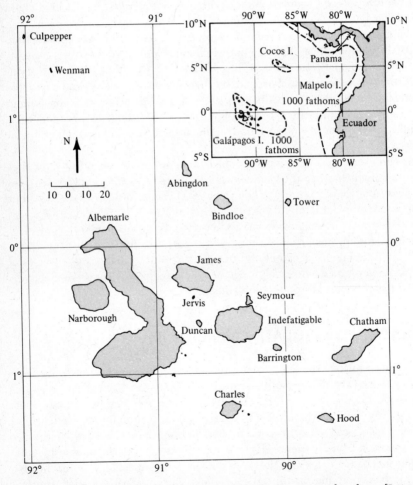

**Figure 7.3.** Two maps of the main islands in the Galápagos Archipelago. [Inset from Lack (1947). Larger map from Bowman (1961), originally published by the University of California Press. Reprinted by permission of the Regents of the University of California.]

of mockingbirds, 2 flycatchers, 2 owls, a hawk, a dove, a cuckoo, a warbler, a martin, and a rail).

This archipelago (16 major islands and a sprinkling of tiny islets) was formed from volcanic eruptions of the ocean floor about 3 to 5 million years ago; thus originally, there were no organisms on the islands, and their entire biota has been derived from mainland species. Because of their remoteness, relatively few different plant and animal stocks have been able to colonize the Galápagos. (However, the position of these islands on the equator presumably makes them particularly vulnerable to invasions from rafts and floating islands carried out to sea in the equatorial current.)

The 13 species of finches are thought to have evolved from a single mainland finch ancestor that reached the islands long ago. (The birds are similar enough to each other that they are classified as a distinct subfamily of finches, endemic to the Galápagos and Cocos islands.) Archipelagos are ideal for geographic isolation and speciation, especially in land birds such as finches that do not readily fly across wide stretches of water. In such effectively isolated populations, different selective pressures on different islands lead to divergent evolution and adaptations; moreover, occasional interchanges between islands result in heightened interspecific competition, which promotes niche diversification. Drepanid honey creepers underwent a similar adaptive radiation in the Hawaiian Islands.

The necessity of geographic isolation and subsequent interisland colonization for the occurrence of speciation and adaptive radiation is nicely demonstrated by the Cocos Island finch, *Pinaroloxias inornata*. Cocos Island is a remote and solitary island several hundred kilometers north of the Galápagos and about the same distance from the mainland (Figure 7.3, inset). Cocos Island supports only one species of finch. There has been no opportunity for geographic isolation or reduced gene flow, and the *Pinaroloxias* gene pool has never split. As might be expected, this finch is a generalist (Smith and Sweatman, 1976), probably with a high degree of phenotypic variability between individuals.

Adaptive radiation of these finches in the Galápagos has produced five different genera that differ in where they forage, how they forage, and what they eat (Figure 7.4). So-called ground finches (*Geospiza*) include six ground-foraging species with broad beaks that eat seeds of different sizes and types as well as flowers of *Opuntia* cactus. The genus *Camarhynchus*, termed tree finches because they forage in trees, contains three insectivorous species with generally somewhat narrower beaks; another species (*Platyspiza crassirostris*) is a vegetarian. Another species, *Cactospiza pallida*, the "woodpecker finch," uses sticks and cactus spines to probe cracks and crevices for insects, much like a woodpecker uses its long pointed tongue. One very distinctive and monotypic genus, the so-called warbler finch, *Certhidea olivacea*, is insectivorous and occurs on almost all the islets and islands in the archipelago and breeds throughout most habitats.

From three to ten species of finches occur on any given island in various combinations, although some species have now gone extinct on some is-

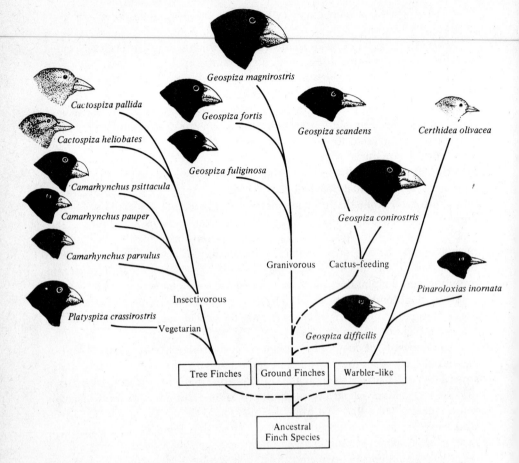

**Figure 7.4.** Phylogeny of the Galápagos finches. [Phylogenetic tree after Lack (1947); head sketches from Grant (1986) after Swarth and Bowman.]

lands (Table 7.1). Beak lengths and depths are highly variable from island to island (Figure 7.5), presumably reflecting different environmental conditions among islands, including interspecific competitive pressures. Figure 7.5 illustrates character displacement in beak depths; the tiny islets of Daphne and Los Hermanos support only one member of a pair of very similar species, either *Geospiza fuliginosa* or *G. fortis*, respectively. On these two small islands, beaks of both species are more similar in beak size than they are on a larger island where the two species occur together in sympatry (Santa Cruz; see upper part of Figure 7.5), where beak depths are completely nonoverlapping, with *fuliginosa* having a small beak (about 6 to 8 mm deep) and *fortis* a larger beak (about 9 to 15 mm deep). Of course, beak dimensions determine in large part the size of the food items the birds eat, but beaks are also used in species recognition.

Larger islands in the Galápagos Archipelago contain a greater variety of habitat types and, as a result, support more species of finches than do smaller

TABLE 7.1  Distributions of Darwin's Finches on Various Major Islands in the Galápagos Archipelago

| Species | Abingdon | Albemarle | Barrington | Bindloe | Charles | Chatham | Culpepper | Duncan | Hood | Indefatigable | James | Jervis | Narborough | Seymour | Tower | Wenman |
|---|---|---|---|---|---|---|---|---|---|---|---|---|---|---|---|---|
| *Geospiza magnirostris* | X | X | X | X | X | – | X | X | – | X | X | X | X | X | X | X |
| *Geospiza fortis* | X | X | X | X | X | X | – | X | – | X | X | X | X | X | – | – |
| *Geospiza fuliginosa* | X | X | X | – | X | X | – | X | X | X | X | X | X | X | – | X |
| *Geospiza difficilis* | X | X | – | X | – | – | X | – | – | X | X | – | – | – | X | X |
| *Geospiza scandens* | X | X | X | X | X | X | – | X | – | X | X | X | X | X | – | – |
| *Geospiza conirostris* | – | – | – | – | – | – | X | – | X | – | – | – | – | – | X | – |
| *Platyspiza crassirostris* | X | X | X | X | X | X | – | X | – | X | X | X | X | X | – | – |
| *Camarhynchus psittacula* | X | – | – | X | X | – | – | X | – | X | X | X | X | X | – | – |
| *Camarhynchus pauper* | – | – | – | – | X | – | – | – | – | – | – | – | – | – | – | X |
| *Camarhynchus parvulus* | – | X | X | – | X | X | – | X | – | X | X | X | X | X | – | – |
| *Cactospiza pallida* | X | X | – | – | – | X | – | X | – | X | X | X | – | – | – | X |
| *Cactospiza heliobates* | – | X | – | – | – | – | – | – | – | – | – | – | X | – | – | – |
| *Certhidea olivacea* | X | X | X | X | X | X | X | X | X | X | X | X | X | X | X | – |
| Total number of species per island | 9 | 10 | 7 | 7 | 9 | 7 | 4 | 9 | 3 | 10 | 10 | 9 | 9 | 8 | 4 | 5 |

*Source:* After Bowman (1961). Originally published by the University of California Press. Reprinted by permission of the Regents of the University of California.

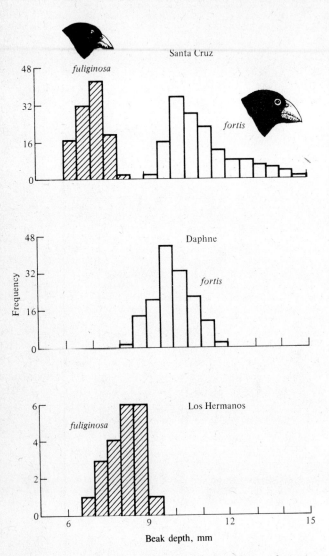

**Figure 7.5.** Histograms of the beak depths of several species of Darwin's finches, genus *Geospiza*, on different islands. In allopatry on the islets of Daphne and Los Hermanos, *G. fortis* and *G. fuliginosa* are more similar in beak size than they are in sympatry on Santa Cruz, where their beak depths are entirely nonoverlapping. [Adapted from Schluter et al. (1985).]

islands. Moreover, the total number of finch species decreases with "average isolation," or the mean distance from other islands, whereas the number of endemic species increases with isolation (Hamilton and Rubinoff, 1963, 1967).

## SELECTED REFERENCES

### Agents of Evolution

Fisher (1930, 1958a, 1958b);   Ford (1964);   Haldane (1932);   Maynard Smith (1958); Wilson and Bossert (1971);   Wright (1931).

### Types of Natural Selection

Kettlewell (1956, 1958);   Mettler and Gregg (1969);   Wilson and Bossert (1971).

### Ecological Genetics

Cockburn (1991);   Dobzhansky (1970);   Ford (1964);   Harvey et al. (1975);   Kettlewell (1956, 1958).

### Allopatric Speciation and Sympatric Speciation

Bush (1975, 1976);   Futuyma (1986);   Mayr (1963);   Otte and Endler (1989);   Patterson (1982);   Prokopy et al. (1982).

### Reproductive Isolating Mechanisms

Alcock (1975);   Futuyma (1986);   Lloyd (1986).

### Darwin's Finches

Abbott et al. (1977);   Bowman (1961);   Grant (1981, 1986);   Hamilton and Rubinoff (1963, 1967);   Lack (1947);   Schluter (1988);   Schluter et al. (1985);   Smith and Sweatman (1976).

**Chapter**
**8**

# Vital Statistics of Populations: Demography

## INTRODUCTION

Although populations are more abstract conceptual entities than cells or organisms and are therefore somewhat more elusive, they are nonetheless real. A gene pool has continuity in both space and time, and organisms belonging to a given population either have a common immediate ancestry or are potentially able to interbreed. Alternatively, a *population* may be defined as a cluster of individuals with a high probability of mating with each other compared to their probability of mating with a member of some other population. As such, Mendelian populations are groups of organisms with a substantial amount of genetic exchange. Such populations are also called *demes*, and the study of their vital statistics is termed *demography*.

In practice, it is often extremely difficult to draw boundaries between populations, except in very unusual circumstances. Certainly, European starlings introduced into Australia are no longer exchanging genes with those introduced into North America, and so each represents a functionally distinct population. Although they might well be potentially able to interbreed, the probability that they will do so is remote because of geographical separation. Such differences

144

also exist at a much finer and more local level, both between and within habitats. For example, starlings in eastern Australia are separated from those in the west by a desert that is uninhabitable to these birds, so they form different populations.

By the preceding definition, organisms reproducing asexually (e.g., a plant that buds off another individual) strictly speaking do not form true populations; there is no gene pool, no interbreeding, and all offspring are essentially identical genetically. However, even such noninterbreeding plants and animals often form collections of organisms with many of the populational attributes of true sexual populations. Most of the animal and many of the plant species that have been studied resort to sexual reproduction at least periodically, and therefore mix up their genes to some extent and form true Mendelian populations. Populations vary in size from very small (a few individuals on a newly colonized island) to very large, such as some wide-ranging and common small insects with populations in the millions. Populations are more usually in the hundreds or thousands. Consideration of both asexual and sexual organisms at the population level often allows us to extend our insight into the activities of individuals in remarkable ways.

An individual's ability to perpetuate its genes in the gene pool of its population, or its reproductive success, represents that individual's *fitness*. Each member of a population has its own relative fitness within that population, which determines in part the fitness of other members of its population; likewise, every individual's fitness is influenced by all other members of its population. Fitness can be defined and understood only in the context of an organism's total environment.

If we consider any continuously varying measurable characteristic, such as height or weight, a population under consideration has a *mean* (average) and a *variance* (a statistical measure of dispersion based on the average of the squared deviations from the mean). Any one individual has only a single value, but the population of individuals has both a mean and a variance. (In fact, we usually estimate true values with the mean and variance of a sample.) These are *population parameters*, characteristics of the population concerned, and they are impossible to define unless we consider a population. Populations also have birth rates, death rates, age structures, sex ratios, gene frequencies, genetic variability, growth rates and growth forms, densities, and so on. Here we examine a variety of such populational characteristics.

## LIFE TABLES AND TABLES OF REPRODUCTION

All sorts of biological phenomena vary in a more or less orderly fashion with age. For example, reproduction begins at puberty and its rate is seldom constant but more usually differs between young versus older adults, and so on. Similarly, the probability of living from one instant to the next is a function of an organism's age as well as the conditions encountered in its immediate environment. Such age-sensitive events are not fixed, of course, but are themselves subject to natural selection and hence vary over evolutionary time. (As one possible example, the age of onset of menarche appears to be decreasing in many human populations.) An age-specific approach thus becomes essential to understand and appreciate many aspects of population biology. Let us now explore such age-related vital statistics of populations.

Two different approaches to sampling populations are depicted in Figure 8.1. The *segment* approach examines all individuals alive over some time interval. A second approach involves following a *cohort* of individuals that entered the population over a particular time interval until no survivors remain. Life tables constructed from these two sorts of samples are termed *vertical* versus *horizontal*, respectively.

Insurance companies employ actuaries to calculate insurance risks. An actuary obtains a large sample of data on some past event and uses them to estimate the average rate of occurrence of a phenomenon. The company then allows itself a suitable profit and margin of safety and sells insurance on the event concerned. Let us consider how an actuary calculates life insurance risks. The raw data consist simply of the average number of deaths at every age in a population: that is, a frequency distribution of deaths by ages (Figure 8.2a). From these values and the age distribution of the population, the actuary calculates age-specific death rates, which are simply the percentages of individuals

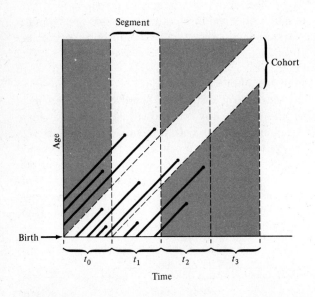

Figure 8.1. A population portrayed as a set of diagonal "life" lines of its various individuals. As time progresses, each individual ages and eventually dies (dots at end of lines). Two types of samples are represented: (1) a *segment* of all the individuals alive during a certain interval of time, and (2) a *cohort*, which consists of all individuals born during some particular time interval. [After Begon and Mortimer (1981) after Skellam.]

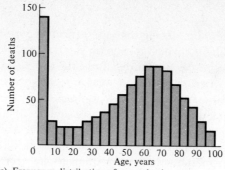

(a) Frequency distribution of age at death

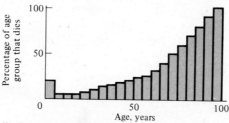

(b) Force of mortality in various age groups

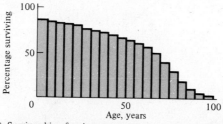

(c) Survivorship of various age groups

**Figure 8.2.** Hypothetical death data of a life insurance actuary (who would treat the sexes separately). (a) The "raw" data consist of a frequency distribution of the age of death of, say, 1000 individuals. For convenience, data are lumped into five-year age intervals. (b) Death rate at age $x$, or $q_x$, expressed as the percentage of each age group that dies during that age interval. (The age distribution of the population at large is needed to compute such age-specific mortality rates.) Values are high in older age groups because they contain relatively few individuals, many of whom die during that age interval. (c) Percentage surviving from an initial cohort of individuals with the preceding age-specific death rates. When divided by 100, these values give the probability that an average newborn will survive to age $x$.

of any age group who die during that age period (Figure 8.2*b*). Death rate at age $x$ is designated by $q_x$, which is sometimes called the "force of mortality" or the *age-specific death rate*. In Figure 8.2, deaths are combined into age classes covering a five-year period. If the population is large and age groups are fine (for instance, consisting of only individuals born on a given day), these curves would be much smoother or more nearly continuous (the distinction between *discrete* and *continuous* events or characteristics will be made repeatedly in this chapter). Demographers begin with discrete age intervals and use the methods of calculus to fit continuous functions to them for estimates at various points within age intervals.

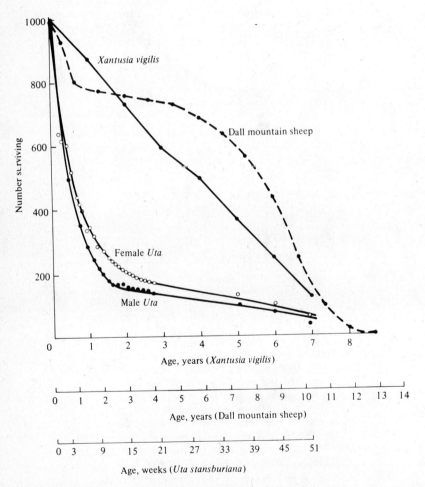

**Figure 8.3.** Several survivorship schedules plotted with an arithmetic vertical axis. (Compare with the more rectangular semilogarithmic plots of Figure 8.4.) Although both types of plots are in common use, logarithmic ones are preferable. In a logarithmic plot, survivorship of the lizard *Xantusia* becomes rectangular (rather than diagonal), whereas that of another lizard, *Uta*, is diagonal (rather than inversely hyperbolic). [After Deevey (1947), Tinkle (1967), and Zweifel and Lowe (1966).]

Still another useful way of manipulating life tables is to calculate age-specific percentage survival (Figure 8.2c). Starting with an initial number or *cohort* of newborn individuals, one calculates the percentage of this initial population alive at every age by sequentially subtracting the percentage of deaths at each age. A smoothed continuous version of survivorship (as in Figures 8.3 and 8.4) is called a *survivorship curve*. The fraction surviving at age x gives the probability that an average newborn will survive to that age, which is usually designated $l_x$.

Ultimately, actuaries are interested in estimating the expectation of further life. How long, on the average, will someone of age x live? For newborn individuals (age 0), average life expectancy is equal to the mean length of life of the cohort. In general, expectation of life at any age x is simply the mean life span remaining to those individuals attaining age x. In symbols,

$$E_x = \frac{\sum_{y=x}^{\infty} l_y}{l_x} \qquad E_x = \frac{\int_x^{\infty} l_y \, dy}{l_x} \qquad (1)$$

where $E_x$ is the expectation of life at age x, and x subscripts age. The equation on the right is a continuous version using the symbolism of integral calculus, whereas the one on the left is an approximation for discrete age intervals (the pivotal age assumption is usually made that x is at the midpoint along the discrete age interval). Calculation of $E_x$ is illustrated in Table 8.1.

In humans, life insurance premiums for males are higher than they are for females because males have a steeper survivorship curve and therefore, at a given age, a shorter life expectancy than females. Figures 8.3 and 8.4 show a variety of survivorship curves, representing the great range they take in natural populations. Rectangular survivorship on a semilogarithmic plot—that is, little mortality until some age and then fairly steep mortality thereafter, as in the lizards *Xantusia vigilis* and *Scincella laterale*, dall mountain sheep, most African ungulates, humans, and perhaps most mammals (Caughley, 1966)— has been called Type I survivorship (Pearl, 1928). Relatively constant death rates with age produce diagonal survivorship curves on semilogarithmic plots as in the lizards *Uta stansburiana* and *Eumeces fasciatus*, the warthog, and most birds; these are classified as Type II curves (actually there are two kinds of Type II curves, representing, respectively, constant risk of death per unit time and constant numbers of deaths per unit time; see Slobodkin, 1962b). Many fish, marine invertebrates, most insects, and many plants have extremely steep juvenile mortality and relatively high survivorship afterward—that is, inverse hyperbolic or Type III survivorship. Of course, nature does not fall into three or four convenient categories, and many real survivorship curves are intermediate between the various "types" previously categorized (as in the palm tree, *Euterpe globosa*; see Figure 8.4c). Moreover, survivorship schedules are not constant but change with immediate environmental conditions. Later, we examine evolution of death rates and old age, but first we must consider the other important populational phenomenon—reproduction.

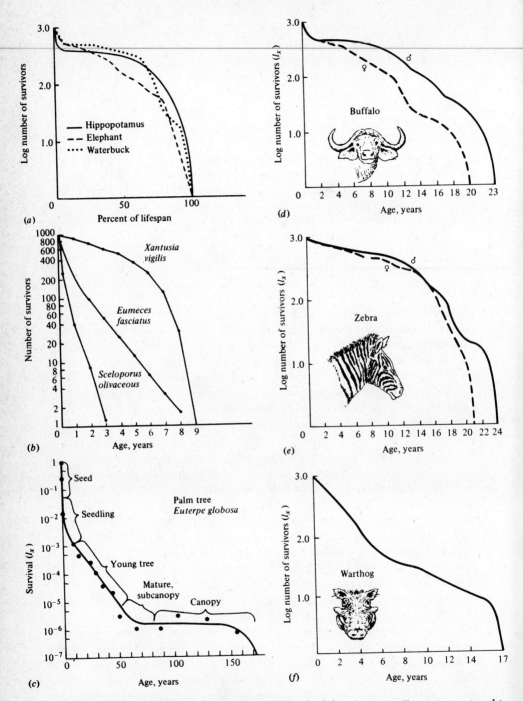

**Figure 8.4.** Semilogarithmic plots of some survivorship schedules. Compare *Xantusia* survivorship in (*b*) with Figure 8.3 which is plotted from the same data. The virtue of a semilogarithmic plot is that a straight line implies equal mortality rates with respect to age. [After Spinage (1972), Van Valen (1975), and Zweifel and Lowe (1966). Spinage by permission of Duke University Press.]

**TABLE 8.1.   Illustration of the Calculation of $E_x$, T, $R_0$ and $\nu_x$ in a Hypothetical Stable Population with Discrete Age Classes, Using Equations (1) to (4)**

| Age (x) | Survivor-ship $l_x$ | Fecundity $m_x$ | Realized Fecundity $l_x m_x$ | Age Weighted by Realized Fecundity $x l_x m_x$ | Expecta-tion of Life $E_x$ (see below) | Reproduc-tive Value $\nu_x$ (see below) |
|---|---|---|---|---|---|---|
| 0 | 1.0 | 0.0 | 0.00 | 0.00 | 3.40 | 1.00 |
| 1 | 0.8 | 0.2 | 0.16 | 0.16 | 3.00 | 1.25 |
| 2 | 0.6 | 0.3 | 0.18 | 0.36 | 2.67 | 1.40 |
| 3 | 0.4 | 1.0 | 0.40 | 1.20 | 2.50 | 1.65 |
| 4 | 0.4 | 0.6 | 0.24 | 0.96 | 1.50 | 0.65 |
| 5 | 0.2 | 0.1 | 0.02 | 0.10 | 1.00 | 0.10 |
| 6 | 0.0 | 0.0 | 0.00 | 0.00 | 0.00 | 0.00 |
| Sums | | 2.2 (GRR) | 1.00 ($R_0$) | 2.78 (T) | | |

Expectation of life:

$E_0 = (l_0 + l_1 + l_2 + l_3 + l_4 + l_5)/l_0 = (1.0 + 0.8 + 0.6 + 0.4 + 0.4 + 0.2)/1.0 = 3.4/1.0 = 3.4$

$E_1 = (l_1 + l_2 + l_3 + l_4 + l_5)/l_1 = (0.8 + 0.6 + 0.4 + 0.4 + 0.2)/0.8 = 2.4/0.8 = 3.0$

$E_2 = (l_2 + l_3 + l_4 + l_5)/l_2 = (0.6 + 0.4 + 0.4 + 0.2)/0.6 = 1.6/0.6 = 2.67$

$E_3 = (l_3 + l_4 + l_5)/l_3 = (0.4 + 0.4 + 0.2)/0.4 = 1.0/0.4 = 2.5$

$E_4 = (l_4 + l_5)/l_4 = (0.4 + 0.2)/0.4 = 0.6/0.4 = 1.5$

$E_5 = l_5/l_5 = 0.2/0.2 = 1.0$

Reproductive value:

$\nu_0 = \frac{l_0}{l_0}m_0 + \frac{l_1}{l_0}m_1 + \frac{l_2}{l_0}m_2 + \frac{l_3}{l_0}m_3 + \frac{l_4}{l_0}m_4 + \frac{l_5}{l_0}m_5 = 0.0 + 0.16 + 0.18 + 0.40 + 0.24 + 0.02 = 1.00$

$\nu_1 = \frac{l_1}{l_1}m_1 + \frac{l_2}{l_1}m_2 + \frac{l_3}{l_1}m_3 + \frac{l_4}{l_1}m_4 + \frac{l_5}{l_1}m_5 = 0.20 + 0.225 + 0.50 + 0.30 + 0.025 = 1.25$

$\nu_2 = \frac{l_2}{l_2}m_2 + \frac{l_3}{l_2}m_3 + \frac{l_4}{l_2}m_4 + \frac{l_5}{l_2}m_5 = 0.30 + 0.67 + 0.40 + 0.03 = 1.40$

$\nu_3 = \frac{l_3}{l_3}m_3 + \frac{l_4}{l_3}m_4 + \frac{l_5}{l_3}m_5 = 1.0 + 0.6 + 0.05 = 1.65$

$\nu_4 = \frac{l_4}{l_4}m_4 + \frac{l_5}{l_4}m_5 = 0.60 + 0.05 = 0.65$

$\nu_5 = \frac{l_5}{l_5}m_5 = 0.10$

The number of offspring produced by an average organism of age $x$ during that age period is designated $m_x$; only those progeny that enter age class zero are counted (age class zero is arbitrary—we could begin a life table at conception, birth, or the age of independence from parental care, depending on which was most convenient and appropriate). Males and females are each credited with one-half of one reproduction for every such offspring produced, so an organism must have two progeny to replace itself. (This procedure makes biological sense in that a sexually reproducing organism passes only half its genome to each of its progeny.) The sum of $m_x$ over all ages, or the total number of offspring that would be produced by an average organism in the absence of mortality, is termed the *gross reproductive rate* (GRR). Like survivorship, patterns of reproduction and fecundities, or $m_x$ schedules, vary widely both with environmental conditions and among different species of organisms. Some, such as annual plants and many insects, breed only once during their lifetime. Others, such as perennial plants and many vertebrates, breed repeatedly. The number of eggs produced and their size relative to the parent also vary over many orders of magnitude. Litter size (usually designated by $B$) refers to the number of young produced during each act of reproduction.

Reproduction may be delayed until fairly late in life, or reproductive activities may begin almost immediately after hatching or birth. The age of first reproduction is usually termed $\alpha$, and the age of last reproduction $\omega$. For an organism that breeds only once, the average time from egg to egg, or the time between generations, termed *generation time (T)*, is simply equal to $\alpha$. But generation time in animals that breed repeatedly is somewhat more complicated. The average time between generations of repeated reproducers can be roughly estimated as $T = (\alpha + \omega)/2$. A more accurate calculation of $T$ is possible by weighting each age by its total realized fecundity, $l_x m_x$, using the following equations (see also Table 8.1):

$$T = \sum_{x=\alpha}^{\omega} x l_x m_x \qquad T = \int_{\alpha}^{\omega} x l_x m_x \, dx \qquad (2)$$

[These equations apply only in a nongrowing population; if a population is expanding or contracting, the right-hand side must be divided by the net reproductive rate, $R_0$ (see next section), to standardize for the average number of successful offspring per individual.] Mean generation time is thus the average age of parenthood, or the *average parental age at which all offspring are born*.

Such age-specific survivorship and fecundity schedules are by no means fixed in ecological time but may fluctuate in temporally variable environments, reflecting the adequacy of current conditions for survival and reproduction.

## NET REPRODUCTIVE RATE AND REPRODUCTIVE VALUE

Clearly, not many organisms live to realize their full potential for reproduction, and we need an estimate of the number of offspring produced by an organism that suffers average mortality. Thus, the *net reproductive rate* $(R_0)$ is defined as the *average number of age class zero offspring produced by an average newborn organism during its entire lifetime*. Mathematically, $R_0$ is simply the product of the age-specific survivorship and fecundity schedules, over all ages at which reproduction occurs:

$$R_0 = \sum_{x=0}^{\infty} l_x m_x \qquad R_0 = \int_{0}^{\infty} l_x m_x \, dx \qquad (3)$$

Alternatively, $\alpha$ and $\omega$ could be substituted for the 0 and $\infty$ limits, since $l_x m_x$ is zero at all nonreproductive ages. Once again, the equation on the left is for discrete age groups, whereas the one on the right is for continuous ages. Table 8.1 illustrates calculation of $R_0$ from a pair of discrete $l_x$ and $m_x$ schedules, and Figure 8.5 diagrams its calculation for continuous ones.

When $R_0$ is greater than 1, the population is increasing; when $R_0$ equals 1, it is stable; and when $R_0$ is less than 1, the population is decreasing. Because of this, the net reproductive rate has also been called the *replacement rate of the population*. A stable population, at equilibrium, with a steep $l_x$ curve must have a correspondingly high $m_x$ curve in order to replace itself (when death rate is high, birth rate must also be high). Conversely, when $l_x$ is high, $m_x$ must be low or else $R_0$ will not equal unity.

Another important concept, first elaborated by Fisher (1930), is reproductive value. To what extent, on the average, do members of a given age group contribute to the next generation between that age and death? In a stable population that is neither increasing nor decreasing, *reproductive value* $(v_x)$ is defined as the *age-specific expectation of future offspring*. Its mathematical definition in a stable population at equilibrium is:

$$v_x = \sum_{t=x}^{\infty} \frac{l_t}{l_x} m_t \qquad v_x \int_x^{\infty} \frac{l_t}{l_x} m_t \, dt \qquad (4)$$

As before, the equation on the left is for discrete age groups and the one on the right for continuous age groups. The term $l_t/l_x$ represents the probability

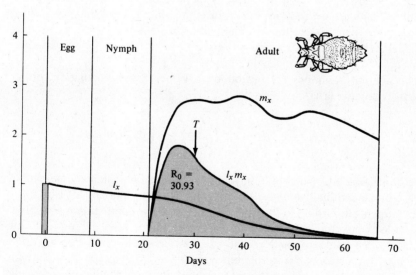

**Figure 8.5.** Continuous $l_x$ and $m_x$ schedules are multiplied together over all ages to obtain the area (shaded) under the $l_x m_x$ product curve of realized fecundity, which is equal to the net reproductive rate, $R_0$. [From F. E. Smith in *Dynamics of Growth Processes*, ed. E. J. Boell, Fig. 1, p. 278. After Evans and Smith (1952) from data for the human louse, *Pediculus humanus*. Copyright © 1954 by Princeton University Press. Reprinted by permission of Princeton University Press.]

of living from age $x$ to age $t$, and $m_t$ is the average reproductive success of an individual at age $t$. Clearly, for newborn individuals in a stable population, $v_0$ is exactly equal to the net reproductive rate, $R_0$. A postreproductive individual has a reproductive value of zero because it can no longer expect to produce offspring; moreover, because natural selection operates only by differential reproductive success, such a postreproductive organism is no longer subject to the direct effects of natural selection. Under many $l_x$ and $m_x$ schedules, reproductive value is maximal around the onset of reproduction and falls off after that because fecundity often decreases with age (but fecundity is subject to natural selection too). Table 8.1 illustrates how to calculate reproductive value in a stable population. Figure 8.6 shows how reproductive value changes with age in a variety of populations.

In populations that are changing in size, the definition of reproductive value is the *present value of future offspring*. Basically, it represents the number of progeny that an organism dying before it reaches age $x + 1$ would have to produce at age $x$ in order to leave as many descendants as it would if it had instead survived to age $x + 1$ and enjoyed average survivorship and fecundity thereafter. In an expanding population (Figure 8.6a), reproductive value of very young individuals is low for two reasons: (1) There is a finite probability of death before reproduction; and (2) because the future breeding population will be larger, offspring to be produced later will contribute less to the total gene pool than offspring currently being born (similarly, in a declining population, offspring expected at some future date are worth relatively more than current progeny because the total future population will be smaller). Thus, present progeny are worth more than future offspring in a growing population, whereas future progeny are more valuable than present offspring in a declining population. This component of reproductive value is tedious to calculate and applies only to populations changing in size.

The general equations for reproductive value in any population, either stable or changing, are:

$$\frac{v_x}{v_0} = \frac{e^{rx}}{l_x} \sum_{t=x}^{\infty} e^{-rt} l_t m_t \qquad \frac{v_x}{v_0} = \frac{e^{rx}}{l_x} \int_{x}^{\infty} e^{-rt} l_t m_t dt \qquad (5)$$

where $e$ is the base of the natural logarithms and $r$ is the instantaneous rate of increase per individual (see later discussion). The exponentials, $e^{rx}$ and $e^{-rt}$, weight offspring according to the direction in which the population is changing. In a stable population, $r$ is zero. Remembering that $e^0$ and $e^{-0}$ equal unity, and that $v_0$ in a stable population is also 1, the reader can verify that equation (5) reduces to (4) when $r$ is zero. Reproductive value does not directly take into account social phenomena such as parental care* or a grandmother's caring for her grandchildren and thereby increasing their probability of survival and successful reproduction.

---

*Defining age class zero as the age of independence from parental care neatly circumvents this problem.

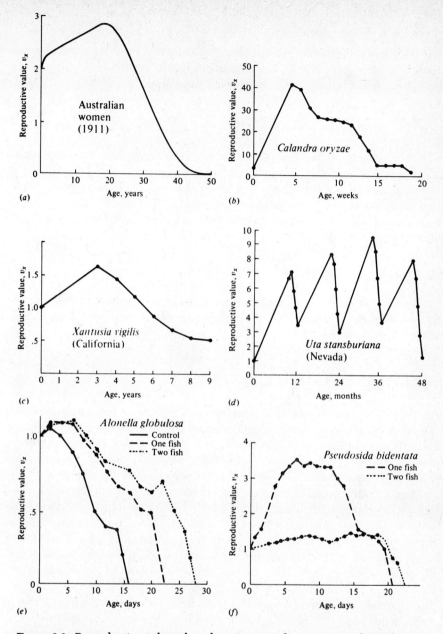

**Figure 8.6.** Reproductive value plotted against age for a variety of populations. (*a*) Australian women, about 1911. [From Fisher (1958a).] (*b*) *Calandra oryzae*, a beetle, in the laboratory. [From data of Birch (1948).] (*c*) *Xantusia vigilis*, a viviparous lizard with a single litter each year. [From data of Zweifel and Lowe (1966).] (*d*) *Uta stansburiana*, an egg-laying lizard that lays several clutches each reproductive season. [From data of Turner et al. (1970).] (*e*) *Alonella globulosa*, a microscopic aquatic crustacean, under three different competitive and predatory regimes in laboratory microcosms. (*f*) *Pseudosida bidentata*, a microscopic aquatic crustacean, under two different laboratory situations. [(*e*, *f*) After Neill (1972).] Age-specific survivorship and fecundity schedules vary with intensity of predation in the two microcrustaceans, generating very different $v_x$ schedules.

It is sometimes useful to partition reproductive value into two components—progeny expected in the immediate future versus those expected in the more distant future. For a nongrowing population,

$$v_x = m_x + \sum_{t=x+1}^{\infty} \frac{l_t}{l_x} m_t \tag{6}$$

The second term on the right-hand side represents expectation of offspring of an organism at age $x$ in the distant future (at age $x + 1$ and beyond) and is known as its *residual reproductive value* (Williams, 1966b). Rearranging equation (6) shows that residual reproductive value ($v_x^*$) is equal to an organism's reproductive value in the next age interval, $v_{x+1}$, multiplied by the probability of surviving from age $x$ to age $x + 1$, or $l_{x+1}/l_x$. In symbols,

$$v_x^* = \frac{l_{x+1}}{l_x} v_{x+1} \tag{7}$$

In summary, any pair of age-specific mortality and fecundity schedules has its own implicit $T$, $R_0$, $r$ (see subsequent discussion), as well as $v_x$ and $v_x^*$ curves.

## STABLE AGE DISTRIBUTION

Another important aspect of a population's structure is its *age distribution* (Figure 8.7), indicating the proportions of its members belonging to each age class. Two populations with identical $l_x$ and $m_x$ schedules, but with different age distributions, will behave differently and may even grow at different rates if one population has a higher proportion of reproductive members.

Lotka (1922) proved that any pair of unchanging $l_x$ and $m_x$ schedules eventually gives rise to a population with a *stable age distribution* (see Figure 8.22, p. 175). When a population reaches this equilibrium age distribution, the percentage of organisms in each age group remains constant. Recruitment into every age class is exactly balanced by its losses due to mortality and aging. Provided $l_x$ and $m_x$ schedules are not changing, a kind of stable age distribution is quickly reached even in an expanding population, in which the per capita rate of growth of each age class is the same (equal to the intrinsic rate of increase per head, $r$, later on), with the consequence that proportions of various age groups also stay constant. Equations (2) to (9) and (12) to (15) assume that a population is in its stable age distribution. However, the intrinsic rate of increase ($r$), generation time ($T$), and reproductive value ($v_x$) are conceptually *independent* of these specific equations, since these concepts can be defined equally well in terms of the age distribution of a population, although they are somewhat more complex mathematically (Vandermeer, 1968). The stable age distribution of a stable population with a net reproductive rate equal to 1 is called the *stationary age distribution*. Computation of the stable age distribution is somewhat tedious, and no more need be said about it here. The interested reader is directed to references at the end of the chapter.

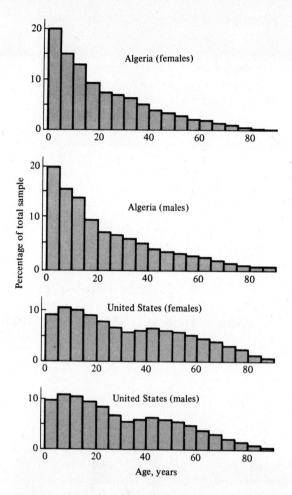

**Figure 8.7.** Age distributions, by sex, in two human populations. The Algerian population is increasing rapidly; the U.S. population is growing more slowly. [After Krebs (1972) after 1968 *Demographic Yearbook of the United Nations*. Copyright. © 1969 by the United Nations. Reproduced by permission.]

# INTRINSIC RATE OF NATURAL INCREASE

Still another useful parameter implicit in every pair of schedules of births and deaths is the *intrinsic rate of natural increase*, sometimes called the *Malthusian parameter*. Usually designated by $r$, it is a measure of the instantaneous rate of change of population size (per individual); $r$ is expressed in numbers per unit time per individual and has the units of 1/time. In a closed population the intrinsic rate of increase is defined as the instantaneous birth rate (per individual), $b$, minus the instantaneous death rate (per individual), $d$. [Similarly, in an open population $r$ is equal to (births + immigration) − (deaths + emigration).] When per capita births exceed per capita deaths ($b > d$), the population is increasing and $r$ is positive; when deaths exceed births ($b < d$), $r$ is negative and the population is decreasing.

In practice, $r$ is somewhat cumbersome to calculate, as its value must be determined by iteration using Euler's implicit equation:

$$\sum_{x}^{\infty} e^{-rx} l_x m_x = 1 \tag{8}$$

where $e$ is the base of the natural logarithms and $x$ subscripts age (derivations of this equation may be found in Mertz, 1970, or Emlen, 1973). Provided that the net reproductive rate, $R_0$, is near one, $r$ can be estimated using the approximate formula

$$r \approx \frac{\log_e R_0}{T} \tag{9}$$

where $T$ is generation time computed with equation (2) (see also May, 1976b). From equation (9) we see that $r$ is positive when $R_0$ is greater than 1, and negative when $R_0$ is less than 1. Because log 1 is zero, an $R_0$ of unity corresponds to an $r$ of zero. Under optimal conditions, when $R_0$ is as high as possible, the maximal rate of natural increase is realized and designated by $r_{max}$. Note that the intrinsic rate of increase is inversely related to generation time, $T$ (see also Figure 8.22, p. 175).

The maximal instantaneous rate of increase per head, $r_{max}$, varies among animals by several orders of magnitude (Table 8.2). Small short-lived organisms, such as the common human intestinal bacterium *Escherichia coli*, have a relatively high $r_{max}$ value, whereas larger and longer-lived organisms, such as humans, have comparatively very low $r_{max}$ values. The components of $r_{max}$ are the instantaneous birth rate per head, $b$, and the instantaneous death rate per head, $d$, under optimal environmental conditions. The evolution of rates of reproduction and death rates is taken up later in this chapter.

A population whose size increases linearly in time would have a constant population growth rate given by

$$\begin{matrix} \text{Growth rate} \\ \text{of population} \end{matrix} = \frac{N_t - N_0}{t - t_0} = \frac{\Delta N}{\Delta t} = \text{Constant} \tag{10}$$

where $N_t$ is the number at time $t$, $N_0$ the initial number, and $t_0$ the initial time. But at any fixed positive value of $r$, the per capita rate of increase is constant, and a population grows exponentially (Figure 8.8). Its growth rate is a function of population size, with the population growing faster as $N$ becomes larger. Suppose you wanted to estimate the rate of change of the population shown in Figure 8.8 at an instant in time, say, at time $t$. As a first approximation, you might look at $N$ immediately before and immediately after time $t$, say, 1 hour before and 1 hour after, and apply the preceding $\Delta N / \Delta t$ equation. But examination of Figure 8.8 reveals that at time $t_1$ (say $t - 1$ hour) the true rate

**TABLE 8.2.  Estimated Maximal Instantaneous Rates of Increase ($r_{max}$, Per Capita Per Day) and Mean Generation Times (in Days) for a Variety of Organisms**

| Taxon | Species | $r_{max}$ | Generation Time $(T)$ |
|---|---|---|---|
| Bacterium | *Escherichia coli* | ca. 60.0 | 0.014 |
| Protozoa | *Paramecium aurelia* | 1.24 | 0.33-0.50 |
| Protozoa | *Paramecium caudatum* | 0.94 | 0.10-0.50 |
| Insect | *Tribolium confusum* | 0.120 | ca.    80 |
| Insect | *Calandra oryzae* | 0.110(.08 − .11) | 58 |
| Insect | *Rhizopertha dominica* | 0.085(.07 − .10) | ca.    100 |
| Insect | *Ptinus tectus* | 0.057 | 102 |
| Insect | *Gibbium psylloides* | 0.034 | 129 |
| Insect | *Trigonogenius globulus* | 0.032 | 119 |
| Insect | *Stethomezium squamosum* | 0.025 | 147 |
| Insect | *Mezium affine* | 0.022 | 183 |
| Insect | *Ptinus fur* | 0.014 | 179 |
| Insect | *Eurostus hilleri* | 0.010 | 110 |
| Insect | *Ptinus sexpunctatus* | 0.006 | 215 |
| Insect | *Niptus hololeucus* | 0.006 | 154 |
| Mammal | *Rattus norwegicus* | 0.015 | 150 |
| Mammal | *Microtus aggrestis* | 0.013 | 171 |
| Mammal | *Canis domesticus* | 0.009 | ca. 1000 |
| Insect | *Magicicada septendecim* | 0.001 | 6050 |
| Mammal | *Homo sapiens* | 0.0003 | ca. 7000 |

is less than, and at time $t_2$ (say, t + 1 hour) is greater than, your straight-line estimate. Differential calculus was developed to handle just such cases, and it allows us to calculate the rate of change at an *instant* in time. As $\Delta N$ and $\Delta t$ are made smaller and smaller, $\Delta N/\Delta t$ gets closer and closer to the true rate of change at time $t$ (Figure 8.8). In the limit, as the $\Delta$'s approach zero, $\Delta N/\Delta t$

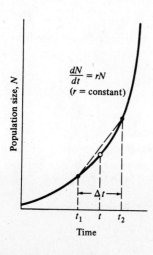

$$\frac{dN}{dt} = rN$$

$(r = \text{constant})$

Population size, $N$

$t_1$   $t$   $t_2$

Time

Figure 8.8.    Exponential population growth under the assumption that the rate of increase per individual, $r$, remains constant with changes in population density. Note that a straight-line estimate of the rate of population growth at time $t$ becomes more and more accurate as $t_1$ and $t_2$ converge; in the limit, as $t_1$ and $t_2$ approach $t$, or $\Delta t \to 0$, the rate of population growth equals the slope of a line tangent to the curve at time $t$ (open circle).

is written as $dN/dt$,* which is calculus shorthand for the instantaneous rate of change of $N$ at $t$. Exponential population growth is described by the simple differential equation

$$\frac{dN}{dt} = bN - dN = (b - d)N = rN \tag{11}$$

where, again, $b$ is the instantaneous birth rate per individual and $d$ is the instantaneous death rate per individual (remember that $r = b - d$).

Using calculus to integrate equation (11) shows that the number of organisms at some time $t$ ($N_t$) under exponential growth is a function of three things: the initial number at time zero ($N_0$), $r$, and the time available for growth since time zero, $t$:

$$N_t = N_0 e^{rt} \tag{12}$$

Here again, $e$ is the base of the natural logarithms. By taking logarithms of equation (12), which is simply an integrated version of equation (11), we get

$$\log_e N_t = \log_e N_0 + \log_e e^{rt} = \log_e N_0 + rt \tag{13}$$

This equation indicates that $\log_e N$ changes linearly in time; that is, a semilog plot of $\log_e N$ against $t$ gives a straight line with a slope of $r$ and a $y$-intercept of $\log_e N_0$.

Setting $N_0$ equal to 1 (i.e., a population initiated with a single organism), after one generation, $T$, the number of organisms in the population is equal to the net reproductive rate of that individual, or $R_0$. Substituting these values into equation (13),

$$\log_e R_0 = \log_e 1 + rT \tag{14}$$

Since log 1 is zero, equation (14) is identical to equation (9).

Another population parameter closely related to the net reproductive rate and the intrinsic rate of increase is the so called *finite rate of increase*, $\lambda$, defined as the rate of increase per individual per unit time. The finite rate of increase is measured in the same time units as the instantaneous rate of increase, and

$$r = \log_e \lambda \quad \text{or} \quad \lambda = e^r \tag{15}$$

In a population without age structure, $\lambda$ is thus identical with $R_0$ [$T$ equal to 1 in equations (9) and (14)].

## EVOLUTION OF REPRODUCTIVE TACTICS

Natural selection recognizes only one currency: successful offspring. Yet even though all living organisms have presumably been selected to maximize their own lifetime reproductive success, they vary greatly in exact modes of repro-

---

*Note that the $d$'s in $dN/dt$ do not refer to death rate, but the entire term represents the instantaneous rate of change in population density, $N$.

duction. Some, such as most annual plants, a multitude of insects, and certain fish like the Pacific salmon, reproduce only once during their entire lifetime. These "big-bang" or *semelparous* reproducers typically exert a tremendous effort in this one and only opportunity to reproduce (in fact, their exceedingly high investment in reproduction may well contribute substantially to their own demise!).

Many other organisms, including perennial plants and most vertebrates, do not engage in such suicidal bouts of reproduction but reproduce again and again during their lifetime. Such organisms have been called *iteroparous* (repeated parenthood). Even within organisms that use either the big-bang or the iteroparous tactic, individuals and species differ greatly in numbers of progeny produced. Annual seed set of different species of trees ranges from a few hundred or a few thousand in many oaks (which produce relatively large seeds—acorns) to literally millions in the redwood (Harper and White, 1974). Seed production may vary greatly even among individual plants of the same species grown under different environmental conditions; an individual poppy (*Papaver rhoeas*) produces as few as four seeds under stress conditions, but as many as a third of a million seeds when grown under conditions of high fertility (Harper, 1966). Fecundity is equally variable among fish; the ocean sunfish, *Mola mola*, is perhaps the most fecund of all vertebrates with a clutch of 200 million tiny eggs. A female codfish also lays millions of relatively tiny eggs. Most elasmobranchs (sharks, skates, and rays), however, produce considerably fewer but much larger offspring. Variability of clutch and litter size is not nearly so great among other classes of vertebrates, but it is still significant. Among lizards, for example, clutch size varies from a fixed clutch of 1 in some geckos and *Anolis* to as many as 40 in certain horned lizards (*Phrynosoma*) and the large *Iguana*. Timing of reproduction also varies considerably among organisms. Due to the finite chance of death, earlier reproduction is always advantageous, all else being equal. Nevertheless, many organisms postpone reproduction. The century plant, an *Agave*, devotes years to vegetative growth before suddenly sending up its inflorescence (some related monocots bloom much sooner). Delayed reproduction also occurs in most perennial plants, many fish such as salmon, a few insects like cicadas, some lizards, and many mammals and birds, especially among large seabirds.

High fecundity early in life may usually be correlated with decreased fertility later on (Figure 8.9). Such fecundity versus age plots may typically be much flatter when early fecundity is lower (Figure 8.9). (This is an excellent example of the principle of allocation.)

Innumerable other examples of the diversity of existing reproductive tactics could be listed. Clearly, natural selection has shaped observed reproductive tactics, with each presumably corresponding in some way to a local optimum that maximizes an individual's lifetime reproductive success in its particular environment. Population biologists would like to understand the factors that influence the evolution of various modes of reproduction.

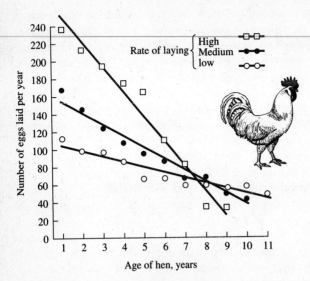

**Figure 8.9.** Age-specific fecundity among three different strains of white Leghorn domestic chickens. Note that fecundity drops off faster with age in birds that lay many eggs early in life, as might be anticipated from the principle of allocation. [Adapted from Romanoff and Romanoff (1949).]

## Reproductive Effort

How much should an organism invest in any given act of reproduction? R. A. Fisher (1930) anticipated this question over 50 years ago:

> It would be instructive to know not only by what *physiological mechanism* a just apportionment is made between the nutriment devoted to the gonads and that devoted to the rest of the parental organism, but also what *circumstances in the life history and environment* would render profitable the diversion of a greater or lesser share of the available resources towards reproduction. [Italics added for emphasis.]

Fisher clearly distinguished between the proximate factor (physiological mechanism) and the ultimate factors (circumstances in the life history and environment) that determine the allocation of resources into reproductive versus nonreproductive tissues and activities. Loosely defined as an organism's investment in any current act of reproduction, *reproductive effort* has played a central role in thinking about reproductive tactics. Although reproductive effort is conceptually quite useful, it has yet to be adequately quantified. Ideally, an operational measure of reproductive effort would include not only the direct material and energetic costs of reproduction but also risks associated with a given level of current reproduction. Another difficulty concerns the temporal patterns of collection and expenditure of matter and energy. Many organisms gather and store materials and energy during periods that are unfavorable for successful reproduction but then expend these same resources on reproduction during a later, more suitable time. The large first clutch of a fat female lizard that has just overwintered may actually represent a *smaller* investment in reproduction than her subsequent smaller clutches that must be produced with considerably diminished energy reserves. Reproductive effort could perhaps be best measured operationally in terms of the effects of various current levels of reproduction upon future reproductive success (see also subsequent discussion).

In spite of these rather severe difficulties, instantaneous ratios of reproductive tissues over total body tissue are sometimes used as a crude first approximation of an organism's reproductive effort (both weights and calories have been used). Thus measured, the proportion of the total resources available to an organism that is allocated to reproduction varies widely among organisms. Among different species of plants, energy expenditure on reproduction, integrated over a plant's lifetime, ranges from near zero to as much as 40 percent (Harper et al., 1970). Annual plants tend to expend more energy on reproduction than most perennials (about 14 to 30 percent versus 1 to 24 percent). An experimental study of the annual euphorb *Chamaesyce hirta* showed that calories allocated to reproduction varied directly with nutrient availability and inversely with plant density and competition (Snell and Burch, 1975).

Returning to Fisher's dichotomy for the apportionment of energy into reproductive versus nonreproductive (somatic) tissues, organs, and activities, we will examine optimal reproductive effort. Somatic tissues are clearly necessary for acquisition of matter and energy; at the same time, an organism's soma is of no selective value except inasmuch as it contributes to that organism's lifelong production of successful offspring. Allocation of time, energy, and materials to reproduction in itself usually decreases growth of somatic tissues and often reduces future fecundity. Increased reproductive effort may also cost by reducing survivorship of the soma; this is easily seen in the extreme case of big-bang reproduction, in which the organism puts everything available into one suicidal bout of reproduction and then dies. More subtle changes in survivorship also occur with minor alterations in reproductive effort (Figure 8.10).

How great a risk should an optimal organism take with its soma in any given act of reproduction? To explore this question, we exploit the concept of residual reproductive value (see p. 156), which is simply the age-specific expectation of all future offspring beyond those immediately at stake. To maximize its overall lifetime contribution to future generations, an optimal organism should weigh the profits of its immediate prospects of reproductive success against the costs to its long-term future prospects (Williams, 1966b). An individual with a high probability of future reproductive success should be more hesitant to risk its soma in present reproductive activities than another individual with a lower probability of reproducing successfully in the future. Moreover, to the extent that present reproduction decreases expectation of further life, it may reduce residual reproductive value directly. For both reasons, current investment in

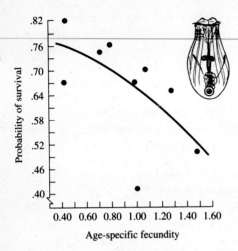

Figure 8.10. Probability of survival decreases with increased age-specific fecundity in the rotifer *Asplanchna*. [From Snell and King (1977).]

reproduction should vary inversely with expectation of future offspring (Figure 8.11).

Several possible different forms for the inverse interaction between reproductive effort and residual reproductive value are depicted in Figure 8.12. Curves in this simple graphical model relate costs and profits in future offspring, respectively, to profits and costs associated with various levels of current reproduction, the latter measured in present progeny. Each curve describes all possible tactics available to a given organism at a particular instant, ranging

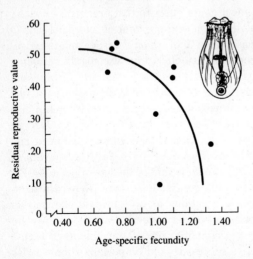

Figure 8.11. Expectation of future offspring declines with current investment in reproduction in laboratory populations of the rotifer *Asplanchna*. [Redrawn from Snell and King (1977).]

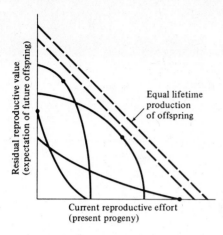

Current reproductive effort
(present progeny)

*y-axis:* Residual reproductive value
(expectation of future offspring)

Equal lifetime
production
of offspring

Figure 8.12. Trade-offs between current reproductive effort and expectation of future offspring at any particular instant (or age). Four curves relate costs in future progeny to profits in present offspring (and vice versa), with a dot marking the reproductive tactic that maximizes total possible lifetime reproductive success. Concave upward curves lead to all-or-none "big-bang" reproduction, whereas convex upward curves result in repeated reproduction (iteroparity). Figures 8.13 and 8.14 depict these trade-offs through the lifetime of a typical iteroparous and a semelparous organism, respectively. [From Pianka (1976b).]

from a current reproductive effort of zero to all-out big-bang reproduction. In a stable population, immediate progeny and offspring in the more distant future are of equivalent value in perpetuation of an organism's genes (see also p. 154); here a straight line with a slope of minus 45° represents equal lifetime production of offspring. Two such lines of equal lifetime success (dashed) are plotted in Figure 8.12. An optimal reproductive tactic exists at the point of intersection of any given curve of possible tactics with the line of equivalent lifetime reproductive success that is farthest from the origin; this level of current reproduction maximizes both reproductive value at that age and total lifetime production of offspring (dots in Figure 8.12). For any given curve of possible tactics, all other tactics yield lower returns in lifetime reproductive success. The precise form of the trade-off between present progeny and expectation of future offspring thus determines the optimal current level of reproductive effort at any given time. Note that concave-upward curves always lead to big-bang reproduction, whereas convex-upward curves result in iteroparity because reproductive value and lifetime reproductive success are maximized at an intermediate current level of reproduction.

Probable trade-offs between immediate reproduction and future reproductive success over the lifetime of an iteroparous organism are depicted in Figure 8.13. The surface in this figure shows the effects of different levels of current fecundity on future reproductive success; the dark dots trace the optimal tactic that maximizes overall lifetime reproductive success. The shadow of this line on the age versus current fecundity plane represents the reproductive schedule one would presumably observe in a demographic study. In many organisms, residual reproductive value first rises and then falls with age; the optimal current level of reproduction will rise as expectation of future offspring declines.

An analogous plot for a semelparous organism is shown in Figure 8.14; here current fecundity also increases as residual reproductive value falls, but the surface for a big-bang reproducer is always concave upward. Exact shapes of the surfaces depicted in these two figures depend on the actual reproductive tactic taken by an organism as well as on immediate environmental conditions

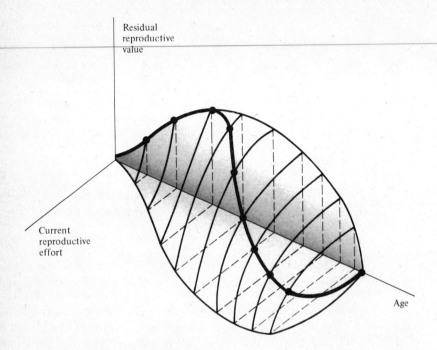

Figure 8.13. During the lifetime of an iteroparous organism, trade-offs between current reproductive effort and future reproductive success might vary somewhat as illustrated, with the dark solid curve connecting the dots tracing the optimal reproductive tactic that maximizes total lifetime reproductive success. The shape of this three-dimensional surface would vary with immediate environmental conditions for foraging, survival, and reproduction, as well as with the actual reproductive tactic adopted by the organism concerned. [From Pianka (1976b).]

for foraging, reproduction, and survival. The precise form of the trade-offs between present progeny and expectation of future offspring is, of course, influenced by numerous factors, including predator abundance, resource availability, and numerous aspects of the physical environment. Unfavorable conditions for immediate reproduction decrease costs of allocating resources to somatic tissues and activities, resulting in lower reproductive effort. (Improved conditions for survivorship, such as good physical conditions or a decrease in predator abundance, would produce similar effects by increasing returns expected from investment in soma.) Conversely, good conditions for reproduction or poor conditions for survivorship result in greater current reproductive effort and decreased future reproductive success.

## Expenditure per Progeny

Not all offspring are equivalent. Progeny produced late in a growing season often have lower probabilities of reaching adulthood than those produced earlier—hence, they contribute less to enhancing parental fitness. Likewise,

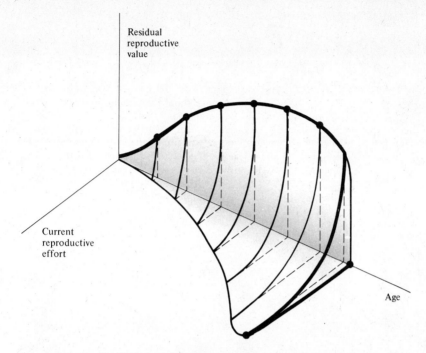

**Figure 8.14.** A plot (like that of Figure 8.13) for a typical semelparous or "big-bang" reproducer. The trade-off surface relating costs and profits in present versus future offspring is always concave upward, and reproduction is all or none. Again, the actual shape of such a surface would reflect the immediate environmental conditions as well as an organism's actual tactic. [From Pianka (1976b).]

larger offspring may usually cost more to produce, but they are also "worth more." How much should a parent devote to any single progeny? For a fixed amount of reproductive effort, average fitness of individual progeny varies inversely with the total number produced. One extreme would be to invest everything in a single very large but extremely fit progeny. Another extreme would be to maximize the total number of offspring produced by devoting a minimal possible amount to each. Parental fitness is often maximized by producing an intermediate number of offspring of intermediate fitness. Here, the best reproductive tactic is a compromise between conflicting demands for production of the largest possible total number of progeny ($r$ selection) versus production of offspring of the highest possible individual fitness ($K$ selection).

A simple graphical model illustrates this trade-off between quantity and quality of offspring (Figures 8.15 and 8.16). In the rather unlikely event that progeny fitness increases linearly with parental expenditure (dashed line $A$ in Figure 8.15), fitness of individual progeny decreases with increased clutch or litter size (the lowermost dashed curve $A'$ in Figure 8.16). However, because parental fitness—the total of the fitnesses of all progeny produced—is flat, no optimal clutch size exists from a parental viewpoint (upper dashed line $A$ in

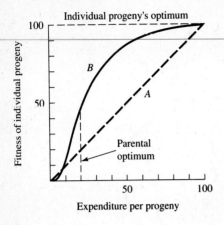

Figure 8.15. Fitness of an individual progeny should generally increase with parental expenditure. Because initial outlays on an offspring usually contribute more to its fitness than subsequent ones, curve *B* is biologically more realistic than line *A*. Note that the parental optimum differs from the optimum for individual progeny, setting up a conflict of interests between the parent and its progeny.

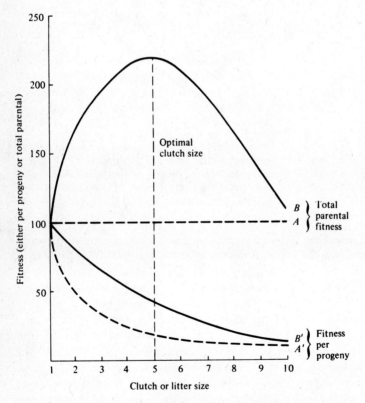

Figure 8.16. Fitness per progeny (*A'* and *B'*) and total parental fitness, the sum of the fitnesses of all offspring produced (*A* and *B*), plotted against clutch and litter size under the assumptions of Figure 8.15. Total investment in reproduction, or reproductive effort, is assumed to be constant. Note that parental fitness peaks at an intermediate clutch size under assumption *B*; optimal clutch size in this example is five.

Figure 8.16). If, however, the biologically plausible assumption is made that progeny fitness increases sigmoidally with parental investment* (curve *B* in Figure 8.15), there is an optimal parental clutch size (peak of uppermost curve *B* in Figure 8.16). In this hypothetical example, parents that allocate only 20 percent of their reproductive effort to each of five offspring gain a higher return on their investment than parents adopting any other clutch size (Figure 8.16). Although this tactic is optimal for parents, it is not the optimum for individual offspring, which would achieve maximal fitness when parents invest everything in a single offspring. Hence, a "parent-offspring conflict" exists (Alexander, 1974; Trivers, 1974; Brockelman, 1975). The exact shape of the curve relating progeny fitness to parental expenditure in a real organism is influenced by a virtual plethora of environmental variables, including length of life, body size, survivorship of adults and juveniles, population density, and spatial and temporal patterns of resource availability. The competitive environment of immatures is likely to be of particular importance because larger, better-endowed offspring should usually enjoy higher survivorship and generally be better competitors than smaller ones.

Juveniles and adults are often subjected to very different selective pressures. Reproductive effort should reflect environmental factors operating upon adults, whereas expenditure per progeny will be strongly influenced by juvenile environments. Because any two parties of the triumvirate determine the third, an optimal clutch or litter size is a direct consequence of an optimum current reproductive effort coupled with an optimal expenditure per progeny (indeed, clutch size is equal to reproductive effort divided by expenditure per progeny). Of course, clutch size can be directly affected by natural selection as well. Recall the example of horned lizards, which are long-lived and relatively *K*-selected as adults, but which because of their tanklike body form have a very large reproductive effort and produce many tiny offspring which must suffer very high mortality (see pp. 112–113).

## Patterns in Avian Clutch Sizes

A good deal of data have now been accumulated demonstrating optimal clutch sizes in birds (see Lack, 1954, 1966, and 1968, for reviews). These elegant studies show that compared with very small and very large clutches, clutches of intermediate size leave proportionately more offspring that survive to breed in the next generation. This is an excellent example of stabilizing selection. Young birds from large clutches leave the nest at a lighter weight (Figure 8.17)

---

*Gains in progeny fitness per unit of parental investment are likely to be greater at lower expenditures per progeny than at higher ones because the proportional increase per unit of allocation is greater at low levels of investment; curves level off at higher expenditures due to the law of diminishing returns.

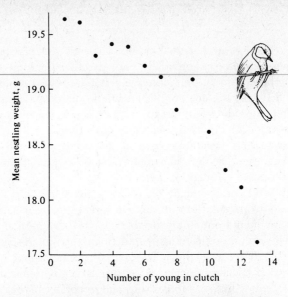

**Figure 8.17.** Average weight of nestling great tits plotted against clutch size, showing that individual young in larger clutches weigh considerably less than those in smaller clutches. [From data of Perrins (1965).]

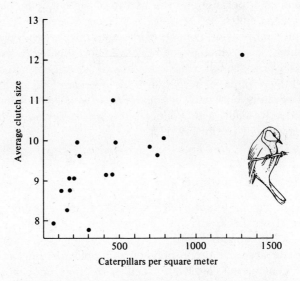

**Figure 8.18.** Average clutch size of great tit populations over 17 years plotted against caterpillar density. Clutches tend to be larger when caterpillars are abundant than when this insect food is sparse. Interestingly, these birds must commit themselves to a given clutch size before caterpillar densities are established—presumably the birds use a fairly reliable physical cue to anticipate food supplies. [After Perrins (1965).]

and have a substantially reduced postfledging survivorship. The optimal clutch apparently represents the number of young for which the parents can, on the average, provide just enough food. Perrins (1965) has good evidence for this in a population of great tits (*Parus major*), which varied their average clutch size from 8 to 12 over a 17-year period, apparently in response to the density of their major food, caterpillars (Figure 8.18).

In European starlings, *Sturnus vulgaris*, clutch size varies from 2 to 8; modal clutch size is 4 or 5 eggs, varying seasonally. Although the total number of chicks actually fledged per nest increases monotonically with clutch size (lower curve, Figure 8.19), mortality among chicks from large clutches during the first three months of life is heavy (upper curve, Figure 8.19). As a result, large clutches do not provide a greater return than smaller clutches, and a clutch of 5 eggs constitutes an apparent optimum.

Clutch size in English chimney swifts, *Apus apus*, varies from 1 to 3 (rarely 4). In these swifts, differential mortality among chicks from clutches of different sizes takes place in the nest before fledging. Swifts capture insect food on the wing and aerial feeding is much better during sunny summers than in cloudy ones (Figure 8.20). Most chicks from clutches of 2 leave the nest in sunny years, but only 1 survives in cloudy years, and the optimal clutch size shifts from 3 to 2 (Figure 8.20). A polymorphism in clutch size persists with mean fledging success over all years being nearly identical (about 1.7 chicks) for clutches of both 2 and 3 (solid line, Figure 8.20).

Wynne-Edwards (1962) interprets the optimal clutch as that which produces a net number of young just replacing the parents during their lifetime of reproduction. As such, his explanation involves group selection (see Chapter 1) because individual birds do not necessarily raise as many young as possible but, rather, produce only as many as are required to replace themselves. Clearly, a "cheater" that produced more offspring would soon swamp the gene pool.

Even within the same widely ranging species, many birds and some mammals produce larger clutches (or litters) at higher latitudes than they do at lower latitudes (Figure 8.21). Such latitudinal increases in clutch size are widespread and have intrigued many population ecologists because of their general occurrence. The following several hypotheses, which are not mutually exclusive, are among those that have been proposed to explain latitudinal gradients in avian clutch sizes. Of course, merely listing hypotheses like this constitutes a kind of a logical "cop-out," but how else can we deal with multiple causality?

**Daylength Hypothesis**    As indicated in Chapter 3, during the late spring and summer, days are longer at higher latitudes than at lower latitudes. Diurnal birds therefore have more daylight hours in which to gather food and thus are able to feed larger numbers of young. However, clutch and litter sizes also increase with latitude in nocturnal birds and mammals, which clearly have a shorter period for foraging.

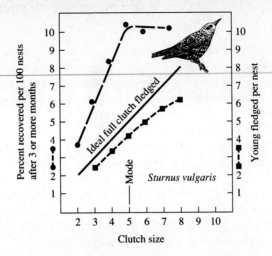

Figure 8.19. Relation of fledging success (lower curve) and of survival to at least 3 months of age (upper curve) in the starling (*Sturnus vulgaris*) as a function of clutch size. Note that although chicks fledged always are rather fewer than the ideal number (diagonal line) determined by the number of eggs laid, increasing the size of the clutch up to eight always increases the number of surviving young per brood at fledging. However, above the modal number of five eggs, such an increase is not reflected in the number of juveniles that reach 3 months of age. Fledging data are from Holland and survival estimates from Switzerland, but comparable data exist for other areas. [After Hutchinson (1978) based on data of Lack (1948).]

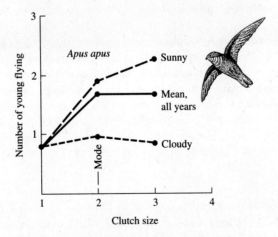

Figure 8.20. Number of young leaving nest as a function of clutch size in swifts (*Apus apus*) in England where the success with more than one egg clearly depends on the weather. [After Hutchinson (1978) after Lack (1956).]

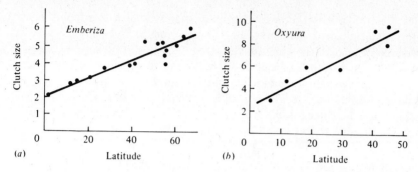

**Figure 8.21.** Graphs of clutch size against latitude for the avian genera (*a*) *Emberiza* and (*b*) *Oxyura*. Each data point represents a species. [After Cody (1966).]

**Prey Diversity Hypothesis**    Due to the very high diversity of insects at lower latitudes, tropical foragers are more confused, less able to form search images, and as a result, are less efficient at foraging than their temperate counterparts (Owen, 1977).

**Spring Bloom or Competition Hypothesis**    Many temperate-zone birds are migratory, whereas few tropical birds migrate. During the spring months at temperate latitudes, there is a great surge of primary production, and insects dependent on these sources of matter and energy rapidly increase in numbers. Winter losses of both resident and migratory birds are often heavy so that spring populations may be relatively small. Hence, returning individuals find themselves in a competitive vacuum with abundant food and relatively little competition for it. In the tropics, wintering migrants ensure that competition is keen all year long, whereas in the temperate zones, competition is distinctly reduced during the spring months. Thus, because birds at higher latitudes are able to gather more food per unit time, they are able to raise larger numbers of offspring to an age at which the young can fend for themselves.

**Nest Predation Hypothesis**    There seem to be proportionately more predators, both individuals and species, in tropical habitats than in temperate ones (see also Chapter 14). Nest failure due to nest predation is extremely frequent in the tropics. Skutch (1949, 1967) proposed that many nest predators locate bird nests by watching and following the parents. Because the parents must make more trips to the nest if they have a large clutch, larger clutches should suffer heavier losses than smaller ones (however, this effect has not yet been demonstrated empirically to my knowledge). A fact in support of this hypothesis is that hole-nesting birds, which are relatively free of nest predators, do not show as great an increase in clutch size with latitude as birds that do not nest in holes (Cody, 1966). Moreover, on tropical islands known to support fewer predators than adjacent mainland areas, birds tend to have larger clutches than they do in mainland populations.

Predators have been implicated as a factor in the evolution of clutch size even in avian species that do not feed their young (nidifugous birds). In Alaskan semipalmated sandpipers, *Calidris pusilla*, ordinary clutches of 4 eggs fledge an average of 1.74 chicks; but clutches artificially raised to 5 fledge only 1 chick, primarily due to heightened predation (Safriel, 1975).

**Hazards of Migration—Residual Reproductive Value Hypothesis**   Latitudinal gradients in clutch or litter size could also be influenced by the trade-off between expectation of future offspring and optimal current investment in reproduction. If hazards of migration or overwintering at high latitudes inevitably result in steeper mortality, expectation of life and residual reproductive value will both be reduced at higher latitudes. This, in turn, would favor an increased effort in current reproduction—and hence larger clutches.

An observation that is somewhat difficult to reconcile with these hypotheses is that clutch size often increases with altitude—as it does in the song sparrow on the Pacific coast (Johnston, 1954). Neither daylength, insect diversity, migratory tendencies, competition, nor predation need necessarily vary altitudinally. Cody (1966) suggested that climatic uncertainty, both instability and unpredictability, may well result in reduced competition at higher elevations.

Using the principle of allocation, Cody merged three of the preceding hypotheses and developed a more general theory of clutch size involving a compromise between the conflicting demands of predator avoidance, competitive ability, and clutch size. His model fits observed facts reasonably well, as one would expect of a more complex model with more parameters.

# EVOLUTION OF DEATH RATES AND OLD AGE

Why do organisms become senile as they grow old? One might predict quite the opposite, since older organisms have had more experience and should therefore have learned how to avoid predators and have more antibodies, and the like. In general, they should be wiser and better adapted, both behaviorally and immunologically. The physiological processes of aging have long been of interest and have received considerable attention, but only fairly recently has the evolutionary process been examined (Medawar, 1957; Williams, 1957; Hamilton, 1966; Emlen, 1970). Here again, Fisher (1930) foreshadowed thought on the subject and paved the way for its development with his concept of reproductive value.

Medawar nicely illustrated the evolution of senescence by setting up an inanimate model. A chemist's laboratory has a stock of 1000 test tubes with a monthly breakage rate of 10 percent. Every month 100 test tubes are broken completely at random and 100 new ones are added to replace them. (Although the rate of breakage is thus deterministic, the model could easily be rephrased in probabilistic or stochastic terms.) All new test tubes are marked with the date of acquisition so that their age (in months) can be determined at a later date. Every test tube has exactly the same probability of survival from any one month to the next: 900/1000 or .9. Thus initially, older test tubes have the same mortality as younger ones and there is no senility. All test tubes are potentially immortal. The probability of surviving two months is the product of the probability of surviving each month separately, or .9 times .9 ($.9^2 = .81$), whereas that of surviving three months is $.9^3$ and that of surviving $x$ months is $.9^x$. After some years, the population of test tubes reaches a stable age distribution with 100 of age 0 months, 90 of age 1, 81 of age 2, 73 of age 3, ..., 28 of age 12, ..., 8 of age 24, ..., 2.25 of age 36, ..., less than one test tube in age groups over 48 months, and so on, totaling 1000 test tubes. (Of course, these numbers are merely the *expected* numbers of tubes of a given age; random sampling and stochastic variations will result in some of the numbers in various age groups being slightly above—and others slightly below—expected values.) Figure 8.22 shows part of the expected stable age distribution, with younger test tubes greatly outnumbering older ones. Virtually no test tubes are over 5 years old, even though individual test tubes are potentially immortal. With time, increased handling results in almost certain breakage.

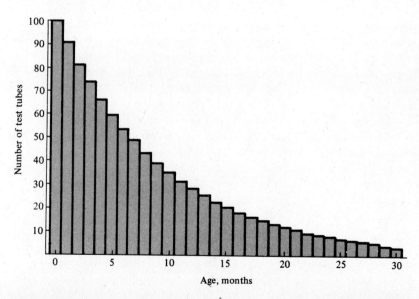

Figure 8.22. Stable age distribution of test tubes with a 10 percent breakage rate per month. Because very few test tubes survive longer than 30 months, the distribution is arbitrarily truncated at this age.

Next, Medawar assigned to each of the 900 test tubes surviving every month an equal share of that month's "reproduction" (i.e., the 100 tubes added that month). Hence, each surviving test tube reproduces one-ninth of one tube a month. Fecundity does not change with age, but the proportions of tubes reproducing does. Younger age groups contribute much more to each month's reproduction than older ones, simply because there are more of them; moreover, as a group their age-specific expectation of future offspring, or total reproductive value, is also higher (Figure 8.22), because as an aggregate their total expectation of future months of life is greater. However, in such an equally fecund and potentially immortal population, the reproductive value of individuals remains constant with age.

Now pretending test tubes have "genes," consider the fate of a mutant whose phenotypic effect is to make its bearer slightly more brittle than an average test tube. This gene is clearly detrimental; it reduces the probability of survival and therefore the fitness of its carrier. This mutant is at a selective disadvantage and will eventually be eliminated from the population. Next, consider the fate of another set of mutant alleles at a different locus that controls the time of expression of the first gene for brittleness. Various alleles at this second locus alter the time of expression of the brittle gene differently, with some causing it to be expressed early and others late. Obviously, a test tube with the brittle gene and a "late" modifier gene is at an advantage over a tube with the brittle gene and an "early" modifier, because it will live longer on the average and therefore produce more offspring. Thus, even though the brittle gene is slowly being eliminated by selection, "late" modifiers accumulate at the expense of "early" modifiers. The later the time of expression of brittleness, the more nearly normal a test tube is in its contribution to future generations of test tubes. In the extreme, after reproductive value decreases to zero, natural selection, which operates only by differential reproductive success, can no longer postpone the expression of a detrimental trait, and it is expressed as senescence. Thus, traits that have been postponed into old age by selection of modifier genes have effectively been removed from the population gene pool. For this reason, old age has been referred to as a "genetic dustbin." The process of selection that postpones the expression of detrimental genetic traits is termed *recession of the overt effects of an allele.*

Exactly analogous arguments apply to changes in the time of expression of beneficial genetic traits, except that here natural selection works to move the time of expression of such characters to early ages, with the result that their bearers benefit maximally from possession of the allele. Such a sequence of selection is termed *precession of the beneficial effects of allele.* We are, in fact, contemplating *age-specific selection.*

In the test tube case, reproduction begins immediately and reproductive value is constant with age. However, in most real populations, organisms are not potentially immortal; furthermore, the onset of reproduction is usually delayed somewhat so that reproductive value first rises and then falls with increasing age (see Figure 8.6a). Thus, individuals at an intermediate age have the highest expectation of future offspring. In this situation, detrimental traits first expressed *after* the period of peak reproductive value can readily be post-poned by selection of appropriate modifiers, but those first expressed *before* the period of maximal reproductive value can be quite another case, particu-larly if they prevent their bearers from reproducing at all. It is difficult to see how selection acting upon the bearer of such a trait could postpone the time of its expression.

In two laboratory studies on *Tribolium* beetles (Sokal, 1970; Mertz, 1975), artificial selection for early reproduction resulted in decreased longevity, demonstrating that senescence does in fact evolve. In Mertz's experiments, selected strains produced more eggs at early ages but total lifetime fecun-dity remained fairly constant, as might be expected due to the principle of allocation.

## JOINT EVOLUTION OF RATES OF REPRODUCTION AND MORTALITY

A species with high mortality obviously must possess a correspondingly high fecundity to persist in the face of its inevitable mortality. Similarly, a very fe-cund organism must on the average suffer equivalently heavy mortality or else its population increases until some balance is reached. Likewise, organisms with low fecundities enjoy low rates of mortality, whereas those with good survivorship have low fecundities. Figure 8.23 shows the inverse relationship

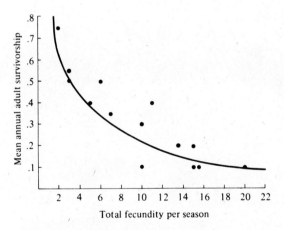

Figure 8.23. Total fecundity per reproductive season plotted against the probability of surviving to a sub-sequent reproductive year for 14 lizard populations. [After Tinkle (1969).] The curve represents popula-tions with a net reproductive rate of one. Deviations from this curve can occur both because $R_0$ is greater or less than one and because of differences in juve-nile survivorship.

between survivorship and fecundity in a variety of lizard populations. Rates of reproduction and death rates evolve together and must stay in some kind of balance. Changes in either of necessity affect the other.

A convenient indicator of an organism's potential for increase in numbers is the maximal instantaneous rate of increase, $r_{max}$, which takes into account simultaneously processes of both births and deaths. Table 8.2 (p. 159) shows the great range of values $r_{max}$ takes. Since actual instantaneous rate of increase averages zero over a sufficiently long period, organisms with high $r_{max}$ values such as *Escherichia coli* are further from realizing their maximal rate of increase than organisms with low $r_{max}$ values such as *Homo sapiens*. In fact, as Smith (1954) points out, $r_{max}$ is a measure of the "harshness" of an organism's average natural environment. As such, it is one of the better indicators of an organism's position along the *r-K* selection continuum (see pp. 189–193). Furthermore, organisms with high $r_{max}$ values usually have much more variable actual rates of increase (and decrease) than organisms with low $r_{max}$ values.

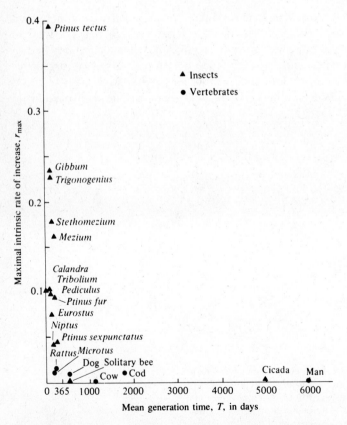

Figure 8.24. Maximal intrinsic rate of increase, $r_{max}$, plotted against mean generation time, *T*, in days, to show the strong inverse hyperbolic relationship between these two population parameters. [From Pianka (1970), partially adapted from Smith (1954).]

The hyperbolic inverse relationship between $r_{max}$ and generation time $T$ is illustrated in Figure 8.24. Small organisms tend to have much higher $r_{max}$ values than larger ones, primarily because of their shorter generation times. Figure 8.25 shows the positive correlation between body size and generation time. The causal basis for this correlation is obvious; it takes time for an organism to attain a large body size. Moreover, delayed reproduction invariably reduces $r_{max}$.

Nevertheless, gains of increased body size must often outweigh losses of reduced $r_{max}$ or large organisms would never have evolved. The frequent tendency of phyletic lines to increase in body size during geological time as evidenced in the fossil record (Newell, 1949) has given rise to the notion of "phyletic size increase." Many advantages of large size are patently obvious, but so are some disadvantages. Certainly, a larger organism is less likely to

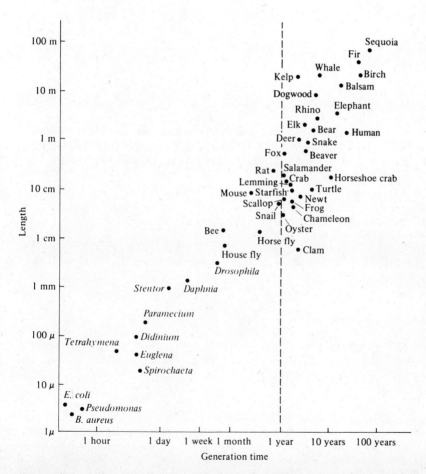

**Figure 8.25.** Log-log plot of organism length against generation time for a wide variety of organisms. [From John Tyler Bonner, *Size and Cycle: An Essay on the Structure of Biology*, Fig. 1, p. 17. Copyright © 1965 by Princeton University Press. Reprinted by permission of Princeton University Press.]

fall victim to predators and would therefore have fewer potential predators and better survivorship. Small organisms are at the mercy of their physical environment, and very slight changes in it can often be devastating; however, larger organisms are comparatively better buffered and therefore better protected. Some disadvantages of large size are that (1) larger organisms require more matter and energy per individual per unit time than smaller ones, and (2) fewer refuges, hiding places, and safe sites exist for large animals than for small ones.

## SELECTED REFERENCES

### Introduction

Begon and Mortimer (1981); Christianson and Fenchel (1977); Cole (1954b, 1958); Dawson and King (1971); Gadgil and Bossert (1970); Harper (1967); MacArthur and Connell (1966); Mettler and Gregg (1969); Poole (1974); Slobodkin (1962b); Wilson and Bossert (1971).

### Life Tables and Tables of Reproduction

Blair (1960); Bogue (1969); Botkin and Miller (1974); Brooks (1967); Caughley (1966); Cole (1965); Deevey (1947); Fisher (1930); Harper and White (1974); Keyfitz and Flieger (1971); Lotka (1925, 1956); Mertz (1970); Pearl (1928); Slobodkin (1962b); Spinage (1972); Zweifel and Lowe (1966).

### Net Reproductive Rate and Reproductive Value

Emlen (1970); Fisher (1930, 1958a); Hamilton (1966); Mertz (1970, 1971a, 1971b); Slobodkin (1962b); Turner et al. (1970); Vandermeer (1968); Williams (1966b); Wilson and Bossert (1971).

### Stable Age Distribution

Emlen (1973); Keyfitz (1968); Krebs (1972); Leslie (1945, 1948); Lotka (1922, 1925, 1956); Mertz (1970); Vandermeer (1968); Williamson (1967); Wilson and Bossert (1971).

### Intrinsic Rate of Natural Increase

Andrewartha and Birch (1954); Birch (1948, 1953); Cole (1954b, 1958); Emlen (1973); Evans and Smith (1952); Fenchel (1974); Fisher (1930); Gill (1972); Goodman (1971); Leslie and Park (1949); May (1976b); Mertz (1970); F. E. Smith (1954, 1963a).

## Evolution of Reproductive Tactics

Alexander (1974);   Ashmole (1963);   Baker (1938);   Bell (1980, 1984a, 1984b);   Brockelman (1975);   Charnov and Krebs (1973);   Chitty (1967b);   Cody (1966, 1971); Cole (1954b);   Dapson (1979);   Fisher (1930);   Gadgil and Bossert (1970);   Goodman (1974);   Grime (1977, 1979);   Gross and Shine (1981);   Harper (1966);   Harper and Ogden (1970);   Harper and White (1974);   Harper et al. (1970);   Hilborn and Stearns (1982);   Istock (1967);   Johnson and Cook (1968);   Johnston (1954);   Klomp (1970);   Lack (1948, 1954, 1956, 1966, 1968, 1971);   Lack and Lack (1951);   Lande (1982);   Mertz (1970, 1971a);   Millar (1973);   Murphy (1968);   Owen (1977);   Partridge and Harvey (1988);   Perrins (1964, 1965);   Pianka (1976b);   Pianka and Parker (1975a);   Romanoff and Romanoff (1949);   Royama (1969);   Safriel (1975);   Salisbury (1942);   Schaffer (1974);   Seger and Brockmann (1987);   Skutch (1949, 1967);   C. C. Smith and Fretwell (1974);   Snell and Burch (1975);   Snell and King (1977);   Stearns (1976, 1989);   Taylor et al. (1974);   Tinkle (1969);   Tinkle et al. (1970);   Trivers (1974);   Wilbur (1977);   Wilbur et al. (1974);   Williams (1966a, 1966b);   Willson (1972b, 1973a);   Wooten (1979);   Wynne-Edwards (1955, 1962).

## Evolution of Death Rates and Old Age

Emlen (1970);   Fisher (1930);   Haldane (1941);   Hamilton (1966);   Medawar (1957); Mertz (1975);   Pearl (1922, 1928);   Snell and King (1977);   Sokal (1970);   Williams (1957);   Willson (1971).

## Joint Evolution of Rates of Reproduction and Mortality

Bonner (1965);   Cole (1954b);   Frank (1968);   Gadgil and Bossert (1970);   Lack (1954, 1966);   Newell (1949);   F. E. Smith (1954);   Tinkle (1969);   Williams (1966a, 1966b).

## Chapter
### 9

# Population Growth
# and Regulation

## VERHULST–PEARL LOGISTIC EQUATION

In a finite world, no population can grow exponentially for very long (Figure 9.1). Sooner or later every population must encounter either difficult environmental conditions or shortages of its requisites for reproduction. Over a long period of time, unless the average actual rate of increase is zero, a population either decreases to extinction or increases to the extinction of other populations.

So far, our populations have had fixed age-specific parameters, such as their $l_x$ and $m_x$ schedules. In this section we ignore age specificity and instead allow $R_0$ and $r$ to vary with population density. To do this, we define *carrying capacity*, $K$, as the density of organisms (i.e., the number per unit area) at which the net reproductive rate ($R_0$) equals unity and the intrinsic rate of increase ($r$) is zero. At "zero density" (only one organism, or a perfect competitive vacuum), $R_0$ is maximal and $r$ becomes $r_{max}$. For any given density above zero density, both $R_0$ and $r$ decrease until, at $K$, the population ceases to grow. A population initiated at a density above $K$ decreases until it reaches the steady state at $K$ (Figure 9.2). Thus, we define $r_a$ (or $dN/dt$ times $1/N$) as the *actual* instantaneous rate of increase; it is zero at $K$, negative above $K$, and positive when the population is below $K$.

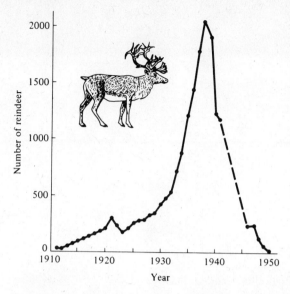

Figure 9.1 In 1911, 25 reindeer were introduced on Saint Paul Island in the Pribolofs off Alaska. The population grew rapidly and nearly exponentially until about 1938, when there were over 2000 animals on the 41-square-mile island. The reindeer badly overgrazed their food supply (primarily lichens) and the population "crashed." Only eight animals could be found in 1950. A similar sequence of events occurred on Saint Matthew Island between 1944 and 1966. [After Krebs (1972) after V. B. Scheffer (1951). The Rise and Fall of a Reindeer Herd. *Science* 73: 356–362.]

The simplest assumption we can make is that $r_a$ decreases linearly with $N$ and becomes zero at an $N$ equal to $K$ (Figure 9.2); this assumption leads to the classical Verhulst-Pearl logistic equation (see Box on p. 187):

$$\frac{dN}{dt} = rN - rN\left(\frac{N}{K}\right) = rN - \frac{rN^2}{K} \tag{1}$$

Alternatively, by factoring out an $rN$, equation (1) can be written

$$\frac{dN}{dt} = rN\left(1 - \frac{N}{K}\right) = rN\left(\frac{K - N}{K}\right) \tag{2}$$

Or, simplifying by setting $r/K$ in equation (1) equal to z,

$$\frac{dN}{dt} = rN - zN^2 \tag{3}$$

The term $rN(N/K)$ in equation (1) and the term $zN^2$ in equation (3) represent the density-dependent reduction in the rate of population increase. Thus, at $N$

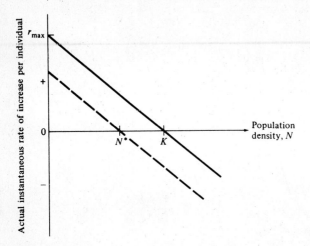

**Figure 9.2** The actual instantaneous rate of increase per individual, $r_a$, decreases linearly with increasing population density under the assumptions of the Pearl-Verhulst logistic equation. The solid line depicts conditions in an optimal environment in which the difference between $b$ and $d$ is maximal. The dashed line shows how the actual rate of increase decreases with $N$ when the death rate per head, $d$, is higher; equilibrium population size, $N^*$, is then less than carrying capacity, $K$. (Compare with Figure 9.5.)

equal to unity (an ecologic vacuum), $dN/dt$ is nearly exponential, whereas at $N$ equal to $K$, $dN/dt$ is zero and the population is in a steady state at its carrying capacity. Logistic equations (there are many more besides the Verhulst-Pearl one) generate so-called *sigmoid (S-shaped)* population growth curves (Figure 9.3). Implicit in the Verhulst-Pearl logistic equation are three assumptions: (1) that all individuals are equivalent—that is, the addition of each new individual reduces the actual rate of increase by the same fraction, $1/K$, at every density (Figure 9.2); (2) that $r_{max}$ and $K$ are immutable constants; and (3) that there is no time lag in the response of the actual rate of increase per individual to changes in $N$. All three assumptions are unrealistic, so the logistic has been strongly criticized (Allee et al., 1949; Smith, 1952, 1963a; Slobodkin, 1962b).

More plausible curvilinear relationships between the rate of increase and population density are shown in Figure 9.4. Note that density-dependent effects on birth rate and death rate are combined by the use of $r$ (these effects are separated later in this section). Carrying capacity is also an extremely complicated and confounded quantity, for it necessarily includes both renewable and nonrenewable resources, which are variables themselves. Carrying capacity almost certainly varies a great deal from place to place and from time to time for most organisms. There is also some inevitable lag in feedback between population density and the actual instantaneous rate of increase. All

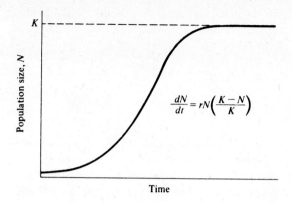

$$\frac{dN}{dt} = rN\left(\frac{K-N}{K}\right)$$

**Figure 9.3** Population growth under the Verhulst-Pearl logistic equation is sigmoidal (S-shaped), reaching an upper limit termed the carrying capacity, $K$. Populations initiated at densities above $K$ decline exponentially until they reach $K$, which represents the only stable equilibrium.

these assumptions can be relaxed and more realistic equations developed, but the mathematics quickly become extremely complex and unmanageable. Nevertheless, a number of populational phenomena can be nicely illustrated using the simple Verhulst-Pearl logistic, and a thorough understanding of it is a necessary prelude to the equally simplistic Lotka-Volterra competition equations, which are taken up in Chapter 12. However, the numerous flaws of the logistic must be recognized, and it should be taken only as a first approximation for

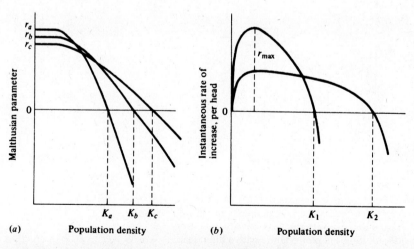

**Figure 9.4** Hypothetical curvilinear relationships between instantaneous rates of increase and population density. Concave upward curves have also been postulated. [From Gadgil and Bossert (1970) and Pianka (1972).]

small changes in population growth, most likely to be valid near equilibrium and over short periods of time (i.e., situations in which linearity should be approximated).

Notice that $r$ in the logistic equation is actually $r_{max}$. The equation can be solved for the *actual* rate of increase, $r_a$, which is a variable and a function of $r$, $N$, and $K$, as follows:

$$r_a = \frac{dN}{Ndt} = r\left(K - \frac{N}{K}\right) = r - \left(\frac{N}{K}\right)r \tag{4}$$

The actual instantaneous rate of increase per individual, $r_a$, is always less than or equal to $r_{max}$ ($r$ in the logistic). Equation (4) and Figure 9.2 show how $r_a$ decreases linearly with increasing density under the assumptions of the Verhulst-Pearl logistic equation.

The two components of the actual instantaneous rate of increase per individual, $r_a$, are the actual instantaneous birth rate per individual, $b$, and the actual instantaneous death rate per individual, $d$. The difference between $b$ and $d$ (i.e., $b - d$) is $r_a$. Under theoretical ideal conditions when $b$ is maximal and $d$ is minimal, $r_a$ is maximized at $r_{max}$. In the logistic equation, this is realized at a minimal density, or a perfect competitive vacuum. To be more precise, we subscript $b$ and $d$, which are functions of density. Thus, $b_N - d_N = r_N$ (which is $r_a$ at density $N$), and $b_0 - d_0 = r_{max}$. When $b_N = d_N$, $r_a$ and $dN/dt$ are zero and the population is at equilibrium. Figure 9.5 diagrams the way in which $b$ and $d$ vary linearly with $N$ under the logistic equation. At any given density, $b_N$ and $d_N$ are given by linear equations

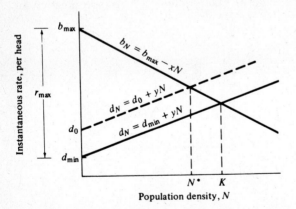

Figure 9.5 The instantaneous birth rate per individual decreases linearly with population density under the logistic equation, whereas the instantaneous death rate per head rises linearly as population density increases. Two death rate lines are plotted, one with a high death rate (dashed line) and one with a lower death rate (solid line). Equilibrium population density, $N^*$, is lowered by either an increased death rate or by a reduced birth rate.

$$b_N = b_0 - xN \tag{5}$$

$$d_N = d_0 + yN \tag{6}$$

where $x$ and $y$ represent, respectively, the slopes of the lines plotted in Figure 9.5 (see also Bartlett, 1960, and Wilson and Bossert, 1971). The instantaneous death rate, $d_N$, clearly has both density-dependent and density-independent components; in equation (6) and Figure 9.5, $yN$ measures the density-dependent component of $d_N$, while $d_0$ determines the density-independent component.

At equilibrium, $b_N$ must equal $d_N$, or

$$b_0 - xN = d_0 + yN \tag{7}$$

Substituting $K$ for $N$ at equilibrium, $r$ for $(b_0 - d_0)$, and rearranging terms,

$$r = (x + y)K \tag{8}$$

or

$$K = \frac{r}{x + y} \tag{9}$$

Note that the sum of the slopes of the birth and death rates $(x + y)$ is equal to $z$, or $r/K$. Clearly, $z$ is the density-dependent constant that is analogous to the density-independent constant $r_{\max}$.

---

### DERIVATION OF THE LOGISTIC EQUATION

The derivation of the Verhulst-Pearl logistic equation is relatively straightforward. First, write an equation for population growth using the actual rate of increase $r_N$:

$$\frac{dN}{dt} = r_N N = (b_N - d_N)N \tag{1}$$

Now substitute the equations for $b_N$ and $d_N$ from equations (5) and (6), above, into (1):

$$\frac{dN}{dt} = [(b_0 - xN) - (d_0 + yN)]N \tag{2}$$

Rearranging terms,

$$\frac{dN}{dt} = [(b_0 - d_0) - (x + y)N]N \tag{3}$$

Substituting in $r$ for $(b - d)$ and, from equation (9) above, $r/K$ for $(x + y)$, multiplying through by $N$, and rearranging,

$$\frac{dN}{dt} = rN - \left(\frac{r}{K}\right)N^2 \tag{4}$$

# DENSITY DEPENDENCE AND DENSITY INDEPENDENCE

Various factors can influence populations in two fundamentally different ways. If their effects on a population do not vary with population density, but the same *proportion* of organisms are affected at any density, factors are said to be *density independent*. Climatic factors often, though by no means always, affect populations in this manner (Table 9.1). If, on the other hand, a factor's effects vary with population density so that the proportion of organisms influenced actually changes with density, that factor is *density dependent*. Density-dependent factors and events can be either positive or negative. Death rate, which presumably often increases with increasing density, is an example of positive or direct density dependence (Figure 9.5); birth rate, which normally decreases with increasing density, is an example of negative or inverse density dependence (Figure 9.6).

Density-dependent influences on populations frequently result in an equilibrium density at which the population ceases to grow. Biotic factors, such as competition, predation, and pathogens, often (though not always) act in this way. Ecologists are divided in their opinions as to the relative importance of density dependence and density independence in natural populations (Andrewartha and Birch, 1954; Lack, 1954, 1966; Nicholson, 1957; Orians, 1962; McLaren, 1971; Ehrlich et al., 1972).

The detection of density dependence can be difficult. In their studies on the population dynamics of *Thrips imaginis* (a small herbivorous insect), Davidson and Andrewartha (1948) found that they could predict population sizes of these insects fairly accurately using only past population sizes and recent climatic conditions. These workers could find no evidence of any density effects; they therefore interpreted their data to mean that the populations of *Thrips* were controlled primarily by density-independent climatic factors. However, a reanalysis of their data shows pronounced density-dependent effects at high densities (Smith, 1961). Population change and population size are strongly inversely correlated, which strongly suggests density dependence.

TABLE 9.1 **Dramatic Fish Kills, Illustrating Density-Independent Mortality**

| Locality | Commercial catch | | Percent decline |
|----------|--------|-------|-----------------|
|          | Before | After |                 |
| Matagorda | 16,919 | 1,089 | 93.6 |
| Aransas | 55,224 | 2,552 | 95.4 |
| Lagunda Madre | 2,016 | 149 | 92.6 |

*Note:* These fish kills resulted from severe cold weather on the Texas Gulf Coast in the winter of 1940.

*Source:* After Odum (1959) after Gunter (1941).

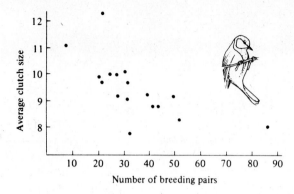

**Figure 9.6** A plot of average clutch size against the density of breeding pairs of English great tits (birds) in a particular woods in a series of years over a 17-year period. [After Perrins (1965).]

Smith also demonstrated a rapidly decreasing variance in population size during the later portion of the spring population increase. Furthermore, these patterns persisted even after partial correlation analysis, which holds constant the very climatic variables that Davidson and Andrewartha considered to be so important. This example illustrates the great difficulty ecologists frequently encounter in distinguishing cause from effect. There is now little real doubt that both density-dependent and density-independent events occur; however, their relative importance may vary by many orders of magnitude from population to population—and even within the same population from time to time as the size of the population changes (Horn, 1968a; McLaren, 1971).

## OPPORTUNISTIC VERSUS EQUILIBRIUM POPULATIONS

Periodic disturbances, including fires, floods, hurricanes, and droughts, often result in catastrophic density-independent mortality, suddenly reducing population densities well below the maximal sustainable level for a particular habitat. Populations of annual plants and insects typically grow rapidly during spring and summer but are greatly reduced at the onset of cold weather. Because populations subjected to such forces grow in erratic or regular bursts (Figure 9.7), they have been termed *opportunistic populations*. In contrast, populations such as those of many vertebrates may usually be closer to an equilibrium with their resources and generally exist at much more stable densities (provided that their resources do not fluctuate); such populations are called *equilibrium populations*. Clearly, these two sorts of populations represent end points of a continuum; however, the dichotomy is useful in comparing different populations. The significance of opportunistic versus equilibrium populations

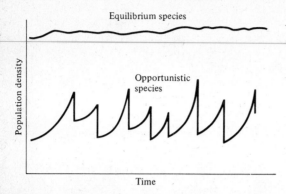

Figure 9.7 Population growth trajectories in an equilibrium species versus an opportunistic species subjected to irregular catastrophic mortality.

is that density-independent and density-dependent factors and events differ in their effects on natural selection and on populations. In highly variable or unpredictable environments, catastrophic mass mortality (such as that illustrated in Table 9.1) presumably often has relatively little to do with the genotypes and phenotypes of the organisms concerned or with the size of their populations. (Some degree of selective death and stabilizing selection has been demonstrated in winter kills of certain bird flocks.) By way of contrast, under more stable or predictable environmental regimes, population densities fluctuate less and much mortality is more directed, favoring individuals that are better able to cope with high densities and strong competition. Organisms in highly rarefied environments seldom deplete their resources to levels as low as do organisms living under less rarefied situations; as a result, the former usually do not encounter such intense competition. In a "competitive vacuum" (or an extensively rarefied environment) the best reproductive strategy is often to put maximal amounts of matter and energy into reproduction and to produce as many total progeny as possible, as soon as possible. Because there is little competition, these offspring often can thrive even if they are quite small and therefore energetically inexpensive to produce. However, in a "saturated" environment, where density effects are pronounced and competition is keen, the best strategy may often be to put more energy into competition and maintenance and to produce offspring with more substantial competitive abilities. This usually requires larger offspring, and because they are energetically more expensive, it means that fewer can be produced.

MacArthur and Wilson (1967) designate these two opposing selective forces $r$ selection and $K$ selection, after the two terms in the logistic equation (however, one should not take these terms too literally, as the concepts are independent of the equation). Of course, things are seldom so black and white; there are usually only shades of gray. No organism is completely $r$-selected or completely

*K*-selected; instead, all must reach some compromise between the two extremes. Indeed, one can think of a given organism as an "*r*-strategist" or a "*K*-strategist" only relative to some other organism; thus statements about *r* and *K* selection are invariably comparative. We think of an *r*-*K* selection continuum and an organism's position along it in a particular environment at a given instant in time (Pianka, 1970, 1972). Table 9.2 lists a variety of correlates of these two kinds of selection.

An interesting special case of an opportunistic species is the *fugitive* species, envisioned as a predictably inferior competitor that is always excluded locally by interspecific competition but persists in newly disturbed regions by virtue of a high dispersal ability (Hutchinson, 1951). Such a colonizing species can persist in a continually changing *patchy* environment in spite of

**TABLE 9.2    Some of the Correlates of *r* and *K* Selection**

|  | *r* selection | *K* selection |
|---|---|---|
| Climate | Variable or unpredictable; uncertain | Fairly constant or predictable; more certain |
| Mortality | Often catastrophic, nondirected, density independent | More directed, density dependent |
| Survivorship | Often Type III | Usually Types I and II |
| Population size | Variable in time, nonequilibrium; usually well below carrying capacity of environment; unsaturated communities or portions thereof; ecologic vacuums; recolonization each year | Fairly constant in time, equilibrium; at or near carrying capacity of the environment; saturated communities; no recolonization necessary |
| Intra- and interspecific competition | Variable, often lax | Usually keen |
| Selection favors | 1. Rapid development<br>2. High maximal rate of increase, $r_{max}$<br>3. Early reproduction<br>4. Small body size<br>5. Single reproduction<br>6. Many small offspring | 1. Slower development<br>2. Greater competitive ability<br>3. Delayed reproduction<br>4. Larger body size<br>5. Repeated reproduction<br>6. Fewer larger progeny |
| Length of life | Short, usually less than a year | Longer, usually more than a year |
| Leads to | Productivity | Efficiency |
| Stage in succession | Early | Late, climax |

*Source:* After Pianka (1970).

pressures from competitively superior species. Hutchinson (1961) used another argument to explain the apparent "paradox of the plankton," the coexistence of many species in diverse planktonic communities under relatively homogeneous physical conditions with limited possibilities for ecological separation. He suggested that temporally changing environments may promote diversity by periodically altering relative competitive abilities of component species, thereby allowing their coexistence.

Recently, McLain (1991) suggested that the relative strength of sexual selection depends on the life history strategy, with $r$-strategists being less likely to be subjected to strong sexual selection than $K$-strategists. Winemiller (1989b, 1992) points out that reproductive tactics among fishes (and probably all organisms) can be placed on a two-dimensional triangular surface in three space with the coordinates: juvenile survivorship, fecundity, and age of first reproduction or generation time (Figure 9.8). This two-dimensional triangular surface has three vertices corresponding to equilibrium ($K$-strategists), opportunistic, and seasonal species. The $r$-$K$ selection continuum runs diagonally across this surface from the equilibrium corner to the opportunistic-seasonal edge. In fish, seasonal breeders exhibit little sexual dimorphism, whereas both opportunistic and equilibrium species display marked sexual dimorphisms (Winemiller, 1992).

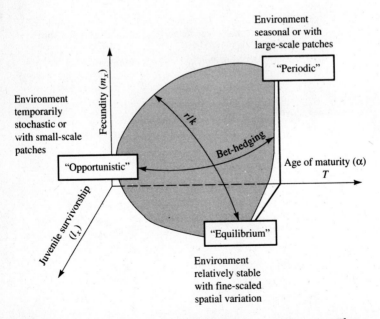

**Figure 9.8** Model for a triangular life history continuum. Three-dimensional representation of reproductive tactics depicting both the $r$-$K$ selection continuum and the bet-hedging axis. [After Winemiller (1992).]

Under situations where survivorship of adults is high but juvenile survival is low and highly unpredictable, there is a selective disadvantage to putting all one's eggs in the same basket, and a consequent advantage to distributing reproduction over a period of time (Murphy, 1968). This sort of reproductive tactic has become known as "bet hedging" (Stearns, 1976) and occurs in both *r*-strategists and *K*-strategists. Winemiller (1992) points out that a bet-hedging axis passes across his triangular surface, from the opportunistic corner endpoint to the edge connecting the seasonal and equilibrium tactics (Figure 9.8).

## POPULATION REGULATION

In the majority of real populations that have been examined, numbers are kept within certain bounds by density-dependent patterns of change. When population density is high, decreases are likely, whereas increases tend to occur when populations are low (Tanner, 1966; Pimm, 1982). If the proportional *change* in density is plotted against population density, inverse correlations usually result (Figure 9.9). Table 9.3 summarizes such data for a variety of populations, including humans (the only species with a significant nonnegative correlation). Such negative correlations are found even in cyclical and erratic populations, such as those considered in the next section.

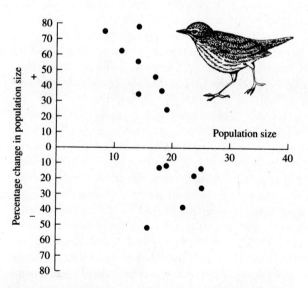

**Figure 9.9** Increases and decreases in population size plotted against population size in the year preceding the increase or decrease for an ovenbird population in Ohio over an 18-year period. [From MacArthur and Connell (1966).]

**TABLE 9.3  Frequencies of positive and negative correlations between percentage change in density versus population density for a variety of populations in different taxa.**

| Taxon | Numbers of populations in various categories | | | | | |
|---|---|---|---|---|---|---|
| | Positive ($p < .05$) | Positive (Not Sig.) | Negative (Not Sig.) | Negative ($p < .10$) | Negative ($p < .05$) | Total |
| Invertebrates (not insects) | 0 | 0 | 0 | 0 | 4 | 4 |
| Insects | 0 | 0 | 7 | 1 | 7 | 15 |
| Fish | 0 | 1 | 2 | 0 | 4 | 7 |
| Birds | 0 | 2 | 32 | 16 | 43 | 93 |
| Mammals | 1[a] | 0 | 4 | 1 | 13 | 19 |
| Totals | 1[a] | 3 | 45 | 18 | 71 | 138 |

[a] *Homo sapiens.*

*Sources:* Tanner (1966) and Pimm (1982).

## POPULATION "CYCLES": CAUSE AND EFFECT

Ecologists have long been intrigued by the regularity of certain population fluctuations, such as those of the snowshoe hare, the Canadian lynx, the ruffed grouse, and many microtine rodents (voles and lemmings) as well as their predators, including the arctic fox and the snowy owl (Elton, 1942; Keith, 1963). These population fluctuations (sometimes called *cycles*, although they should not be) are of two types: voles, lemmings, and their predators display roughly a four-year periodicity; hare, lynx, and grouse have approximately a ten-year cycling time. Lemming population eruptions and the fabled, but very rare, suicidal marches of these rodents into the sea have frequently been popularized (and even staged for movie production!) and are therefore all too "well known" to the lay population.

The tantilizing regularity of these fluctuations in population density presents ecologists with a "natural experiment"—hopefully one that can provide some general insights into factors influencing population densities. Many different hypotheses for the explanation of population cycles have been offered, and the literature on them is extensive (see references at end of chapter). Here, as elsewhere in ecology, it is often extremely difficult or even impossible to devise tests that separate cause from effect, and many of the putative causes of population cycles may in fact merely be side effects of the cyclical changes in the populations concerned. Several of the currently more popular hypotheses, which are not necessarily mutually exclusive, are outlined in subsequent sections of this chapter. Descriptions and discussion of others, such as the "sunspot" and the "random peaks" hypotheses, can be found in the references. Bear in mind that two or more of these hypothetical mechanisms could act together in any given situation.

**Stress Phenomena Hypothesis**  At the extremely high densities that occur during population peaks of voles, a great deal of fighting occurs among these rodents. Body sizes fluctuate with population density, such that animals are larger at peak densities. The so-called stress syndrome is manifested by the animals, their adrenal weights increase, and they become extremely aggressive—so much so that successful reproduction is almost completely curtailed. Eventually "shock disease" may set in and large numbers of animals may die off, apparently because of the physiological stresses on them. Christian and Davis (1964) review evidence pertaining to this rather mechanistic hypothesis.

Some plausible extensions to the stress hypothesis can be made using the ideas of optimal reproductive tactics. Recall that, because current offspring are "worth more" in an expanding population, there is an advantage to early and intense reproduction (high reproductive effort). However, as a population ceases to grow and enters into a decline, the opposite situation arises, favoring little or no current reproduction. Also, if, as seems highly likely, juvenile survivorship diminishes as population density increases, the profits to be gained from reproduction would also decrease. Curtailment of present reproduction and total investment in aggressive survival activities could repay an individual that survived the crash with the opportunity for "sweepstakes" reproductive success!

**Predator–Prey Oscillation Hypothesis**    In simple ecological systems, predator and prey populations can oscillate because of the interaction between them (see also Chapter 15). When the predator population is low, the prey increase, which then allows predators to increase—although this increase lags behind that of the prey. Eventually, predators overeat their prey and the prey population begins to decline; but because of time lag effects, the predator population continues to increase for a period, driving the prey to an even lower density. Finally, at low enough prey densities, many predators starve and the cycle repeats itself. However, prey populations oscillating for reasons other than predation pressures obviously constitute cyclical food supplies for their predators, which should in turn lag behind and oscillate with prey availability. Short of a predator removal experiment, it is thus extremely difficult to determine whether or not changes in prey populations are causally related to changes in predator population density. (Prey populations sometimes fluctuate regularly even in the absence of predators.)

**Epidemiology—Parasite Load Hypothesis**    Under this hypothesis, parasitic epidemics spread through oscillating populations at peaks, causing massive mortality. But because parasites fail to find their hosts very efficiently at low densities, most hosts are unparasitized and the population increases again until another parasite epidemic brings it back under control.

**Food Quantity Hypothesis**    Arctic hare populations sometimes oscillate without lynx populations, perhaps due to a predator–prey "cycle" involving themselves as predators and their own food plants as prey. Under this hypothesis, dense hare populations decrease the quantity of suitable foods, which in turn causes a decline in the hare population. In time the plants recover, and after a lag period hares again increase. Clearly, lynx could entrain to such a cycle.

**Nutrient Recovery Hypothesis**    According to this hypothesis (Pitelka, 1964; Schultz, 1964, 1969), one reason for the periodic decline of rodent populations (especially lemmings) is that the quality of their plant food changes in a cyclical way. During a lemming "high," the ground is blanketed with lemming fecal pellets and many important chemical elements such as nitrogen and phosphorus are tied up and unavailable to growing plants. In the cold arctic tundra, decomposition of fecal materials takes a long time. During this period, lemmings decline due to inadequate nourishment. After a lapse of a few years, feces are decomposed and their nutrients recycled once again, having been taken up by plants. Because their plant food is now especially nutritious, lemmings increase in numbers and the cycle repeats. Schultz (1969) has evidence of such cyclical changes, but whether they are causing, or merely side effects of, the lemming population fluctuations has not been established. The nutrient recovery mechanism does not apply to microtine cycles in general, for Krebs and DeLong (1965) provided supplemental food to a declining population of *Microtus*, but this failed to reverse the decline. More unambiguous experiments like this one are needed to assess the importance of various mechanisms of population control.

**Other Food Quality Hypotheses** Freeland (1974) proposed a related hypothesis based on changes in the relative abundances of toxic versus palatable food plants due to preferential grazing pressures by voles. He suggests that voles graze back competitively superior palatable plants during rodent outbreaks, which allows the competitively inferior unpalatable plants species to spread. Thus, rodent diets must shift to toxic plants, and voles do not fare as well at high densities as they do at low ones.

A similar process, but one involving induced chemical defenses of plants, has been suggested for snowshoe hares by Pease et al. (1979) and Bryant (1980a, 1980b). According to this hypothesis, heavily grazed plant individuals respond to intense herbivory by producing heavily protected new growth that, in turn, constitutes relatively poor fodder for snowshoe hares.

**Genetic Control Hypothesis** This hypothesis, credited to Chitty (1960, 1967a), explains population fluctuations in terms of changing genetic composition of the population concerned. During troughs, the animals experience little competition and are relatively *r*-selected, whereas at peaks competition is intense and they are more *K*-selected. Thus, directional selection, related to population density, is always occurring; modal phenotypes are never the most fit individuals in the population, and in each generation the gene pool changes. The population always lags somewhat behind the changing selective pressures and so no stable equilibrium exists. Some evidence exists for such genetic changes in populations of *Microtus* (Tamarin and Krebs, 1969), but here again, cause and effect are extremely difficult to disentangle.

A number of interesting observations have been made on various fluctuating populations that are relevant to some of the foregoing hypotheses.

Fenced populations of voles reach much higher population densities than do unfenced ones (Krebs et al., 1969) and, as a consequence, overgraze their food supplies more seriously than unfenced natural populations. Fencing presumably does not alter predation levels, but prevents emigration of both juvenile and subordinate animals, thereby raising local density. Krebs (personal communication) suggests that fencing alters spacing behavior and dispersal, hence restricting "self-regulation" of the population. Such an argument borders on group selection but may be compatible with classical Darwinian selection at the level of the individual if carefully rephrased.

Population fluctuations are sometimes out of synchrony within fairly close proximity of one another. In one such situation marked individuals from a population nearing its peak and about to decline were transferred to a nearby population that was just beginning to increase (Krebs, unpublished, as reported in Putnam and Wratten, 1984). Transplanted animals peaked and declined as observed in their own host populations but did not adopt the behavior of the increasing alien population into which they were introduced. These observations suggest that factors intrinsic to the animals, rather than extrinsic phenomena, must be involved.

According to Hanski et al. (1991), specialized mammalian predators (small mustelid weasels) maintain or enhance the cyclicity of microtine rodent populations, whereas generalist predators (larger mammals and hawks) stabilize rodent populations, particularly at lower latitudes. These authors argue that small mustelids are major predators at high latitudes in Scandinavia, but that generalist predators become increasingly important as latitude decreases. Both the amplitude and the length of microtine cycles increase with latitude in Scandinavia (Hanski et al., 1991), as predicted by theory.

Some of the various factors, and how they change with lemming population size over a typical "cycle," are summarized in Figure 9.10.

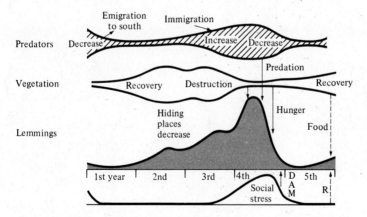

**Figure 9.10.** A schematic representation of various factors involved in the population "cycle" of lemmings. Events and factors with a negative influence on the population are shown with solid arrows; those with a beneficial influence depicted with a dashed arrow. D, disease; A, aggression; M, migration; and R, recovery of health. [From Itô (1980).]

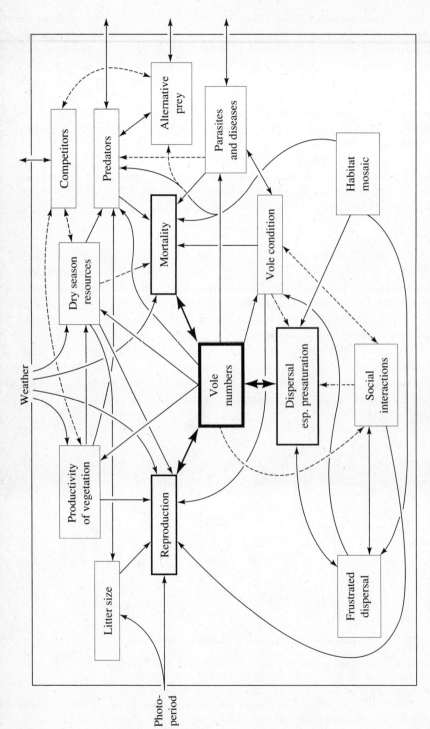

**Figure 9.11** Another schematic representation of factors known (solid arrows) or suspected (dashed arrows) of influencing numbers of California voles. [From Lidicker (1988).]

One must always be wary of oversimplification and "single-factor thinking"; most or even all of the preceding hypothetical mechanisms could work in concert (Figure 9.11) to produce observed population "cycles" (Lidicker, 1988). The extreme difficulty of separating cause from effect, already illustrated, plagues much of ecology. Simple tests that actually refute hypotheses are badly needed. For scientific understanding to progress rapidly and efficiently, a logical framework of *refutable* hypotheses, complete with alternatives, is most useful. However, while such a single-factor approach may work quite satisfactorily for systems exhibiting simple causality, it has proven to be distressingly ineffective in dealing with ecological problems where multiple causality is at work. Once again, one of the major dilemmas in ecology seems to be finding effective ways to deal with multiple causality.

## METAPOPULATIONS

Most species are broken up into many subpopulations, isolated from one another to varying degrees, depending on barriers to dispersal and proximity. Such a set of all possible populations is termed a *metapopulation* (Figure 9.12). Levins (1969) proposed the following simple mathematical model to describe the dynamics of metapopulations:

$$\frac{dp}{dt} = mp(1 - p) - ep$$

where $p$ denotes the fraction of habitat patches occupied by the species, $e$ is the rate of local extinction, and $m$ represents the rate of colonization of empty patches. An equilibrium is reached when $p$ is equal to $1 - e/m$. This model assumes that all patches of habitat are identical, and that the species is either present or absent. Moreover, all local populations are assumed to have the same probability of going extinct, and the rate of colonization of empty patches is proportional to the fraction of occupied patches ($p$) and to the fraction of empty patches ($1 - p$). More complex models have been developed that relax many of these restrictive simplifying assumptions.

For example, Hanski (1982) added a term to the extinction part of the equation which causes the probability of going extinct to fall as the number of patches occupied increases, such that

$$\frac{dp}{dt} = mp(1 - p) - ep(1 - p)$$

A stochastic version of Hanski's model predicts that species fall into two types: regionally common and abundant (*core species*), which are well-spaced out in niche space, and the opposite, rare species that are present only in patches (so-called *satellite species*).

Processes of population turnover, extinction, and reestablishment of new populations constitute the study of metapopulation dynamics. Populations can go extinct within a particular area, but by surviving in an adjacent patch of

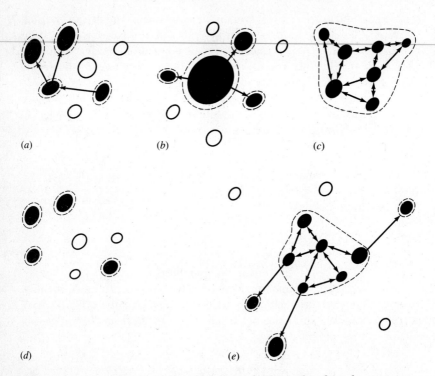

Figure 9.12 Five different kinds of metapopulations. Closed circles represent habitat patches (filled = occupied; unfilled = vacant). Dashed lines indicate boundaries of "populations." Arrows represent migration events (colonizations). (*a*) Levins' metapopulation. (*b*) Source-boundary metapopulation. (*c*) "Patchy" population. (*d*) Non-equilibrium metapopulation [differs from (*a*) in that no re-colonization occurs]. (*e*) An intermediate case that combines both case (*b*) and (*c*). [Adapted from Harrison (1991).]

habitat, still survive in the overall landscape. Animals with active habitat selection, such as many species of birds and lizards, can reach an ecological and evolutionarily stable equilibrium between "source" and "sink" habitats, with periodic dispersal from the former to the latter maintaining the species locally (Pulliam, 1988). The evolution of dispersal ability holds central focus in such systems. An appropriate mixture of spatial patchiness coupled with disturbance and dispersal can promote coexistence in competitive systems. The metapopulation concept is closely linked to ideas in landscape ecology (see pp. 375–376), although the much needed bridge between the two new concepts has not yet been made.

## SELECTED REFERENCES

### Verhulst–Pearl Logistic Equation

Allee et al. (1949);  Andrewartha and Birch (1954);  Bartlett (1960);  Beverton and Holt (1957);  Chitty (1960, 1967a, 1967b);  Clark et al. (1967);  Cole (1965);  Ehrlich and Birch (1967);   Errington (1946, 1956);   Fretwell (1972);   Gadgil and Bossert (1970);   Gibb (1960);   Green (1969);   Grice and Hart (1962);   Hairston et al. (1960); Horn (1968a);   Hutchinson (1978);   Krebs (1972);   McLaren (1971);   Murdoch (1966a, 1966b, 1970);   Nicholson (1933, 1954, 1957);   Pearl (1927, 1930);   Slobodkin (1962b); F. E. Smith (1952, 1954, 1963a);   Solomon (1949, 1972);   Southwood (1966);   M. H. Williamson (1971);   Wilson and Bossert (1971).

### Density Dependence and Density Independence

Andrewartha (1961, 1963);   Andrewartha and Birch (1954);   Brockelman and Fagen (1972);   Davidson and Andrewartha (1948);   Ehrlich et al. (1972);   Gunter (1941); Horn (1968a);   Hutchinson (1978);   Lack (1954, 1966);   McLaren (1971);   Nicholson (1957);   Odum (1959);   Orians (1962);   Pianka (1972);   F. E. Smith (1961, 1963b).

### Opportunistic versus Equilibrium Populations

Ayala (1965);   Charlesworth (1971);   Clarke (1972);   Dobzhansky (1950), Force (1972); Gadgil and Bossert (1970);   Gadgil and Solbrig (1972);   Grassle and Grassle (1974); Grime (1979);   Hutchinson (1951, 1961);   Itô (1980);   King and Anderson (1971); Lewontin (1965);   Luckinbill (1979);   MacArthur (1962);   MacArthur and Wilson (1967);   McLain (1991); Menge (1974);   Murphy (1968);   Pianka (1970, 1972);   Roughgarden (1971);   Seger and Brockmann (1987);   Stearns (1976);   Wilson and Bossert (1971); Winemiller (1989b, 1992).

### Population Regulation

Lack (1954, 1966);   MacArthur and Connell (1966);   Pimentel (1968);   Pimm (1982); F. E. Smith (1961, 1963a, 1963b);   Solomon (1972);   St. Amant (1970);   Tanner (1966).

### Population "Cycles": Cause and Effect

Bryant (1980a, 1980b);   Bryant et al. (1983);   Chitty (1960, 1967a);   Christian and Davis (1964);   Cole (1951, 1954a);   Elton (1942);   Freeland (1974);   Gilpin (1973); Hanski et al. (1991);   Hilborn and Stearns (1982);   Itô (1980);   Keith (1963, 1974); Krebs (1964, 1966, 1970, 1978);   Krebs and DeLong (1965);   Krebs and Myers (1974); Krebs et al. (1971);   Krebs et al. (1969);   Lidicker (1988);   Pease et al. (1979); Pitelka (1964);   Platt (1964);   Putman and Wratten (1984);   Schaffer and Tamarin (1973);   Schultz (1964, 1969);   Sinclair et al. (1993);   Tamarin and Krebs (1969); Wellington (1960);   Williams (1966a);   Wynne-Edwards (1962).

### Metapopulations

Bengtsson (1989);   Boorman and Levitt (1973);   Caswell and Cohen (1991);   DeAngelis and Waterhouse (1987);   Ebenhard (1991);   Gilpin and Hanski (1991);   Gotelli (1991); Hanski (1982);   Hanski and Gilpin (1991);   Harrison (1991);   Harrison et al. (1988);   Hassell and May (1990);   Hastings (1991);   Hastings and Wolin (1989); Levins (1969, 1970);   Pulliam (1988);   Pulliam and Danielson (1991);   Sjogren (1991); Verboom et al. (1991).

## Chapter
## *10*

# Sociality

## USE OF SPACE: HOME RANGE AND TERRITORIALITY

Most habitats consist of a spatial-temporal mosaic of many different, often intergrading, elements, each with its own complement of organisms and other resources. Because of this extensive environmental heterogeneity, an individual's exact location is often a major determinant of its immediate fitness. Members of a prey species that are well protected from their predator(s) in one environmental patch type may be extremely vulnerable to the same predator(s) in another patch. Natural selection, by favoring those individuals that select better microhabitats, should produce a correlation between preference for a given patch type and fitness within it. The density of other individuals in a particular patch, of course, strongly influences the suitability of a given patch. Moreover, because most animals use several to many microhabitats for different purposes or at different times of day, the fitness relations of the use of space are usually quite complex.

Organisms can be spaced about the landscape in two extreme ways. They may occur in groups (*clumped* or *contagious spatial distributions*) or individuals may be evenly spread out (*dispersed spatial distributions*). Intermediate between these extremes are *random spatial distributions*, in which organisms are spread randomly over the landscape. Statistical techniques have been developed to quantify the spatial relationships of individuals in a population; one such technique uses the ratio of variance to mean in the numbers of individuals per quadrat. When this ratio is unity, the distribution of organisms in the quadrats fits the Poisson distribution and the organisms are randomly distributed with respect to the quadrats. Ratios less than unity indicate dispersion, whereas those greater than unity reflect clumped distributions. Dispersed spatial distributions are generally indicative of competition and *K* selection; however, random and clumped distributions, in themselves, indicate little about factors influencing the distribution of the organisms concerned.

Organisms vary widely in their degree of mobility (frequent ecological synonyms are *motility* and *vagility*). Some, such as terrestrial plants and sessile marine invertebrates like barnacles, spend their entire adult life at one spot, with their gametes (or larvae) being the dispersal stages. Others, like earthworms and snails, although vagile, seldom move very far. Still others, such as the Monarch butterfly, some sea turtles, and whales, as well as migratory birds, regularly move distances of many kilometers during their lifetimes.

Returns on many species of banded birds have shown that, even after migrating thousands of kilometers, individuals often return to the same general area where they were raised (this is known as *philopatry*). Similarly, fruit flies *(Drosophila)* labeled with radioactive tracers do not usually move very far. Such restricted movements presumably allow individuals to become genetically adapted to local conditions. Two extreme types of populations are distinguished, although once again there are all degrees of intermediates: *viscous populations*, in which individuals do not usually move very far, and *fluid populations*, in which individuals cover great distances. In viscous populations there is little gene flow and great genetic variability can occur from place to place, whereas the opposite is true of fluid populations. [Note that kin selection (pp. 222–225) is much more likely to occur in viscous populations than in fluid ones—colonizing species with high dispersal abilities are likely to have more fluid populations than equilibrium species or climax species.] At a local level, however, inbreeding in a viscous population may lead to reduced genetic variability.

Important differences occur between organisms in the way they use space. The distinction between two-dimensional and three-dimensional patterns of utilization is fundamental (one-dimensional use of space may also be approximated, as along the crest of a sand ridge or the shore of a stream, lake, or ocean). Thus, we find certain ecological similarities between organisms as diverse as plankton, pelagic fish, flying insects, many birds, and bats—all of which live in a three-dimensional world.

The area or volume over which an individual animal roams during the course of its usual daily wanderings and in which it spends most of its time is the animal's *home range*. Often home ranges of several individuals overlap; home ranges are not defended and are not used to the exclusion of other animals. In contrast, *territories* are defended and used exclusively by an individual, a pair, a family, or a small inbred group of individuals. Nonoverlapping territories normally give rise to dispersed spacing systems and are invariably indicative of competition for some resource in short supply.

Several different kinds of territories are classified by the functions they serve. Many seabirds, such as gulls, defend only their nest and the area immediately adjacent to it; that is, they have a *nesting territory*. Some male birds and mammals, such as grouse and sea lions, defend territories used solely for breeding, termed *mating territories*. By far the most widespread type of territory, however, is the *feeding territory*, which occurs in a few insects, some fish, numerous lizards, many mammals, and most birds.

For territoriality to evolve, some resource must be in short supply and that resource must also be defendable (Figure 10.1). Food items are generally not defendable, since most animals eat their prey as soon as they encounter it. However, the space in which prey occurs can often be defended with a reasonable amount of effort. Sometimes even space is not easily defended, especially when food items are very sparse or extremely mobile; under such circumstances, feeding territories cannot be evolved because costs of defense exceed benefits that could be gained (Brown, 1964). An optimal territory size exists where the difference between benefit gained and cost of defense is greatest (Figure 10.2). Often birds that nest in colonies (i.e., seabirds and swallows) defend nesting territories, but because their prey are very mobile and thus indefensible, they have no feeding territories but feed in flocks. Territoriality is particularly prominent in insectivorous and carnivorous birds, probably largely as a result of the energetic economy of flight and their consequent great mobility, which makes territorial defense economically feasible. Typically, males of these birds set up territories during the early spring, often even before wintering females return. During this period many disputes over territories occur and there is much fighting among males. Once breeding has begun, however,

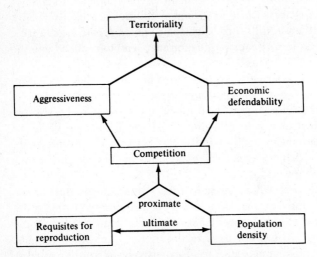

**Figure 10.1** Diagrammatic representation of the various factors influencing the evolution of territoriality. [From Brown (1964).]

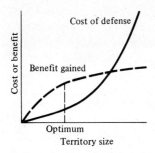

Figure 10.2 Cost of territorial defense increases monotonically with territory size, but net profits gained decrease above a certain size. At the optimum territory size, the difference between benefit and cost is maximal.

disputes over boundaries are usually greatly reduced, with males advertising that they are still "on territory" only briefly during the morning and evening hours.

Male ovenbirds recognize individual neighbors by their territorial songs; Weedon and Falls (1959) played tape-recorded songs back in different places and at different rates and times, and watched responses of various males to playbacks. When a tape recording of a nonneighbor's call is played from a neighboring territory (after the original neighbor has been removed), a male ovenbird reacts by responding vigorously and singing frequently and loudly. The male thus recognizes the substitution of calls, because the same male does not react strongly to the call of his original neighbor. This technique was used to study the function of the territorial song of the white-throated sparrow (Falls, 1969); a greater frequency of playback elicits a stronger response, suggesting that highly motivated birds sing more often than birds that are less likely to win a territorial encounter.

There are definite advantages to recognizing territorial calls of established neighbors and not reacting strongly to them; it would be a waste of time and energy to respond vigorously to such calls once territorial boundaries have been well established. Both individuals stand only to benefit from a "gentleman's agreement" over where such boundaries lie.

Possession of home ranges and territories also serves other important functions. By becoming familiar with a small local area, an animal can learn (1) when and where food is likely to be found, (2) the locations of safe retreats from predators, and (3) in some cases, when and where those predators are likely to be encountered. Thus, resident individuals will normally have distinct advantages over nonresidents (so-called *vagrants*).

An enormous literature on territoriality exists; the interested reader can find an entry to it through the references at the end of this chapter.

# SEX

Most organisms employ sexual reproduction, although many plants and invertebrates use it only infrequently. The evolutionary origin and selective advantage(s) of sexual reproduction remain major unresolved problems in biology

(Williams, 1975). Sexual reproduction itself remains an enigma to students of evolution because organisms engaging in sex perpetuate the genetic materials of another organism. In contrast, an organism that reproduces asexually transmits only its own genes to a clone of its offspring genetically identical to itself, thereby leaving twice as many copies of its genes. In effect, reproducing sexually reduces a reproductive organism's contribution to each of her progeny by a full one-half, meaning that to perform as well as an asexual organism, a sexual form must produce twice as many progeny. Sexual reproduction has arisen at least twice. Presumably, it first came into existence in bacteria billions of years ago in the primieval seas. In bacteria, sex involves exchanging genetic material but does not necessarily result in immediate reproduction. A more elaborate form of sexual reproduction arose later in protists that involved the evolution of diploidy as well as a complex reduction division (meiosis) and production of haploid gametes with only one set of chromosomes. This form of sexual reproduction has persisted to the present day through the evolution of more complex organisms such as ourselves. Diploidy may have evolved as a sort of "fail-safe" mechanism: when there are two copies of the genetic material, if an error is made, a "good" accurate backup copy still exists.

One plausible idea for the origin of sex is a predation hypothesis. Early organisms that consumed others could have simply adopted some of their prey's genetic material and put it to use to their own ends. According to this view, the predator incorporated some of its prey's loci into its own genome, thereby immediately acquiring the ability to synthesize some useful gene products and hence enhancing its own immediate performance and fitness. For example, evidence is overwhelming that components of the prokaryotes, bacteria and blue-green algae, have been incorporated into eukaryotic higher organisms as cell organelles (chloroplasts and mitochondria).

Once gametes evolved, there was a distinct advantage to producing two distinct types: one large and nutritious, but sedentary, gamete that would support early development (eggs, oocytes, ovules) and another more mobile, but smaller, gamete that carries little more than genetic material (pollen, sperm). Such a specialization of function is superior to the presumed primitive state in which the two gametes are similar in size and function (*isogamy*). The situation in which gametes adopt different functions is termed *anisogamy*. Anisogamy gives rise to an asymmetry that results in an interesting fundamental, yet inescapable, "conflict of interests" between males and females (Trivers, 1972), the basis of sexual selection (Darwin, 1871).

Numerous varieties of sexual reproduction exist. Perhaps the finest of all is facultative sexuality, as seen in water fleas (Cladocera): these aquatic microcrustaceans abandon sex completely during the relatively constant summer months to form all-female clones, with each producing only genetically identical daughters (all females possess two full sets of their mother's chromosomes). With the onset of winter, females produce meiotic eggs that develop into haploid males with only one set of their mother's chromosomes. These males inseminate haploid eggs of females in all clones, which then produce a special overwintering resting egg through sexual reproduction.

Some organisms are also hermaphroditic, including simultaneous hermaphrodites (in which one individual has both male and female gonads at the same time—as in many invertebrates such as earthworms and many plants) and sequential hermaphrodites. Among certain marine fishes, some species are males when young but then change sex to become females as they grow older and larger, whereas other species of coral reef fish are females when small and become males as they get older and larger. Sex change in such fish is under social control. At least one sequentially hermaphroditic plant species may be able to switch back and forth from being either male or female, and vice versa. In the most familiar organisms, most vertebrates and some plants, the sexes are separate.

Sexual processes allow the genes in a gene pool to be mixed up each generation and recombined in various new combinations; genetic variability is thus generated by sexual reproduction. The potential rate of evolution of a sexual population is far greater than that of a group of asexual organisms, simply because a variety of beneficial mutations are readily combined into the same individual in a sexual species. But a rapid potential rate of evolution is seldom of as much immediate advantage to an individual organism as is a doubly high rate of reproduction. Sexual reproduction is certainly very basic in diploid organisms and is doubtless an ancient and primitive trait. Considered from an individual's perspective, however, sex is expensive, because an individual's genes are thereby mixed with those of another organism and hence each of its offspring carries only half of its genes (i.e., heritability is halved). In contrast, a female reproducing asexually (including parthenogenesis) duplicates only her own genome in each of her offspring. Even Fisher (1930) suggested that sex could conceivably have evolved for the benefit of the group by way of some non-Darwinian form of group selection. Strangely enough, although many temporary losses of sexuality have been secondarily evolved, relatively few known organisms seem to have completely lost the capacity to exchange their genes with those of other organisms for any geologically long period. All female, unisexual species (well known in fish and lizards) are presumably short lived on the geological time scale. (See Box on "Virgin Birth in Human Females?" on p. 208.)

One brave evolutionist has concluded that sex in higher vertebrates is maladaptive (Williams, 1975, p. 109). Evolutionary benefits of genetic recombination and increased variability must more than offset the disadvantage of one organism perpetuating another's genes. In animals with biparental care, two parents can usually raise twice as many progeny as a single parent, offsetting the cost of sex. One possible advantage to an individual is the "good genes" hypothesis: by reproducing sexually, an organism can mix its genes with other desirable genes, thereby enhancing the fitness of its progeny (of course, this can work both ways, for by mating with a less fit partner, an organism would tend to diminish its own fitness). Another idea is that competition between siblings is reduced by the formation of a variety of types under sexual reproduction (in contrast, cloned offspring should interfere maximally with one another because they are genetically identical and hence require similar resources). If heterozygosity in itself confers increased fitness, however, sexual reproduction can clearly be advantageous to individuals.

# VIRGIN BIRTH IN HUMAN FEMALES?

Parthenogenetic reproduction could occur among human females yet remain unnoticed. Indeed, such a woman could have a husband and be totally unaware of her own condition. She would have only daughters, each of whom would carry only her genes, which would almost certainly increase in the gene pool, at least over the short term. Is there any evidence for this? Claims of reproduction without males are not to be expected from nunneries, but neither have any emanated from prisons where women are kept isolated from men. Parthenogenesis in humans may seem far-fetched, but 50 years ago no one suspected that parthenogenesis could occur in any vertebrate: now all-female species have been documented in fish, amphibians, reptiles, and birds (all major orders of vertebrates except mammals).

In the mid-1950s, the British medical journal *Lancet* published an editorial pointing out that it could be difficult to establish suitable criteria for recognition of parthenogenesis in humans. This set into motion a train of events that led to an interesting, if too limited, scientific examination. The *Sunday Pictorial* newspaper asked mothers who believed that they had produced a child by virgin birth to come forward. Two different mechanisms exist by which a female could reproduce without contact with a male: (1) budding from somatic cells of the mother or incomplete disjunction during meiosis of gametogenic cells, or (2) autofertilization. In the first situation, mother and daughter would be perfect clones, genetically identical (like identical twins). In the second process, the mother would have to produce a sperm that would inseminate her own egg. Mother and daughter would not be genetically identical, although the daughter would possess a subset of her mother's genes, possibly being homozygous at some loci where her mother was heterozygous.

The newspaper article unfortunately mentioned that such children would have to be daughters (it would have been interesting to see whether or not any sons were claimed, but, if so, they could not possibly be parthenoforms). Ultimately, 19 women presented themselves along with their daughters as examples of "virgin birth." Eleven of these did not profess that no father existed, but were under the mistaken impression that the search was for a hymen intact after conception (but long since broken in birth).

The remaining eight pairs were examined by Balfour-Lynn (1956), who blood typed mothers and daughters and found antigens present in six daughters that were absent in their mothers, clear evidence of genetic differences. In another pair, the mother had blue eyes and the daughter brown eyes, indicating genetic differences. In the single remaining case, "Mrs. Alpha and daughter," there was apparent genetic identity in blood groups and several other genetically determined traits, including electrophoretic analysis of serum. The probability of such a close match between a mother and daughter produced by heterosexual reproduction was less than one chance in a hundred ($P < .01$).

As a final check, reciprocal skin grafts were carried out. The transplant from daughter to mother was rejected (shed) in about 4 weeks, while the one from mother to daughter remained healthy for 6 weeks before it was removed. Balfour-Lynn (1956) considered these skin-graft results obscure, but Beatty (1967) interpreted them to mean that the daughter possessed antigens not present in the mother and, therefore, that the results could not confirm parthenogenetic reproduction. Autoimmune responses are known that result in rejection of grafts of one's own skin. Clearly, the jury is still out on this intriguing question: further studies like this one should be undertaken. By now, "Mrs. Alpha's daughter" may well have daughters of her own that could be tested by modern techniques such as DNA fingerprinting.

It may be no accident that many parthenoform unisexuals have a biparental origin, arising from the hybridization of two bisexual parental species. Of course, in such a situation, clonal reproduction maintains and perpetuates heterozygosity perfectly, even better than sexual reproduction would.

## SEX RATIO

In populations of many dioecious organisms, there are approximately equal numbers of males and females. The *sex ratio* is defined as the proportion of males in the population. To be more precise, we distinguish the sex ratio at conception, or the *primary sex ratio,* from that at the end of the period of parental care, the *secondary sex ratio.* The sex ratio of newly independent nonbreeding animals (e.g., as recently fledged birds) is the *tertiary sex ratio,* whereas that of the older breeding adult population is the *quaternary sex ratio.*

Some have asked "Why have males?" Why are various sex ratios often near equality (i.e., 50:50, or 0.5)? Darwin (1871) speculated that sex ratios of 1:1 might benefit groups by minimizing intrasexual fighting over mates. Other workers have reasoned that since one male can easily serve a number of females, it might "be better for the *species*" if the population sex ratio were biased in favor of females, because this would increase the total number of offspring produced. Similarly, males are sometimes viewed as supernumerous and therefore "dispensable." Such interpretations invoke naive group selection, and it is preferable to look for an explanation of sex ratio in terms of selection at the level of the individual.

Here again, Fisher (1930) first solved the problem. He noted that in sexually reproducing diploid species, exactly half the genes (more precisely, half those on autosomal chromosomes) come from males and half from females each and every generation, no matter what the population sex ratio. This statement merely asserts that every individual organism has a mother and a father, but its implications regarding the sex ratio are extensive. The reproductive success of all females must therefore be equal to that of all males—hence, if the numbers of one sex are lower than those of the other sex, then, on average, an individual of the underrepresented sex leaves more descendents than an individual of the overrepresented sex. In effect, individuals of the sex in short supply become more valuable.

**Ronald Fisher**

Fisher concluded that, at equilibrium, an optimal organism should allocate exactly half its reproductive effort to producing progeny of each sex; thus, if

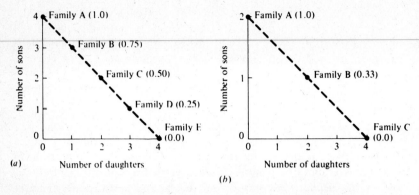

**Figure 10.3** Two hypothetical situations illustrating possible family structures without a sexual dimorphism in costs to the parent (a) and with such a sexual differential (b). Parents are assumed to invest a given fixed amount in reproduction. In (a) sons cost parents the same amount as daughters and the optimal family sex ratio (provided that the population is at equilibrium) is 0.50, at which sex ratio, parental investment is equalized on offspring of the two sexes. In (b) male progeny cost their parents twice as much as female progeny, and expenditure on offspring of the two sexes is equalized at a family sex ratio of 0.33 (again, provided that the population at large is near equilibrium). Table 10.1 and the text develop the reasons that parental investment should be divided equally between offspring of each sex when the population is at equilibrium.

each male offspring costs approximately as much to produce as each female offspring, the optimal family sex ratio is near 50:50, provided that the population is in or near equilibrium. Note that Fisher's argument does *not* depend on competition for mates in any way, as it assumes that each male has the same probability of mating as every other male (likewise all females are assumed to be of equivalent fitness).

Figure 10.3 and Table 10.1 illustrate Fisher's principle in two hypothetical populations. In the first case, there is no sexual dimorphism in energy demands of progeny, and the parental investment argument can be translated directly into numbers; thus, this case leads to an optimal family sex ratio at equilibrium of 0.5. This is not true in the second case, where there is an investment differential such that the individual offspring of one sex require twice as much parental expenditure as individuals of the other sex; this case leads to an optimal family sex ratio at equilbrium of either 0.33 or 0.67, depending on which sex is more expensive to produce. In both cases, the optimal family sex ratio differs when the population sex ratio deviates from the optimal family sex ratio at equilibrium. In such a circumstance, families producing the sex in deficit (compared to the equilibrium sex ratio) have a selective advantage; this results in excess production of the underrepresented sex, which then forces the population sex ratio to the equilibrium sex ratio.* Sex ratio is a special case

*The notion of such "unbeatable" tactics has been extended and generalized by Maynard Smith (1976), who terms them *evolutionarily stable strategies*, or ESSs.

**TABLE 10.1    Comparison of the Contribution to Future Generations of Various Families in Case *a* and Case *b* of Figure 10.3 in Populations with Different Sex Ratios**

| Case *a* | Number of males | Number of females |
|---|---|---|
| Initial population | 100 | 100 |
| Family *A* | 4 | 0 |
| Family *C* | 2 | 2 |
| Subsequent population (sum) | 106 | 102 |

$C_A = 4/106 = 0.03773$
$C_C = 2/106 + 2/102 = 0.03846$ (family C has a higher reproductive success)

| Case *b* | Number of males | Number of females |
|---|---|---|
| Initial population | 100 | 100 |
| Family *A* | 2 | 0 |
| Family *B* | 1 | 2 |
| Subsequent population | 103 | 102 |

$C_A = 2/103 = 0.01942$
$C_B = 1/103 + 2/102 = 0.02932$ (family B is more successful)

| | Number of males | Number of females |
|---|---|---|
| Initial population | 100 | 100 |
| Family *B* | 1 | 2 |
| Family *C* | 0 | 4 |
| Subsequent population | 101 | 106 |

$C_B = 1/101 + 2/106 = 0.02877$
$C_C = 4/106 = 0.03773$ (family C is more successful than family B)

Natural selection will favor families with an excess of females until the population reaches its equilibrium sex ratio (below).

| | Number of males | Number of females |
|---|---|---|
| Initial population | 100 | 200 |
| Family *B* | 1 | 2 |
| Family *C* | 0 | 4 |
| Subsequent population | 101 | 206 |

$C_B = 1/101 + 2/206 = 0.01971$
$C_C = 4/206 = 0.01942$ (family B now has the advantage)

*Note:* The contribution of family $x$ is abbreviated $C_x$.

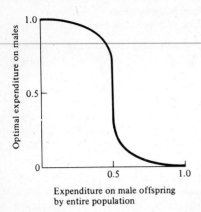

Figure 10.4 The optimal family expenditure on sons varies with expenditure on male offspring in the population at large.

of *frequency-dependent selection,* which arises whenever the selective value of a trait changes with its frequency of occurrence. Figure 10.4 shows the way in which optimal family sex ratio varies with population sex ratio when the two sexes are equally expensive to produce. This plot is more general in that it shows how optimal investment of a family varies as average expenditure in the population at large changes. Note that when average expenditure on offspring of the two sexes by the entire population is near equilibrium (1:1), a relatively broad range of family strategies is near optimal; as overall population expenditure deviates more from this equilibrium, the optimal family tactic rapidly converges on producing families containing only the more valuable underrepresented sex.

Thus, the only factor that can influence the primary and secondary sex ratios is a sexual difference in costs of progeny to the parents. A special case is differential mortality of the sexes *during the period of parental care* (Figure 10.5). Because differential energetic requirements and differential mortality after this period do not interact with parental expenditure, they cannot directly alter either the primary or secondary sex ratios unless their effects are also manifest during the period of parental care. Sexual dimorphisms, or physiological, morphological, or behavioral differences between the sexes, are of paramount importance in any discussion of sex ratio. A great variety of ecological factors can influence sexual dimorphism, especially sexual selection and mating systems, which are complexly intertwined (see next section).

Most reptiles do not have sex chromosomes, and sex is determined by the environmental temperature at which eggs are incubated. Presumably environmental determination of sex is the ancestral condition and genetic determination of sex evolved at least twice (females are the heterogametic sex in birds, whereas males are heterogametic in mammals).

Parental manipulation of sex ratio has been demonstrated in eusocial insects and in zebra finches (Trivers and Hare, 1976; Burley, 1981, 1986). In mate choice experiments, zebra finches banded with red bands are preferred over those banded with green bands. In long-term breeding experiments, parental birds with one parent banded with either an attractive

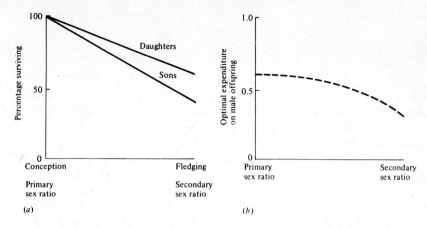

**Figure 10.5** When there is differential mortality between offspring of the two sexes during the period of parental care, parental expenditure on the two sexes can only be equalized through skewed primary and secondary sex ratios. For example, if sons die off more rapidly than daughters (*a*), the optimal primary sex ratio will be biased toward sons, whereas the optimal secondary sex ratio is biased in favor of daughters (*b*). The line curves because the proportion of males to females decreases more and more rapidly as surviving daughters accumulate.

or an unattractive leg band produced more progeny of the same sex or of the opposite sex, respectively (Burley, 1986).

## SEXUAL SELECTION AND MATING SYSTEMS

Fisher (1958a) quotes an unnamed "modern" biologist as having asked the question, "Of what advantage could it be *to any species* for the males to struggle for the females and for the females to struggle for the males?" (italics added). As Fisher points out, this is really a pseudoquestion because the fundamental units of natural selection are individuals rather than species. The question thus reflects the attitude of a group selectionist. In the next few pages, we see exactly why this is not a biologically meaningful question.

Given that an organism is to mix its genes with those of another individual (i.e., it is to reproduce sexually), just which other individual those genes are mixed with can make a substantial difference. An organism mating with a very fit partner by virtue of associating its genes with "good" genes passes its own genes on to future generations more effectively than another genetically identical individual (twin) mating with a less fit partner. Thus, those members of any population that make the best matings leave a statistically greater contribution to future generations. As a result, *within* each sex there is competition for the best mates of the opposite sex; this leads to the *intrasexual* component of *sexual selection*. Intrasexual selection usually generates antagonistic and aggressive interactions between members of a sex, with those individuals best able to dominate other individuals of their own sex being at a relative advantage. Often, direct physical battle is unnecessary and mere gestures (or various other signals of "strength") are enough to determine which individual "wins" an encounter. This makes some selective sense, for if the outcome of a fight is relatively certain, little if anything can be gained from actually fighting; in fact, there is some disadvantage due to the finite risk of injury to both contestants. Similar considerations apply in the defense of territories.

Maynard Smith (1956) convincingly demonstrated mating preferences in laboratory populations of the fruit fly *Drosophila subobscura*. Females of these little flies usually mate only once during their lifetime and store sperm in a seminal receptacle. Males breed repeatedly. Maynard Smith mated genetically similar females to two different strains of males, one inbred (homozygous) and one outbred (heterozygous), and collected all eggs laid by these females during their entire lifetimes. A similar total number of eggs were laid by both groups of females, but the percentage of eggs that hatched differed markedly. Females mated to inbred males laid an average of only 264 fertile eggs each, whereas those bred to outbred males produced an average of 1134 fertile eggs per female (and hence produced over four times as many viable offspring). Maynard Smith reasoned that there should therefore be strong selection for females to mate with outbred rather than with inbred males. When virgin females were placed in a bottle with outbred males, mating occurred within an hour in 90 percent of the cases; however, when similar virgins were offered inbred males, only 50 percent mated during the first hour. Both kinds of males courted females vigorously and repeatedly attempted to mount females, but outbred males were much more successful than inbred ones. By carefully watching their elaborate courtship behavior, Maynard Smith discovered that inbred males responded more slowly than outbred males to the rapid side-step dance of females. Presumably as a result of this lagging, females often rejected advances of inbred males and flew away before being inseminated. These observations clearly show that females exert a preference as to which males they will accept. Similar mating preferences almost certainly exist in most natural populations, although they are usually difficult to demonstrate.

In an elegant study of mate choice among captive feral pigeons, Burley and Moran (1979) and Burley (1981b) demonstrated clear mating preferences among these monogamous birds. Pair bonds were first broken and pigeons kept isolated by sex for several months. Then a "chooser" bird was offered a "choice pair" of potential mates of the opposite sex (Figure 10.6). Pigeons in such a choice pair were tethered to prevent direct physical contact and differed from one another in a given phenotypic aspect, such as age, plumage color, or past reproductive experience. Choosers selected mates with reproductive experience over inexperienced birds and tended to reject very old potential mates in favor of younger ones. A clear hierarchy in preference for plumage traits was also evident; "blue" birds were almost invariably preferred over "ash red" birds, and among female choosers, "blue check" males were chosen over "blue bar" males (either blue phenotype was in turn preferred over "ash red").

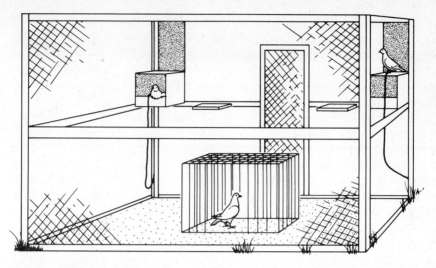

Figure 10.6 Experimental setup for chooser-choice pair experiments. Choice pair birds are tethered next to nest boxes in opposite corners of the large pen. The chooser bird, of the opposite sex, is kept in the small cage until it is fully aware of both of the members of the choice pair. [From Burley and Moran (1979).]

Over evolutionary time, natural selection operates to produce a correlation between male fitness and female preference, because those females preferring the fittest males associate their own genes with the best male genes and, therefore, produce the fittest sons (female fitness is correlated with male preference for the same reason). This can be termed the "sexy son" phenomenon.

As a result of such mating preferences, populations have *breeding structures*. At one extreme is inbreeding, in which genetically similar organisms mate with one another *(homogamy)*; at the other extreme is outbreeding, in which unlikes mate with each other *(heterogamy)*. Outbreeding leads to association of unlike genes and thus generates genetic variation. Inbreeding produces genetic uniformity at a local level, although variability may persist over a broader geographic region. Both extremes represent nonrandom breeding structures; randomly mating *panmictic* populations described by the Hardy-Weinberg equation of population genetics lie midway between them. However, it is very doubtful that any natural population is truly panmictic.

Animal populations also have *mating systems*. Most insectivorous birds and carnivorous birds and mammals are *monogamous*, with a pair bond between one male and one female. In such a case, both parents typically care for the young. *Polygamy* refers to mating systems in which one individual maintains simultaneous or sequential pair bonds with more than one member of the opposite sex. There are two kinds of polygamy, depending on which sex maintains multiple pair bonds. In some birds, such as marsh wrens and yellow-headed blackbirds, one male may have pair bonds with two or more females at the same time *(polygyny)*. Much less common is *polyandry*, in which one female has simultaneous pair bonds with more than one male; polyandry occurs in a few bird species, such as some jacanas, rails, and tinamous. In some species, a male has several short pair bonds with different females in sequence; typically, each such pair bond lasts only long enough for completion of copulation and insemination. This occurs in a variety of birds (including some grackles, hummingbirds, and grouse) and mammals (many pinnipeds and some ungulates). Finally, an idealized mating system (perhaps, more appropriately, a lack of a mating system) is *promiscuity*, in which each organism has an equal probability of mating with every other organism. True promiscuity is extremely unlikely and probably nonexistent; it would result in a panmictic population. It may be approached in some invertebrates such as certain polychaete worms and crinoid echinoderms, which shed their gametes into the sea, or in terrestrial plants that release pollen to the wind, where they are mixed by currents of water and air. However, various forms of chemical discrimination of gametes—and therefore mating preferences—probably occur even in such sessile organisms.

The intersexual component of sexual selection (that occurring between the sexes) is termed *epigamic selection*. It is often defined as "the reproductive advantage accruing to those genotypes that provide the stronger heterosexual stimuli," but it is also aptly described as "the battle of the sexes." Epigamic selection operates by mating preferences. Of prime importance is the fact that what maximizes an individual male's fitness is not necessarily coincident with what is best for an individual female, and vice versa. For example, in most vertebrates, individual males can usually leave more genes under a polygynous mating system, whereas an individual female is more likely to maximize her reproductive success under a monogamous or polyandrous system. Sperm are small, energetically inexpensive to produce, and are produced in large numbers. As a result, vertebrate males have relatively little invested in each act of reproduction and can and do mate frequently and rather indiscriminately (i.e., males tend toward promiscuity). Vertebrate females, on the other hand, often or usually have much more invested in each act of reproduction because eggs or offspring are usually energetically expensive. Because these females have so much more at stake in each act of reproduction, they tend to exert much stronger mating preferences than males and to be more selective as to acceptable mates. By refusing to breed with promiscuous and polygynous males, vertebrate females can sometimes "force" males to become monogamous and to contribute their share toward raising the offspring. In effect, polygyny is the outcome of the battle of the sexes when the males win out (patriarchy), whereas polyandry is the outcome when females win out (matriarchy). Monogamy is a compromise between these two extremes. Under a monogamous mating system, a male must be certain that the offspring are his own; otherwise, he might expend energy raising offspring of another male (note that females do not have this problem). No wonder monogamous males jealously guard their females against stolen copulations! Nevertheless, cuckoldry is not infrequent. Certainty of paternity is a serious problem for males, but females can be confident that their progeny are indeed their own (female parentage is certain). On the other hand, females mated monogamously are vulnerable to desertion once reproduction is underway.

Let us now examine ecological determinants of mating systems. Some assert that sex ratios "drive" mating systems; under such an interpretation, polygyny arises when males are in short supply and polyandry occurs when there are not enough females to go around. According to this explanation, many species are monogamous simply *because* sex ratios are often near equality. In fact, quite the reverse is true, with sexual selection and mating systems indirectly and directly determining sexual dimorphisms and hence various sex ratios. In many birds and some mammals, floating populations of nonbreeding males exist. These can be demonstrated by simply removing breeding individuals; typically, they are quickly replaced with younger and less experienced animals (Hensley and Cope, 1951; Stewart and Aldrich, 1951; Orians, 1969b).

Only 14 of the 291 species (5 percent) of North American passerine birds are regularly polygynous (Verner and Willson, 1966). Some 11 of these 14 (nearly 80 percent) breed in prairies, marshes, and savanna habitats. Verner and Willson suggested that in these extremely productive habitats, insects are continually emerging and thus food supply is rapidly renewed; as a result, several females can successfully exploit the same feeding territory. However, a similar review of the nesting habitats used by polygynous passerines in Europe (which also constitute about 5 percent of the total number of species) showed no such prevalence toward grassland or marshy habitats (Haartman, 1969). Indeed, for elusive reasons, Haartman suggested that closed-in, safe nests were a more important determinant of polygynous mating systems than were breeding habitats. Crook (1962, 1963, 1964, 1965) has suggested that among African weaverbirds, monogamy is evolved when food is scarce and both parents are necessary to raise the young, whereas polygyny evolves in productive habitats with abundant food where male assistance is less essential. This argument, of course, ignores entirely "the battle of the sexes" (epigamic selection).

One of the most complete field studies of polygyny to date is that of Verner (1964), who studied long-billed marsh wrens in Washington state. Male wrens build dummy nests (not used by females) scattered around their territories; during courtship, females are escorted around a male's entire territory and shown each dummy nest (this presumably allows females to assess the quality of a male's territory). Some males possessed two females (one male had three), whereas males on adjacent territories had only one female or none at all. Territories of bigamous and trigamous males were not only larger than those of bachelors and monogamous males, but they also contained more emergent vegetation (where female wrens forage). Verner reasoned that a female must be able to raise more young by pairing with a mated male on a superior territory than by pairing with a bachelor on an inferior territory even though she obtains less help from her mate. This has since been demonstrated in red-winged blackbirds (Figure 10.7). Verner noted that the evolution of polygyny depends on males being able to defend territories containing enough food to support more than one female and her offspring; this condition for the evolution of polygyny requires fairly productive habitats. Female wrens are antagonistic toward each other, and as a result, males cannot make a second mating until their first female is incubating; a temporal staggering of the females is produced

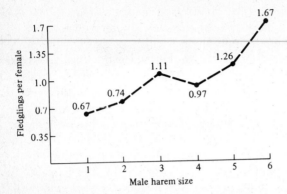

Figure 10.7 Reproductive success of female red-winged blackbirds mated with males having different harem sizes. [After Alcock (1975). *Animal Behavior.* From data of Holm (1973). Copyright ©1973 by the Ecological Society of America.]

(Verner, 1965). Building on Verner's work and studies on blackbirds, Verner and Willson (1966) defined the *polygyny threshold* as the minimum difference in habitat quality of territories held by males in the same general region that is sufficient to favor bigamous matings by females (Figure 10.8).

Polygyny is much more prevalent in mammals than in birds, presumably because in most mammals females nurse their young and, at least among herbivorous species, males can do relatively little* to assist females in raising the young (such species typically have a pronounced sexual dimorphism). A notable exception is carnivorous mammals that are often monogamous during the breeding season, with males participating in feeding both the female and the young (typically sexual dimorphisms are slight in such species). Similarly, most carnivorous and insectivorous birds are monogamous, and males can and do gather food for nestlings. Often, sexual dimorphism in such species is slight, and those that are dimorphic are usually migratory (sexual dimorphism may promote rapid pairing as well as species recognition). Birds whose young are well developed at hatching *(precocial* as opposed to *altricial* birds) typically have little male parental care and are frequently polygynous, with pronounced sexual dimorphisms.

Given a population sex ratio near equality and a monogamous mating system, every individual (both male and female) has a substantial opportunity to breed and to pass on its genes; however, with an equal population sex ratio and polygyny, a select group of the fittest males makes a disproportionate number of matings. Dominant battle-scarred males of the northern sea lion, *Eumetopias jubata,* that have "won" the rocky islets where most copulations occur, often

---

*Why male mammals do not lactate remains an unresolved evolutionary question (Daly, 1979).

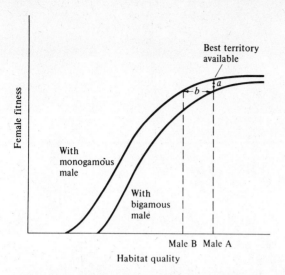

**Figure 10.8** Graphical model of the conditions necessary for the evolution of polygyny. Reproductive success of females is correlated with the quality of a male's territory: females select mates that give them the highest individual fitness. Distance *a* is the difference in fitness between a female mated monogamously and a female mated bigamously in the same environment. Distance *b* is the polygyny threshold, or the minimum difference in territory quality held by males in the same region that is sufficient to favor bigamous matings by females. [After Orians (1969b).]

have harems of 10 to 20 females. Under such circumstances, these males sire most progeny and their genes constitute half the gene pool of the subsequent generation. Because those heritable characteristics making them good fighters and dominant animals are passed on to their sons, contests over the breeding grounds may be intensified in the next generation. It is essentially an "all-or-none" proposition in that only the winning males are able to perpetuate their genes. As a result of this intense competition between males for the breeding grounds, intrasexual selection has favored a striking sexual dimorphism in size. Whereas adult females usually weigh less than 500 kilograms, adult males may weigh as much as 1000 kilograms. Sexual dimorphism in size is even more pronounced in the California sea lion, *Zalophus californianus*, where females attain weights of only about 100 kilograms, whereas males reach nearly 500 kilograms. Among 13 species of these pinniped mammals, sexual size dimorphisms are more profound in species that have larger harems (Figure 10.9). Presumably, the upper limit on such a differential size escalation is set by various other ecological determinants of body size, such as predation pressures, foraging efficiency, and food availability.

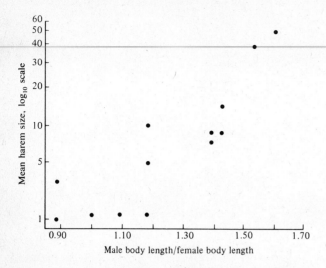

Figure 10.9 Average harem size in various species of pinniped sea mammals increases with the degree of sexual size dimorphism. [From Trivers (1985) after LeBoufe.]

Somewhat analogous situations occur among various polygynous birds. Many species of grouse exhibit multiple short pair bonds, lasting only long enough for copulation, with males displaying their sexual prowess in groups termed "leks." Dominant, usually older, males occupy the central portion of the communal breeding grounds and make a disproportionate percentage of the matings. Receptive females rush past peripheral males to get to central males for copulation. Strong sexual dimorphisms in size, plumage, color, and behavior exist in many grouse. In addition to intrasexual selection, epigamic selection operating through female choice can also produce and maintain sexual dimorphisms; usually both types of sexual selection occur simultaneously, and it is often difficult to disentangle their effects. Indeed, by choosing to mate with gaudy and conspicuous males, females have presumably forced the evolution of some bizarre male sexual adaptations, such as the long tails of some male birds of paradise (this is known as *runaway sexual selection*). Certain bower birds have avoided becoming overly gaudy (and hence dangerously conspicuous) by evolving a unique behavioral adaptation; males build highly ornamented bowers that are used to attract and to court females and that signal the male's intersexual attractiveness. Frequently, if not usually, the same sexual characteristics (such as size, color, plumage, song, behavior) advertise both intrasexual prowess and intersexual attractiveness. This makes evolutionary sense because an individual's overall fitness is determined by its success at coping with both types of sexual selection, which should usually be positively correlated; moreover, economy of energy expenditure is also obtained by consolidation of sexual signals.

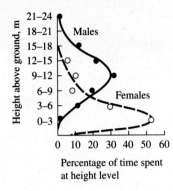

Figure 10.10 Male red-eyed vireos tend to forage higher than females. [Adapted from P. Williamson (1971). Copyright ©1971 by the Ecological Society of America.]

Extravagant male traits such as a peacock's tail could be exploited by females as indicators of the male's ability to survive despite his handicap (Zahavi, 1975, 1977). This "handicap hypothesis" was extended by Hamilton and Zuk (1982), who suggest that females might also assess a male's resistance to parasites by the brightness or elaborateness of his epigamic colors or display.

Sexual dimorphisms sometimes serve still another ecological function, by reducing niche overlap and competition between the sexes (Figure 10.10). In certain island lizards (Schoener, 1967, 1968a) and some birds (Selander, 1966), strong sexual dimorphisms in the feeding apparatus (jaws and beaks, respectively) are correlated with differential utilization of food resources.

## FITNESS AND AN INDIVIDUAL'S STATUS IN ITS POPULATION

An organism's fitness is determined by the interaction between its phenotype and the totality of its environment. In $K$-selected organisms, fitness is determined largely by the biotic environment, especially the individual's status within its own population; however, fitness of $r$-selected organisms may often depend less on the biotic environment and be more strongly influenced by the physical environment.

Compared with breeding individuals, members of a nonbreeding floating population have very low immediate fitnesses (although their fitness rises if they are able to breed later). Even within a breeding population, various individuals may often differ substantially in fitness; for instance, among long-billed marsh wrens, males vary greatly in their individual fitness, with bigamous and trigamous males being more fit than monogamous ones. Variance in the reproductive success of males is typically much greater than that of females, providing the context for intense sexual selection. Predators tend to crop mainly the "excess" members of bobwhite quail and muskrat populations (Errington, 1956, 1963); presumably, resident prey individuals know their own territories and home ranges—making them more difficult to capture than vagrant individuals with a less stable status in the population.

# SOCIAL BEHAVIOR AND KIN SELECTION

A wide variety of ecological phenomena have been interpreted as having been evolved for the benefit of the population rather than for the benefit of individuals. Clutch size, sex ratio, and sexual selection have already been discussed; predator alarm calls and so-called "prudent predation" are discussed in Chapter 15, and "selection at the level of the ecosystem" is briefly considered in Chapter 16. Another broad area latent with opportunities for interpretation of events and phenomena as having arisen for the advantage of a group is social behavior. Why, for example, does a worker honey bee sacrifice her own reproduction for the good of the colony? As is well known, a worker bee will even give up her own life in defense of the hive. To find the probable answer to this question, we must first develop some basic considerations and definitions.

True altruism occurs only when an individual behaves so that it suffers a net loss while a neighbor (or neighbors) somehow gains from its loss (Table 10.2). Obviously selfish behavior will always be of selective advantage; the problem is to explain the occurrence of apparently altruistic behavior such as that of the worker honey bee.

The very existence of social behavior implies that *individuals* living in cooperating groups are in fact leaving more genes in the population gene pool than hermits (Hamilton, 1964). Thus, social behavior may be expected to evolve when distinct advantages are inherent in group participation. To develop his thesis, Hamilton defines *kin selection* as selection operating between closely related individuals to produce cooperation. *Inclusive fitness* includes an individual's own personal fitness (its own progeny) plus the offspring of its relatives appropriately discounted by their degree of relatedness. As an extreme example, an individual should theoretically sacrifice its own life if it can thereby save the lives of more than two siblings, each of which shares half its genes. Such behavior furthers its own genotype even more effectively than living to reproduce itself. This is not true altruistic behavior because the individual making the "sacrifice" actually gains more than it loses. Kin selection operates at a much more subtle level. Closely related relatives are much more likely to benefit from such pseudoaltruistic behavior than distantly related ones; for the latter to occur, loss to the benefactor should be very small or the number of distant relatives benefited must be large. Thus, the condition necessary for evolution of such pseudoaltruistic behavior is that total gain(s) to relative(s) must exceed the loss to the pseudoaltruist. Parental care is, of course, a special case of kin selection.

Here once again, ideas were foreshadowed by Fisher (1930) in his consideration of the evolution of distastefulness and warning coloration in insects. Many distasteful or even poisonous insects, especially the larvae of some moths and butterflies, are brightly colored; vertebrate predators, especially birds, quickly learn not to eat such warningly colored insects. Fisher noted the difficulty of explaining the origin of this gaudy coloration since new ultraconspicuous mutants would be expected to suffer the first attacks of inexperienced predators and therefore be at a selective disadvantage to less gaudily colored genotypes.

TABLE 10.2    **The Four Possible Situations Involving an Individual's Behavior and Its Influence on a Neighbor**
Kin selection occurs when an individual actually gains more than it loses (due to the advantage the individual obtains in perpetuating its genes among its neighbors who are relatives). Thus it is pseudoaltruistic behavior. True altruistic behavior, in which an individual actually loses while its neighbor gains, is virtually unknown (except, perhaps, in humans); moreover, group selection is necessary for its evolution.

|  | Neighbor(s) gain | Neighbor(s) lose |
|---|---|---|
| Individual gains | Pseudo-altruistic behavior (kin-selection) | Selfish behavior (selected) |
| Individual loses | True altruistic behavior (counter-selected) | Mutually disadvantageous behavior (counter-selected) |

He suggested that benefits to siblings on the same branch accruing under a gregarious family system could favor evolution of warning coloration; hence, such ultraconspicuous mutants would be "pseudoaltruists." Fisher even made his argument quantitative by noting that, although each sibling shares only half the individual's genes, their numbers ensure that the total number of shared genes benefited greatly exceeds the number destroyed in the individual's own genome. Such a gene for gaudiness raises inclusive fitness, but it cannot spread unless it is already present among relatives: hence the mutant has to have already spread before it can be selected.

Two orders of insects are of particular interest because they have evolved eusociality—namely, the hymenoptera and isoptera. Members of the insect order Hymenoptera, which includes ants, bees, wasps, and hornets, often form colonies and exhibit apparent altruistic social behavior. Social behavior is thought to have arisen independently many times among hymenoptera, which have a peculiar haplodiploid genetic system; males are produced asexually and are haploid. Thus, all sperm produced by any given male are genetically identical (barring somatic mutation) and carry that male's entire genome. (Males have no father but do have a grandfather.) Females are normal diploids, with each ovum carrying only one-half their genomes. In some hymenopteran species, a queen is thought to mate just once during her entire lifetime and store one male's sperm in a spermatotheca; thus, all her progeny have the same father. A result of this strange genetic system is that sisters are more closely related to one another than mothers are to their own daughters; the former share three-fourths of their genes, the latter only one-half. Hence, one would predict that worker bees should help raise their sisters to reproductive maturity in preference to mating themselves, which in fact they often do. Interestingly, workers can and sometimes do lay haploid male-producing eggs. Males share fewer genes with both siblings and progeny, and never work for the benefit of the colony.

Honey bees mate during nuptial flights in the air. Multiple matings by females have been shown to occur by releasing drones with marked sperm. A phenomenon known as sperm precedence occurs in many insects, where the first sperm in is the last sperm out (the last male to copulate gets the first progeny). The implications of multiple mating on the relatedness among colony members merits consideration. Relatedness will remain high as long as the queen is using a given male's sperm, but relatedness will diminish when she switches between the sperm of two males. Bees have long been known to have periods of unrest during which queens are killed and replaced by one of their own royal brood. Such "mutinies" would be expected to occur precisely when queens are switching between the sperm of two different drones (this prediction has not yet been demonstrated conclusively).

Termites (order Isoptera) also form highly organized colonies, complete with a queen and "king," various secondary reproductives, and both male and female workers with distinct castes. But because both sexes of termites have a normal diploid genetic system, the kin selection argument is inadequate to explain the evolution of sociality in termites. Several hypotheses have been offered for this apparent evolutionary enigma. One involves cyclic inbreeding (Bartz, 1979, 1980). According to this argument, the primary king and queen in a termite colony are unrelated to one another, but each is itself a highly homozygous product of intense inbreeding. As a result, both the maternal and paternal copies of each of their genomes are virtually identical. This situation results in a very high degree of relatedness among their progeny since each offspring receives almost exactly the same set of genes. Termites are highly dependent on one another for continual replenishment of their intestinal protozoa. (These endosymbionts, which produce the cellulases that enable termites to digest wood, are lost with each molt and must be replenished continually during the lifetime of an individual.) Presumably, a pair of termites (the king and queen) maximize their own reproductive success by producing many nonreproductive progeny (workers), which in turn allow production of many successful reproductive offspring (new kings and queens). However, natural selection among workers should operate to release them from such parental "control." Doubtlessly, termites also enjoy other advantages from group cooperation, such as protection from predators and the elements (note that this is also an advantage for hymenopterans).

A detailed long-term study of helpers at the nest in colonially nesting white-fronted bee-eaters in Kenya (Emlen and Wrege, 1988, 1989) demonstrated that helpers tended to assist close relatives much more often than distant kin or nonrelatives (Table 10.3), providing strong evidence for kin selection.

Another form of pseudoaltruism, termed *reciprocal altruism* (Trivers, 1971), does not require genetic affinity or kin selection to operate. In reciprocal altruism, some behavioral act incurs a relatively minor loss to a donor but provides a recipient with a large gain; thus, two entirely unrelated animals can both benefit from mutual assistance. An example that could perhaps be explained by reciprocal altruism is the posting of sentinels. A sentinel crow spends a brief time period sitting in a tree watching for predators while the rest of the flock forages; in turn, it receives continual sentinel protection from the remainder of the flock during the much longer period of time in which it forages. (Such crows may not be sentinels at all but may merely have had their fill of food to eat; their squawk could be a signal intended to inform the predator that it has been detected; see also pp. 321–322.) Reciprocity is, of course, absolutely essential for the evolution and maintenance of this sort of altruistic behavior, unless kin selection is also operating to promote it.

Kin selection and reciprocal altruism are appealing concepts that facilitate explanation of the evolution of social behavior by natural selection at the level of individuals. However, neither mechanism is readily verified by observation. Future empirical work in these areas, although difficult, will be of some interest.

TABLE 10.3 **Helpers at the Nest in White Fronted Bee Eaters in Kenya**

| Breeders | $r^*$ | No. cases | % cases |
|---|---|---|---|
| Father × Mother | 0.5 | 78 | 44.8 |
| Father × Stepmother | 0.25 | 17 | 9.8 |
| Mother × Stepfather | 0.25 | 16 | 9.2 |
| Son × Nonrelative | 0.25 | 18 | 10.3 |
| Brother × Nonrelative | 0.25 | 12 | 6.9 |
| Grandfather × Grandmother | 0.25 | 5 | 2.9 |
| Half brother × Nonrelative | 0.13 | 3 | 1.7 |
| Uncle × Nonrelative | 0.13 | 2 | 1.1 |
| Grandmother × Nonrelative | 0.13 | 1 | 0.6 |
| Grandson × Nonrelative | 0.13 | 1 | 0.6 |
| Great grandfather × Nonrelative | 0.13 | 1 | 0.6 |
| Nonrelative × Nonrelative | 0.0 | 20 | 11.5 |
| TOTAL | | 174 | 100.0 |

*Coefficient of Relatedness.

*Source:* Adapted from Emlen and Wrege (1988).

# THE EVOLUTION OF SELF-DECEIT

Recent studies with young children have shown that the leaders among children are those who can lie more convincingly. Interestingly, the correlation between leadership and ability to be dishonest did not hold among adult women, although it does among adult males. These startling new discoveries have obvious implications in politics!

Deceit of others is fairly straightforward but self-deceit is an extremely interesting phenomenon worth closer scrutiny. We like to think that we perceive the world around us accurately. In a series of experiments, voices of human subjects were taped and played back to subjects who were wired up to a polygraph "lie detector." The polygraph measures electrical conductance across the skin's surface based on perspiration. Our skins, and presumably our subconscious minds, virtually always recognize tape recordings of one's own voice accurately. However, these experiments demonstrated two different forms of self-deception. Sometimes, the conscious response of a subject to hearing his or her own voice is "No, that's not me," but electrical conductance in the subject's own skin shows otherwise (the subconscious recognizes his or her own voice). Other times, a subject asserts "Yes, that's me!" to the voice of another person, but his or her skin indicates that the subconscious knows otherwise (i.e., the truth). People who fail to recognize their own voices and who project themselves into another person's voice typically have poor self-images. Presumably, these people were being quite "honest" and genuinely felt that they were giving correct answers. In both situations, these persons' conscious minds were deceiving themselves while their subconscious retained accurate information.

What possible adaptive value could such complex self-deception have? The rather startling possibility is that self-deceit makes one a more effective "liar," enabling one to persuade other humans of some misinformation (imagine the benefits in politics and litigation!). Indeed, the evolution of the subconscious mind itself would seem to be a necessary precursor for self-deceit to even become possible. It is an open question as to what the subconscious mind actually does with its veritable treasure trove of accurate perceptions. My guess is that such information is exploited to maximize the reproductive success of individuals (*without* their own "knowledge")! The profound implications of self-deceit for politics are disturbing.

# SELECTED REFERENCES

## Use of Space: Home Range and Territoriality

Brown (1964, 1969); Brown and Orians (1970); C. R. Carpenter (1958); F. L. Carpenter (1987); Falls (1969); Howard (1920); Hutchinson (1953); Kohn (1968); McNab (1963); Menge (1972a); Morse (1971); Orians and Horn (1969); Orians and Willson (1964); Pielou (1969); C. C. Smith (1968); Tinbergen (1957); Weedon and Falls (1959).

## Sex

Bull (1983); Bull and Harvey (1989); Charnov (1982); Darwin (1871); Fisher (1930); Fricke and Fricke (1977); Shapiro (1980); Stearns (1987); Trivers (1972); Williams (1971, 1975).

## Sex Ratio

Burley (1981a, 1986); Charnov (1982); Darwin (1871); Fisher (1930, 1958a); Hamilton (1967); Kolman (1960); Maynard Smith (1976); Trivers and Hare (1976); Verner (1965).

## Sexual Selection and Mating Systems

Bayliss (1978, 1981); Blumer (1979); Burley (1977, 1981a, 1981b); Burley and Moran (1979); Crook (1962, 1963, 1964, 1965, 1972); Daly (1979); Darwin (1871); Dawkins and Carlisle (1976); Downhower and Armitage (1971); Emlen (1968b); Emlen and Oring (1977); Erckmann (1983); Fisher (1930, l958a); Fricke and Fricke (1977); Gross and Shine (1981); Haartman (1969); Hamilton (1961); Hamilton and Zuk (1982); Handford and Mares (1985); Hensley and Cope (1951); Holm (1973); Houde and Endler (1990); Howard (1974); Hrdy (1981); Kirkpatrick (1982); Kirkpatrick and Ryan (1991); Kolman (1960); Lack (1968); Lightbody and Weatherhead (1987, 1988); Margulis and Sagan (1986); Maynard Smith (1956, 1958, 1971); Orians (1969b, 1972, 1980b); Payne (1979); Perrone and Zaret (1979); Pleszczynska (1978); Ralls and Harvey (1985); Ridley (1978); Ryan (1990); Schoener (1967, 1968a); Selander (1965, 1966, 1972); Shapiro (1980); Smouse (1971); Stewart and Aldrich (1951); Thornhill (1976); Thornhill and Alcock (1983); Trivers (1972, 1985); Trivers and Willard (1973); Verner (1964, 1965); Verner and Engelsen (1970); Verner and Willson (1966); Werren et al. (1980); Whittingham et al. (1992); Wiley (1974); Williams (1966a, 1971, 1975); Willson and Burley (1983); Willson and Pianka (1963); Wittenberger (1976); Xia (1992); Zahavi (1975, 1977).

## Fitness and an Individual's Status in Its Population

Errington (1946, 1956, 1963); Fretwell (1972); C. C. Smith (1968); Verner (1964, 1965); Verner and Engelsen (1970); Wellington (1957, 1960).

## Social Behavior and Kin Selection

Alexander (1974); Bartz (1979, 1980); Brown (1966, 1975); Cole (1983); Crook (1965); Dawkins (1976); Eberhard (1975); Emlen and Wrege (1988, 1989); Evans (1977); Fisher (1930, 1958a); Greenberg (1979); Hamilton (1964, 1967, 1970, 1971, 1972); Hardin (1982); Horn (1968b); Hughes (1988); Maynard Smith (1964); Michod (1982); Myles and Nutting (1988); Oster and Wilson (1979); Price and Maynard Smith (1973); Sherman (1977); Trivers (1971, 1974, 1985); Trivers and Hare (1976); Wallace (1973); Wiens (1966); E. O. Wilson (1971, 1975); Wynne-Edwards (1962).

## The Evolution of Self-Deceit

Gur and Sackeim (1979); Trivers (1985).

**Chapter**
**11**

# Interactions Between Populations

## DIRECT INTERACTIONS

The traditional approach to population interactions has been to consider just the direct pairwise interactions. In this simplistic view of things, two populations may or may not affect each other; if they do, the influence may be beneficial or adverse. By designating a detrimental effect with a minus, no effect with a zero, and a beneficial effect with a plus, all possible population interactions can be conveniently classified. When neither of two populations affects the other, the interaction is designated as (0, 0). Similarly, a mutually beneficial relationship is (+, +) and a mutually detrimental one is (−, −). Other possible interactions are (+, −), (−, 0), and (+, 0), making a total of six fundamentally different ways in which populations can interact (Table 11.1).

   *Competition* (−, −) takes place when each of two populations affects the other adversely. Typically, both require the same resource(s) that is (are) in short supply; the presence of each population inhibits the other. If the resource is another population (a prey species), competition is indirect and mediated by means of resource depression—this type of competition is termed *exploitation* competition. Other kinds of competition also occur. For example, competition can also be direct, as in agonistic encounters such as allelopathy or interspe-

229

TABLE 11.1   **Summary of the Various Sorts of Direct Pairwise Interactions That May Occur Between Two Populations**

| Type of interaction | Species A | B | Nature of the interaction |
|---|:---:|:---:|---|
| Competition | − | − | Each population inhibits the other |
| Predation, parasitism, Batesian mimicry | + | − | Population A—the predator, parasite, or mimic—kills or exploits members of population B—the prey, host, or model |
| Neutralism | 0 | 0 | Neither population affects the other |
| Mutualism, Müllerian mimicry | + | + | Interaction is favorable to both (can be obligatory or faculative) |
| Commensalism | + | 0 | Population A, the commensal, benefits, whereas B, the host, is not affected |
| Amensalism | − | 0 | Population A is inhibited, but B is unaffected |

*Source:* Adapted from Odum (1959) after Haskell (1947).

cific territoriality (known as *interference competition*). *Predation* (+, −) occurs when one population affects another adversely but benefits itself from the interaction. Usually, a predator kills its prey and consumes part or all of the prey organism. (Exceptions include lizards losing their tails to predators and plants losing their leaves to herbivores.) *Parasitism* (+, −) is essentially identical to predation, except that the host (a member of the population being adversely affected) is usually not killed outright but is exploited over some period of time. Thus, parasitism can in some ways be considered a "weak" form of predation; Batesian mimicry (p. 323) and herbivory could be placed here as well. Interactions that benefit both populations (+, +) are classified as *mutualisms*. In some mutualisms, the association is obligatory (neither population can exist without the other), but in others the interaction is facultative because it is not an essential condition for survival of either population (Müllerian mimicry, p. 323, falls under this heading). When one population benefits while the other is unaffected, the relationship is termed *commensalism* (+, 0). *Amensalism* (−, 0) is said to occur when one population is affected adversely by another but the second is unaffected. *Neutralism* (0, 0) occurs when the two populations do not interact and neither affects the other in any way whatsoever; it is thus of little ecological interest. True neutralism is likely to be very rare or even nonexistent in nature, because there are probably indirect interactions between all the populations in any given ecosystem, although their significance may be minimal. Three of the six population interactions—competition, predation, and mutualisms—are of overwhelming importance; an entire chapter is devoted to competition and predation. Mutualisms are considered in this chapter.

# COMPLEX POPULATION INTERACTIONS

Interactions among populations often become quite intricate, particularly in diverse communities. Colwell (1973) studied interactions among four species of nectar-feeding birds, four species of flowering plants, and two species of mites (Figure 11.1) in Costa Rica. Three hummingbirds, *Colibri thalassinus*, *Eugenes fulgens*, and *Panterpe insignis*, compete for nectar and are pollinating vectors for the plants. Flowers of various species differ in corolla lengths and are visited differentially by hummingbirds that differ in beak lengths. Only *Eugenes*, which has the longest beak, can reach the nectar of *Centropogon talamancensis* through its very long corolla (Figure 11.1). However, *Panterpe* hummingbirds steal nectar from this *Centropogon* species by piercing the base of flower corollas. The fourth species of bird, *Diglossa plumbea*, is a nectar thief that obtains nectar from all four plant species by breaking their corolla bases. Two species of mites live within flowers of different plant species, moving

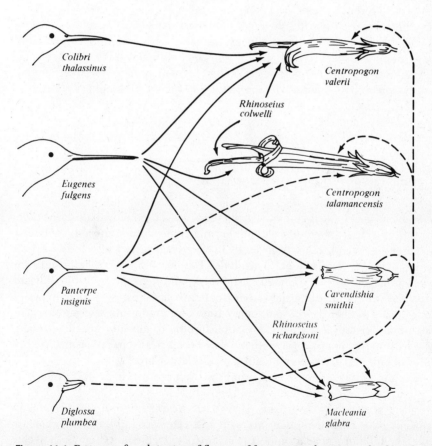

**Figure 11.1** Patterns of exploitation of flowers of four species by various birds and mites. Broken lines indicate illegitimate visits by nectar thieves, which pierce the base of the corolla. [From Colwell (1973). Copyright ©1973 by The University of Chicago Press.]

among flowers by climbing up beaks and riding in hummingbird nostrils. Both mite species are found on *Panterpe* and *Eugenes* hummingbirds, which make legitimate pollinating visits through the corollas of flowers of three and four species of plants, respectively. One mite, *Rhinoseius colwelli*, is restricted to flowers of the two species of *Centropogon* and never occurs in either *Macleania* or *Cavendishia* flowers. The second species of mite, *R. richardsoni*, is found only in *Macleania* and *Cavendishia* flowers and never occurs in *Centropogon* flowers. Significantly, avian pollinator visits are structured so that transfers between *Macleania* and *Cavendishia* flowers are frequent, whereas transfers between flowers of these two species and those of *Centropogon* are much more infrequent. Experimental introductions of mites into flowers without mites showed that both species can live and reproduce successfully in the flowers normally occupied only by the other species. Adult male mites are extremely aggressive, particularly in interspecific encounters, and Colwell observed male *R. colwelli* killing *R. richardsoni*. Over evolutionary time, adult male mites may have reinforced the observed species-specific separation on flowers of different species by killing mites of the other species when these made the mistake of leaving their hummingbird carriers to invade a flower of the wrong species, which contained adult males of the other species of mite. Among these ten species, then, interactions include intense interference competition (between the two species of mites), exploitation competition (among nectarivorous birds), facultative mutualism (between hummingbird pollinators and plants), parasitism (between plants and nectar thieves), and commensalism (between mites and their hummingbird carriers).

## MUTUALISTIC INTERACTIONS AND SYMBIOTIC RELATIONSHIPS

*Symbiosis* means living together. Usually the term is used only to describe pairs of organisms that live together without harming one another, thereby excluding parasitism $(+, -)$ and amensalism $(-, 0)$, in which one party is affected adversely (see Table 11.1 for explanation of symbols). Hence, symbiotic relationships include mutualism $(+, +)$, commensalism $(+, 0)$, and neutralism $(0, 0)$. Obligate mutualisms can be distinguished from facultative ones. These various types of interactions can change in evolutionary time and grade into one another. Although mutualism is a symmetric relationship, there may nevertheless usually be an asymmetry in costs versus benefits to each of the parties concerned (a conflict of interests arises even in mutualistic relationships!).

As pointed out earlier, true neutralism is uninteresting as well as uncommon and therefore need not be considered. However, mutualism and commensalism are fairly widespread, particularly in diverse communities. Many organisms have formed mutualisms with ants (DeVries, 1991a, 1992). For example, the bullhorn *Acacia* supports a colony of ants, feeding them both nectar and protein, which in turn protect the plant from a wide variety of herbivores (when ants are removed by poisoning them, these plants are quickly defoliated). Likewise, caterpillars in many different Lepidopteran families have evolved

close mutualisms with ants to defend themselves against parasites and predators (Pierce, 1985). These caterpillars "sing" to the ants as well as feed them a
nutritious protein-rich diet (DeVries, 1991b, 1992). Numerous other cases are
also known. In many legumes such as mesquite, root nodules house bacteria
that fix atmospheric nitrogen. Mycorrhizae, or fungal roots, supply mineral
nutrients to host plants but in return gain organic carbon from the host. Many
deep-sea fish harbor bioluminescent bacteria, exploiting their light-emission
abilities to the fish's own ends in the dark depths of the oceans. Certain types
of algae are endozoic, actually living inside the cells of animal hosts, particularly
coelenterates such as corals and *Hydra*. In these situations, algal photosynthate
is shared with the host. Some invertebrate "hosts" actually digest away most
of the alga, retaining ("kidnapping"?) just the chloroplasts, which continue to
photosynthesize inside the animal! Perhaps the ultimate in mutualistic interactions concerns the intriguing theory of endosymbiosis; very strong evidence
exists that certain of the cell organelles found in higher organisms (eukaryotes),
particularly chloroplasts and mitochondria, are actually the remnants of symbiotic prokaryotic organisms (Ehrman, 1983; Margulis, 1970, 1974, 1976) that
have been permanently incorporated into the eukaryotes.

Some birds ride on the backs of water buffalo (the bird obtains food while
the mammal is freed of many insect pests); other small birds pick between
the teeth of crocodilians (the bird obtains food while the reptile gets its teeth
cleaned). Certain ants exploit aphids for the latter's honeydew, tending their
herds of hemipterans much as a shepherd watches over his flock. Other species
of ants and termites actually cultivate fungi for food.

An African bird known as the honey guide has formed a unique alliance
with the honey badger or ratel (a large skunklike mammal); the honey guide
locates a beehive and leads the honey badger to it, whereupon the mammal
tears open the bee's nest and eats its fill of honey and bee larvae. Later the bird
has its meal of beeswax and larvae. The honey guide can find beehives with
relative ease but cannot open them, whereas the ratel is in just the opposite
situation; cooperation clearly increases the efficiency of both species.

In marine environments, certain species of labrid fish are "cleaners," maintaining cleaning stations where other species of larger fish come to be cleaned
of ectoparasites and bacteria, sometimes lining up rather like cars at a gas
station. Interspecific displays are used in recognition. Cleaner fish are conspicuous and brightly colored. Interestingly, an unrelated fish species has evolved
in another family, the saber-toothed blenny, which mimics the cleaner fish, but
brings woe to the unsuspecting large fish (these blennies *eat* the vascularized
gill tissue of the large fish!).

Because most land plants cannot move, they often exploit animals both
for pollination and for seed dispersal (some rely on wind, too). Seeds of many
fruits pass unharmed through the intestines of herbivores and germinate to
grow a new plant from the droppings of the animal dispersing agent. Colorful
flowers with nectar and brightly colored fruits can only be interpreted as having been evolved to attract appropriate animals. Here, as in plant-herbivore
interactions, a high degree of plant-animal specificity has arisen. Animals that
pollinate a particular plant are referred to as *pollinating vectors*. For example,

in Central America different species of male euglossine bees are highly specific to particular species of tiny epiphytic orchids; male bees travel long distances between orchids. Different bee species are attracted by different orchid fragrances (Dressler, 1968), as can be shown by putting out "baits" of artificially synthesized orchid "fragrances." (These male bees do not obtain nectar from the orchids they visit, but only obtain orchid products that the insects use for production of their own pheromones to attract females.) These bees are probably necessary for, and may have allowed the evolution of, the great diversity of tropical orchids, many of which are evidently quite rare and far apart. Such specificity of pollinating vectors assures that the plant's pollen is transmitted to the ovules of its own species. Whereas female euglossine bees are not as specific to the plant species they pollinate as males, individual females travel distances up to 23 kilometers (Janzen, 1971a) and regularly move long distances between sparsely distributed plants in gathering nectar and pollen; thus, they probably promote outcrossing among tropical plants at low densities. Indeed, Janzen suggests that such "traplining" by female bees may actually permit the very existence of plant species forced to very low densities by factors such as competition and predation on their seeds and seedlings.

Some pollinators, such as *Heliconius* butterflies (Gilbert, 1972), obtain amino acids from the pollen of plants they pollinate. Because production of nectar and pollen (and fruit) requires matter and energy, attracting animal pollinators (and seed dispersers) has its costs to the plant. Nectar and fruits are usually rich in sugars and other carbohydrates but contain relatively little protein; in contrast, pollen and seeds contain considerably greater amounts of nitrogen and other limiting materials. Due to the frequent scarcity of such vital nutrients, carbohydrates are presumably cheaper for a plant to produce than amino acids and proteins. Thus, pollen-eating pollinators presumably cost a plant considerably more than strict nectar feeders. Returns from visiting a flower (or eating a fruit) must be great enough to an animal pollinator or seed disperser to make it worthwhile, yet small enough that the animal will travel the distance necessary to disperse the pollen or seeds. This intricate energetic interplay between plants and their pollinators is reviewed by Heinrich and Raven (1972).

Obligate mutualisms are less common than facultative ones, probably because both populations depend completely on the relationship and neither can survive without the other. A very high degree of interdependency occurs between figs and the fig wasps that pollinate them (wasp eggs are laid inside fig fruits where larvae develop). There are hundreds of species of figs, each with its own species of wasp (this is a good example of tight, or species-specific, coevolution). Similarly, termites cannot themselves produce enzymes to digest the cellulose in wood, but by harboring in their intestines a population of protozoans that can make such enzymes, the insects are able to exploit wood successfully as a food source. Neither termite nor protozoan could survive without the other. These intestinal endosymbionts are passed on from one generation of termites to the next through exchange of intestinal contents (see p. 224). Large grazing mammals have a rumen in their gut system, an anaerobic chamber that houses endosymbiotic protozoans and bacteria, which similarly assist in digestion. Another putative example of mutualism is lichens, which are composed of a fungus and an alga; the fungus provides the supportive tissue, whereas the alga performs photosynthesis. (Algae of some lichens can be grown without the fungi.)

Commensalism occurs when one population is benefited but the other is unaffected $(+, 0)$. Small epiphytes such as bromeliads and orchids, which grow on the surfaces of large trees without obvious detriment to the tree, might be an example. A well-documented case of commensalism is the association between cattle egrets and cattle (Heatwole, 1965). These egrets follow cattle that are grazing in the sun and capture prey (crickets, grasshoppers, flies, beetles, lizards, frogs) that move as cattle approach. The number of cattle egrets associated with cattle is strongly dependent on the activities of the cattle; thus, Heatwole observed fewer egrets than expected on a random basis near resting cattle, but nearly twice as many egrets as expected (if the association were entirely random) accompanied cattle that were actively grazing in the sun. Since the birds seldom take prey (such as ticks and other ectoparasites) directly from the bodies of the cattle, the mammals probably benefit little from their relationship with egrets. Moreover, egret-feeding rates and feeding efficiency are markedly higher when these birds are associated with cattle (Table 11.2).

**TABLE 11.2    Various Aspects of the Association of Cattle Egrets with Cattle**

| Category | Number of cattle | Percent cattle | Number of associated egrets | |
|---|---|---|---|---|
| | | | Expected | Observed |
| Grazing in sun | 735 | 39.1 | 239 | 439 |
| Grazing in shade | 55 | 2.9 | 18 | 21 |
| Standing in sun | 146 | 7.8 | 48 | 46 |
| Standing in shade | 257 | 13.7 | 84 | 17 |
| Lying in sun | 503 | 26.8 | 164 | 69 |
| Lying in shade | 143 | 7.6 | 47 | 17 |
| Walking | 39 | 2.1 | 13 | 3 |
| Total | 1878 | 100.0 | 612 | |

| | Mean number per minute | Number of times count was higher than for opposite egret | Percent of times count was higher than for opposite egret |
|---|---|---|---|
| FEEDINGS, $N = 84$ | | | |
| Associated | 2.34 | 58 | 69 |
| Nonassociated | 1.71 | 26 | 31 |
| STEPS, $N = 62$ | | | |
| Associated | 20.1 | 7 | 11 |
| Nonassociated | 32.1 | 55 | 89 |
| FEEDINGS/STEP, $N = 59$ | | | |
| Associated | 0.129 | 52 | 88 |
| Nonassociated | 0.051 | 7 | 12 |

*Note:* Upper box shows the numbers of egrets associated with cattle engaged in different activities. Lower box shows feeding rates, steps taken per prey item (energy expended in foraging), and feeding efficiencies of egrets associated with and not associated with cattle.

*Source:* From Heatwole (1965).

Because of their plus-plus and symmetric nature, mutualisms exhibit positive feedback and hence can run away—for this reason, they are destabilizing unless the intraspecific negative self-damping is stronger than the interspecific positive mutualistic effects. Mutualistic relationships are easily modeled with equations similar to the Lotka-Volterra competition equations simply by changing the signs of the alphas ($K$'s have also been changed to $X$'s since they no longer represent maximal densities), as follows:

$$\frac{dN_1}{dt} = r_1 N_1 \left( \frac{X_1 - N_1 + \alpha_{12} N_2}{X_1} \right) \tag{1}$$

$$\frac{dN_2}{dt} = r_2 N_2 \left( \frac{X_2 - N_2 + \alpha_{21} N_1}{X_2} \right) \tag{2}$$

Equilibrium conditions are described by a pair of linear equations and are shown graphically in Figure 11.2. Populations reach equilibrium at density $X_1$ or $X_2$ in the absence of the other species, and each population's equilibrium density is increased by increasing the density of the other species. If both $X_1$ and $X_2$ are positive and if $\alpha_{12}$ and $\alpha_{21}$ are chosen so that isoclines cross, the joint equilibrium is stable. More realistic, but also more complex, cost-benefit models of mutualism are discussed by Roughgarden (1975), Vandermeer and Boucher (1978), Dean (1983), Wolin and Lawlor (1984), Wolin (1985), and Post et al. (1985).

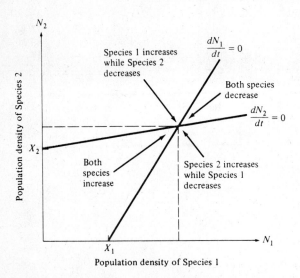

Figure 11.2 Isoclines for two species with a mutually beneficial interaction and a stable joint equilibrium (see text).

# INDIRECT INTERACTIONS

Superimposed on direct pairwise interactions, more subtle *indirect* interactions are mediated through other members of the community concerned. Darwin anticipated the concept of indirect interactions and gave as an example the interactions among cats, field mice, humblebees (bumblebees), and red clover. The bees pollinate clover, but field mice raid bee nests and eat bee larvae. Lots of clover grows around villages, presumably because cats keep mice populations down, allowing bumblebees to flourish, which in turn assists clover. Darwin's staunch defender Huxley carried Darwin's example further and noted that spinsters (who have lots of cats) facilitate Britain's naval prowess because strong sailors must be well fed and British beef thrives on clover. Here we have a long string with a pathlength of seven: spinsters → cats → mice → bees → clover → beef → sailors → naval prowess!

Five different sorts of indirect interactions involving three or four different species' populations are depicted in Figure 11.3. Pointed arrows indicate beneficial effects, whereas circle-headed "arrows" depict detrimental interactions.

Two consumers sharing a common prey may compete indirectly through classical exploitative competition (resource depression). Two prey species may appear to compete because, if either increases, a shared predator also increases, which operates to the detriment of the other prey population—Holt (1977) called this *apparent competition*. Three species' populations at three different trophic levels result in what has been termed a *food-chain mutualism* (such vertical interactions have also been called *cascading effects* or *cascades*).

Indirect Interactions

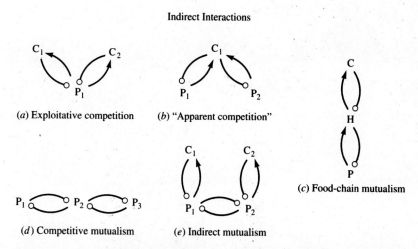

(*a*) Exploitative competition    (*b*) "Apparent competition"

(*c*) Food-chain mutualism

(*d*) Competitive mutualism    (*e*) Indirect mutualism

Figure 11.3  (*a*) Indirect competition through classical exploitative competition resulting from resource depression. (*b*) Apparent competition—Holt (1977). (*c*) Food-chain mutualism. (*d*) Competitive mutualism—Pianka (1981). (*e*) Facilitation, or indirect mutualism—Vandermeer et al. (1985). [Adapted from Pianka (1987), with thanks to J. H. Brown.]

The plant and carnivore are indirect mutualists because the plant generates herbivores, which constitute food for the carnivore. The carnivore reduces herbivory, which benefits the plants. An example was provided by Power et al. (1985): fish-eating bass feed upon herbivorous minnows in pools of an Oklahoma creek. When bass were removed (pools were fenced to keep these predators out) and minnow densities raised, the standing crop of algae diminished. With the readdition of bass, minnows retreated to shallow water and algal densities increased significantly over the next two weeks. Three species' populations at the same trophic level, arranged so that one species ($P_2$) is sandwiched between two others (referred to as *horizontal interactions*), can also result in indirect mutualism. Populations $P_1$ and $P_3$ are indirect mutualists because each inhibits the other's competitor $P_2$. Such a situation can also arise even when $P_1$ and $P_3$ are actually weak competitors, so long as competitive interactions with $P_2$ are strong (this has been called *competitive mutualism*—Pianka, 1981b). An alternate depiction of how indirect competitive mutualism arises is shown in Figure 11.4. A four-species system results in an indirect "mutualism" (perhaps best termed *facilitation*—Vandermeer et al., 1985). In this case, populations $C_1$ and $C_2$, which do not interact directly but consume different prey species, interact indirectly because their prey compete: if consumer $C_1$ increases, its prey

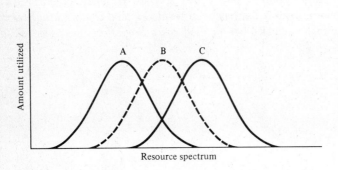

**Figure 11.4** Conditions that can lead to competitive mutualism between species. Species A and C overlap moderately in their utilization of resources, so that, in isolation, these two species are potentially weak competitors. However, both species overlap more extensively with a third species, Species B, and hence potentially experience intense competition with Species B. Since each species, A and C, exerts a stronger influence on Species B than they do on one another, when all three species occur together, each exerts strong competition on Species B, thus reducing the intensity of competition between Species B and the other. Resulting *indirect* effects between Species A and C, as mediated through Species B, are benefical (each species reduces the fitness of a strong competitor of the other species). Hence the net interaction between Species A and C changes qualitatively in the presence or absence of Species B. [From Pianka (1981).]

$P_1$ decreases, which in turn reduces the competition with $P_2$, hence allowing an increase in this second prey population ($P_2$) and providing more food for consumer $C_2$. Numerous other sorts of indirect interactions are also possible.

An indirect effect can be defined mathematically as the product of all the various direct effects along a directed series of links, or a *pathway*, in which no species node is passed through more than once (Lane, 1985). Such a path product represents the indirect effect between two nodes, which may also be connected by a direct effect. Typically, the longer the pathway by which an indirect effect is mediated, the longer is the time lag required for the effect to be transmitted from one node to another. Thus, indirect effects typically take longer to occur than direct effects. Positive indirect effects can arise both by means of mutualistic links and by means of products of an *even* number of negative links. If, however, there are an *odd* number of negative links in a pathway, the overall indirect effect is negative.

Indirect effects are usually weaker than direct effects. However, because there are many more indirect effects than direct ones in a given system, the former can assume paramount importance even though they are weak. Indirect effects may actually oppose the direct effects, and if their overall effects are intense enough, the overall net effect of one population on another, sometimes termed the *community effect*, can actually be reversed. Although this sort of double thinking seems circuitous and complex, it may prove to be vital to understanding community organization; direct interactions in opposition to indirect interactions would moderate each other, leaving a target species only weakly affected.

Indeed, an interaction between any given pair of populations depends vitally on the complex network of other interactions within which the pair concerned is embedded. Indirect effects render interpretation of simple experiments and observations extremely difficult.

## SELECTED REFERENCES

### Direct Interactions

Haskell (1947, 1949);  Krebs (1972);  Levins (1968);  MacArthur (1972);  MacArthur and Connell (1966);  MacArthur and Wilson (1967);  Odum (1959, 1971).

### Complex Population Interactions

Colwell (1973);  Seifert and Seifert (1976).

### Mutualistic Interactions and Symbiotic Relationships

Addicott (1985);  Allee (1951);  Allee et al. (1949);  Axelrod and Hamilton (1981); Boucher (1985);  Boucher et al. (1982);  Briand and Yodzis (1982);  Colwell (1973); Dean (1983);  DeVries (1991a, 1991b, 1992);  Dressler (1968);  Ehrman (1983); Gilbert (1971, 1972, 1979);  Heatwole (1965);  Heinrich and Raven (1972);  Janzen

(1966, 1967, 1971a, 1971b);   Margulis (1970, 1974, 1976);   May (1982);   Pierce (1985); Post et al. (1985);   Roughgarden (1975);   Seifert and Seifert (1976);   Vandermeer (1980);   Vandermeer and Boucher (1978);   Whittaker (1970);   Wolin (1985);   Wolin and Lawlor (1984).

## Indirect Interactions

Bender et al. (1984);   J. H. Brown and Ojeda (1987);   J. H. Brown et al. (1986); Darwin (1859);   Holt (1977);   Kerfoot and Sih (1987);   Lane (1985);   Lawlor (1979, 1980);   Levine (1976);   Patten (1983);   Pianka (1980, 1981b, 1987);   Power et al. (1985);   Strauss (1991);   Vandermeer (1980);   Vandermeer and Boucher (1978);   Vandermeer et al. (1985);   D. S. Wilson (1986).

## Chapter
## 12

# Competition

## MECHANISMS OF COMPETITION

Competition occurs indirectly when two or more organismic units use the same resources and when those resources are in short supply. Such a process involving resource depression or depletion has been labeled *consumptive* or *exploitative* competition. Competition may also occur through more direct interactions such as in interspecific territoriality or in the production of toxins, in which case it is termed *interference* competition. Competition for space, as in the rocky intertidal, often mediated by who gets there first, has been termed *preemptive* competition. Interaction between the two organismic units reduces the fitness or equilibrium population density of each. This can occur in several ways. By requiring that an organismic unit expend some of its time and/or matter and/or energy on competition, or the avoidance of competition, a competitor may effectively reduce the amounts left for maintenance and reproduction. By using up or occupying some of a scarce resource, competitors directly reduce the amount available to other organismic units. Mechanisms of interference competition, such as interspecific territoriality, will be favored by natural selection only when there is a potential for overlap in the use of limited resources at the outset (i.e., when exploitative competition is potentially possible).

*Intraspecific* competition, considered in Chapter 9, is competition between individuals belonging to the same species, and usually to the same population.

*Interspecific* competition occurs between individuals belonging to different species and is of chief interest in the present chapter. Because of its symmetry, it is always advantageous, *when possible*, for either party in a competitive relationship to avoid the interaction; competition has therefore been an important evolutionary force that has led to niche separation, specialization, and diversification. If, however, avoidance of a competitive interaction is impossible, natural selection may sometimes favor convergence.

Competition is not an on-off process; rather, its level presumably varies continuously as the ratio of demand over supply changes. Thus, there is little, if any, competition in an ecological vacuum, whereas competition is keen in a fully saturated environment. All degrees of intermediates exist.

## LOTKA-VOLTERRA COMPETITION EQUATIONS

Competition was placed on a fairly firm, albeit greatly oversimplified, theoretical basis over 50 years ago by Lotka (1925) and Volterra (1926a, 1926b, 1931). Their equations describing competition have strongly influenced the development of modern ecological theory and nicely illustrate a mathematical model of an important ecological phenomenon. They also helped to develop a number of very useful concepts, such as competition coefficients, the community matrix, and diffuse competition, that are conceptually independent of the equations.

The Lotka-Volterra competition equations are a modification of the Verhulst-Pearl logistic equation and share its assumptions. Consider two competing species, $N_1$ and $N_2$, with carrying capacities $K_1$ and $K_2$ in the absence of one another. Each species also has its own maximal instantaneous rate of increase per head, $r_1$ and $r_2$. The simultaneous growth of the two competing species occurring together is described by a pair of differential logistic equations:

$$\frac{dN_1}{dt} = r_1 N_1 \left( \frac{K_1 - N_1 - \alpha_{12} N_2}{K_1} \right) \tag{1}$$

$$\frac{dN_2}{dt} = r_2 N_2 \left( \frac{K_2 - N_2 - \alpha_{21} N_1}{K_2} \right) \tag{2}$$

where $\alpha_{12}$ and $\alpha_{21}$ are competition coefficients; $\alpha_{12}$ is a characteristic of Species 2 that measures its competitive inhibition (per individual) on the Species 1 population; and $\alpha_{21}$ is a similar characteristic of Species 1 that measures its inhibitory effects on Species 2. Competition coefficients are subscripted to show at a glance which population is affected and which is having the effect; thus, $\alpha_{12}$ measures the inhibitory effect of one $N_2$ individual on the growth of the $N_1$ population, whereas $\alpha_{21}$ represents the effect of one $N_1$ individual on the $N_2$ population. In the absence of any interspecific competition [$\alpha_{12}$ or $N_2$ equal zero in equation (1); $\alpha_{21}$ or $N_1$ equal zero in equation (2)], both populations grow sigmoidally according to the Verhulst-Pearl logistic equation and reach an equilibrium population density at their carrying capacity.

**TABLE 12.1    Summary of the Four Possible Cases of Competition Under the Lotka-Volterra Competition Equations**

| | Species 1 can contain Species 2 ($K_2/\alpha_{21} < K_1$) | Species 1 cannot contain Species 2 ($K_2/\alpha_{21} > K_1$) |
|---|---|---|
| Species 2 can contain Species 1 ($K_1/\alpha_{12} < K_2$) | Case 3: Either species can win | Case 2: Species 2 always wins |
| Species 2 cannot contain Species 1 ($K_1/\alpha_{12} > K_2$) | Case 1: Species 1 always wins | Case 4: Neither species can contain the other; stable coexistence |

By definition, the inhibitory effect of each individual in the $N_1$ population on its own population's growth is $1/K_1$ (see also Chapter 9); similarly, the inhibition of each $N_2$ individual on the $N_2$ population is $1/K_2$. Likewise, from inspection of equations (1) and (2), the inhibitory effect of each $N_2$ individual on the $N_1$ population is $\alpha_{12}/K_1$, and the inhibitory effect of each $N_1$ individual upon the $N_2$ population is $\alpha_{21}/K_2$. Competition coefficients are normally, though not always (see subsequent discussion), numbers less than 1. The outcome of competition depends on the relative values of $K_1$, $K_2$, $\alpha_{12}$, and $\alpha_{21}$. Four possible cases of competitive interaction correspond to different combinations of values for these constants (Table 12.1).

To see this, we ask at what density of $N_1$ individuals is the $N_2$ population held exactly at zero, and vice versa? In other words, what density of each species will always prevent the other from increasing? By inspection, note that at an $N_2$ of $K_1/\alpha_{12}$, $N_1$ can never increase and that when $N_1$ reaches $K_2/\alpha_{21}$, $N_2$ can never increase [substitute these values in equations (1) and (2)].

Therefore, in the absence of the other species, populations of both species increase at any density below their own carrying capacity and decrease at any value above it. In the presence of $K_1/\alpha_{12}$ individuals in the $N_2$ population, $N_1$ decreases at all densities; and in the presence of $K_2/\alpha_{21}$ individuals in the $N_1$ population, $N_2$ decreases at every density.

Recall that, under the Verhulst-Pearl logistic equation, $r_a$ decreases linearly with increasing $N$, reaching a value of zero at density $K$ (see Figure 9.2). Exactly the same relationships hold for the Lotka-Volterra competition equations, except that here a family of straight lines relate $r_1$ to $K_1$ and $r_2$ to $K_2$; each line corresponds to a different population density of the competing species (Figure 12.1$a$ and 12.1$b$).

The $r$ axis is omitted in Figure 12.2 and $N_1$ is simply plotted against $N_2$. Points in this $N_1 - N_2$ plane thus correspond to different proportions of the two species and to different population densities of each. Setting $dN_1/dt$ and $dN_2/dt$ equal to zero in equations (1) and (2), and solving, gives us equations for the boundary conditions between increase and decrease for each population:

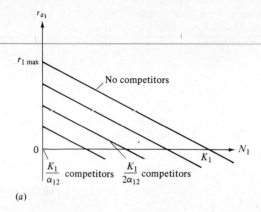

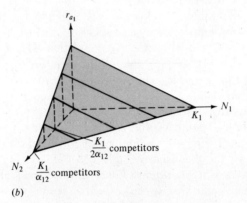

Figure 12.1. Two plots showing how the actual instantaneous per capita rate of increase $(r_{a_1})$ varies with the densities of a population $(N_1)$ and its competitor $(N_2)$ under the Lotka-Volterra competition equations. (a) Two-dimensional graph with four lines, each of which represents a given density of competitors (compare with Figure 9.2). (b) Three-dimensional graph with an $N_2$ axis showing the plane on which each of the four lines in (a) lie. At densities of $N_1$ and $N_2$ above $K_1$ and $K_1/\alpha_{12}$, respectively, the plane continues with $r_{a_1}$ becoming negative [compare with (a)].

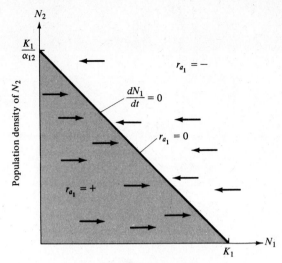

Figure 12.2. A plot identical to that of Figure 12.1*b*, but only the $N_1$-$N_2$ plane is shown at $r_{a_1}$ equal to zero. The line represents all equilibrium conditions $(dN_1/dt = 0)$ at which the $N_1$ population just maintains itself $(r_{a_1} = 0)$. In the shaded area below the line, $r_{a_1}$ is positive and the $N_1$ population increases (arrows); above the isocline, $r_{a_1}$ is always negative and the $N_1$ population decreases (arrows). An exactly equivalent plot could be drawn on the same axes for the competitor, $N_2$, except that the intercepts of the $N_2$ isocline $(dN_2/dt = 0)$ are $K_2$ and $K_2/\alpha_{21}$ and arrows parallel the $N_2$ axis rather than the $N_1$ axis. See also Figure 12.3.

$$\frac{K_1 - N_1 - \alpha_{12}N_2}{K_1} = 0 \quad \text{or} \quad N_1 = K_1 - \alpha_{12}N_2 \tag{3}$$

$$\frac{K_2 - N_2 - \alpha_{21}N_1}{K_2} = 0 \quad \text{or} \quad N_2 = K_2 - \alpha_{21}N_1 \tag{4}$$

These two linear equations are plotted in Figure 12.3, giving $dN/dt$ isoclines for each species; below these isoclines populations of each species increase, above them they decrease. Thus, the lines represent equilibrium population densities or saturation values; neither species can increase when the combined densities of the two lie above its isocline.

The four cases of competitive interaction are summarized in Table 12.1 and shown graphically in Figure 12.3. Only one combination of values (Case 4) leads to a stable equilibrium of the two species; this case arises when neither species is able to reach densities high enough to eliminate the other—that is, when both $K_1 < K_2/\alpha_{21}$ and $K_2 < K_1/\alpha_{12}$. Implicit in these inequalities are

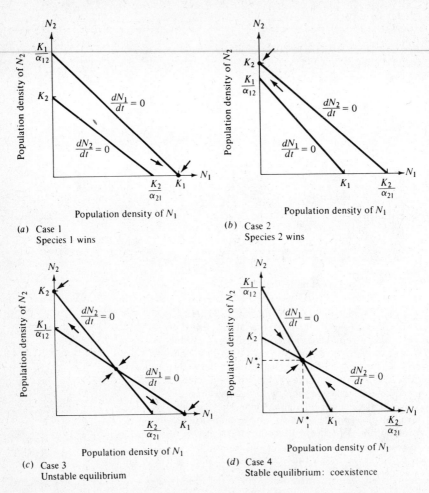

(a) Case 1
Species 1 wins

(b) Case 2
Species 2 wins

(c) Case 3
Unstable equilibrium

(d) Case 4
Stable equilibrium: coexistence

**Figure 12.3.** Plots like that of Figure 12.2, but with isoclines of both species superimposed upon one another. (a) Case 1: The $N_1$ isocline lies above the $N_2$ isocline and Species 1 always wins in competition. The only stable equilibrium (dot) is at $N_1 = K_1$ and $N_2 = 0$. (b) Case 2: The reverse, in which Species 2 is the superior competitor and always excludes Species 1. Here the only stable equilibrium (dot) is at $N_2 = K_2$ and $N_1 = 0$. (c) Case 3: Each species is able to contain the other (Table 12.1); that is, each inhibits the other population's growth more than its own. Three possible equilibria exist (dots), but the joint equilibrium of both species (where the two isoclines cross) is unstable. Alternate stable equilibria are $N_2 = K_2$ and $N_1 = 0$ or $N_1 = K_1$ and $N_2 = 0$. Depending on initial proportions of the two species, either can win. (d) Case 4: Neither species can contain the other, but both inhibit their own population growth more than that of the other species. Only one equilibrium exists at $N_1^*$ and $N_2^*$; both species thus coexist at densities below their respective carrying capacities.

the conditions necessary for coexistence. Each population must inhibit its own growth more than that of the other species. For this to occur, if $K_1$ equals $K_2$, $\alpha_{12}$ and $\alpha_{21}$ must be numbers less than 1. When carrying capacities are not equal, $\alpha_{12}$ or $\alpha_{21}$ can take on a value greater than 1 and coexistence may still be possible as long as the product of the two competition coefficients is less than 1 ($\alpha_{12}\alpha_{21} < 1$) and the ratio of $K_1/K_2$ falls between $\alpha_{12}$ and $1/\alpha_{21}$ (see MacArthur, 1972, p. 35). Population sizes at the joint equilibrium ($N_1^*$ and $N_2^*$) of the two species in Figure 12.3d are below their respective carrying capacities $K_1$ and $K_2$; thus, in competition, neither population reaches densities as high as it does without competition. None of the other three cases lead to stable coexistence of both populations and hence they are of less interest. However, in Case 3 an unstable equilibrium exists, with each species inhibiting the *other's* growth rate more than its own; in this case, the outcome of competition depends entirely on the initial proportions of the two species.

It is sometimes useful to rearrange equations (1) and (2) by multiplying through the bracketed term by $r_1N_1$ or $r_2N_2$, respectively:

$$\frac{dN_1}{dt} = r_1N_1 - r_1\frac{N_1^2}{K_1} - r_1N_1\alpha_{12}\frac{N_2}{K_1} \tag{5}$$

$$\frac{dN_2}{dt} = r_2N_2 - r_2\frac{N_2^2}{K_2} - r_2N_2\alpha_{21}\frac{N_1}{K_2} \tag{6}$$

These equations can be written in simpler form by consolidating constants:

$$\frac{dN_1}{dt} = r_1N_1 - z_1N_1^2 - \beta_{12}N_1N_2 \tag{7}$$

$$\frac{dN_2}{dt} = r_2N_2 - z_2N_2^2 - \beta_{21}N_1N_2 \tag{8}$$

where $z_1$ and $z_2$ are equal to $r_1/K_1$ and $r_2/K_2$, respectively; similarly, $\beta_{12}$ is $z_1\alpha_{12}$ and $\beta_{21}$ is $z_2\alpha_{21}$. In equations (7) and (8), the first term to the right of the equals sign is the density-independent rate of population increase, whereas the second and third terms measure intraspecific self-damping and interspecific competitive inhibition of this rate of increase, respectively.

The Lotka-Volterra equations can also be written in a more general form for a community composed of $n$ different species:

$$\frac{dN_i}{dt} = r_iN_i\left(\frac{K_i - N_i - \sum_{j \neq i}^{n}\alpha_{ij}N_j}{K_i}\right) \tag{9}$$

where the $i$'s and $j$'s subscript species and range from 1 to $n$. (Note that all $\alpha_{ii}$'s are 1.) At steady state, $dN_i/dt$ must equal zero for all $i$, and equilibrium population densities are given by an equation analogous to equations (3) and (4):

$$N_i^* = K_i - \sum_{j \neq i}^{n} \alpha_{ij} N_j \qquad (10)$$

Notice that the more competitors a given species has, the larger the term $(\sum \alpha_{ij} N_j)$ becomes, and the further that species' equilibrium population size is from its $K$ value; this accords well with biological intuition. The total competitive effect of the remainder of the community on a particular population is known as *diffuse competition* (see also pp. 277–278).

Implicit in the Lotka-Volterra competition equations are a number of assumptions; some can be relaxed, although the mathematics rapidly become unmanageable. Maximal rates of increase, competition coefficients, and carrying capacities are all assumed to be *constant* and immutable; they do not vary with population densities, community composition, or anything else. As a result, all inhibitory relationships within and between populations are strictly linear, and every $N_1$ individual is identical, as is every $N_2$ individual. However, similar results can be reached without assuming linearity even by a purely graphical argument (Figure 12.4). Response to changes in density is instantaneous. In addition, the two species are not allowed to diverge; thus, the environment is assumed to be completely unstructured or homogeneous. Competition in a patchy environment has also been modeled mathematically (Skellam, 1951; Levins and Culver, 1971; Horn and MacArthur, 1972; Slatkin, 1974; Levin, 1974).

In real populations, rates of increase, competitive abilities, and carrying capacities *do* vary from individual to individual, with population density, com-

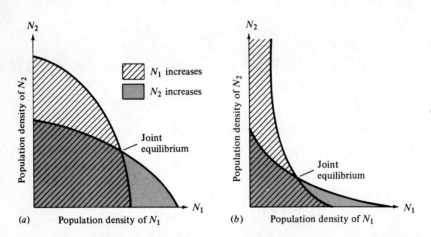

**Figure 12.4.** Nonlinear isoclines representing graphically the conditions for stable coexistence of two competitors. Concave upward isoclines have been observed in bottle experiments with *Drosophila* flies (Ayala, Gilpin, and Ehrenfeld, 1973) and have also emerged from some innovative models of competition such as those of Schoener (1973, 1977b).

munity composition, and in space and time. Indeed, temporal variation in the environment may often allow coexistence by continually altering the competitive abilities of populations inhabiting it. Time lags are doubtless of some importance in real populations. Finally, a heterogeneous environment may allow real competitors to evolve divergent resource utilization patterns and to reduce interspecific competitive inhibition.

Almost all theory built on the Lotka-Volterra competition equations deals only with equilibrium conditions. But real ecological systems (and portions thereof) may often be partially unsaturated, which in itself could allow coexistence of otherwise competitively intolerant populations. For instance, either predation or density-independent rarefaction might hold down population levels and thereby decrease competition. However, increasing rarefaction in the Lotka-Volterra equations does not reduce the intensity of competitive inhibition, except by reduction of population densities. A more realistic model might incorporate the ratio of instantaneous demand to supply; one such possibility might be to make competition coefficients variables and functions of the combined densities of both populations. In such a nonlinear system, at saturation values ($K_1$, $K_2$, or $N_1^*$ plus $N_2^*$ in Figure 12.3$d$), demand/supply is unity and the $\alpha$'s would take on maximal values, but as a perfect competitive vacuum is approached, these same $\alpha$'s would become vanishingly small.

The Lotka-Volterra competition equations play an extremely prominent role in modern ecological theory (Levins, 1968; MacArthur, 1968, 1972; Vandermeer, 1970, 1972; May, 1976a); however, their numerous biologically unrealistic assumptions underscore the inadequacy of existing competition theory. Although these equations have been overworked and overextended by enthusiastic theorists, they have nevertheless contributed substantially to the development of many important ecological concepts, such as the actual rate of increase, $r$ and $K$ selection, intraspecific and interspecific competitive abilities, diffuse competition, competitive communities, and the community matrix, all of which in fact are *independent* of the equations. Thus, these equations help to provide a useful conceptual framework.

## COMPETITIVE EXCLUSION

How does one population drive another to extinction? In Cases 1, 2, and 3 of Figure 12.3, one species ultimately eliminates the other entirely when the two come into competition and the system is allowed to go to saturation; we say then that competitive exclusion has occurred. Consider an ecological vacuum inoculated with small numbers of each species. At first, both populations grow nearly exponentially at rates determined by their respective instantaneous maximal rates of increase. As the ecological vacuum is filled, the actual rates of increase become progressively smaller and smaller. Both populations are infinitely unlikely to have exactly the same rates of increase, competitive abilities, and carrying capacities. Hence, as the ecological vacuum is filled, a time must come when one population's actual rate of increase drops to zero, while the other's rate of increase is still positive. This situation represents a turning point in competition, for the second population now increases still further and its competitive inhibition of the first is further intensified, reducing the actual rate of increase of the first population to a negative value. The first population is now declining, while the second is still increasing; barring changes in competitive parameters, competitive exclusion (extinction of the first population) is only a matter of time. This process has been demonstrated experimentally (Figure 12.5).

Some rather strong statements concerning competitive exclusion have been made. Among them is a hypothesis called the competitive exclusion "principle"; two species with identical ecologies cannot live together in the same place at the same time. Ultimately, one must edge out the other; complete ecological overlap is impossible. The corollary is that if two species coexist, there must be ecological differences between them. Since any two organismic units are infinitely unlikely to be *exactly* identical, mere observation of ecological differences between species does not constitute "verification" of the hypothe-

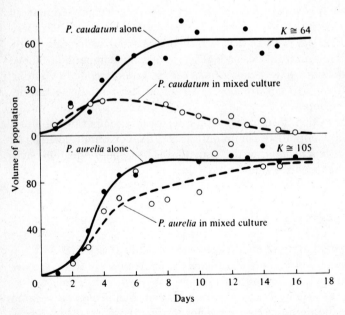

**Figure 12.5.** Competitive exclusion in a laboratory experiment with two protozoans, *Paramecium caudatum* and *P. aurelia*. [From Gause (1934), reprint ed. Copyright © 1964 by The Williams and Wilkins Co., Baltimore, Md.] Other experiments with *P. bursaria* and *P. caudatum* resulted in coexistence of both species at densities below their carrying capacities.

sis. Untestable hypotheses like this one are of little scientific utility and are gradually forgotten by the scientific community.

However, the competitive exclusion "principle" has served a useful purpose by emphasizing that some ecological difference may be necessary for the coexistence of competitive communities in *saturated* environments. Ecologists have gone on to ask more penetrating and dynamic questions: How much ecological overlap can two species tolerate and still coexist? How does this maximal tolerable overlap vary as the ratio of demand to supply changes? How high must migration rates be for competitively inferior fugitive species to persist in spatially and temporally varying patchy habitats? Can temporally changing competitive abilities lead to coexistence?

## BALANCE BETWEEN INTRASPECIFIC AND INTERSPECIFIC COMPETITION

Intraspecific and interspecific competition probably often have opposite effects on a population's tolerance, as well as on its use of resources and its phenotypic variability. To see this, consider an idealized case with no interspecific competition and look at the effects of intrapopulational competition on the population's use of a resource or habitat. We will consider the former a resource continuum and the latter a habitat gradient, although an exactly parallel argument is easily developed for discrete resources and habitats (Figure 12.6). Individuals should spread themselves out more or less evenly along such a continuum or gradient. If all resources along the continuum are not being utilized to approximately the same extent, those individuals using relatively unused portions should encounter less intense intraspecific competition and, as a result, presumably will often have higher individual fitnesses. Hence, we would predict that, in using such a continuous resource in its entirety, individuals should behave so as to equalize the ratio of demand to supply along the continuum, which in turn equalizes the level of intraspecific competition. [MacArthur (1972) calls this the "principle of equal opportunity."]

Using the same argument, consider now the manner in which an *expanding* population makes use of a resource continuum or a habitat gradient (Figure 12.7). (Again, the argument applies equally well to resource categories that are not continuous but discrete; see Figure 12.6.) The first individuals will no doubt select those resources and/or habitats that are optimal in the absence of competition. However, as the density of individuals increases, competition among them reduces benefits to be gained from these optimal resources and/or habitats and favors deviant individuals that use less "optimal," but also less hotly contested, resources and/or habitats. By these means, intraspecific competition can often act to increase the variety of resources and habitats utilized by a population.

Interspecific competition, on the other hand, generally tends to *restrict* the range of habitats and resources a population uses, because different species

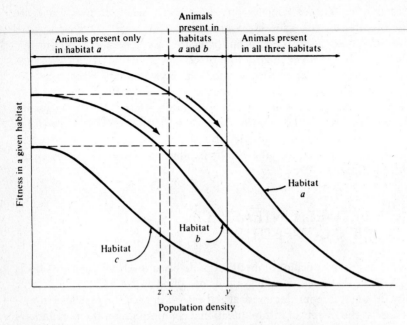

**Figure 12.6.** A graphical model of density-dependent habitat selection. Fitnesses of individuals plotted against population density in three hypothetical habitats, *a*, *b*, and *c*. Fitness decreases as habitats are filled in all three habitats, but at any given density, fitness is highest in habitat *a*, intermediate in habitat *b*, and lowest in habitat *c*. Habitat *a* is preferred until densities reach level *x*, at which point it pays to invade habitat *b* at low densities. Habitat *c*, however, remains vacant. As habitats *a* and *b* continue to fill (arrows), it eventually pays to colonize the least desirable habitat *c* (when densities in *a* and *b* reach densities of *y* and *z*, respectively). For fitness to remain constant as all three habitats are filled, population density must always be lowest in habitat *c*, intermediate in habitat *b*, and highest in the best habitat, *a*. [Modified after Fretwell and Lucas (1969).]

normally have differing abilities at exploiting habitat types and harvesting resources. Individuals using marginal habitats presumably cannot compete as effectively against members of another population as can individuals exploiting more "optimal" habitats. Thus, in most communities, any given population is "boxed in" by other populations that are superior at exploiting adjacent habitats. Indeed, because these two forces oppose each other, one could even hypothesize that, at equilibrium, total intraspecific competition should be exactly balanced by total interspecific competition. In fact, this assertion is not quite correct because inherent genetic and physiological limitations must also restrict the range of habitats and resources used by an organism.

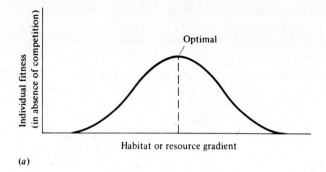

(a)

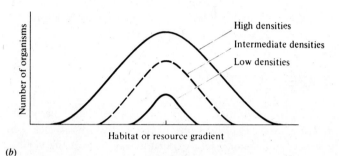

(b)

**Figure 12.7.** Diagrammatic representation of how an expanding population might use a gradient of habitats or resources. (*a*) Individual fitness as a function of the habitat or resource gradient in the absence of any intraspecific competition. (*b*) At low densities, most individuals select near-optimal environmental conditions, but as density increases, intraspecific competition for more optimal habitats (or resources) increases, favoring individuals that exploit less optimal, but also less hotly contested, habitats or resources. Variety of habitats or resources exploited thus increases with increasing population density.

## EVOLUTIONARY CONSEQUENCES OF COMPETITION

Many of the long-term consequences of competition have been mentioned in Chapters 1 and 5; others are treated in Chapters 14 and 15. For instance, natural selection in saturated environments (*K* selection) favors competitive ability. A raft of presumed populational products of intraspecific competition over evolutionary time include rectangular survivorship, delayed reproduction, decreased clutch size, increased size of offspring, parental care, mating systems, dispersed spacing systems, and territoriality. Perhaps the most far-reaching evolutionary effect of interspecific competition is *ecological diversification*, also termed *niche separation*. This, in turn, has made possible, and has led to, the development of complex biological communities. Another presumed result of both intraspecific and interspecific competition is increased efficiency of utilization of resources in short supply.

Although the concept of competition is a central theme in much of modern ecological theory, it has proven to be surprisingly difficult to study in the field and is still poorly understood as an actual phenomenon. Some ecologists consider competition among the most important of ecological generalizations, yet others maintain that it is of little utility in understanding nature. At least three possible reasons, not necessarily mutually exclusive, for this difference in viewpoint have been suggested: (1) competition in nature is often elusive and very difficult to study and to quantify; (2) ecologists with an evolutionary approach might often consider competition more important than workers concerned with explaining more immediate events (Orians, 1962);* and (3) there may be a natural dichotomy of relatively $r$-selected and relatively $K$-selected organisms, especially in terrestrial communities (Pianka, 1970).

## LABORATORY EXPERIMENTS

Competition is often fairly easily studied by direct experiment, and many such studies have been made. Gause (1934) was one of the first to investigate competition in the laboratory; his classic early experiments on protozoa verified competitive exclusion. He grew cultures of two species of *Paramecium* in isolation and in mixed cultures, under carefully controlled conditions and nearly constant food supply (Figure 12.5). "Carrying capacities" and respective rates of population growth of each species when grown separately and in competition were then calculated (competition coefficients are also readily computed from such data). Interestingly enough, the protozoan with the highest maximal instantaneous rate of increase per individual (*P. caudatum*) was the inferior competitor, as expected from considerations of $r$ and $K$ selection.

An effect of environment on the outcome of competition was demonstrated by Park (1948, 1954, 1962) and his many colleagues. Working with two species of flour beetles (*Tribolium*), these investigators showed that, depending on conditions of temperature and humidity, either species could eliminate the other (Table 12.2). The outcome of competition between two beetle species can be reversed by a protozoan parasite (Park, 1948). In early experiments, the outcome of competition under particular environmental conditions could not always be predicted in a set of environments termed the "indeterminate zone." More recently, however, this zone has been substantially reduced by taking into account the reproductive values and genotypes of beetles (Lerner and Ho, 1961; Park et al., 1964). Under some environmental conditions, cultures that had begun with a numerical preponderance of a given species always resulted in the extermination of the other species (Figure 12.8), thereby

---

*For a discussion of the difference between the "proximate" and the "ultimate" approaches to biological phenomena, see p. 11.

**TABLE 12.2 Outcome of Competition Between Two Species of Flour Bettles, *Tribolium confusum* and *T. castaneum***

Many Replicates of Laboratory Experiments at Different Temperatures and Humidities

| Temperature (°C) | Relative humidity (%) | Climate | Single species numbers | Mixed species (% wins) confusum | Mixed species (% wins) castaneum |
|---|---|---|---|---|---|
| 34 | 70 | Hot-moist | confusum = castaneum | 0 | 100 |
| 34 | 30 | Hot-dry | confusum > castaneum | 90 | 10 |
| 29 | 70 | Warm-moist | confusum < castaneum | 14 | 86 |
| 29 | 30 | Warm-dry | confusum > castaneum | 87 | 13 |
| 24 | 70 | Cold-moist | confusum < castaneum | 71 | 29 |
| 24 | 30 | Cold-dry | confusum > castaneum | 100 | 0 |

*Source:* From Krebs (1972) after Park (1954).

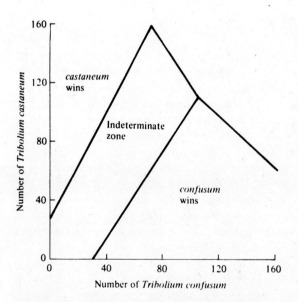

Figure 12.8. Outcome of competition between laboratory strains of two beetles, *Tribolium confusum* and *T. castaneum*, is a function of initial densities of the species; an initial preponderance of either species increases the likelihood of its winning. The "indeterminate zone" represents initial conditions under which either species can win. Note the similarity with Case 3 in Figure 12.3c. [From Krebs (1972) after Neyman, Park, and Scott. Originally published by the University of California Press; reprinted by permission of The Regents of the University of California.]

constituting an empirical verification of Case 3 of the Lotka-Volterra equations (Figure 12.3c).

In a laboratory study of competitive interactions among ciliate protozoans, Vandermeer (1969) cultured each of four species separately and in all possible pairs. These experiments allowed estimation of $r$'s, $K$'s, and $\alpha$'s. Observed pairwise interactive effects were similar to those actually observed when all four species were grown together in mixed culture—an indication that higher-order interactions among these species were slight. However, several similar studies (Hairston et al., 1968; Wilbur, 1972; Neill, 1974) suggest strong interactive effects among species, with the competitive effects between any two species depending strongly on the presence or absence of a third.

Competition and competitive exclusion have now been demonstrated in laboratory experiments on a wide variety of plants and animals. Potential flaws in many of these investigations are that, for practical reasons, they are carried out in constant and simple environments, almost invariably on small, often relatively $r$-selected, organisms that may not encounter high levels of competition regularly under natural circumstances.

## EVIDENCE FROM NATURE

Competition is notoriously difficult to demonstrate in natural communities, but a variety of observations and studies suggest that it does indeed occur regularly in nature and that it has been important in molding the ecologies of many species of plants and animals. Even if competition did not occur on a day-to-day basis, it could nevertheless still be a significant force; active avoidance of interspecific competition in itself implies that competition has occurred sometime in the past and that the species concerned have adapted to one another's presence. Also it might be difficult to find competition actually occurring in nature because inefficient competitors should be eliminated by competitive exclusion and therefore might not normally be observable. We might not expect to find abundant evidence of competition in small, short-lived organisms, such as insects and annual plants, but would look for it in larger, longer-lived organisms, such as vertebrates and perennial plants.

Ecologists have several different sorts of evidence, much of which is circumstantial, suggesting that competition either has occurred or is occurring in natural populations. These include: (1) studies on the ecologies of closely related species living in the same area; (2) character displacement; (3) studies on "incomplete" floras and faunas and associated changes in niches, or *niche shifts*; and (4) taxonomic composition of communities.

Closely related species, especially those in the same genus, or "congeneric" species, are often quite similar morphologically, physiologically, behaviorally, and ecologically. As a result, competition is intense between pairs of such species that live in the same area, known as *sympatric congeners*; selection may be strong to render their ecologies more different or to lead to ecological separation. Many groups of closely related sympatric species have been studied, and almost without exception, detailed investigations on relatively *K*-selected organisms have revealed subtle but important ecological differences between such species. Usually, the differences are of one or more of three basic types: (1) the species exploit different habitats or microhabitats (differential spatial utilization of the environment); (2) they eat different foods; or (3) they are active at different times (differential patterns of temporal activity). Such ecological differences are known as *niche dimensions* because they are important in defining a species' role in its community and its interactions with other species.

Many examples of clear-cut habitat and microhabitat differences could be cited. MacArthur (1958) studied spatial utilization patterns in five species of sympatric warblers (genus *Dendroica*) by noting the time spent in precise locations by foraging individuals of each species. Each species has its own unique pattern of exploiting the forest (Figure 12.9).

Overdispersion in general and territoriality in particular are indicative of competition in that they reduce its intensity. Overdispersion is widespread in plants, as is territoriality in vertebrate populations. Indeed, interspecific territoriality has been documented in many species of birds (Orians and Willson, 1964) and probably occurs in other taxa as well; thus, both intraspecific and interspecific competition have led to territorial behavior.

Differences in time of activity among ecologically similar animals can effectively reduce competition, *provided that resources differ at different times.* This is true in situations where resources are rapidly renewed, because the resources available at any one instant are relatively unaffected by what has happened at previous times. Perhaps the most obvious type of temporal separation is that between day and night; animals active during the daytime are *diurnal*, those active at night are *nocturnal*. Examples of pairs apparently separated by such temporal differences are hawks and owls, swallows and bats, or grasshoppers and crickets. Patterns of activity within the course of the day alone also differ, with some species being active early in the morning, others at midday, and so on. Seasonal separation of activity also occurs among some animals, such as certain lizards. In many animals, daily time of activity changes seasonally (Figure 12.10).

Dietary separation among closely related animal species has been shown repeatedly. For example, Table 12.3 on p. 260 shows that several sympatric congeneric species of the marine snail genus *Conus* (commonly called "cone shells") eat distinctly different foods (Kohn, 1959). Similarly, three species of stoneflies eat prey of different sizes (Figure 12.11, p. 260). Among desert lizards, diets of several sympatric species are composed predominantly of ants, termites, other lizards, and plants (Pianka, 1966b, 1986a). Similar cases of food differences among related sympatric species are known in many birds and mammals.

Simultaneous differences in the use of space, time, and food have also been documented for some sympatric congeneric species. In lizards of the genus *Ctenotus* (Pianka, 1969), for instance, seven sympatric species forage at different times, in different microhabitats, and on different foods. Frequently,

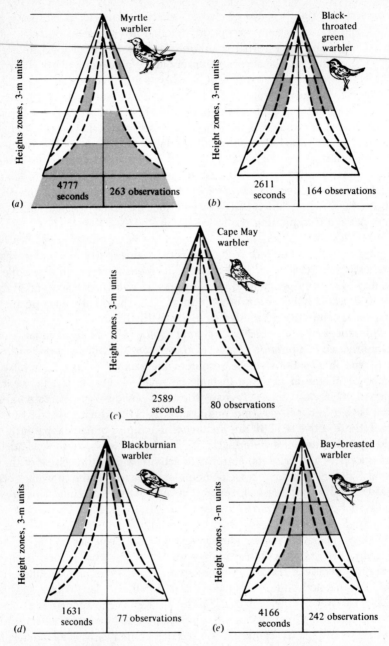

**Figure 12.9.** Differential use of parts of trees in a coniferous forest by five sympatric species of congeneric warblers (genus *Dendroica*). Shading indicates parts of trees in which foraging activities of each species are most concentrated. The right side of each schematic tree represents use based on the total number of birds observed (sample size given below each "tree"); the left side represents use based on total number of seconds of observation (again, sample size is given at the bottom of each "tree"). [After MacArthur (1958). By permission of Duke University Press.]

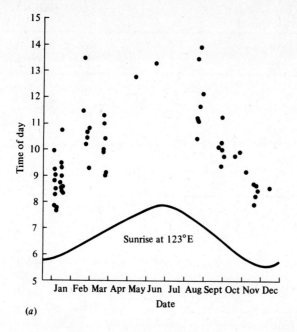

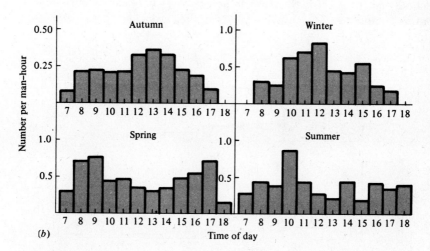

**Figure 12.10.** Seasonal changes in daily activity patterns in two species of Australian desert lizards. (*a*) A small blue-tailed skink, *Ctenotus calurus*; each dot represents one lizard. [From Pianka (1969). By permission of Duke University Press.] (*b*) The "military dragon," an agamid, *Ctenophorus isolepis*; numbers observed per man-hour of observation at different times of day in each season. [After Pianka (1971b).]

TABLE 12.3 **Major Foods (Percentages) of Eight Species of Cone Shells, *Conus*, on Subtidal Reefs in Hawaii**

| Species | Gastro-pods | Entero-pneusts | Nereids | Eunicea | Tere-bellids | Other polychaetes |
|---|---|---|---|---|---|---|
| *flavidus* | | 4 | | | 64 | 32 |
| *lividus* | | 61 | | 12 | 14 | 13 |
| *pennaceus* | 100 | | | | | |
| *abbreviatus* | | | | 100 | | |
| *ebraeus* | | | 15 | 82 | | 3 |
| *sponsalis* | | | 46 | 50 | | 4 |
| *rattus* | | | 23 | 77 | | |
| *imperialis* | | | | 27 | | 73 |

*Source:* From data of Kohn (1959). By permission of Duke University Press.

in such cases, pairs of species with high overlap in one niche dimension have low overlap along another, presumably reducing competition between them.

The phenomenon of *character displacement*, which refers to increased differences between species where they occur together, is also evidence that competition occurs in nature. Sometimes two widely ranging species are ecologically more similar in the parts of their ranges where each occurs alone without its competitor (i.e., an *allopatry*) than they are where both occur together

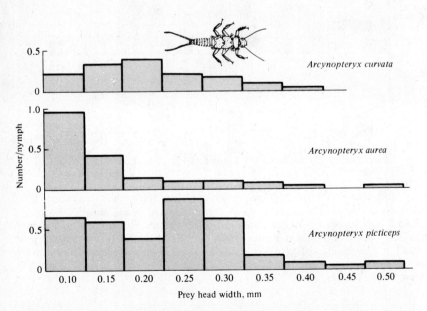

Figure 12.11. Size distributions of dipteran prey among three sympatric species of stoneflies in the genus *Arcynopteryx*. [From Sheldon (1972).]

(in *sympatry*). This sort of ecological divergence can take the form of morphological, behavioral, and/or physiological differences. One way in which character displacement occurs is in the size of the food-gathering or "trophic" apparatus, such as mouthparts, beaks, or jaws. Prey size is usually strongly correlated with the size of an animal's beak or jaw as well as with its structure (Figure 12.12). Although the evidence is circumstantial (see Grant, 1972, for a review), character displacement in either body size or the size of the trophic apparatus is thought to have occurred in some lizards, snails (Figure 12.13), birds, mammals, and insects, presumably separating food niches. Such niche shifts in the presence of a potential competitor suggest that each population has adapted to the other by evolving a means to reduce interspecific competition.

Morphological character displacement in the size of mouthparts need not evolve if the populations concerned have diverged in other ways; hence, it is expected only in situations where both competitors occur side by side, exploiting identical microhabitats (i.e., true *syntopy*). Animals that forage in different microhabitats, such as the warblers of Figure 12.9, have adapted to one another primarily by means of behavioral, rather than morphological, character displacement.

Size differences between closely related sympatric species have been implicated as being necessary for coexistence (Hutchinson, 1959; MacArthur, 1972; Schoener, 1965, 1983), and even in the "assembly" of communities (Case

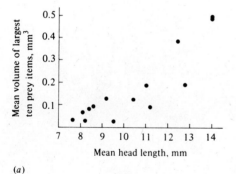

(a)

(b)

Figure 12.12. Two plots of prey size versus predator size. (a) Plot of the average volume of the ten largest prey items (in cubic millimeters) against mean head length in 14 species of lizards (genus *Ctenotus*). [After Pianka (1969). By permission of Duke University Press.] (b) Average prey weight plotted against mean body weight among 13 species of hawks (log-log plot). [After Schoener (1968). By permission of Duke University Press.]

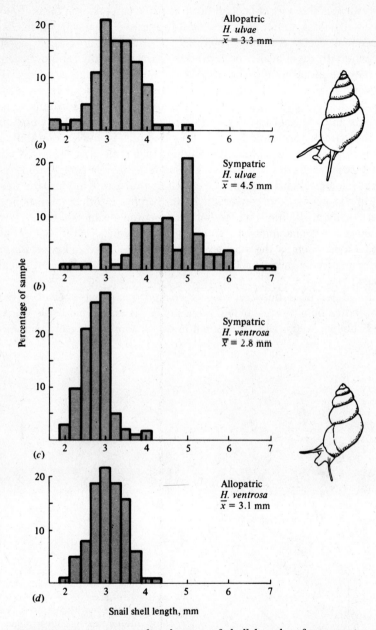

**Figure 12.13.** Frequency distributions of shell lengths of two species of mud snails in Denmark. The two species, *Hydrobia ulvae* and *H. ventrosa*, are similar in size where each occurs alone in allopatry [(*a*) and (*d*)]; however, their shell lengths are markedly divergent in the area of sympatry [(*b*) and (*c*)]. [After Fenchel (1975).]

et al, 1983), although there has been considerable dispute over the statistical validity of these patterns (Grant, 1972; Horn and May, 1977; Grant and Abbott, 1980). There may be a definite limit on how similar two competitors can be and still avoid competitive exclusion; character displacement in average mouthpart sizes is often about 1.3, and the length ratio of 1.3 has been suggested as a crude estimate of just how different two species must be to coexist syntopically. (For body mass, a ratio of 2 corresponds to the length ratio of 1.3.)

Perhaps the most thorough analysis of such "Hutchinsonian" ratios is that provided for the world's bird-eating hawks by Schoener (1984), who computed size ratios among all possible pairs and triplets of the 47 species of accipiter hawks. Frequency distributions of expected size ratios were generated for all possible combinations of species, which were then compared with the much smaller number of existing accipiter assemblages. Schoener found a distinct paucity of low size ratios among real assemblages, strongly suggesting size assortment.

**G. E. Hutchinson**

Clearly, the preceding argument applies only in competitive communities, and ecological overlap, or a greater similarity between syntopic species, can presumably be greater in unsaturated habitats where competition is reduced.

Another type of evidence for competition comes from studies on so-called "incomplete" biotas, such as islands, where all of the usual species are not present (see also Chapter 2). Those species that invade such areas often expand their niches and exploit new habitats and resources that are normally exploited by other species on areas with more complete faunas. On the island of Bermuda, for example, considerably fewer species of land birds occur than on the mainland, with the three most abundant being the cardinal, catbird, and white-eyed vireo. Crowell (1962) found that, compared with the mainland, these three species are much more abundant on Bermuda and that they occur in a wider range of habitats. In addition, all three have somewhat different feeding habits on the island, and one species at least (the vireo) employs a greater variety of foraging techniques.

Mountaintops represent islands on the terrestrial landscape just as surely as Bermuda is an island in the ocean, and they often show similar phenomena. For example, two congeneric species of salamanders, *Plethodon jordani* and *P. glutinosus*, occur in sympatry on mountains in the eastern United States. In sympatry, the two species are altitudinally separated, with *glutinosus* occurring at lower elevations than *jordani*; vertical overlap between the two species never exceeded 70 meters (Hairston, 1951). On mountaintops where *jordani* occurs, *glutinosus* is restricted to lower elevations, whereas on adjacent mountains that lack *jordani*, *glutinosus* is found at higher elevations, often right up to the peak.

Niche expansion under reduced interspecific competition has been termed *ecological release*. Further evidence of competition stems from a corollary of ecological release; when mainland forms are introduced on to islands, native species are frequently driven to extinction, presumably through competitive exclusion. Thus, many birds that once occurred only on Hawaii became extinct shortly after the introduction of mainland birds such as the English sparrow and the starling (Moulton and Pimm, 1986). Similar extinctions have apparently occurred in the Australian marsupial fauna (e.g., the Tasmanian wolf, but also many species of mid-sized marsupials) with the introduction of placental mammal species (e.g., the dingo dog and the European fox). Of course, fossil history is replete with cases of natural invasions and subsequent extinctions. The simplest and most plausible explanation for many of these observations is that surviving species were superior competitors and that niche overlap was too great for coexistence. Before natural selection could produce character displacement and niche separation, one species had become extinct. Elton (1958) discusses many other examples of ecological invasions among both plants and animals.

Another sort of observation, involving the taxonomic composition of communities, has been used to try to assess whether or not competition has been an important force in nature. Because closely related species should be strong competitors, one might predict that fewer pairs of congeneric species will occur within any given natural community than would be found in a completely random sample from the various species and genera occurring over a broader geographic area. Such a paucity of sympatric congeners, if observed, would suggest that competitive exclusion occurs more often among congeneric species than in more distantly related ones. This test was applied to many communities by Elton (1946), who was well aware of the problem of defining a "community." Frequently, where there are two abutting communities, each supports its own member of a congeneric pair, and such pairs must be excluded wherever possible. Despite this potential bias toward an increased proportion of congeneric species pairs, Elton found fewer congeners than expected on a strictly random basis. Elton's analysis has since been shown to be incorrect by Williams (1964), who gave a corrected statistical approach to the problem. Using this correct technique, Terborgh and Weske (1969) calculated the expected number of congeneric species pairs of Peruvian birds in seven habitats (Figure 12.14). They found that the four habitats richest in total number of species had greater than the expected number of congeneric pairs, thus refuting any increased incidence of competitive exclusion among congeners in this particular avifauna. I also failed to find any consistent impoverishment in the number of congeneric species pairs in a number of lizard communities (Pianka, 1973). More analyses of this sort on a broad range of taxa in different communities might be worthwhile. Some field experiments on competition are briefly considered in Chapter 14.

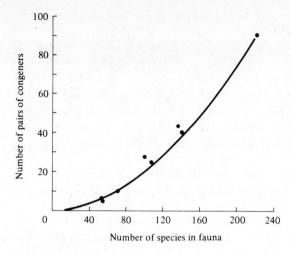

Figure 12.14. Observed (points) and expected (curve) numbers of pairs of congeneric bird species in seven Peruvian habitats. The combined avifauna of all seven areas contained 92 different pairs of congeners among a total of 221 species (plotted at the uppermost right-hand corner). Random subsamples of this total fauna would contain the numbers of congeneric pairs indicated by the curve. If competitive exclusion occurred more frequently among congeners, the observed number of such pairs (points) would fall below the expected curve. [From Terborgh and Weske (1969). By permission of Duke University Press.]

## OTHER PROSPECTS

Although competition is presumably central to numerous ecological processes and phenomena, current understanding of competitive interactions remains inadequate from both theoretical and empirical points of view. Possibilities abound for significant work. Clearly, the great temporal and spatial heterogeneity of the real world demands a dynamic approach to competitive interactions.

Models that depart from the notion of competitive communities at equilibrium with their resources are of interest. The notion of a competition coefficient itself may be somewhat illusory and may often obscure the real mechanisms and dynamics of competition. Even so, theory could be improved substantially simply by treating alphas as *variables* in both ecological and evolutionary time. For example, the actual shapes of resource utilization curves might be allowed to change (subject to an appropriate constraint, such as holding the area under the curves constant), either in ecological time by behavioral release or in evolutionary time through directional selection that favors deviant phenotypes. Because the competitive effects between any given pair of species are often sensitive to the presence or absence of a third species, theory is needed on *interactive* competition coefficients.

An attractive alternative to competition coefficients is to measure the intensity of interactions between species by the sensitivity of each species' own density to changes in the density of the other. As such, these interactions are represented mathematically by partial derivatives ($\partial N_i / \partial N_j$ and $\partial N_j / \partial N_i$ terms); if the presence of species $j$ is detrimental to species $i$, $\partial N_i / \partial N_j$ is negative, whereas a beneficial interaction has a positive sign. Note that this approach involves population dynamics and that it can be applied to prey-predator and symbiotic interactions as well as to competitive ones.

Great potential also exists for further development of theory on diffuse competition. Consider two communities with similar numbers of species but different guild structures. In the first, several distinct clusters of competing species have strong competitive interactions among themselves but weak interactions with members of other guilds. In the second community, all members interact more or less equally and diffuse competition is more intense. (Differences in the degree of niche dimensionality would produce such a difference between communities; see also pp. 277–280.) What effects will such differences in degree of competitive "connectedness" have upon various community-level properties like stability? Will maximal tolerable overlap be less in the second community? To what extent do the resources available force guild structure? Ecologists are currently seeking answers to such questions.

Prospects for future empirical work are even brighter, although certainly more difficult and challenging. Well-designed and executed experiments or perturbations of equilibrium densities, which introduce or remove species, will certainly allow partial quantification of competitive effects, but as Schoener (1974a, 1983) points out, in themselves they probably will not provide much insight into the actual mechanisms of competition. As previously indicated, such experiments will not easily tease apart direct effects from the indirect interactions mediated by other members of ecological communities (Bender et al., 1984). Clever empirical work will probably provide greater insights into competitive mechanisms than further theoretical explorations. Unfortunately, however, the precise nature of these crucial observations is difficult to anticipate.

## SELECTED REFERENCES

### Mechanisms of Competition

Birch (1957);   Brian (1956);   Crombie (1947);   Elton (1949);   Hazen (1964, 1970); Miller (1967);   Milne (1961);   Milthorpe (1961);   Pianka (1976a).

### Lotka-Volterra Competition Equations

Andrewartha and Birch (1953);   Bartlett (1960);   Haigh and Maynard Smith (1972); Horn and MacArthur (1972);   Levin (1974);   Levins (1966, 1968);   Levins and Culver (1971);   Lotka (1925);   MacArthur (1968, 1972);   May (1976a);   Neill (1974);

Pielou (1969); Schoener (1973, 1976b); Skellam (1951); Slatkin (1974); Slobodkin (1962b); Strobeck (1973); Vandermeer (1970, 1973, 1975); Volterra (1926a, 1926b, 1931); Wangersky and Cunningham (1956); Wilson and Bossert (1971).

## Competitive Exclusion

Bovbjerg (1970); Cole (1960); DeBach (1966); Gause (1934); Hardin (1960); Jaeger (1971); MacArthur and Connell (1966); Miller (1964); Patten (1961); Pianka (1972).

## Balance Between Intraspecific and Interspecific Competition

Connell (1961a, 1961b); Fretwell (1972); Fretwell and Lucas (1969); MacArthur (1972); MacArthur et al. (1972).

## Evolutionary Consequences of Competition

Collier et al. (1973); Connell (1961a, 1961b); Grant (1972); MacArthur (1972); MacArthur and Wilson (1967); Orians (1962); Pianka (1970); Ricklefs and Cox (1972).

## Laboratory Experiments

Gause (1934, 1935); Gill (1972); Hairston et al. (1968); Harper (1961a, 1961b); Krebs (1972); Lerner and Ho (1961); Neill (1972, 1974, 1975); Neyman et al. (1956); Park (1948, 1954, 1962); Park et al. (1964); Vandermeer (1969); Wilbur (1972).

## Evidence from Nature

Beauchamp and Ullyott (1932); Bender et al. (1984); Bovbjerg (1970); Brown and Wilson (1956); Case and Bender (1981); Case et al. (1983); Cody (1968, 1974); Colwell and Fuentes (1975); Crowell (1962); Dayton (1971); Dunham (1980); Elton (1946, 1949, 1958); Fenchel (1974, 1975); Fenchel and Kofoed (1976); Gadgil and Solbrig (1972); Grant (1972, 1986); Grant and Abbott (1980); Hairston (1951); Horn and May (1977); Huey et al. (1974); Hutchinson (1959); Kohn (1959, 1968); MacArthur (1958, 1972); Moulton and Pimm (1986); Orians and Willson (1964); Pianka (1966b, 1969, 1971b, 1973, 1974, 1975, 1986); Pittendrigh (1961); Schluter (1988); Schluter et al. (1985); Schoener (1965, 1968a, 1974a, 1974b, 1975b, 1982, 1983, 1984); Terborgh and Weske (1969); Vaurie (1951); Werner (1977); Werner and Hall (1976); Williams (1964); Wilson and Keddy (1986a, 1986b).

## Other Prospects

Ayala et al. (1973); Bender et al. (1984); Horn and MacArthur (1972); Lawlor (1979, 1980); Levins and Culver (1971); MacArthur (1968, 1972); May (1973, 1976a); Pianka (1976a, 1981b, 1987); Pomerantz (1980); Roughgarden (1974a, 1976a); Schoener (1973, 1976b); Wiens (1977).

# Chapter
## 13

# The Ecological Niche

## HISTORY AND DEFINITIONS

The concept of the niche pervades all of ecology; were it not for the fact that the ecological niche has been used in so many different ways, ecology might almost be defined as the study of niches. Many aspects of the niche have already been considered, and some others are examined here.

Among the first to use the term niche was Grinnell (1917, 1924, 1928). He viewed the niche as the functional role and position of an organism in its community. Grinnell considered the niche essentially a behavioral unit, although he also emphasized it as the ultimate distributional unit (thereby including spatial features of the physical environment). Later Elton (1927) defined an animal's niche as "its place in the biotic environment, *its relations to food and enemies*" (his italics) and as "the status of an organism in its community." Furthermore, he said that "the niche of an animal can be defined to a large extent by its size and food habits." Others, such as Dice (1952), use the term to refer to a subdivision of habitat; thus, Dice states that "the term [niche] does not include, except indirectly, any consideration of the function the species serves in the community." Clarke (1954) distinguished two separate meanings for the term niche, the "functional niche" and the "place niche." Clarke noted that different species of animals and plants fulfill different functions in the ecological complex and that the same functional niche may be filled by quite different species in different geographical regions. The idea of "ecological equivalents" was first stressed by Grinnell in 1924 (see also pp. 362–364).

One of the most influential treatments of niche is that of Hutchinson (1957a). Using set theory, he treats the niche somewhat more formally and defines it as the total range of conditions under which the individual (or pop-

268

ulation) lives and replaces itself. Hutchinson's examples for niche coordinates are nonbehavioral and have thus emphasized the niche as a place in space rather like a microhabitat or the "habitat niche" of Allee et al. (1949). This emphasis is unfortunate to the extent that it tends to exclude the "behavioral niche" from consideration. Hutchinson's distinction between the fundamental and the realized niche (pp. 272–274) is one of the most explicit statements that an animal's potential niche is seldom fully utilized at a given moment in time or a particular place in space. This distinction has proven useful in clarifying the roles of other species, both competitors and predators, in determining the niche of an organism.

Odum (1959) defined the ecological niche as "the position or status of an organism within its community and ecosystem resulting from the organism's structural adaptations, physiological responses, and specific behavior (inherited and/or learned)." He emphasized that "the ecological niche of an organism depends not only on where it lives but also on what it does." The place an organism lives, or where one would go to find it, is its habitat. For Odum, the habitat is the organism's "address," whereas the niche is its "profession." Weatherley (1963) suggested that the definition of niche be restricted to "the nutritional role of the animal in its ecosystem, that is, its relations to all the foods available to it." However, some ecologists prefer to define the term niche more broadly and to subdivide it into components such as the "food niche" or the "place niche."

Because concepts of the ecological niche have taken on so many different forms, it is often difficult to be sure exactly what a particular ecologist means when this entity is invoked. No one denies that there is a broad zone of inter-action between the traditional entities of "environment" and "organismic unit"; the major problem is to specify precisely in any given case just what subset of this enormous subject matter should be considered the "ecological niche." Considerable effort and insight have gone into separating and distinguishing the organism from its environment. Indeed, one can argue fairly plausibly that it constitutes a step backward to confound these two concepts again. Some therefore avoid using the word "niche" altogether and insist that we can get along perfectly well without it.

Following earlier terminology, the *ecological niche* is defined as *the sum total of the adaptations of an organismic unit,* or as all of the various ways in which a given organismic unit conforms to its particular environment. As with environment, we can speak of the niche of an individual, a population, or a species. The difference between an organism's environment and its niche is that the latter concept includes the organism's capacity for exploiting its environment and involves the ways in which an organism actually interfaces with and *uses* its environment.

The niche concept has gradually become linked to the phenomenon of in-terspecific competition, and it is increasingly becoming identified with patterns of resource utilization. Niche relationships among competing species are fre-quently visualized and modeled with bell-shaped utilization curves along a con-tinuous resource gradient, such as prey size or height above ground (Figure 13.1).

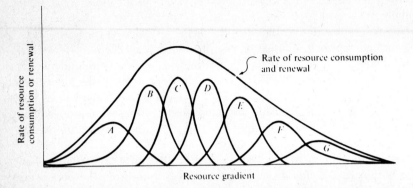

**Figure 13.1** Niche relationships among members of competitive communities are usually modeled with bell-shaped utilization curves along a resource spectrum, such as height above ground or prey size. Among the seven hypothetical species shown, those toward the tails have broader utilization curves because their resources renew more slowly. In such an assemblage, consumers are at equilibrium with their resources and the rate of resouce consumption is equal to the rate of renewal along the entire resouce gradient.

Emphasis on resource use is operationally tractable and has generated a rich theoretical literature on niche relationships in competitive communities. In the remainder of this chapter we consider in detail various aspects of this theory, including niche breadth, niche overlap, and niche dimensionality.

## THE HYPERVOLUME MODEL

Building on the law of tolerance, Hutchinson (1957a) and his students constructed an elegant formal definition of niche. When the tolerance or fitness of an organismic unit is plotted along a single environmental gradient, bell-shaped curves usually result (Figure 13.1). Tolerances for two different environmental variables can be plotted simultaneously (Figure 13.2). In Figure 13.3, hypothetical tolerances for three different variables are plotted in a three-dimensional space. Adding each new environmental variable simply adds one more axis and increases the number of dimensions of the plot by one. Actually, Figure 13.3 represents a four-dimensional space, with the three axes shown representing the three environmental variables, while the fourth axis* represents reproduc-

---

*This one cannot be shown directly but is implicit in the figure in the same way that Figure 13.2a is implicit in Figure 13.2b.

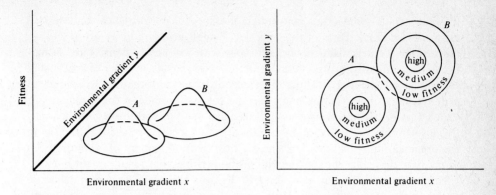

**Figure 13.2** Two plots of the fitnesses of two organismic units, A and B, versus their position along two environmental gradients, x and y. (a) A three-dimensional plot with a fitness axis. (b) A two-dimensional plot with the fitness axis omitted; low, medium, and high fitness represented by contour lines.

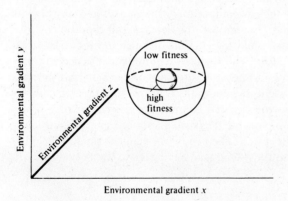

**Figure 13.3** A plot (like that of Figure 13.2b) of fitness along three different environmental gradients, x, y and z, showing zones of low and high fitness. A four-dimensional plot with a fitness axis analogous to Figure 13.1a is implicit in this graph.

tive success or some other convenient measure of performance that we will call *fitness density*. Parts of this space with high fitness density are relatively optimal for the organism concerned; those with low fitness density are suboptimal. Conceptually, this process can be extended to any number of axes, using $n$-dimensional geometry. Thus, Hutchinson defines an organism's niche as an $n$-dimensional hypervolume enclosing the complete range of conditions under which that organism can successfully replace itself (Hutchinson's "niche" may be closer to my definition of environment). All variables relevant to the life of the organism must be included, and all must be independent of each other. An immediate difficulty with this model of the niche is that not all environmental variables can be nicely ordered linearly. To avoid this problem and to make the entire model more workable, Hutchinson translated his $n$-dimensional hypervolume formulation into a set theory mode of representation. Unfortunately, fitness density attributes of the $n$-dimensional model are lost in the conversion to a set theory model.

Hutchinson designates the entire set of optimal conditions under which a given organismic unit can live and replace itself as its *fundamental niche*, which can then be represented as a set of points in environmental space. The fundamental niche is thus a hypothetical, idealized niche in which the organism encounters no "enemies" such as competitors or predators and in which its physical environment is optimal. In contrast, the actual set of conditions under which an organism exists, which is always less than or equal to the fundamental niche, is termed its *realized niche*. The realized niche takes into account various forces that restrict an organismic unit, such as competition and perhaps predation. The fundamental niche is sometimes referred to as the *precompetitive* or *virtual* niche, whereas the realized niche is the *postcompetitive* or *actual* niche (however, this terminology neglects factors other than competition—such as predation—that might restrict the occupied region of the fundamental niche). These two concepts are thus somewhat analogous to the notions of $r_{max}$ and $r_a$.

## NICHE OVERLAP AND COMPETITION

Niche overlap occurs when two organismic units use the same resources or other environmental variables. In Hutchinson's terminology, each $n$-dimensional hypervolume includes part of the other, or some points in the two sets that constitute their realized niches are identical. Overlap is complete when two organismic units have identical niches; there is no overlap if two niches are completely disparate. Usually, niches overlap only partially, with some resources being shared and others being used exclusively by each organismic unit.

Hutchinson (1957a) treats niche overlap in a simplistic way, assuming that the environment is fully saturated and that niche overlap cannot be tolerated for any period of time; hence, competitive exclusion must occur in the overlapping parts of any two niches. Competition is assumed to be intense and to result in survival of only a single species in contested niche space. Although this simplified approach has its shortcomings, it is useful to examine each of the logically possible cases (Figure 13.4) before considering niche overlap and competition in a more realistic way. First, two fundamental niches could be identical, corresponding exactly to one another, although such ecological identity is unlikely.

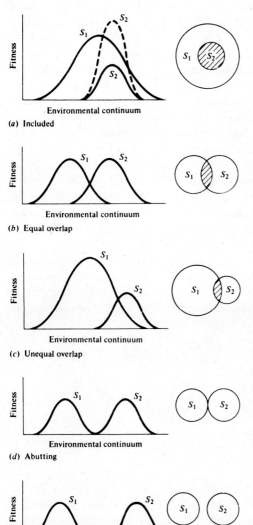

**Figure 13.4** Various possible niche relationships, with fitness density models on the left and set theoretic ones on the right. (*a*) An included niche. The niche of Species 2 is entirely contained within the niche of Species 1. Two possible outcomes of competition are possible: (1) if Species 2 is superior (dashed curve), it persists and Species 1 reduces its utilization of the shared resources; (2) if Species 1 is superior (solid curves), Species 2 is excluded and Species 1 uses the entire resource gradient. (*b*) Overlapping niches of equal breadth. Competition is equal and opposite. (*c*) Overlapping niches of unequal breadth. Competition is not equal and opposite because Species 2 shares more of its niche space than Species 1 does. (*d*) Abutting niches. No direct competition is possible, but such a niche relationship can arise from competition in the past and be indicative of the avoidance of competition, as in interference competition. (*e*) Disjunct niches. Competition cannot occur and is not even implicit in this case.

In this most improbable event, the competitively superior organismic unit excludes the other. Second, one fundamental niche might be completely included within another (Figure 13.4a); given this situation, the outcome of competition depends on the relative competitive abilities of the two organismic units. If the one with the included niche is competitively inferior, it is exterminated and the other occupies the entire niche space; if the former organismic unit is competitively superior, it eliminates the latter from the contested niche space. The two organismic units then coexist, with the competitively superior one occupying a niche included within the niche of the other. Third, two fundamental niches may overlap only partially, with some niche space being shared and some used exclusively by each organismic unit (Figure 13.4b and 13.4c). In this case, each organismic unit has a "refuge" of uncontested niche space and coexistence is inevitable, with the superior competitor occupying the contested (overlapping) niche space. Fourth, fundamental niches might abut one another (Figure 13.4d); although no direct competition can occur, such a niche relationship may reflect the avoidance of competition. Finally (Figure 13.4e), if two fundamental niches are entirely disjunct (no overlap), there can be no competition and both organismic units occupy their entire fundamental niche. Figure 13.5 illustrates the distinction between the fundamental and the realized niche for an organismic unit with six competitors.

A major shortcoming of the foregoing discussion is that, in nature, niches often do overlap yet competitive exclusion does not take place. Niche overlap in itself obviously need not necessitate competition. Overlap in habitats used may simply indicate that competitors have diversified in other ways. Should resources not be in short supply, two organismic units can share them without detriment to one another. In fact, extensive niche overlap may often be correlated with *reduced* competition, just as disjunct niches may frequently indicate avoidance of competition in situations where it could potentially be severe (such as in cases of interspecific territoriality). For these reasons, the ratio of demand to supply, or the degree of saturation, is of vital concern in the relationship between ecological overlap and competition. Indeed, much current research is designed to clarify, both theoretically and empirically, the

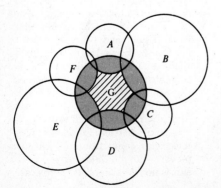

Figure 13.5 Set theory model of the fundamental niche (stippled and cross hatched) of Species G and its realized niche (cross hatched), which is a subset of the fundamental niche, after competition and complete competitive displacement due to six superior competitors, species *A, B, C, D, E,* and *F.*

relationship between competition and niche overlap; ecologists are now asking questions such as "How much niche overlap can coexisting species tolerate?" and "How does this maximal tolerable niche overlap vary with the degree of saturation?" (Figure 13.6).

Competition is the conceptual backbone of much current ecological thought. Nonetheless, competition remains surprisingly elusive to study in the field and hence is still poorly understood (probably because avoidance of competition is always advantageous when possible). Precise mechanisms by which available resources are divided among members of a community must be known before determinants of species diversity and community structure can be understood fully. Resource partitioning among coexisting species, or niche segregation, has therefore attracted considerable interest (for reviews, see Lack, 1971; MacArthur, 1972; Schoener, 1974a, 1986; Pianka, 1976b, 1981).

The basic raw data for analysis of niche overlap is the resource matrix, which is simply an $m$ by $n$ matrix indicating the amount (or rate of consumption) of each of $m$ resource states utilized by each of $n$ different species. When coupled with data on resource availability, utilization can be expressed with "electivities" (Ivlev, 1961), which measure the degree to which consumers actually utilize resources disproportionately to their supply. From such a matrix, one can generate an $m$ by $n$ matrix of overlap between all pairs of species, with the ones on the diagonal and values less than unity as off-diagonal elements. Overlap is sometimes equated with competition coefficients (alphas) because overlap is much easier to measure. Again, the caveat: overlap need not result in competition unless resources are in short supply. Extensive overlap may be possible when there is a surplus of resources (low demand/supply ratio), whereas maximal tolerable overlap may be much less in more saturated environments (see Figure 13.6). Because the principle of equal opportunity dictates that the ratio of demand over supply be constant along any *particular* resource gradient, intensity of competition should be directly proportional to the actual overlap observed along any given resource spectrum (see Figures 12.7 and 13.1). Patterns of niche overlap along different resource axes or between different communities must be compared with caution.

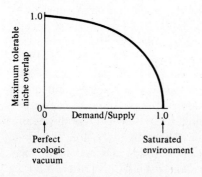

Figure 13.6 The niche overlap hypothesis predicts that maximum tolerable niche overlap must decrease as the intensity of competition increases. Maximal niche overlap in a saturated environment may not be zero, as figured, but some positive quantity related to the character displacement ratio of 1.3 (see page 263). [From Pianka (1972). Copyright ©1972 by The University of Chicago Press.]

## NICHE DYNAMICS

Realized niches of most organisms change both in time and from place to place as physical and biotic environments vary. Temporal niche changes can be considered at two levels: (1) on a short-term basis (i.e., in an ecological time scale), usually the life of a single individual or at most a few generations, and (2) on a long-term basis, over evolutionary time and many generations (see pp. 290–291). Thus, the realized niche can be thought of as an ever-changing subset of the fundamental niche or, in the $n$-dimensional hypervolume model, as a pulsing hypervolume bounded by the hypervolume corresponding to the fundamental niche.

Some organisms, particularly insects, have entirely disjunct, nonoverlapping niches at different times in their life histories: caterpillars and butterflies, maggots and flies, tadpoles and toads, planktonic larval but sessile adult barnacles, and aquatic larval versus terrestrial adult insects (mosquitos, stoneflies, dragonflies, etc.). In all these cases, a drastic and major modification of an animal's body plan at metamorphosis allows a pronounced niche shift. Niches of other organisms change more gradually and continuously during their lifetimes. Thus, juvenile lizards eat smaller prey than do adults and are often active earlier in the day at lower environmental temperatures (their smaller size and greater surface-to-volume ratio facilitates faster warming). Incomplete metamorphosis, in which an insect changes rather gradually with each molt (as in grasshoppers), is similar.

An organism's immediate neighbors in niche space, or its potential competitors, can (but need not) exert strong influences upon its ecological niche. Whereas realized niches of relatively $r$-selected organisms are determined primarily by their physical environments, realized niches of more $K$-selected organisms are perhaps more strongly influenced by their biotic environments. Selective pressures and niches may vary during an individual's life. Thus, in temperate zones, early spring is a time when annual plants are relatively $r$-selected; later in the season they become progressively more $K$-selected (Gadgil and Solbrig, 1972). Even within a given species, some individual organisms may be more $r$-selected than others, such as populations in different microhabitats or parts of a species' geographic range, or individuals at different positions in the rocky intertidal.

Theoretically, reduced interspecific competition should often allow niche expansion. In an attempt to observe this, Crowell (1962) examined and compared ecologies of three species of birds on Bermuda with those of mainland populations. Considerably fewer species of land birds occur on the island than on the mainland; the three most abundant are the cardinal, catbird, and white-eyed vireo. On Bermuda, these three species have very dense populations; however, in mainland habitats where there are many more species of birds— and thus, a greater variety of interspecific competitors—densities of these same three species are usually considerably lower. Although there are, of course, inevitable differences between the habitats of Bermuda and mainland North America, the striking difference between avifaunas should nevertheless have a major effect on the ecologies of the birds. The catbird and cardinal have generally more restricted place and foraging niches on the island (perhaps available niches are more restricted), but the white-eyed vireo has expanded both its place niche and its foraging niche. All three species nest at a wider variety of heights on Bermuda.

## NICHE DIMENSIONALITY

Although the $n$-dimensional hypervolume model of the niche is extremely attractive conceptually, it is abstract and rather difficult to apply to the real world. Indeed, to construct such a hypervolume, we would have to know essentially everything about the organism concerned. Because we can never know *all* factors impinging upon any organismic unit, the fundamental niche must remain an abstraction. Even realized niches of most organisms have so many dimensions that they defy quantification. When considering relatively $K$-selected organisms, the number of niche dimensions can be limited to those on which competition is effectively reduced. Competition is often avoided by differences in microhabitats exploited, foods eaten, and times of activity, and so the effective number of niche dimensions can be reduced to three—place, food, and time. We can think of a saturated community as occupying some volume in a space with these three dimensions; thus, a community is something like a three-dimensional jigsaw puzzle, each piece being one species occupying only a part of the overall volume.

Most niche theory is framed in terms of a single niche dimension. Each species then has only two neighbors in niche space, and overlap or similarity matrices contain many zeros and only two positive entries on the off-diagonal per row. However, real species of plants and animals seldom sort themselves out on a single resource axis; instead, pairs of species usually show moderate niche overlap along two or more niche dimensions. Complementarity of niche dimensions often occurs, with pairs that have high overlap along one niche dimension overlapping little on another dimension, and vice versa (Figure 13.7). Multidimensional niche relationships can be complex. One feature of increased niche dimensionality is that niches may be overlapping or identical along one axis and yet be separated or even disjunct along another (Figure 13.8). Thus, an observer oblivious to the first niche dimension in Figure 13.8 would consider species pair A and F, pair E and B, and pair D and C as completely overlapping, when in fact they are partially or entirely separated along the unknown dimension!

In a one-dimensional niche space, any given niche can be bounded only on two sides, whereas there can be many more neighbors in a two-dimensional niche space, and still more in three or more dimensions. As the effective number of niche dimensions rises, the potential number of neighbors in niche space increases more or less geometrically. As dimensionality increases, overlap matrices contain fewer off-diagonal elements of zero, and the variance in observed overlap usually falls both within rows and over the entire matrix. Hence, niche

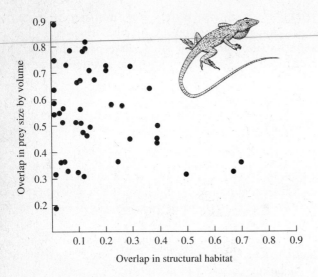

**Figure 13.7** Overlap in prey size plotted against overlap in structural microhabitat among various species of *Anolis* lizards on the island of Bimini. Pairs with high dietary overlap tend to exploit different structural microhabitats; conversely, those with high spatial overlap exhibit relatively little overlap in prey sizes eaten. [From Schoener (1968). Copyright ©1968 by the Ecological Society of America.]

dimensionality strongly affects the *potential* for "diffuse" competition arising from the total competitive effect of all interspecific competitors (MacArthur, 1972). The overall effect of relatively weak competitive inhibition per species summed over many other species could well be as strong or even stronger than much more intense competitive inhibition (per species) by fewer competing species. A population with many niche dimensions has the potential for many immediate neighbors in niche space, which can intensify diffuse competition.

Imagine that height above ground and prey size are two such critical niche dimensions that species use differentially and thereby avoid or reduce interspecific competition. Analysis of resource utilization and niche separation along more than a single niche dimension should ideally proceed through estimation of proportional *simultaneous* utilization of all resources along each separate niche dimension. These define a three-dimensional resource matrix, with each entry representing the probability of capture of a prey item of a given size category at a particular height interval by each species present. Obtaining such multidimensional utilization data is extremely difficult, however, because most animals move and integrate over both space and time. Accurate estimates of an animal's true use of a multidimensional niche space could only be obtained by continually monitoring an individual's use of all resources. (Even then, the degree to which prey individuals move between microhabitats will affect competition in obscure but vitally important ways!) Because such continual observation is often extremely tedious or even impossible, one usually approximates from

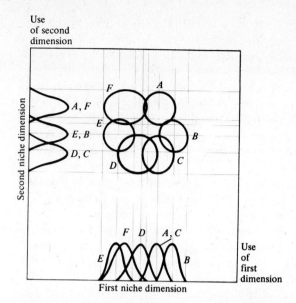

Figure 13.8 Hypothetical niches of six species differing along two niche dimensions. Although overlap is broad or complete along either single dimension alone (bell-shaped curves), niches overlap minimally or not at all when both niche dimensions are considered (circles and ellipses).

separate unidimensional utilization distributions (Figures 13.8 and 13.9). Just as the three-dimensional shape of a mountain cannot be accurately determined from two of its silhouettes viewed at right angles, these "shadows" do not allow accurate inference of the true multidimensional utilization.

The question of the degree of dependence or independence of dimensions becomes critical. Provided that niche dimensions are truly independent, with

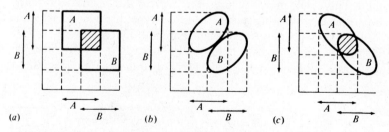

Figure 13.9 Three different cases of niche overlap on two resource dimensions with identical unidimensional projections (double-headed arrows). In (a) niche dimensions do not interact and unidimensional projections accurately reflect multidimensional conditions (dimensions are truly independent and the niche axes are orthogonal). In (b) and (c) niche dimensions are partially interdependent and in (b), unidimensional shadows are quite misleading.

rey of any size being equally likely to be captured at any height, overall multidimensional utilization is simply the product of the separate unidimensional utilization functions (May, 1975b). Under perfect independence, the probability of capture of prey item $i$ in microhabitat $j$ is then equal to the probability of capture of item $i$ times the probability of being in microhabitat $j$. Unidimensional estimates of various niche parameters (including overlap) along component niche dimensions may then simply be multiplied to obtain multidimensional estimates. However, should niche dimensions be partially interdependent (Figure 13.9), there is no substitute for knowledge of true multidimensional utilization. True multidimensional overlap can vary greatly depending on the exact form of this dependence (see Figure 13.9). In the extreme case of complete dependence (e.g., if, prey of each size are found only at one height), there is actually only a single niche dimension, and a simple average provides the best estimate of true utilization. Moreover, the arithmetic average of estimates of unidimensional niche overlap obtained from two or more separate unidimensional patterns of resource use actually constitutes an *upper bound* on the true multidimensional overlap (May, 1975b).

It is difficult (or virtually impossible) to evaluate the degree of interdependence of niche dimensions for many species. However, in relatively sedentary species, the degree to which foods eaten are influenced by microhabitat can sometimes be assessed. One such study of a sedentary legless lizard showed that most species and castes of termites are eaten in fairly similar proportions by lizards taken from different microhabitats, suggesting that these two niche dimensions are largely independent (Huey et al., 1974).

In reviewing major factors leading to ecological isolation among birds, Lack (1971) concluded that the most important were differences in geographic range, habitat, and foods eaten. Schoener (1974b) reviewed patterns of resource partitioning in over 80 natural communities, ranging from simple organisms such as slime molds through various mollusks, crustaceans, insects, and other arthropods to various members of the five classes of vertebrates, including lizards. He identified and ranked five resource dimensions by degree of importance in niche segregation: macrohabitat, microhabitat, food type, time of day, and seasonality of activity. Schoener concludes that habitat dimensions are generally more important in separating niches than food-type dimensions, which in turn tend to be important more often than temporal dimensions. Terrestrial poikilotherms partition food by being active at different times of day relatively often compared with other animals.* Predators partition resources by diurnal

---

*Use of time of activity as a niche dimension can be justified in several ways. If resources are rapidly renewed, exploitative competition cannot occur unless individuals are active within a fairly short time of each other (otherwise, resources may be replenished during the interval separating the species). When microhabitat and food categories are crude and changing in time, temporal differences in activity may in fact be associated with differential use of space and food resources that would not be reflected without much greater precision in recognition of appropriate nontemporal resource states. Thus, a microhabitat such as "open sun" clearly changes with the daily march of temperature; similarly a given crude prey category such as "ants" may usually lump a series of prey species with temporal segregation. Hence, it is often both useful and appropriate to treat time as a niche dimension.

differences in time of activity more frequently than do other groups, ar
tebrates segregate less by seasonal activity differences than do nonverteb
Schoener also found that segregation by food type is more important for
mals feeding on large foods relative to their own size than it is among anin
that feed on relatively small items.

# NICHE BREADTH

## Specialization versus Generalization

Some organisms have smaller niches than others. *Niche breadth*, also called
*niche width* and *niche size*, can be thought of as the extent of the hypervolume
representing the realized niche of an organismic unit. For example, a koala,
*Phascolarctos cinerius*, eats only leaves of certain species of *Eucalyptus* and
thus has a more specialized food niche than the Virginia opossum, *Didelphis
virginianus*, which is a true omnivore that eats nearly anything. Statements
about niche breadths must invariably be comparative; we can only say that
a given organismic unit has a niche that is narrower or broader than that of
some other organismic unit. Highly specialized organisms like the koala usually, though not always, have narrow tolerance limits along one or more of their
niche dimensions. Often such specialists have very specific habitat requirements, and as a result, they may not be very abundant. In contrast, organisms
with broad tolerances are typically more generalized, with more flexible habitat requirements, and are usually much more common. Thus, specialists are
often relatively rare, whereas generalists are more abundant. However, rare
organisms may frequently occur in clumps so that their local density need not
necessarily be low.

The only currency of natural selection is differential reproductive success.
This fact raises a question: If specialization involves becoming less abundant,
why have organisms become specialized at all? Since generalized organisms
can usually exploit more food types, occupy more habitats, and build up larger
populations, they might be expected by their very numbers to out-reproduce
more specialized, competing members of their own population and thereby
swamp the population gene pool. The answer to this apparent dilemma lies
in the old adage that a jack-of-all-trades is a master of none. More specialized
individuals are presumed to be more efficient on their own ground than are
generalists.

Under what conditions will a jack-of-all-trades win in a competition with a
more specialized species? MacArthur and Levins (1964, 1967) considered this
question and developed the following model. First, imagine an ant-eating lizard
in an environment that contains only a single food resource type, colonies of
ants 3 mm in length, and a variable population of ant-eating lizards that exploit
the insect food resource. Assume that ants are eaten whole and that lizards
differ only in the size of their jaws, forming a fairly continuous phenotypic
spectrum. Some phenotypes will be well adapted to use 3-mm ants and very
effective at harvesting them; others will be less efficient, either because their

ws are too large or too small. Next, consider the same phenotypic spectrum of lizards in another "pure" environment, this one composed solely of colonies of 5-mm ants. Almost certainly the best-adapted phenotype will differ from that in the 3-mm ant environment (Figure 13.10a), and the phenotype most efficient at using 5-mm ants will be one with a larger mouth. Now, consider a *mixed* ant colony with equal numbers of 3- and 5-mm ants (say, two castes) in a *homogeneous* mixture. Which phenotype will be optimal in this new mixed environment? Assuming that the lizards encounter and use the two ant sizes in

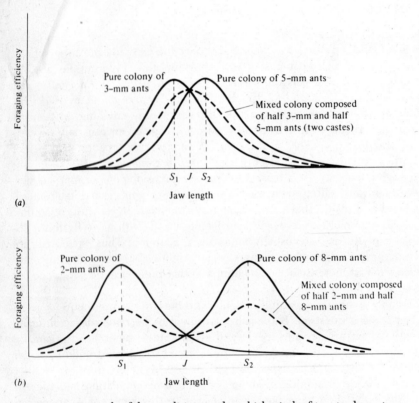

Figure 13.10 Example of the conditions under which a jack-of-two-trades outperforms two specialists. Foraging efficiency, measured in useful calories gathered per unit of foraging time, is plotted against a phenotypic spectrum of jaw lengths for a hypothetical population of ant-eating lizards. (a) Solid curves: performance of various lizard phenotypes in "pure" environments containing only 3- and 5-mm ants, respectively. Dashed curve: performance of various phenotypes in an environment containing a homogeneous mixture of equal numbers of both 3- and 5-mm ants (say, two castes). The jack-of-both-trades, $J$, has a higher foraging efficiency than either specialist, $S_1$ or $S_2$, which are the most efficient foragers in the respective "pure" environments. (b) Foraging efficiencies of the various lizard phenotypes when the difference between the two ant types is greater, say, 2- versus 8-mm ants. In this case, performance in the mixed environment is bimodal, and the two specialists outperform the jack-of-both-trades. [After MacArthur and Connell (1966).]

exactly equal proportions, the relationship between phenotype and harvesting effectiveness must be exactly intermediate between the two similar relationships in pure environments (Figure 13.10*a*). The dashed line can thus be drawn in Figure 13.10*a* midway between the first two (if the two resources were not in exactly equal proportions, this new line would simply be closer to one or the other of the original lines); depending on the shapes of the curves and the distance between them, this new line can take either unimodal (Figure 13.10*a*) or bimodal (Figure 13.10*b*) shape. In the former case, the phenotype of highest harvesting efficiency is intermediate between the best "pure 3-mm ant eater" and the best "pure 5-mm ant eater," and this "jack-of-both-trades" (probably the phenotype that could best exploit 4-mm ants) is competitively superior. In the latter case, because the two types of ants are very different in size, the two phenotypes with high harvesting effectiveness are separated from one another by intermediate phenotypes with *lower* efficiencies at exploiting a mixture of 2- and 8-mm ants, and the two specialists will eliminate the jack-of-both-trades.

Effects of interspecific competition on niche breadth are complex and under different conditions may actually favor either niche contraction or niche expansion. Thus, a competitor may reduce food availability in some microhabitats but leave prey densities in other microhabitats unaltered, effectively reducing expectation of yield in some patches but not others. A competitor that is an optimal forager should restrict its patch utilization to those with higher expectation of yield, thereby decreasing the breadth of its place niche. Conversely, a more generalized competitor that reduces food availability more or less equally in all microhabitats by reducing the overall level of prey availability can force its competitor to expand the range of resources it uses, thereby increasing the breadth of its food niche. In a food-sparse environment, an optimal forager simply cannot afford to bypass as many potential prey items as it can in a food-dense environment; therefore, more suboptimal prey must be eaten in the food-sparse habitat. Reduced interspecific competition is often accompanied by an increase in the range of habitats a species uses, but marked changes in the variety of foods eaten with changes in interspecific competition seem to be much less common (MacArthur, 1972).

Theory on optimal foraging predicts that dietary niche breadth should generally expand as resource availability decreases (Emlen, 1966, 1968a; MacArthur and Pianka, 1966; Schoener, 1971; MacArthur, 1972; Charnov, 1973, 1976a, 1976b; Stephens and Krebs, 1986). In an environment with a scant food supply, a consumer cannot afford to bypass many inferior prey items because mean search time per item encountered is long and expectation of prey encounter is low. In such an environment, a broad niche maximizes returns per unit expenditure, promoting generalization. In a food-rich environment, however, search time per item is low since a foraging animal encounters numerous potential prey items; under such circumstances, substandard prey items can be bypassed because expectation of finding a superior item in the near future is high. Hence, rich food supplies are expected to lead to selective foraging and narrow food niche breadths.

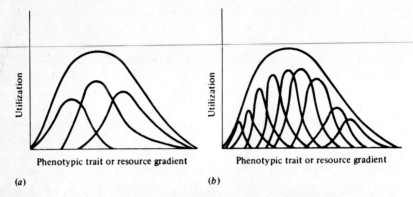

Figure 13.11 Diagrammatic representation of two populations differing in the within-phenotype and between-phenotype components of niche breadth. (*a*) A few generalized individuals exploit most of the resource gradient. (*b*) A population with the same overall niche breadth consists of numerous different, more specialized individuals. The within-phenotype component is greater in (*a*) than in (*b*); the between-phenotype component of niche breadth is greater in (*b*) than in (*a*).

Two fundamental components of niche breadth are distinguished: the "between-phenotype" versus "within-phenotype" components (Van Valen, 1965; Orians, 1971; Roughgarden, 1972, 1974b, 1974c). A population with a niche breadth determined entirely by the between-phenotype component would be composed of specialized individuals with no overlap among them in resources used; a population composed of pure generalists with each member exploiting the entire range of resources used by the total population would have a between-phenotype component of niche breadth of zero and a maximal within-phenotype component. Clearly, real populations will lie somewhere between these two extremes (Figure 13.11).

## Optimal Use of Patchy Environments

Environments that are decidedly discontinuous—that is, heterogeneous environments consisting of a patchwork of rather different resources—are termed *patchy environments*; those with similar or well-mixed resources are called *homogeneous* or *uniform environments*. Frequently, a particular organism in a heterogeneous environment exploits only a part of the environmental mosaic, and different organisms are specialized to use each of the patch types. However, as the degree of difference between patches decreases, the advantage of being generalized increases. Thus, we distinguish again two different extremes in the way in which organisms can utilize their environments. When an organism in the course of its daily activities encounters and uses resources in the same proportions in which they actually occur (i.e., does not select resources or patches in the environment), the organism is said to utilize its environment in a *fine-grained* manner (Levins, 1968). Organisms that spend disproportionate amounts of time in different patches are said to use their environment in a *coarse-grained* manner. An animal that encounters its prey in the actual proportions in which they occur (i.e., in a fine-grained manner), through selection of particular prey types, can exploit this fine-grained environment in a coarse-grained way. Fine-grained utilization of the environment may be forced on an organism in situations where the size of the patches in the environmental mosaic (the *grain size* of the environment) is small compared to the organism's own size. Thus, larger animals, all else being equal, tend to encounter the world in a more fine-grained way than do smaller ones. For example, a meadow and an adjacent forest each has its own herbivorous mouse (often a *Microtus* and a *Peromyscus*, respectively), whereas a deer exploits both the meadow and the forest, using the former at night for feeding and the latter by day for sleeping and retreat. Small animals such as insects generally exploit their environments in a more coarse-grained way than do larger animals such as vertebrates. Sessile animals and plants, as individuals, must tend toward coarse-grained utilization of space simply by virtue of their immobility. Because individual sessile organisms may be distributed widely in an environmental mosaic, however, populations of them may approximate fine-grained utilization of the spatial environment. Statements about environmental grain size or the degree of coarse-grained versus fine-grained utilization of resources are usually comparative; when one says that a given animal "lives in a coarse-grained environment," this is relative to some other animal.

Consider now the ways in which insectivorous birds could forage in a mixed forest containing several different species of trees and a wide range of different types of insects (MacArthur and MacArthur, 1961). Assume that some degree of coarse-grained utilization is necessary. A bird could forage only in one species of tree and fly from each such tree to another; much time and energy would then be expended in flying between trees and over unsuitable ones. Alternatively, birds could eat only one category of insect food (perhaps a particular spectrum of insect sizes), capturing these wherever they occur in any of the various tree species. In the course of searching out its particular insect prey, the bird would doubtless encounter a variety of other prey types (or sizes). The first strategy would presumably lead to different species of birds being specialized to each of the various tree species; the latter would result in each bird species having its own particular range of prey types. A third way in which birds could exploit such a patchy environment is to compromise on both of the preceding ways and to specialize as to exactly *where* in the trees they forage and *how* they feed. Thus, a bird might select a layer in the forest and forage through this layer, taking whatever prey are available within broad limits, feeding in different trees, and consuming many different prey types. In this case, different species of birds would differ as to where and how they forage, with each species exploiting a "natural feeding route" in the forest. A compromise strategy such as this is usually a more efficient way for birds to exploit a patchy environment than is specializing as to places or prey exploited; most birds have indeed developed their own unique patterns as to where and how they forage. Kinglets and titmice tend to forage in the crowns of trees; other species such as many warblers (see also Figure 12.9) forage in other parts of trees. Many woodpeckers exploit crevices in tree trunks in an ascending spiral to some height and then fly or glide down to a low point on the trunk of a nearby tree and repeat the process. Only in cases of extreme *concentrations* of food does it pay to specialize on a particular food type. Many parrots are quite specialized as to fruits and nectars they eat; when these foods are encountered, they are usually very dense and in superabundance. Such exceptionally rich energy sources are worth searching out; once one is located, dividends are relatively great.

Consider now a model for utilization of patchy environments based on optimization of an animal's time budget, although the model applies equally well to an energy budget (MacArthur and Pianka, 1966). Assume that environmental resources, say, prey species, are encountered by a foraging animal in the same proportions in which they actually occur or that the environment is fine-grained (this assumption is relaxed later). The animal is able to select from the available array of prey types and can *use* this fine-grained environment in a coarse-grained way. For convenience, we also assume that the animal either does or does not eat any given kind of prey; that is, no one type is eaten only part of the time it is encountered. What number of different kinds of prey will provide the animal with maximal return per unit time (or expenditure)? Total foraging time, per item eaten, can be broken down into two components: time spent on search (search time) versus that spent on pursuit, capture, and eating (pursuit time). A fine-grained environment is searched for all types of food simultaneously, whereas prey are pursued, captured, and eaten singly. Prey types are ranked from those providing the highest harvest per unit time (or energy) to those of lowest yield—that is, from the prey species whose capture requires the least expenditure per calorie assimilated to that requiring the most. Of course, the diet includes the most rewarding item; as an animal expands its diet to include progressively less-rewarding kinds of prey, more and more acceptable items are encountered and search time (or energy) per prey item *decreases*. However, as new and varied prey, often hard to catch or to swallow, are added to the diet, pursuit time (or energy) per item eaten will usually *increase*. So long as the time or energy saved in reduced search is greater than the increase in time or energy expended in increased pursuit, the diet should be enlarged to include the next, less-rewarding, kind of prey. At some point, losses accompanying further enlargement will balance or exceed gains, thereby marking the optimal diet.

Though no general statements about an animal's diet are possible from this model, some testable *comparative* predictions can be made. For instance, search time per item eaten should be less in a food-dense environment than in a food-sparse one. However, pursuit time, a function of the relative abilities of predator and prey and of the variety of prey types and predator escape mechanisms, should be little altered by changes in food density per se. As a result, productive environments should be used in a more specialized way than less productive ones. Similarly, animals that spend little effort searching for their prey should be more specialized than those with higher ratios of time spent in search to time spent on pursuit. Should food be scarce, foraging animals are unlikely to bypass potential prey, whereas during times or in areas with abundant food, individuals may be more selective and restrict their diets to better food types. Thus, a low expectation of finding prey, or a high mean search time per item, demands generalization; a higher expectation of locating prey items (short mean search time per item) allows some degree of specialization.

So far, the environment in this model has been homogeneous and fine-grained. Let us now include patchy environments in which various patches contain different arrays of prey. Types of patches are now ranked in order of decreasing *expectation of yield* or from the patch in which the most calories are likely to be obtained from prey per unit expenditure to that with the least return. The two components of the animal's time (or energy) budget are now "hunting time" per item captured—time spent *within* suitable patches (hunting time is equivalent to the "foraging time" previously discussed, or to the sum of search time plus pursuit time)—versus the "traveling time" per item caught—time spent traveling *between* suitable patches. Time spent traveling between patches decreases as an animal expands the number of different kinds of patches on its itinerary; time spent hunting within patches must increase as the itinerary is enlarged to include an increased variety of patches. Optimal use of a patchy environment therefore depends on the rate of decrease in traveling time (per prey item) relative to the rate of increase in hunting time (per prey item) associated with expanding the number of different patch types exploited. Suppose that food density is suddenly increased in all patches; *both* traveling time and hunting time, per item, will then be reduced. Because only search time decreases with increased food density, animals that expend greater amounts of energy on search (searchers) will have their hunting time reduced more than those that spend relatively more energy on pursuit (pursuers). Hence, under high food densities, pursuers should restrict the variety of patches they use more than searchers.

Consider now the effects of patch size on the optimal number of patch types exploited. Envision two environments differing only in the sizes of their patches and not at all in the proportions or qualities of the various patch types. Hunting time per item in identical but different-sized patches is the same in both environments because patch quality is unaltered. However, traveling time per item *decreases* as patch size increases because distance between patches varies linearly with the linear dimension of a patch, whereas hunting area or volume within a patch varies as its square or cube, depending on whether the animal exploits space in two or three dimensions, respectively. Since larger patches offer smaller traveling time per unit of hunting time, they can be used in a more specialized way than can smaller patches. In the extreme, as patches become vanishingly small compared to the size of the organism, patch selection is impossible and completely fine-grained utilization must take place. Similarly, patches that are very large compared to an organism approximate a one-patch environment and an animal can (or must) spend all of its time in that one patch.

Thus, smaller or less mobile animals should use fewer different patches than larger or more mobile ones.

Competitors should normally reduce the density of some types of prey in some patches. Any prey item worth exploiting in the absence of competition is still worth eating if there is competition for it, but this is not true of patterns of patch exploitation. The decision whether or not to forage within a given patch type depends on an organism's expectation of yield in that patch. Should food within a given patch type become scarce (due to competitors or some other factor), inclusion of that patch type in the itinerary increases mean hunting time per item sharply and reduces efficiency. Thus, the presence of competitors in some patches should cause an optimal predator to restrict the variety of patch types it exploits.

## The Compression Hypothesis

When faced with more intense competition from another species, many organisms restrict their utilization of shared microhabitats or other resources. These adjustments are those that take place in ecological time, during the lifetime of the organism concerned. The fact of such niche contractions, coupled with theoretical considerations, has led to the so-called *compression hypothesis* (MacArthur and Wilson, 1967), which states that as more species invade a community, place niches are compressed while food niches either remain constant or expand (Figure 13.12). Any prey item worth eating should be acceptable no matter what the intensity of competition, but an animal must choose the places it forages on the basis of its *expectation* of yield, which will usually be markedly decreased in some patches of habitat by heightened competition. Thus, the compression hypothesis predicts that during short-term, nonevolutionary time, habitats used should shrink with increased competition, while the range of foods eaten should expand or remain the same. Moreover, should competitors reduce overall levels of available foods more or less equally among all patches on a species' itinerary, food niche expansion will be favored.

## Morphological Variation–Niche Breadth Hypothesis

A species can be a generalist in two ways: (1) a population can contain a variety of different phenotypes, each using a smaller range of resources than the overall population, or (2) each individual in a population can itself be relatively flexible and generalized, with the resources utilized by any individual being similar to those exploited by the entire population. Roughgarden (1972) refers to these two components of niche breadth as the between-phenotype and within-phenotype components of niche breadth (see Figure 13.11). By allowing different phenotypes to exploit different resources, phenotypic variability within a population should thus increase the overall range of resources exploited by the population. Moreover, by reducing niche overlap among members of a population, between-phenotype niche expansion might be expected to reduce the average degree of interphenotypic competition.

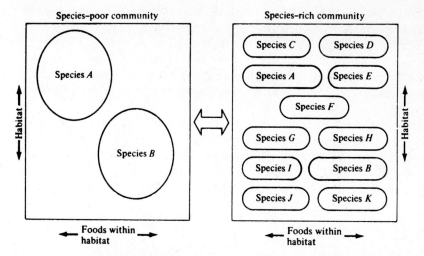

**Figure 13.12** Diagrammatic illustration of the compression hypothesis. As more species invade a habitat, interspecific competition forces any single species to decrease the range of habitats it exploits; however, the range of foods eaten should either increase or remain more or less constant. Conversely, if a species invades a habitat that is impoverished with species, reduced interspecific competition should often allow an expansion in the variety of habitats (or microhabitats) exploited. The compression hypothesis applies only in ecological time and does not address evolutionary niche shifts. [After MacArthur and Wilson (1967). *The Theory of Island Biogeography.* Copyright ©1967 by Princeton University Press. Reprinted by permission of Princeton University Press.]

Island species, freed from some interspecific competition, often tend to exploit a wider range of habitats than mainland species do—a phenomenon often referred to as *ecological release.* Van Valen (1965) reasoned that reduced competition from other species should also favor an increased morphological variability because this would promote niche expansion. He postulated that island species should often be morphologically more variable than their mainland counterparts; moreover, Van Valen found evidence for just such an increased phenotypic variability in five out of six species of birds known to have broader niches on certain islands. Morphological variation in length of wing, tail, tarsus, and bill in some Mexican insular bird populations is lower than in mainland ones, however (Grant, 1967). In a later study, Grant (1971) found no consistent trends in tarsal length variation between mainland and island populations of Mexican birds. He speculates that, under spatially uniform environmental conditions, selection favors little variation among individuals in feeding ecology and associated morphology by acting strongly against individuals that depart widely from the average phenotype, but that under spatially varied (patchy) conditions, the opposite may be true.

Soulé and Stewart (1970) restate and somewhat reverse the hypothesis; generalized, broad-niched species should be phenotypically and morphologically more variable than more specialized and more narrow-niched species (they call this the niche-variation hypothesis). However, Soulé and Stewart were unable to find any evidence that generalized African bird species such as crows are in fact morphologically more variable than more specialized species. Van Valen and Grant (1970) point out that the broad niche of these crows could well be primarily due to within-phenotype flexibility in resource utilization and that one need not necessarily expect great morphological variability in such a situation.

Also of interest here is the fact that, although tropical species are often considered to be more specialized than temperate species, bill dimensions of some tropical birds are at least as variable as those of some North temperate species (Willson, 1969). More information is needed before the relationships between morphological variation and niche breadth can be adequately assessed.

## EVOLUTION OF NICHES

Niche changes over evolutionary time are rather difficult to document, although their occurrence cannot be disputed. As new species arise from the fission of existing ones through the process of speciation, new niches come into existence. Life on the earth almost certainly arose in aquatic environments, and early organisms were doubtless very small and simple. During the evolutionary history of life over geological time, organisms have become more and more complex and diversified, and the earth has been filled with an overwhelming variety of plants and animals. Some taxonomic groups of organisms, such as dinosaurs, have gone extinct and been replaced by others. Major breakthroughs in the body plans of organisms periodically open up new adaptive zones and allow bursts of evolution of new and diverse species, termed *adaptive radiations*. A major force that has led to niche separation and diversification is interspecific competition. Thus, the first terrestrial organisms found themselves in a wide open ecological and competitive vacuum, freed from competition with aquatic organisms, and they rapidly radiated into the many available new terrestrial niches. Similarly, evolution of endothermy and aerial exploitation patterns have allowed major adaptive radiations; flight has evolved independently at least four times, in insects, reptiles, birds, and mammals. Often, evolutionary interactions between two or more taxa have had reciprocal effects upon one another; thus, the origin and radiation of flowering plants (angiosperms) in the Mesozoic presumably allowed insects to diversify widely, whereas species specificity of pollinating insects in turn may well have allowed considerable diversification of plants. Indeed, Whittaker (1969) has suggested that organic diversity is self-augmenting.

## PERIODIC TABLE OF NICHES

In chemistry, the urge of scientists to order and classify natural phenomena resulted in the well-known periodic table of the elements, which allowed chemists to predict unknown elements and their chemical properties and led to our understanding of electron shells. Some ecologists wonder whether something like a "periodic table of niches" might be possible. Of course, nothing about ecological niches is quite so simple or discrete as the number of electrons in the outer shell of a chemical element, but most aspects of niches have many more dimensions and are more continuous. Some patterns described earlier may be used to construct a very primitive periodic table of niches (Figure 13.13). Thus, trophic niches repeat themselves in organisms of different sizes that are relatively more or less $r$- and $K$-selected. An aphid is more like a lemming and a mantid more like a weasel, in their food niche, whereas in terms of body size and position on the $r$-$K$ selection continuum, the aphid and mantid are relatively alike, as are the lemming and the weasel. Niche dimensions other than those used in Figure 13.13, such as diurnal and nocturnal time of activity, could also be used to construct similar but different periodic tables. A periodic table of niches for aquatic organisms would differ somewhat from that for terrestrial ones, especially in that there are very few relatively $K$-selected aquatic primary producers. Perhaps one day, as the young science of ecology matures, we will be able to construct something analogous to, but much more complex than, the periodic table of the elements that will order niches, allow predictions, and improve our understanding of the elusive ecological niche.

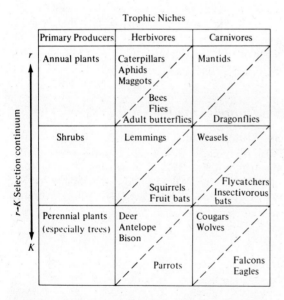

Figure 13.13 A crude "periodic table of niches" of terrestrial organisms, with examples. Dashed diagonal lines separate herbivores and carnivores that exploit space in two and three dimensions. Many other niche dimensions, such as time of activity, might profitably be used in such classifications.

# SELECTED REFERENCES

## History and Definitions

Allee et al. (1949);   Clarke (1954);   Colwell and Fuentes (1975);   Dice (1952);   Elton (1927);   Gaffney (1975);   Grinnell (1917, 1924, 1928);   Hutchinson (1957a);   Levins (1968);   MacArthur (1968, 1972);   McNaughton and Wolf (1970);   Odum (1959, 1971); Parker and Turner (1961);   Pianka (1976a);   Ross (1957, 1958);   Savage (1958);   Udvardy (1959);   Van Valen (1965);   Vandermeer (1972b);   Weatherley (1963);   Whittaker and Levin (1975);   Whittaker et al. (1973).

## The Hypervolume Model

Green (1971);   Hutchinson (1957a, 1965);   Maguire (1967, 1973);   Miller (1967); Shugart and Patten (1972);   Vandermeer (1972b);   Warburg (1965).

## Niche Overlap and Competition

Arthur (1987);   Case and Gilpin (1974);   Colwell and Futuyma (1971);   Hespenhide (1971);   Horn (1966);   Huey et al. (1974);   Hutchinson (1957a);   Inger and Colwell (1977);   Ivlev (1961);   Klopfer and MacArthur (1960, 1961);   Lock (1971); MacArthur (1957, 1960a, 1972);   MacArthur and Levins (1964, 1967);   May (1974, 1975b);   May and MacArthur (1972);   Miller (1964);   Orians and Horn (1969);   Pianka (1969, 1972, 1973, 1974, 1976a, 1976b, 1981b);   Pielou (1972);   Roughgarden (1972, 1974a, 1974b, 1974c);   Sale (1974);   Schoener (1968a, 1970, 1974a, 1986); Schoener and Gorman (1968);   Selander (1966);   Smouse (1971);   Terborgh and Diamond (1970);   Turelli (1981);   Vandermeer (1972b);   Willson (1973b).

## Niche Dynamics and Niche Dimensionality

Cody (1968);   Colwell and Fuentes (1975);   Colwell and Futuyma (1971);   Crowell (1962);   Gadgil and Solbrig (1972);   Huey et al. (1974);   Inger and Colwell (1977); Lack (1971);   Levins (1968);   MacArthur (1964, 1972);   MacArthur et al. (1972); MacArthur and Pianka (1966);   MacArthur and Wilson (1967);   MacMahon (1976); May (1975b);   Pianka (1973, 1974, 1975, 1976a);   Pianka et al. (1979);   Schoener (1974b).

## Niche Breadth

### Specialization versus Generalization

Charnov (1973, 1976a, 1976b);   Emlen (1966, 1968a);   Fox and Morrow (1981); Futuyma and Moreno (1988);   Greene (1982);   King (1971);   MacArthur (1972); MacArthur and Connell (1966);   MacArthur and Levins (1964, 1967);   MacArthur and Pianka (1966);   Orians (1971);   Roughgarden (1972, 1974a, 1974b, 1974c); Schoener (1971);   Stephens and Krebs (1986);   Van Valen (1965);   Willson (1969).

### Optimal Use of Patchy Environments

Emlen (1966, 1968a);   Hutchinson and MacArthur (1959);   Kamil and Sargent (1982);   Kamil et al. (1986);   King (1971);   Levins (1968);   MacArthur (1972); MacArthur and Levins (1964);   MacArthur and MacArthur (1961);   MacArthur and Pianka (1966);   Schoener (1969a, 1969b, 1971);   Stephens and Krebs (1986).

### The Compression Hypothesis

Crowell (1962);   MacArthur (1972);   MacArthur et al. (1972);   MacArthur and Pianka (1966);   MacArthur and Wilson (1967);   Schoener (1974b).

### Morphological Variation–Niche Breadth Hypothesis

Grant (1967, 1971);   Orians (1974);   Rothstein (1973);   Roughgarden (1972); Soulé and Stewart (1970);   Van Valen (1965);   Van Valen and Grant (1970);   Willson (1969).

## Evolution of Niches

Arthur (1987);   Hutchinson (1965);   Lawlor and Maynard Smith (1976);   MacArthur (1968, 1972);   MacArthur and Levins (1964, 1967);   Pianka (1993);   Roughgarden (1975, 1976);   Whittaker (1969, 1972).

## Periodic Table of Niches

Pianka (1986a, 1993).

Chapter
14

# Experimental Ecology

## DESIGN OF EXPERIMENTS

Experiments are designed to test hypotheses, typically based on simple causality. Ecological experiments are usually difficult to perform, even without insistence on proper statistical procedures. Because random and directed changes will probably be taking place during the time course of an experiment, a so-called *control*, a nearby untreated area, is essential for comparison with the manipulation. Such control plots should be randomly intermixed with treatment plots, both in space and in time. Moreover, both the control and the experiment must be replicated enough times to convince a statistician that there is a significant difference between the controls and the experiments. If such replicates are not truly independent of one another, the experimenter is committing "pseudoreplication" (Hurlbert, 1984). Few experiments in ecology can meet these standards, and many are not even replicated. Moreover, multiple causality and indirect effects make the unexpected results of many experiments difficult to interpret.

## ECOLOGICAL EXPERIMENTS

Although difficult, various manipulations of natural populations involving species removals, additions, and transplants have been informative. Terrestrial plants and relatively immobile animals, such as marine invertebrates in the rocky intertidal, are best suited for such studies.

A potent technique for studying the effects of predation in nature is to exclude predators from an area and then compare this experimental site with an adjacent "control" area, which is similar but unaltered, with normal access for predators. Hopefully, resulting differences between the two areas can best be ascribed to predation. Paine (1966) performed such a predator removal experiment along the rocky intertidal seacoast of the Olympic peninsula in Washington State. When the major top predator, the sea star *Pisaster ochraceus*, was removed, the number of species remaining was drastically reduced. Control areas with *Pisaster* supported some 15 species of marine invertebrates, but the area without the starfish had only 8 species. The rocky intertidal is a space-limited system. In the absence of predation, more efficient occupiers of space (so-called *competitive dominants*), especially the bivalve *Mytilus californianus* (a mussel), dominated the area. Presumably, by preying on *Mytilus*, *Pisaster* continually open up new patches that are rapidly colonized by fugitive species that are less efficient competitors for space. Thus, by reducing the intensity of competition at lower trophic levels, this starfish predator allows the coexistence of otherwise competitively incompatible species. Such *keystone* predators often have a powerful impact on community structure.

Competition for space between two species of barnacles, *Balanus balanoides* and *Chthamalus stellatus*, was studied by Connell (1961a, 1961b) along a rocky Scottish seacoast. These two crustaceans occupy sharply defined horizontal zones, as do most sessile organisms in the marine intertidal, with *Chthamalus* occupying an upper, and *Balanus*, a lower zone (Figure 14.1). Larvae of both species settle and attach over a wider vertical range than the zone occupied by adults. By manipulation of wire cages to exclude a snail predator (*Thais*) and by periodic removal of barnacles, Connell elucidated the various forces causing the zonation. He demonstrated that, although *Chthamalus* loses in competition with *Balanus*, adults persist in a narrow band because *Chthamalus* are more tolerant of physical desiccation than *Balanus*. Connell suggests that the lower limit of distribution of intertidal organisms is usually determined mainly by biotic factors, such as competition with other species and predation, whereas the upper limit is more often set by physical factors, such as dry conditions prevailing during low tides.

In another such field experiment, all individuals of the large intertidal sea star *Pisaster ochraceus* were removed from a small island reef and added to another similar reef, while a third undisturbed nearby reef was monitored as a "control" (Menge, 1972b). A smaller starfish *Leptasterias hexactis*, whose diet overlaps broadly with that of *Pisaster* (thus a potential competitor for food), was present on all three reefs. The average weight of individual *Leptasterias* increased significantly with removal of *Pisaster* and decreased with its addition, while the size of control *Leptasterias* remained constant (Figure 14.2). Estimated standing crops of the two species of sea stars varied inversely over a series of study areas (Figure 14.3), further implicating interspecific competition.

In another field manipulation, Dunham (1980) studied competition between two species of rock-dwelling iguanid lizards at Big Bend National Park

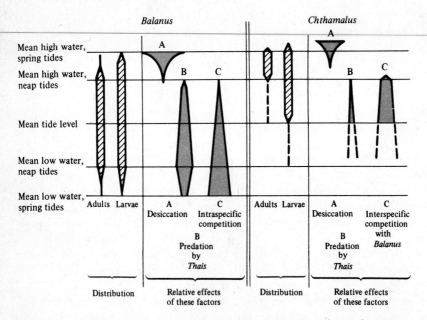

Figure 14.1 Vertical distributions of larvae and adults of two barnacle species, *Balanus balanoides* and *Chthamalus stellatus*, in the rocky intertidal of Scotland. Relative intensities of various limiting factors are represented diagrammatically by widths of shaded areas. [After Connell (1961b). By permission of Duke University Press.]

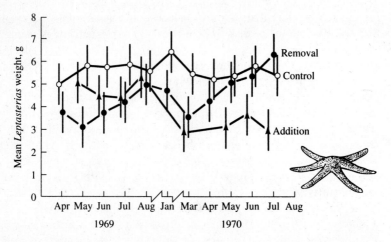

Figure 14.2 Response of individual *Leptasterias* (wet weights) to the removal or addition of *Pisaster* sea stars, with control. [From Menge (1972).]

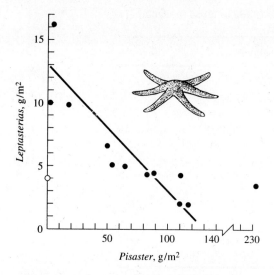

**Figure 14.3** Inverse correlation between the biomasses of *Pisaster* and *Leptasterias* across ten sites in Puget Sound. [After Menge (1972).]

in Texas using similar, but reciprocal, removal experiments. In two dry years, food supplies were apparently scant and removal of the larger lizard species (*Sceloporus merriami*) had numerous significant effects on the smaller species (*Urosaurus ornatus*), including increases in density, feeding success, growth rates, lipid levels, and prehibernation body weights. Treatments did not differ from controls during the two wet years, when insect food resources were presumably superabundant. Only one effect of the smaller species on the larger one was evident in these removal experiments—*Sceloporus* survival was significantly higher in one of the two dry years. Clearly, competition is not reciprocal and varies in intensity from year to year.

Perhaps one of the more ambitious experiments ever undertaken in ecology was reported by Munger and Brown (1981) and Brown et al. (1986). These investigators undertook labor-intensive manipulations to tease apart the interactions occurring among seed plants and their seed-eating predators, the heteromyid rodents and ants, in the Chihuahuan Desert in extreme southeastern Arizona. At a reasonably homogeneous flatland desert site, twenty-four 50 × 50 meter (0.25 hectare) plots were fenced off. Small holes (1.9 centimeters in diameter) in these fences allowed ingress and egress for small rodents but physically excluded all kangaroo rats (*Dipodomys*). (These fences were likened to "semi-permeable membranes.") Control plots had large 6.5-centimeter diameter holes, which allowed free movement of all rodents including *Dipodomys*. Rodents were removed by live trapping and ants by poisoning. In addition, four seed-addition treatments allowed inferences about the effects of seeds on seed predators. Eight plots had holes in their fences, eight were poisoned, and eight were seed-addition treatments, as follows. Eleven different experimental

TABLE 14.1  **Experimental Design in Study on Seed Predation in the Chihuahuan Desert**

| Plots | Treatment |
|---|---|
| 11,14 | Controls |
| 6,13 | Seed addition, large seeds, constant rate |
| 2,22 | Seed addition, small seeds, constant rate |
| 9,20 | Seed addition, mixed seeds, constant rate |
| 1,18 | Seed addition, mixed seeds, temporal pulse |
| 5,24 | Rodent removal, *Dipodomys spectabilis* (largest kangaroo rat) |
| 15,21 | Rodent removal, all *Dipodomys* (kangaroo rat) species |
| 7,16 | Rodent removal, all seed-eating rodents |
| 8,12 | *Pogonomyrmex* harvesting ants |
| 4,17 | All seed-eating ants |
| 3,19 | All *Dipodomys* plus *Pogonomyrex* ants |
| 10,23 | All seed-eating rodents plus all seed-eating ants |

*Source:* From Brown et al. (1986).

manipulations were each replicated once at random (i.e., in two plots). The remaining two plots were untreated "controls" (Table 14.1).

Rodent removals resulted in an increase in *Pheidole xerophila*, a small ant species. Seed addition resulted in increases in rodent densities. Removal of large rodents resulted in an increase in the densities of smaller rodents. On plots where all species of rodents were removed, large-seeded species of plants gained at the expense of small-seeded species. In the short term, rodents and ants competed for seeds. However, over the long term (treatments were monitored for 10 years!), ants indirectly benefited rodents via facilitation: rodents prefer large seeds and ants prefer small seeds, but because large-seeded plants compete with small-seeded plants, the presence of rodents indirectly benefits ants. The direct impact of ants on rodents, however, appears to be negligible.

Hundreds of such manipulative field experiments have now been undertaken, both to demonstrate the importance of competition and predation, although their design is too often less than perfect (Hairston, 1989; Underwood, 1986). Schoener (1983) reviewed the extensive literature on competition experiments, which include both aquatic and terrestrial as well as plant and animal studies; a full 90 percent of the 164 studies and 76 percent of the species examined exhibit evidence of interspecific competition. In a similar review of predation removal-addition experiments, Sih et al. (1985) reported even higher percentages of studies demonstrating the significant impact of predators.

Such experiments suffer from at least two major shortcomings: first, failure to detect a niche shift in response to removal of a potential competitor may simply be an indication that the target population is not ecologically flexible. [The idea that species have become evolutionarily locked into particular adaptive zones due to competition in the historical past has been termed "the ghost of competition past" by Rosenzweig (1979) and others.] Likewise, predator removal experiments can fail in systems with donor control where predators consume only prey that would die anyway from other causes. Second, and perhaps of greater interest and importance, the extent to which indirect effects mediated through other members of a community network act in opposition to the direct pairwise interaction is not easily determined (Bender et al., 1984). Indeed, Sih et al. (1985) reported surprisingly high numbers of "unexpected" results, which could well stem from indirect interactions (Pimm, 1991).

## A DEFAUNATION EXPERIMENT

An interesting ecological experiment was performed by Simberloff and Wilson (1970, and included references). After censusing the entire arthropod faunas of several very tiny mangrove islets in the Florida Keys, all arthropods were exterminated by fumigation with a deadly gas (methyl bromide). The vegetation survived relatively unscathed, but even insect eggs were killed. The process of recolonization was then monitored over a two-year period. Arthropods rapidly recolonized; within a mere 200 days, numbers of species on the islets had stabilized (Figure 14.4). Although turnover rates remained quite high throughout the experimental period, numbers of species on the islets remained relatively constant over a period of nearly two years, providing strong evidence that the islands have indeed reached an equilibrium. Two islands, $E1$ and $E2$, seem to

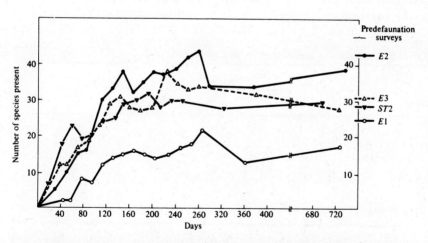

**Figure 14.4** Colonization curves (number of arthropod species through time) of four tiny mangrove islets in the Florida keys following complete extermination of the arthropod fauna, which resulted in relatively little damage to the vegetation. Numbers of species on each islet first rose rapidly, then soon leveled off to species densities fairly close to those before defaunation (plotted on the right vertical axis). [After Simberloff and Wilson (1970). By permission of Duke University Press.]

have reached equilibrium at a slightly lower species density after defaunation. The apparent depression in the number of species at equilibrium may indicate that the species that reinvaded these islands interfere with one another more than the members of the original communities did; alternatively, this reduction in species density may be due to the fact that new immigrants are not adapted to exploit the island's resources as well as were the original inhabitants. In any case, these results demonstrate that the actual composition of an island's fauna may itself partially determine the equilibrium number of species an island will support.

An instructive analysis of these data was undertaken by Heatwole and Levins (1972), who tallied up the numbers of species of insects in various (somewhat arbitrary) trophic categories before and after defaunation (Table 14.2). The apparent return to a similar state could be evidence of resiliency and suggests that an equilibrium trophic structure might exist. However, the species pool of potential invaders contains many more species of herbivorous insects, ants, and other carnivores than scavengers, detritovores, wood borers, or parasites, so that this apparent "return to an equilibrium state" could also arise as an artifact of sampling (Simberloff, 1976; May, 1981).

Wilson (1969) suggests that an island community may experience a sequence of several distinct sorts of equilibria through time (Figure 14.5). First, a "noninteractive" equilibrium in the number of species present may be reached even before component populations come to equilibrium demographically and with one another. Then, because competitive and predatory interactions are intensified as populations saturate the island with individuals, a second "interactive" equilibrium is reached. Both stages should occur relatively rapidly. Wilson (1969) envisions two other types of equilibria, which would require considerably greater time spans to attain. As various species go extinct and others

TABLE 14.2 **Evidence for Stability of Trophic Structure?**

| | Trophic classes | | | | | | | | |
|---|---|---|---|---|---|---|---|---|---|
| Island | H | S | D | W | A | C | P | ? | Total |
| E1 | 9(7) | 1(0) | 3(2) | 0(0) | 3(0) | 2(1) | 2(1) | 0(0) | 20(11) |
| E2 | 11(15) | 2(2) | 2(1) | 2(2) | 7(4) | 9(4) | 3(0) | 0(1) | 36(29) |
| E3 | 7(10) | 1(2) | 3(2) | 2(0) | 5(6) | 3(4) | 2(2) | 0(0) | 23(26) |
| ST2 | 7(6) | 1(1) | 2(1) | 1(0) | 6(5) | 5(4) | 2(1) | 1(0) | 25(18) |
| E7 | 9(10) | 1(0) | 2(1) | 1(2) | 5(3) | 4(8) | 1(2) | 0(1) | 23(27) |
| E9 | 12(7) | 1(0) | 1(1) | 2(2) | 6(5) | 13(10) | 2(3) | 0(1) | 37(29) |
| Total | 55(55) | 7(5) | 13(8) | 8(6) | 32(23) | 36(31) | 12(9) | 1(3) | 164(140) |

*Note:* The table is after Heatwole and Levins (1972). The islands are labeled in Simberloff and Wilson's (1969) original notation, and on each the fauna is classified into the trophic groups: herbivore (H); scavenger (S); detritus feeder (D); wood borer (W); ant (A); predator (C); parasite (P); class undetermined (?). For each trophic class, the first figures are the number of species before defaunation, and the figures in parentheses are the corresponding numbers after recolonization. The total number of different species encountered in the study was 231 (the simple sum 164 + 140 counts some species more than once).

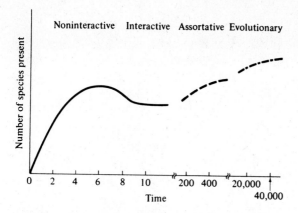

**Figure 14.5** Hypothetical sequence of community equilibria on an island through time. The time scale is arbitrary and intended only to emphasize the vastly greater time spans required for the assortative and evolutionary equilibria (see text). [From Wilson (1969).]

invade, the composition of an island's biota may gradually change until a certain set of species from the available species pool is reached that is composed of species with particularly low extinction rates; Wilson (1969) terms this the "assortative" equilibrium. Finally, given a much longer time span, the component species could actually evolve minimal extinction rates and an "evolutionary" equilibrium might be reached.

## PHYLOGENY AND THE MODERN COMPARATIVE METHOD

Evolution of phenotypes (or species) in a range of different environments is analogous to the response of individuals in a nested sequence of experimental treatments (Harvey and Pagel, 1991). The analogue to an experiment begins by subjecting a population of individuals to the same treatment at the base of a monophyletic tree (Figure 14.6). Subpopulations are divided and redivided sequentially after varying intervals of time. Between divisions, each subpopulation is subjected to a particular environmental "treatment" which differs from that experienced by other subpopulations.

Under this analogy, however, no record is kept of the "treatments" administered to various subpopulations. An estimate of the probable historical record of "treatments" and "responses" can be reconstructed, however, from appropriate information on extant lineages, using parsimony, as follows. We assume that past environments for a given lineage tend to be similar to present environments ("treatments"). Organisms tend to occupy habitats similar to those

| First state: | 1 | 2 | 1 | 1 | 2 | 2 | 2 | 2 |
|---|---|---|---|---|---|---|---|---|
| Second state: | 1 | 2 | 1 | 1 | 2 | 2 | 2 | 1 |
| Environment: | A | B | C | D | E | F | G | H |
| Phenotype: | a | b | c | d | e | f | g | h |

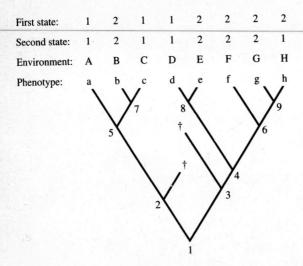

Figure 14.6 A phylogenetic tree can be viewed as an analogue to a nested experiment. At the basal node 1, a population is subdivided into two subpopulations, which are subjected to different environmental "treatments." At each subsequent node, subpopulations are redivided and subjected to new varying "treatments." Probable states at ancestral nodes can be estimated from states of existing extant species at the tips of the phylogenetic branches. [Adapted from Harvey and Pagel (1991).]

occupied by their ancestors. Extant populations contain information about their evolutionary history because the structure of the phylogenetic tree is known (Figure 14.6). For example, the ancestral subpopulation at node 4 is likely to have had condition 2 at both states because a majority of descendant subpopulations exhibit that condition. However, node 5 would be more likely to have exhibited condition 1 as judged from the prevalent condition in its descendants. Note also that the states of extant subpopulations tend to covary across environments ("treatments"), making the condition of either state a good predictor of the condition of the other state. Phenotypes $a$ and $d$ have probably undergone convergent evolution, as have phenotypes $b$ and $e$, and both $b$ and $e$ have converged with $f$ and $g$.

# SELECTED REFERENCES

## Design of Experiments

Fisher (1935);   Hairston (1989);   Hurlbert (1984);   Krebs (1989);   Underwood (1986).

## Ecological Experiments

Bender et al. (1984);   Brown et al. (1986);   Connell (1961a, 1961b);   Dunham (1980); Hairston (1989);   Lubchenco (1978, 1980);   Menge (1972a, 1972b);   Menge and Menge (1974);   Munger and Brown (1981);   Paine (1966);   Pimm (1991);   Rosenzweig (1979);   Salzburg (1984);   Schoener (1983);   Sih et al. (1985);   Underwood (1986).

## A Defaunation Experiment

Heatwole and Levins (1972);   May (1981);   Simberloff (1976);   Simberloff and Wilson (1970);   Wilson (1969).

## Phylogeny and the Modern Comparative Method

Brooks and McLennan (1991);   Felsenstein (1985, 1988);   Garland (1992);   Garland et al. (1991, 1992);   Harvey and Pagel (1991);   Harvey and Purvis (1991);   Huey and Bennett (1987);   Ridley (1983).

*Chapter*
## 15

# Predation and Parasitism

## PREDATION

Predation is often readily observed and easily studied, and neither its existence nor its importance in nature is doubted. It is directional in the sense that one member of the pair (the predator) benefits from the association while the other (the prey) is affected adversely. In contrast, competition is a symmetric process in that both species are affected adversely and, where possible, each tends to evolve mechanisms whereby the relationship with the other is avoided.

Individual predators that are better able to capture prey should have more resources at their disposal and should therefore normally be more fit than those that are less proficient at capturing prey. Hence, natural selection acting on the predator population tends to increase the predator's efficiency at finding, capturing, and eating its prey. However, members of the prey population that are better able to escape predators should normally be at a selective advantage within the prey population. Thus, selection on the prey population favors new adaptations that allow prey individuals to avoid being found, caught, and eaten. Obviously, these two selective forces oppose one another; as the prey become more adept at escaping from their predators, the predators in turn evolve more efficient mechanisms for capturing them. Hence, in the evolution of a prey-predator relationship, the prey evolves so as to dissociate itself from the interaction, while the predator continually maintains the relationship. Long-term evolutionary escalations of this sort have resulted in some rather intricate and often exceedingly complex adaptations. Consider, for instance, the complex social hunting behavior of lions and wolves; the long sticky tongues and accurate aim of some fish, toads, and certain lizards; the folding fangs and venom-injection apparatus of viperine snakes; spiders and their webs; the deep-sea angler fish; and snakes such as boas that suffocate their prey by constriction. Other examples include the rapid and very accurate strikes of predators as diverse as praying mantids, dragonflies, fish, lizards, snakes, mammals, and birds. Prey have equally elaborate predator escape mechanisms, such as the posting of sentinels, predator alarm calls, background color matching, and thorns. Many prey organisms recognize their predators at some distance and employ appropriate avoidance tactics well before the predator gets close enough to make a kill; this behavior in turn has forced many predators to hunt by ambush.

One of my favorite examples of the joint adaptations of a predator and its prey is provided by the starling and the peregrine falcon. The peregrine is a magnificent bird-eating hawk whose hunting behavior must be seen to be fully appreciated. These falcons take other birds as large as themselves; nearly all prey is captured on the wing and in the air. Peregrines have exceedingly keen vision; foraging individuals climb high up into the sky and move across country. When a potential prey is sighted flying along below, the peregrine closes its wings and dives or "stoops." To make its ambush most effective, the falcon often "comes out of the sun" at its prey. Diving peregrines have been estimated to reach speeds of over 300 kilometers per hour (nearly 100 meters per second!). Most prey are killed instantly by the sudden jolt of the peregrine's talons. Large prey are allowed to fall to the ground and are eaten there, but smaller items may be carried away in the air. (Little wonder that falconers and their dogs find it extremely difficult to get small birds to fly when a peregrine is "waiting on" overhead! On occasion, game birds allow themselves to be overtaken on the ground by dogs in preference to taking to the air and risking the falcon's deadly stoop.) Starlings normally fly in loose flocks, but when a peregrine is sighted, often at a considerable distance, they quickly assume a very tight formation (Figure 15.1). Tight flocking is a response specific to the peregrine and is not employed with other hawks. Falcons are much less likely to attack a tight flock than a single bird; indeed, "stragglers" slightly outside of the starling flock formation are often taken by peregrines. Presumably, the falcon itself could be injured if it were to stoop into a tight flock. Thus, even a predator as effective as the peregrine falcon has had its hunting efficiency impaired by appropriate behavioral responses of its prey.

There is an important difference between predation on animals and predation on plants. In most animals, predation is an all-or-nothing proposition in that the predator kills the prey outright and consumes most or all of it; however, when plants are eaten, usually only a portion of the plant is consumed by its predator. Hence, predation on plants (herbivory) is more like parasitism among animals. But even partial predation must often reduce a prey individual's ability

**Figure 15.1** Starlings normally fly in loose flocks (left), but when under attack by a peregrine falcon, they fly as close together as possible (right).

to survive and to reproduce. Because of this fundamental difference, though, selective pressures on animals to avoid being eaten may be stronger than they are on plants. Nevertheless, plants have evolved elaborate antipredator devices (see pp. 329–332).

## PREDATOR–PREY OSCILLATIONS

In terms of their population dynamics, plus-minus interactions are perhaps most interesting because they display a natural tendency to oscillate (May, 1973, 1981; Toft, 1986). Yet these interactions can be stabilized by allowing certain sorts of realistic self-damping to occur. In some ways, however, the theory of predation has lagged behind that for competition; perhaps its asymmetry makes it more difficult to model. Lotka (1925) and Volterra (1926a, 1926b; 1931) wrote a simple pair of equations to model prey and predator populations:

$$\frac{dN_1}{dt} = r_1 N_1 - p_1 N_1 N_2 \tag{1}$$

$$\frac{dN_2}{dt} = p_2 N_1 N_2 - d_2 N_2 \tag{2}$$

where $N_1$ is prey population density, $N_2$ is population density of the predator, $r_1$ is the instantaneous rate of increase of the prey population (per head), $d_2$ is the death rate of the predator population (per head), and $p_1$ and $p_2$ are predation constants. The term $p_1N_1$ represents the functional response, describing the consumption response of individual predators to changes in prey density, a constant in these simple equations. The $p_2N_1$ term in equation (2) represents the numerical response, describing the way in which prey are converted into new predator individuals. Each population is limited by the other and there are no self-limiting density effects (i.e., no second-order $N_1$ or $N_2$ terms). Thus, in the absence of the predator, the prey population expands exponentially and the rate of increase of the prey population is potentially unlimited. The product of the densities of the two species, $N_1N_2$, reflects the number of contacts between them; after multiplication by the constant, $p_2$, this term becomes the maximal rate of *increase* of the predator population ($p_2N_1N_2$). The same term multiplied by the constant, $p_1$, appears with a negative sign in the prey equation and acts to *decrease* the rate of growth of the prey population.

The equations are solved by setting $dN/dt$ equal to zero, factoring out the appropriate $N$ to get the actual rate of increase, and setting this $r_a$ equal to zero. These algebraic manipulations show that the prey reaches an equilibrium population density when the predator's density is $r_1/p_1$; similarly, the predator is at equilibrium when the prey's density is $d_2/p_2$ (Figure 15.2). Thus, each species' isocline corresponds to a particular (constant) density of the other species, and again there is no self-damping term such as the $-zN^2$ term in the competition equations. Below some threshold prey density, predators always decrease, whereas above that threshold they increase; similarly, prey increase

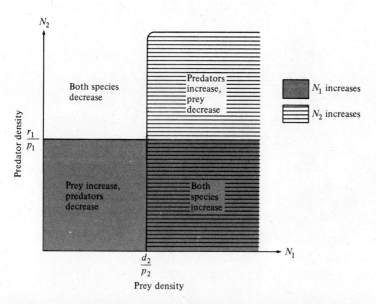

Figure 15.2 Prey and predator isoclines for the Lotka-Volterra prey-predator equations (see text).

below a particular predator density but decrease above it (Figure 15.2). A joint equilibrium exists where the two isoclines cross, but prey and predator densities do not converge on this point. Rather, any given initial pair of densities results in oscillations of a certain magnitude. Initial densities near the joint equilibrium point result in repeating oscillations of low amplitude; initial densities farther from the joint equilibrium point generate oscillations of greater amplitude. Thus, this pair of differential equations has a periodic solution, with the population densities of both prey and predator changing cyclically and out of phase over time. The amplitude of the fluctuations depends on initial conditions. Mathematically, a system of such repeating and undamped oscillations is termed *neutrally stable*. Neutral stability probably does not exist in the biological world since most individuals and populations encounter either self-regulation or density-dependent feedback.

Addition of a simple self-damping term $(-zN_1^2)$ to the prey equation results either in a rapid approach to equilibrium or in damped oscillations, both of which lead eventually to the joint equilibrium (Figure 15.3). However, a self-damping term for the predator should include the prey's density as a determinant of the predator's carrying capacity. Perhaps a more realistic (although mathematically less tractable) pair of simple equations for modeling the prey-predator relationship is:

$$\frac{dN_1}{dt} = r_1N_1 - z_1N_1^2 - \beta_{12}N_1N_2 \tag{3}$$

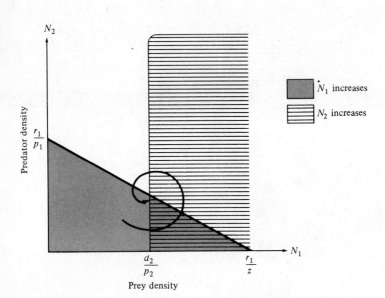

Figure 15.3 Prey and predator isoclines with self-damping in the prey population. Population densities converge on the stable joint equilibrium.

$$\frac{dN_2}{dt} = \gamma_{21}N_1N_2 - \beta_2\frac{N_2^2}{N_1} \tag{4}$$

The prey equation is the simple Lotka-Volterra competition equation, but the predator equation has a new twist in that competitive inhibition of the predator population is now a function of the relative densities of predator and prey. Thus, inhibition of the predator population increases both with increased predator density and with decreased prey density. Notice also that the predator population cannot increase unless there are some prey. However, even though this pair of equations overcome some of the faults of previous pairs, they are still unrealistic in at least one important way. Imagine a situation in which there are more prey than the predator population can possibly exploit; in such a case, growth rate of the predator cannot be simply proportional to the product of the two densities as in equation (4), but some sort of threshold effect must be taken into account.

Equations like the preceding ones entirely omit many important subtleties from the prey-predator interaction. For instance, Solomon (1949) distinguished two separate components of the way in which predators respond to changes in prey density. First, individual predators capture and eat more prey per unit time as prey density increases until some satiation threshold is reached, above which the number of prey taken per predator is more or less constant (Figures 15.4 and 15.5a); second, increased prey density raises the predator's population size and a greater number of predators eat an increased number of prey (Figure 15.5b). Solomon termed the former the *functional response* and the latter the *numerical response* of the predator. Three types of functional responses are recognized, representing pure forms among a continuum of possibilities (Figure 15.4). [Equations (1) and (3) model a Type 1 linear functional response without a ceiling.] Note that a predator's functional response can allow regulation of prey density without an increase in predator numbers (no numerical response). Using the "systems" approach that relies on continued feedback between observation and model, Holling (1959a, 1959b, 1966) developed elaborate models of predation incorporating both the functional and the numerical responses as well as other parameters, including various time lags and hunger level. These models are more realistic and descriptive than the others (see

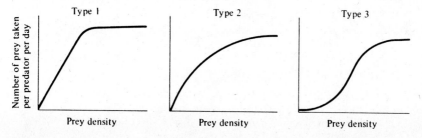

Figure 15.4 Three types of functional responses. [After Holling (1959a).]

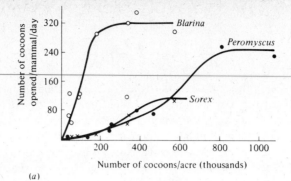

(a)

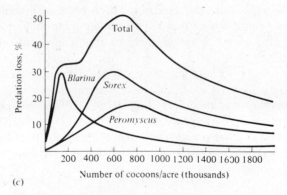

(b)

(c)

**Figure 15.5** (a) Number of cocoons (prey) eaten per mammal per day by three small mammals plotted against the density of their prey (the so-called functional response). (b) Density of each of the three mammal predators plotted against prey density (the so-called numerical response of the predators). (c) Combined functional plus numerical responses of each predator species and the total, which represents the overall intensity of predation on the prey population, as a function of prey density. [From Holling (1959a).]

previous discussion), but they are also more complex and restricted. Obviously, a realistic model of prey-predator relationships *must* be quite complex!

A simple graphical model of the prey-predator interaction was developed by Rosenzweig and MacArthur (1963), who reasoned somewhat as follows. In the absence of predators, the maximum equilibrium population density of the prey is $K_1$, the prey's carrying capacity. Similarly, some lower limit on prey density is likely to exist, below which contacts between individuals are too rare to ensure reproduction and the prey population thus decreases to extinction. Likewise, at any given density of prey, there must be some maximal predator density that can just be supported without either an increase or a decrease in the prey population. Using these arguments, a prey isocline ($dN_1/dt = 0$) can be drawn in the $N_1 - N_2$ plane (Figure 15.6) similar to those drawn earlier in Figures 15.2 and 15.3. As long as the prey isocline has but a single peak, the exact shape of the curve is not important to the conclusions that can be derived from the model. Above this line, prey populations decrease; below it they increase. Next, consider the shape of the predator isocline ($dN_2/dt = 0$). Below some threshold prey density, individual predators cannot gather enough food to replace themselves and the predator population must decrease; above this threshold prey density, predators will increase. For simplicity, first assume (this assumption can be relaxed) that there is little interaction or competition between predators, as would occur when predators are limited by some factor other than prey availability. Given this assumption, the predator isocline should look somewhat like that shown in Figure 15.7a. If there is competition between predators, higher predator densities will require denser prey

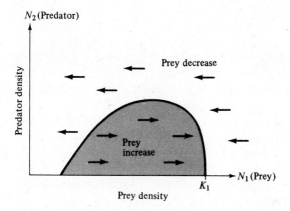

Figure 15.6 Hypothetical form of the isocline of a prey species ($dN_1/dt = 0$) plotted against densities of prey and predator. Prey populations increase within the shaded region and decrease above the line enclosing it. Prey at intermediate densities have a higher turnover rate and will support a higher density of predators without decreasing.

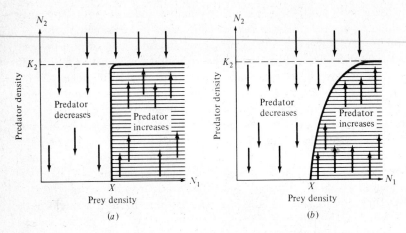

**Figure 15.7** Two hypothetical predator isoclines. (*a*) Below some threshold-prey density, $X$, individual predators cannot capture enough prey per unit time to replace themselves. To the left of this threshold-prey density, predator populations decrease; to the right of it, they increase provided that the predators are below their own carrying capacity, $K_2$ (i.e., within the cross hatched area). So long as predators do not interfere with one another's efficiency of prey capture, the predator isocline rises vertically to the predator's carrying capacity, as shown in (*a*). (*b*) Should competition between predators reduce their foraging efficiency at higher predator densities, the predator isocline might slope somewhat like the curve shown. More rapid learning of predator escape tactics by prey through increased numbers of encounters with predators would have a similar effect.

populations for maintenance and the predator isocline will slope somewhat as in Figure 15.7*b*. In both examples, the carrying capacity of the predator is assumed to be set by something other than prey density. Only one point in the $N_1 - N_2$ plane represents a stable equilibrium for both species—the point of intersection of the two isoclines (where $dN_1/dt$ and $dN_2/dt$ are both zero). Consider now the behavior of the two populations in each of the four quadrants marked $A$, $B$, $C$, and $D$ in Figure 15.8. In quadrant $A$, both species are increasing; in $B$, the predator increases and the prey decreases; in $C$, both species decrease; and in $D$, the prey increases while the predator decreases. Arrows or vectors in Figure 15.8 depict these changes in population densities.

Relative magnitudes of the changes in the population densities of prey and predator determine another important property of this model—that is, whether or not a stable equilibrium exists. There are three cases, corresponding to vectors that (1) spiral inward, (2) form a closed circle, or (3) spiral outward (Figure 15.8*a*, 15.8*b*, 15.8*c*). These three cases correspond to damped oscillations, oscillations of neutral stability, and oscillations increasing in amplitude until a limit cycle is reached, respectively (Figure 15.8*a*, 15.8*b*, 15.8*c*). Such oscillations of prey and predator could perhaps produce population "cycles" like those of lemmings and their predators (pp. 194–199). Given time, the case with damped oscillations will reach its equilibrium value at which neither prey

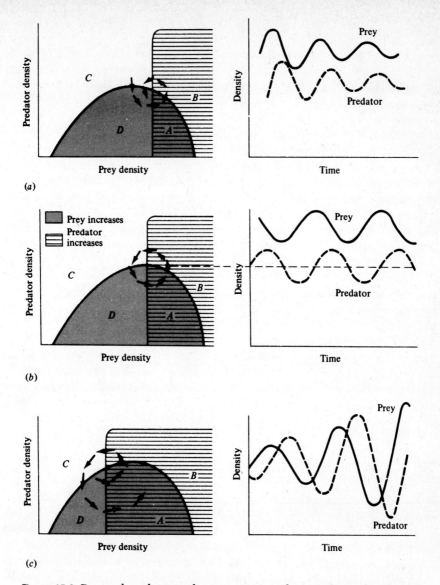

**Figure 15.8** Prey and predator isoclines superimposed upon one another to show stability relationships. (*a*) An inefficient predator that cannot successfully exploit its prey until the prey population is near its carrying capacity. Vectors spiral inward, prey-predator population oscillations are damped, and the system moves to its joint stable equilibrium point (where the two isoclines cross). (*b*) A moderately efficient predator that can begin to exploit its prey at some intermediate density. Vectors here form a closed ellipse, and populations of prey and predator oscillate in time with neutral stability, as in Figure 15.2. (*c*) An extremely efficient predator that can exploit very sparse prey populations near their limiting rareness. Vectors now spiral outward and the amplitude of population oscillations increases steadily until a limit cycle is reached, often leading to the extinction of either the predator or both the prey and the predator. Such a cyclical interaction can be stabilized by providing the prey with a refuge from predators. [After MacArthur and Connell (1966).]

nor predator population densities change; this case corresponds to a predator that is relatively *inefficient* at gathering prey (the predator cannot even begin to exploit the prey population until prey are fairly near their own carrying capacity). Similarly, the case that produces oscillations of increasing amplitude corresponds to a very *efficient* predator, which can exploit the prey population nearly down to its limiting rareness. Such an overly efficient predator should rapidly exterminate its prey (and thus, go extinct itself unless alternative prey are available); little wonder that increasing oscillations are never observed in nature! Because of predator escape tactics of prey, many (or most) real predators are probably relatively inefficient, tending to crop only those prey present in excess of a substantial prey population (Errington, 1946). Damped oscillations also result from competition among predators, producing a sloping predator isocline (Figure 15.7*b*). Hence, the case with damped oscillation is probably the most realistic reflection of nature.

Individual predators that reproduce successfully at low prey densities will normally outcompete and replace less efficient individuals that require higher prey densities; hence, natural selection acting on the predator moves its isocline to the left and *reduces* the stability of the interacting system.

However, selection operating in favor of those prey individuals best able to escape predators opposes the action of selection on the predators and forces the predator isocline to the right (presumably it also raises the prey isocline); thus, natural selection on the prey population tends to *increase* the stability of the system. Indeed, unless the prey is one step ahead of the predator, the latter can be expected to overeat its prey and take both populations to extinction.

## "PRUDENT" PREDATION AND OPTIMAL YIELD

Some have suggested that an intelligent predator should crop its prey so as to maximize the prey's turnover rate and therefore the predator's yield. Such a "prudent" predator would maintain the prey population at the density that gives the maximum rate of production of new prey biomass. In terms of the prey isoclines depicted in Figure 15.8, this prey density of "optimal yield" corresponds to the density at the peak of the prey isocline. Humans have the capacity to be such a prudent predator; indeed, optimal yield has long been a goal in management of exploited populations in fisheries biology. But do other, less intelligent predators also maximize their yield? A truly "prudent" predator should prey preferentially upon those prey individuals with low growth rates and low reproductive values and leave those with rapid growth rates and higher reproductive values alone. In fact, predators often do take the aging and decrepit prey individuals, which are frequently easy to catch, while younger, more vigorous ones escape.

However, there is a potential flaw in the prudent predation interpretation, provided that several predator individuals encounter the same prey items. If, say, young juicy prey are less experienced and easier to catch, an individual predator who "cheated" and ate them would be likely to leave more genes than the prudent genotypes that did not exploit this food supply; as a result, non-prudence would become incorporated into the gene pool and spread. Exactly the same considerations apply to a competing species that is able to use the prey individuals in question. Hence, we would expect prudence to evolve only in a situation where a single predator has exclusive use of a prey population; perhaps some feeding territories are examples.

Another, much more likely, explanation for the occurrence of apparent "prudent" predation in nature can be made in terms of the prey organisms themselves. As was pointed out earlier, the intensity of natural selection is directly proportional to expectation of future offspring (reproductive value); thus, one might predict that individual prey with high reproductive values would have more to gain from escaping predators than would those with lower reproductive values. After one's expectation of future offspring has dropped to zero, nothing further can be gained from being able to escape a predator and predator avoidance cannot be evolved. Thus, many cases of apparent "prudent" predation may well be simply part and parcel of old age; the evolution of senescence has wide significance! Viewed in this way, the susceptibility of prey to predation should be inversely related to their reproductive value.

## SELECTED EXPERIMENTS AND OBSERVATIONS

Predation is readily studied in the laboratory and, under certain favorable circumstances, in the field. Gause (1934) studied a simple prey-predator system in the laboratory using two microscopic protozoans, *Paramecium caudatum* and *Didinium nasutum*. *Didinium* are voracious predators on *Paramecium*. When both species were placed together in a test tube containing clear medium (which supports a culture of bacteria, the food for *Paramecium*), *Didinium* overate its food supply, exterminated it, and then starved to death itself (Figure 15.9a). When some sediment was added to the medium (making it "heterogeneous" rather than "homogeneous"), providing a refuge or "safe site" for the prey, *Didinium* went extinct but the *Paramecium* population recovered (Figure 15.9b). In a third experiment (Figure 15.9c), Gause introduced new individuals of each species at regular time intervals; such "immigrations" resulted in two complete cycles of prey and predator. In other experiments, using *Paramecium aurelia* as the predator and a yeast, *Saccharomyces exiguus*, as prey, Gause (1935) obtained nearly three complete cycles.

In an experiment like Gause's with *Paramecium* and *Didinium* (but with *P. aurelia* as prey), Luckinbill (1973) showed that an unstable prey-predator interaction could be stabilized simply by adding methyl cellulose to the medium: this slowed down movements of both prey and predator and reduced the rate of contact between them.

Huffaker (1958) performed similar laboratory experiments on two species of mites, using oranges as the plant food for the system. One mite was herbivorous, eating the oranges; the second mite was a predator on the herbivorous one. In simple systems with oranges close together and evenly spaced, the predator simply overate its prey and both species became extinct. Increasing distances between oranges only lengthened the time required for extinction

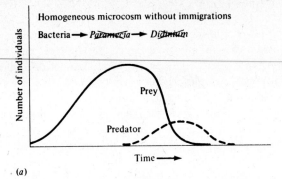

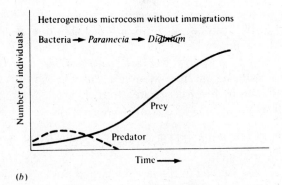

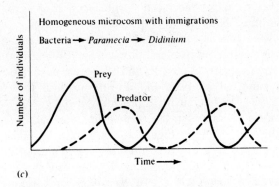

**Figure 15.9** Three laboratory prey-predator experiments with protozoans. (*a*) In a simple homogeneous microcosm without immigration of new prey or predators, the predator quickly overeats and exterminates its prey and then all predators themselves starve to death. (*b*) In a more heterogeneous system, the predator goes extinct and the prey population recovers to expand to its carrying capacity. (*c*) Even in a homogeneous microcosm, immigration of prey and predators results in both populations oscillating in time. [From Gause (1934), reprint ed. Copyright © 1964 by The Williams and Wilkins Co., Baltimore, Md.]

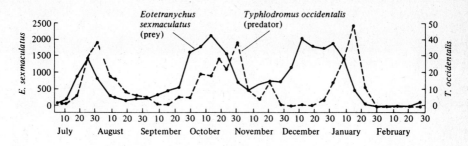

Figure 15.10 Fluctuations in the population size of a predatory mite, *Typhlodromus occidentalis*, and its prey (another mite), *Eotetranychus sexmaculatus*, in a spatially heterogeneous environment. [After Huffaker (1958).]

but did not allow coexistence. However, by introducing barriers to dispersal and making the system still more complex, Huffaker obtained three complete prey-predator cycles (Figure 15.10). Thus, environmental heterogeneity increased the stability of the system of a predator and its prey. In addition, these experiments illustrate the existence of prey-predator oscillations as predicted by theory.

In a very revealing (although more complex) laboratory investigation, Utida (1957) examined both competition and predation simultaneously. His system was composed of three species: a beetle (*Callosobruchus chinensis*) as prey and two species of predatory wasps as competitors (*Neocatolaccus mamezophagus* and *Heterospilus prosopidis*). The prey beetle was provided with a continually renewed food supply. Both species of wasps, which have similar life histories, were dependent on the beetle population as a common food source. Population densities of the three species fluctuated widely and erratically (Figure 15.11), but after four years, some 70 generations later, all three were still coexisting. Populations of the two predatory wasps tended to fluctuate out of phase with one another. Analysis showed that *Heterospilus* was more efficient at finding and exploiting the beetle population when it was at low densities, but *Neocatolaccus* was more efficient at high prey densities (Figure 15.12); thus, the

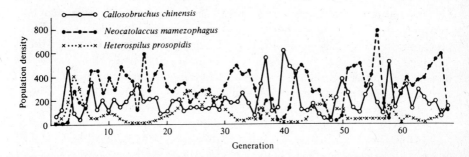

Figure 15.11 Changes in the population sizes of a beetle host (*Callosobruchus*) and two predatory wasps over a four-year period. Although fluctuations appear erratic, all three species survived the entire period of four years (some 70 generations). [After Utida (1957).]

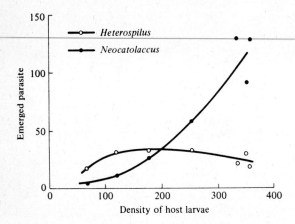

**Figure 15.12** Number of parasitic wasps emerging from larvae of the host beetle plotted against the density of these larvae. *Neocatolaccus* is a more successful parasite at high host densities, whereas *Heterospilus* has the advantage at lower densities of host larvae. [After Utida (1957).]

competitive advantage shifted between the two wasps as the density of beetle prey changed in time. Utida (1957) thought that fluctuations of the beetle population were caused both by the effects of the two wasp predators as well as by density-dependent changes in the rate of reproduction of the beetle itself. Thus, this system provides an example of coexistence of two competitors on a single resource due to changing abundance of that resource. The stability of the system is apparently a result of biotic interactions.

## EVOLUTIONARY CONSEQUENCES OF PREDATION: ESCAPE TACTICS

Generalized predators feeding on a variety of prey obviously must be adapted to cope with a wider variety of predator escape tactics than more specialized predators that normally deal with fewer types of prey. By diverging in their "strategies" of predator avoidance over evolutionary time, individuals of two (or more) prey species can make it increasingly difficult for a single predator to capture and exploit both prey types efficiently; thus, prey evolution can "force" a predator to restrict the range of foods it eats. Such evolution of a diversity of escape tactics among prey, especially morphological diversity, has been termed *aspect diversity* (Rand, 1967; Ricklefs and O'Rourke, 1975; Schall and Pianka, 1980).

Antipredator devices are extremely varied. Some mechanisms for predator escape are quite simple and straightforward, but others may be exceedingly intricate and subtle. As an example of the former, many lizards dig special escape tunnels in their burrow systems that come up near the surface and allow the lizard to break out should a predator corner it underground.

Behavior and anatomy often make animals difficult to detect and to follow; such cryptic adaptations can involve sound, smell, color, pattern, form, posture, and movement. Concealing or cryptic coloration is widespread and often depends on appropriate behavior; to hide itself, an animal must select the proper background and orient itself correctly. Some moths normally align the dark markings of their wings to be parallel with cracks and crevices of the tree bark substratum. Nearly all diurnal animals and some nocturnal ones are countershaded, with their dorsal (upper) parts darker than their ventral (lower) parts. Lighting from above casts shadows below; in a countershaded animal, these balance dorsoventrally, reducing contrast and producing a neutral density—the net effect, of course, is to make the animal more difficult to see. Countershading occurs in most insects, fish, amphibians, lizards, snakes, birds, and many mammals. A counterexample that proves the point is provided by a few animals that are normally belly up in nature, such as the "upside-down" catfish *Synodontis nigriventris* and certain moth larvae; these animals are darker ventrally than they are dorsally! Because a successful predator must also be inconspicuous in order to catch its prey, crypticity is equally important to predators as it is to prey.

Many insects resemble the parts of the plants on which they live, especially leaves, twigs, thorns, or bark; leaf butterflies and walking sticks are familiar examples. Both a green and a brown color phase often occur in such cryptically shaped animals. For example, females of two southern grasshoppers, *Syrbula admirabilis* and *Chortophaga viridifasciata*, have green and brown color phases (strangely, males are almost always brown!). Green females predominate in wetter, greener habitats; but in immediately adjacent drier and browner areas, the brown form is most prevalent (Otte and Williams, 1972). Determination of a female's color is not under strict genetic control but is developmentally flexible in response to local conditions.

Actual demonstrations that coloration differences and background-color matching have selective value are unfortunately rather scarce. The best documented example is that of the moth, *Biston betularia*, in England. This moth, along with several hundred other species, has evolved rapidly during the last century in response to human modification of its habitat. In the 1800s, *Biston* were pale colored, spending their daytime hours on pale, lichen-covered tree trunks. However, with the buildup of industry and concomitant air pollution, the lichens have died and tree trunks in some areas have been covered with a layer of soot and grime, becoming quite dark. In early collections, black moths (melanics) were very rare, but they have become increasingly more common. Indeed now these melanistic varieties comprise the vast majority of moth populations in polluted areas. This phenomenon of directional selection, termed *industrial melanism*, has also taken place in the United States and in Europe in other species of moths. In an elegant series of experiments, Kettlewell (1956) made reciprocal transfers of pale moths from a nonpolluted woods with melanic moths from a polluted

TABLE 15.1   **Numbers of Typical and Melanic Marked Moths *(Biston betularia)* Released and Recaptured in a Polluted Woods Near Birmingham and an Unpolluted Woods Near Dorset[a]**

|  | Polluted woods | Unpolluted woods |
| --- | --- | --- |
| Numbers of marked moths released |  |  |
| Typical | 64 | 496 |
| Melanic | 154 | 473 |
| Number recaptured |  |  |
| Typical | 16 (25%) | 62 (12.5%) |
| Melanic | 82 (53%) | 30 (6.3%) |

[a]The wild population in the polluted woods was 87 percent melanic.

*Source:* From data of Kettlewell (1956).

area (Table 15.1). These moths, along with resident moths occurring at each locality, were marked with a tiny inconspicuous paint spot beneath their wings, and attempts were made to recapture them on later days. As expected, pale moths had lower survivorship in the polluted woods and melanic moths had lower survivorship in clean, lichen-covered forests. Moreover, Kettlewell actually observed foraging birds catching mismatched moths!

A black lava flow on white desert sands in New Mexico provides a "natural" experiment that strongly suggests that background-color matching has evolved and is adaptive (Benson, 1933). This lava flow is completely surrounded by white sandy areas and has presumably been stocked mostly with animals derived from those that live on the white sands. Two closely related pocket mice live in the area: *Perognathus intermedius ater* is nearly pitch black and occurs only on the lava; *P. apache gypsi* is pale white and lives only on the white sands.

Some animals are covered with blotches of different colors, which tend to break up their shape and to make them more difficult to see. Good examples are rattlesnakes and boa constrictors; often, these snakes look so much like a pile of dead leaves that they may go undetected. This type of coloration, especially prevalent in larger animals that have difficulty in finding hiding places, such as leopards, tigers, and giraffe, has been termed *disruptive coloration*.

A form of adaptive coloration that is not necessarily cryptic is called *flash coloration*. Many inconspicuous insects, including some butterflies and grasshoppers, are extremely cryptic when at rest; but when disturbed they fly away and reveal brightly colored underwings (often red, yellow, or orange). These insects thus suddenly become extremely visible and conspicuous, catching the predator's eye. When they land, they close their wings and quickly move away from the spot on which they landed. As anyone who has chased grasshoppers knows, it is very difficult to keep track of the position of such an animal. (Squid and octopi employ a remotely similar strategy when they squirt out their "ink," leaving a dense cloud in the water. Typically, the prey animal immediately changes both color and course, becoming pale and swimming at right angles to its original direction of flight, thereby evading the predator.) A rather derived kind of flash coloration occurs in some butterflies and moths that have large owl-like eyes on their underwings. These eyes are normally hidden by the upper pair of wings. When a small bird approaches, the upper wings are suddenly twitched aside, revealing an owl-like face beneath. Some small birds are apparently so startled that they fly away, leaving the insect alone! Some such eyespots are perpetually on display (Stradling, 1976). Similarly, the yellow "eyes" on some large green caterpillars may make them resemble green tree snakes.

Another, smaller type of eyespot actually invites the attack of a predator. Many predators instinctively go for the eyes of their prey, because eyes are usually one of the most vulnerable parts of an animal and the loss of vision readily incapacitates prey (lions and wolves have found another "Achilles heel" on large ungulates—they simply hamstring the animal). Many species of butterflies possess small "fake" eyespots along the periphery of their wings that may actually invite attack. Eyespots painted on such butterflies in places without eyespots are damaged when the animals are released and recaptured, apparently from being pecked at by birds (Sheppard, 1959). Thus, the butterfly obtains a second chance at escape by luring the predator's attack away from its own eyes. Behavioral adaptations may sometimes serve similar functions; certain snakes raise their blunt tails and wave them around in a very headlike manner, occasionally actually making short menacing lunges with their tails at the threatening predator. Should the unwary predator grab the snake by this "head," the serpent still has its real head free to bite back.

Many birds and mammals have various sorts of alarm signals, which, when given, warn other animals that a predator is in the immediate vicinity. Beaver warn one another of danger by slapping their tails loudly against the surface of the water. Similarly, rabbits in a warren (a colony of related individuals using the same burrow system) often "thump" with their hind feet to signal the approach of a predator. The white underside of the tail of some rabbits and ungulates (such as the white-tailed deer) is raised when the animal flees from a predator, perhaps serving as a warning signal to nearby animals. Prairie dogs, many primates, and many foraging bird flocks, such as crows, frequently post sentinels (see also p. 225) that watch for predators from a good vantage point and warn the group should one appear in the distance. Among birds, alarm calls in response to the presence of bird-eating hawks are especially prevalent. Typically, these faint shrill whistles are extremely difficult for a vertebrate predator to locate; often, they are similar in widely different bird species, presumably having converged over evolutionary time.

Because different species may benefit from the call of one individual, hawk alarm calls appear to be somewhat altruistic. However, warning calls are normally used only during the nesting season and are therefore best interpreted as having arisen through kin selection. The ventriloquial nature of the call, coupled with the fact that the caller has already seen the predator and therefore knows exactly where the danger lies, ensures that the risk to the bird giving the alarm is slight. Likewise, the fact that banded birds often return to breed in the same area where they themselves were raised means that birds in any given area will be related and share many genes, which in turn ensures that the total gain to the many relatives of the pseudoaltruist will frequently be quite large. If hawk alarm calls have evolved through kin selection, one would predict that the frequency of use of such calls should be inversely related to the distance the caller is from its own nest and immediate relatives (no one has yet demonstrated this empirically). The fact that these calls work across species lines is easily explained because individuals that recognize warning signals of other sympatric species as alarms should be at a relative advantage within their own population. The convergence of hawk alarm calls can be explained by either or both of two mechanisms. The first concerns the ventriloquial properties of the call. The number of ways in which an alarm call can be loud enough to function as a warning, and yet still be difficult to locate, are decidedly limited. Thus, convergence could be a simple result of physical constraints inherent in the system. A second, equally likely, possibility is that natural selection has favored convergence because it facilitates interspecific recognition of alarm calls. Thus, individuals tending to produce call variants more like those of another species would benefit because they would also be more likely to recognize the calls of the other species (as such, call convergence would be very similar to Müllerian mimicry, discussed later in this chapter).

There are several alternative hypotheses for alarm calls. The calls may not be intended as "alarms" at all, but as signals intended to inform the predator that it has been detected and that it not need try to capture the caller (p. 225). Charnov and Krebs (1975) suggest that the caller could decrease its own chance of being captured by alerting other individuals, because the predator's attention would be drawn away from the caller and toward the large numbers scurrying for cover. Still another argument for a selfish caller is based on the fact that many predators tend to stay nearby and return to an area where they have successfully captured prey in the past; to the extent that alarm calls prevent a predator from catching any prey at all, they may serve to keep the predator on the move and hence out of the immediate area of the caller. Some birds that forage in mixed-species flocks in the tropics issue false alarm calls when they see a juicy prey item, which distracts the attention of other birds, allowing the selfish caller to capture the prey item (Munn, 1986).

Yet another evolutionary consequence of predation is *warning coloration;* unpalatable or poisonous animals have often evolved bright colors that advertise their distastefulness. Such markings are called *aposematic,* which translates "away signal." These animals are usually colored with the same conspicuous colors we use for signs along highways: reds, yellows, and blacks and whites. Examples of animals with warning coloration are bees and wasps, monarch butterflies, coral snakes, skunks, and certain brightly colored poisonous frogs and salamanders. Signals that warn potential predators may also involve pattern, posture, smell, or sound. A rattlesnake's buzzing presumably serves to warn other larger animals such as the American bison not to come too near (unfortunately for the rattler, however, this warning only attracts human attention, which usually results in the snake's demise). Experiments have shown that avian and lizard predators learn to avoid distasteful prey. In the case of poisonous prey, this learning may actually become incorporated into the gene pool and manifested as an "instinctive" avoidance.

*Mimicry* is an interesting sidelight of warning coloration that nicely demonstrates the power of natural selection. An organism that commonly occurs in a community along with a poisonous or distasteful species can benefit from a resemblance to another species that displays warning colors, even though the "mimic" itself is nonpoisonous and possibly quite palatable. False warning coloration is termed *Batesian mimicry* after its discoverer. Many species of harmless snakes mimic poisonous snakes; in Central America some harmless snakes are so similar to poisonous coral snakes that only an expert can distinguish the mimic from the "model." Similarly, harmless flies and clearwing moths often mimic bees and wasps, and palatable species of butterflies mimic distasteful species. Batesian mimicry is disadvantageous to the model because some predators will encounter palatable or harmless mimics and thereby take longer to learn to avoid the model. The greater the proportion of mimics to models, the longer is the time required for predator learning, and the greater the number of model casualties. In fact, if mimics became more abundant than models, predators might not learn to avoid the prey item at all but might actively search out model and mimic alike. For this reason, Batesian mimics are usually much less abundant than their models; also, these mimics are frequently polymorphic and mimic several different model species.

A different kind of mimicry occurs when two species, both distasteful or dangerous, mimic one another; this phenomenon is termed *Müllerian mimicry.**  Both bees and wasps, for example, are usually banded with yellows and blacks. Because potential predators encounter several species of mimics more frequently than a single species, they learn faster to avoid them, and the relationship is actually beneficial to both prey species (Benson, 1972). The resemblance need not be so close as it must be under Batesian mimicry because neither species actually deceives the predator; instead, each only reminds the predator of its dangerous or distasteful properties. Müllerian mimicry is beneficial to all parties including the predator; such mimics can be equally common and are rarely polymorphic.

Plants, being sessile, cannot use many of the escape techniques of animals and are obviously and decidedly more limited in the ways they can deter potential predators. A plant with a patchy or spotty distribution in time or space may escape some predation simply by virtue of its unpredictable availability; thus, an annual that is here today but gone tomorrow may be more difficult for herbivores to find and use than an evergreen perennial that is always relatively available (see subsequent discussion in this chapter). Some plants, especially perennials, have evolved morphological adaptations, such as hairs, spines, and hooks (Gilbert, 1971), that discourage many herbivores quite directly. By far the most widespread predator deterrent of plants is what might be termed

---

*Actually, the dichotomy between Batesian and Müllerian mimicry is somewhat artificial in that these two sorts of mimicry grade into one another.

chemical warfare. A great variety of secondary chemical substances occur in plants that are not known to serve any direct physiological function for their possessors. Many are evidently not breakdown products of larger molecules due to metabolic processes and wastes, but are secondary substances produced by active synthesis from smaller molecular precursors. Such secondary chemical substances often contain nitrogen and other elements that are available to the plant only in limited supply; moreover, it takes energy to produce these chemicals. Clearly, there are definite costs to the plant in production of herbivore repellents. Nearly a century ago the German botanist Stahl (1888) suggested that these secondary substances might reduce the plant's palatability to herbivores. Stahl's prediction has now been amply verified for many different plant-herbivore systems.

Agriculturalists and plant breeders have produced many strains that are highly resistant to normal herbivores. Genetic varieties, or morphs, of plants in nature have been shown to be differentially palatable to herbivores; thus, Jones (1962, 1966) found that a number of herbivores, ranging from insects and snails to *Microtus* (a mammal), preferred a "noncyanogenic" morph of the plant *Lotus corniculatus* over a "cyanogenic" one. Tannins have been implicated as the agent that repels some herbivores; oak leaf tannin significantly reduces the growth rate of larvae of the moth *Operopthera brumata* (Feeny, 1968). Frequently, herbivores eat only relatively new growth and do not utilize older parts of plants, presumably because the latter contain tannins and other repellent chemicals (Feeny, 1970). Other chemical substances believed to protect plants from animals and fungi include essential oils and resins, alkaloids, terpenes, and terpenoids. The latter two classes of compounds have especially penetrating odors and tastes; sesquiterpenes are fatal to sheep. Wild herbivores that have evolved alongside poisonous plants would be unlikely to eat them, whereas domestic sheep and cattle will eat many poisonous fodders.

The argument presented at the beginning of this section suggests that sympatric plant species may usually gain by evolving qualitatively different antiherbivore secondary chemical defenses. As a result, plants in general are not continuous resources but constitute a spectrum of qualitatively distinct, discrete food types. Such divergent antiherbivore chemistries force the evolution of herbivore specialists by reducing the efficiency of generalized herbivores. Because related plants have similar secondary chemistries, phylogenies of certain herbivorous insects (particularly butterflies) closely parallel the phyletic relationships of their host plants (Ehrlich and Raven, 1964; Benson et al., 1975).

## PARASITISM

Although the process of parasitism is similar to that of predation in terms of the signs of the interaction coefficients between the two populations, predation differs from parasitism in that members of the species affected detrimentally (the *host*) are seldom killed outright but may live on for some time after becoming parasitized. Batesian mimicry and herbivory can be considered as special cases of parasitism. So-called *parasitoid* insects, such as ichneumonid wasps, which lay their eggs in or on another host insect (larvae develop inside the host, consuming and eventually killing it), merge into more traditional predators.

Parasitism has been defined using several criteria (Kennedy, 1975): (1) the parasite is physiologically dependent on its host; (2) the parasite usually has a higher reproductive potential than its host (for successful dispersal and infection, parasites typically must be exceedingly fecund); (3) parasites are capable of killing highly infected hosts; (4) the infection process tends to produce an overdispersed distribution of parasites within the host population. In addition, parasites are typically substantially smaller than their hosts and usually have a much shorter generation time than do their hosts.

Several fundamentally different classes of parasites may be distinguished. *Ectoparasites*, such as ticks and mites, exploit the outer surface(s) of their hosts, whereas *endoparasites** live inside their host organisms (these are often called pathogens or disease agents). A spectrum can be envisioned ranging from microparasites to macroparasites to parasitoids to predators. A wide variety of important ecological characteristics change along this continuum (Table 15.2).

Still another very different but interesting type of parasitism is social parasitism, which includes thievery, slavery, and *brood parasitism*. Best known in birds, brood parasitism occurs when members of one species exploit another for parental care, usually by means of deceiving the hosts into believing that they are raising their own progeny (cowbirds are the most familiar example).

As in predator-prey interactions, parasites and their hosts are typically engaged in an antagonistic evolutionary interaction due to the inherent basic conflict of interests. Selection should always favor resistant host individuals (Haldane, 1941). To the extent that natural selection favors evolution of reduced parasite virulence (see also subsequent discussion in this chapter), parasitic interactions may evolve gradually toward commensalisms and ultimately even become mutualistic interactions.

$$\text{Parasitism} \rightarrow \text{Commensalism} \rightarrow \text{Mutualism}$$
$$(+, -) \quad \leftarrow \quad (+, 0) \quad \leftarrow \quad (+, +)$$

Of course, selection could also proceed in the opposite direction (reverse arrows). Such changes may also occur during ecological time as during the ontogeny of parasites.

Most parasites are very specialized. Many endoparasites have become intimately dependent on their hosts; tapeworms have lost their digestive systems and viruses consist of little but "naked" genetic material (indeed, they are not even really "alive" outside of their hosts!). Elaborate life cycles involving intermediate hosts and various infective stages are prevalent.

---

*These include *microparasites* such as viruses, bacteria, plus parasitic protozoans, and *macroparasites*, largely various "worms" such as acanthocephalans, many nematodes, trematodes, and tapeworms (Anderson and May, 1979).

**TABLE 15.2  Comparison of Certain Ecological Characteristics That Vary Along a Parasite-Predator Spectrum**

| Characteristic | Microparasite | Macroparasite | Parasitoid | Predator |
|---|---|---|---|---|
| Body size | Much smaller than hosts | Smaller than hosts | Mature stages similar in size | Larger than prey |
| Intrinsic rate of population growth | Much faster than hosts | Faster than hosts | Comparable but slightly slower | Usually slower than prey |
| Interaction with host individuals in natural populations | One host usually supports several populations of different species | One host supports a few to many individuals of different species | One host can support several individuals | Many prey items are eaten by each predator |
| Effect of the interaction on the host individual | Mildly to fairly deleterious | Variable; not too virulent to definitive; can be intermediate | Eventually fatal | Usually immediately fatal |
| Stability of the interaction between trophic levels | Intermediate | High | Intermediate | Usually low |
| Ability to regulate the lower trophic level | Moderate | Low | Fairly high | High |

*Source:* Modified from Anderson and May (1982) and Toft (1986).

A degree of specificity typically arises in parasite-host interactions, and many parasites have evolved species-specific host requirements. An interesting ramification of the very intimate interaction between parasites and their hosts is *molecular mimicry*. In this phenomenon, antigenic determinants of parasitic origin (known as *eclipsed antigens*) resemble host antigens to such an extent that they do not elicit the formation of host antibodies. This, of course, allows the parasite to dwell safely inside the host's tissues in a more or less uncontested fashion, protected from the host's immune response. Parasites exploit many other innovative adaptations to confound their hosts; trypanosomes simply switch antigens—every time the host mounts an immune response, the parasite "sheds" its previous coat and puts on another (Turner, 1980). Some tapeworms secrete copious amounts of a sticky substance composed of proteins and sugars. Host antibodies get gummed up in this "glycocalyx"; some tapeworms produce such vast amounts of glycocalyx that the host's antibodies can never find the worm. Clearly, this host defense tactic costs tapeworms a substantial amount.

Among the many subtle aspects of parasite biology is *host-altered behavior*, in which parasites actually cause increased vulnerability of infected hosts to their own parasites or predators (Holmes and Bethel, 1972; Moore, 1985). A simple example is a rabid animal's biting other animals and thereby passing on the rabies viral infection. (A venereal disease causing increased sexual activity or promiscuity would be similar.) A somewhat more complex example involves the lancet fluke *Dicrocoelium dentriticum*, whose primary host is sheep; metacercariae of this trematode infect ants as intermediate hosts. One or two metacercariae encyst in subesophageal ganglia of an ant's brain, which affects the behavior of the ant, causing it late in the day to climb to the tip of a grass stem and to clamp its mandibles closed and hang there overnight. As temperatures fall, the ant's jaws become locked and it cannot move. This behavior places the infected ant in the position where it is most likely to be inadvertently ingested by a grazing sheep early the next morning.

A prevalent conception among parasitologists is that very virulent parasites represent recently evolved parasite-host interactions, whereas more benign parasites are indicative of older, more ancient associations. Although this assertion can be disputed, evidence exists for the evolution of reduced virulence. One of the best examples concerns a carcinogenic myxoma virus, which produces a mild nonlethal disease in its natural host, the South American cottontail rabbit *Sylvilagus brasilensis*. In 1950, this virus was used as a biological control agent in Australia in an effort to eradicate the introduced European rabbit *Oryctolagus cuniculus*. Fleas and mosquitos transmit the virus between infected rabbits. Initially, this virus proved exceedingly virulent in *Oryctolagus*, causing better than a 99 percent kill among infected rabbits. Of course, selection was intense for increased resistance among rabbits. Even so, a less virulent strain of myxoma virus rapidly became established in the field. Apparently, the exceedingly virulent strain often killed its host outright even before being transmitted to other rabbits, whereas less virulent strains were more likely to persist long enough to be passed on to other hosts. In other situations, such as when a parasite finds itself engaged in a race against its host's immune response, selection may actually favor increased virulence.

Parasites can have rather profound effects on the ecology of their hosts. Effects of the malarial parasite *Plasmodium mexicanus* on members of a population of *Sceloporus occidentalis* lizards were examined in California by Schall (1982). About one-third to 40 percent of the male lizards were infected, whereas only 15 to 30 percent of the female lizards had malaria (percentage infected increases with age). Blood hemoglobin levels were lower in parasitized lizards than in unparasitized ones. When lizards were at rest, oxygen consumption rates did not differ between the two groups, but during maximal activity, infected animals had significantly lower metabolic rates. Moreover, both the capacity for aerobic metabolism and running stamina (as measured in an oval track) were reduced in infected lizards as compared to controls. Parasitized lizards also tended to have smaller fat reserves and smaller clutch sizes than noninfected ones. Effects of infection by malarial parasites might be subtle and hosts might appear healthy, but these parasites seem to reduce host fitness substantially.

In another study, Schall (1992) demonstrated that another species of malarial parasite allowed coexistence of two species of Caribbean *Anolis* lizards (in the absence of the parasite only one species of lizard occurs, but if this species of lizard is parasitized, the other lizard species can coexist with it).

Like predators, parasites can affect community structure, although sometimes in obscure ways. Recall that the outcome of interspecific competition between two species of flour beetles could be reversed by a protozoan parasite (Park, 1948). Avian malaria may have contributed to the extinction of some members of the Hawaiian avifauna.

# EPIDEMIOLOGY

Disease transmission lends itself quite naturally to quantitative treatment (Anderson, 1981). Among the phenomena one can examine are (1) the percentage of hosts that are susceptible, infected, or immune; (2) the rate of spread of the pathogen under different conditions (particularly the density of the host population and the frequency of infection); and (3) the extent to which density-dependent probability of infection regulates host population growth. The stability of the interaction and the evolution of host resistance and disease severity are also of some interest.

Smallpox epidemics in human populations were modeled mathematically by Bernoulli over two centuries ago. Epidemiological models typically make the simplifying assumption that host population size is constant, and examine the dynamics of parasitism, usually in terms of the proportion of hosts infected. Two rate parameters are critical: the rate of transmission of the disease from infected to susceptible hosts and the rate at which infected hosts recover to become immune. A critical parameter is the basic reproductive rate of the infection. Does each infection, on average, produce more than one new infection (leading to an epidemic) or fewer than one? In such a simple epidemiological mathematical model, two equilibria exist: one with no infection and the other with constant but dynamically renewing proportions of hosts in each of three states, susceptible, infected, and immune. Interestingly, which of these two equilibria exists depends both on the two rate parameters and on what is termed the threshold host population size, or the critical density of hosts necessary for parasites to replace themselves and to spread. In small host populations, parasites cannot infect new hosts rapidly enough to survive, whereas an epidemic may take off with the same parameters in a larger host population. Such epidemiological models suggest that vaccination efforts should be more intensive in urban areas than in rural ones (country folks are less likely to get ill than city slickers!). With no lag in transmission and without immunity, the time course of an epidemic is sigmoidal (Figure 15.13). Such a system is accurately modeled with a simple differential equation of the form:

$$\frac{dY}{dt} = \beta XY \tag{5}$$

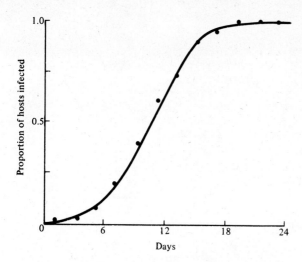

**Figure 15.13** The time course of an epidemic is typically sigmoidal, with the rate of new infection reaching its maximum when about half the population is infected, while the other half is vulnerable. [From Anderson (1981) after Stivens.]

where $Y$ represents the number of infected host individuals and $X$ is the number susceptible to the pathogen. With such a linear functional response, the parasite spreads slowly at first and its rate of spread is maximized when half the host population is infected and the other half is vulnerable to infection. In more realistic and more complex models with inoculation time lags and immunity, interactions between parasites and their hosts can generate cycles of epidemics, which, of course, are all too familiar to parents and pediatricians.

## COEVOLUTION

In its broadest sense, *coevolution* refers to the joint evolution of two (or more) taxa that have close ecological relationships but do not exchange genes and in which reciprocal selective pressures operate to make the evolution of either taxon partially dependent on the evolution of the other (Ehrlich and Raven, 1964). Thus, coevolution includes most of the various forms of population interaction, from competition to predation to mutualism. A nice example of tight coevolution between pinworm parasites and their primate hosts is shown in Figure 15.14.

The term coevolution is also often used in a more restricted sense to refer primarily to the interdependent evolutionary interactions between plants and animals, especially their herbivores and pollinators. A plant may evolve a secondary chemical substance that deters the vast majority of predators, but if a particular herbivore can in turn evolve a physiological means of coping with the

Enterobius species                                                    Primate hosts

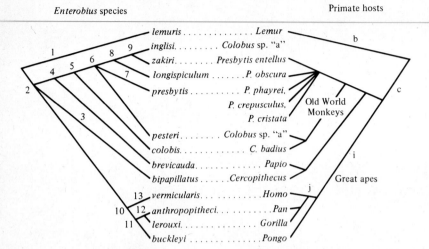

**Figure 15.14** Parallel phylogenies of the nematode genus *Enterobius* and their primate hosts. [From Mitter and Brooks (1983) after Brooks and Glen (1982).]

chemical deterrent, it can thereby obtain an uncontested food supply. Through this kind of coevolution, many herbivores have become strongly specialized on a single species or a few closely related species of plants. Thus, *Drosophila pachea* is the only species of fruit fly that can exploit the "senita" group of cacti; these plants produce an alkaloid that is fatal to the larvae of all other fruit flies, but *D. pachea* has apparently evolved a means of detoxifying this chemical (Kircher et al., 1967).

In many cases such specialized herbivores even use a plant's toxic chemicals (often quite pungent) as cues in locating and selecting their host plants. Some herbivores, such as the monarch butterfly, actually sequester plant poisons (cardiac glycosides in this case), which in turn make the herbivore unpalatable or even poisonous to its own potential predators.

Danaid butterflies and certain moths make double use of polyuridine alkaloids—these noxious chemicals are not only sequestered by larvae and adults and used for antipredator purposes, but are also exploited as chemical precursors in the synthesis of pheromones important for mate attraction. An arginine mimic, l-canavanine, present in many legumes, ruins protein structure in most insects; however, a bruchid beetle has evolved metabolic machinery that enables it to utilize plants containing canavanine.

Attempts have been made to generalize about the coevolution of herbivores and plant antiherbivore tactics (Orians, 1974). Feeny (1975) argues that rare or ephemeral plant species are hard for herbivores to find and hence are protected by escape in time and space; moreover, he asserts, such plant species should evolve a diversity of qualitatively different, chemically inexpensive, defenses that should constitute effective evolutionary barriers to herbivory by nonadapted generalized herbivores that are most likely to find such "cryptic" plants. However, these same secondary chemicals will be only minimal ecological barriers to adapted specialized herbivores, against which the plant's primary antiherbivore tactic is escape in time and space (i.e., not being found). In contrast, Feeny reasons that abundant or persistent plant species cannot prevent herbivores from finding them either in ecological or evolutionary time. Such "apparent" plant species appear to have evolved more expensive quantitative defenses, including tough leaves of low nutrient or water content which contain large amounts of relatively nonspecific chemicals such as tannins (Table 15.3). Feeny points out that such plant defenses should pose a significant ecological barrier to herbivores, although perhaps only a weak evolutionary barrier unless supplemented with qualitative chemical defenses (some plants have both).

Cates and Orians (1975) develop somewhat different but related predictions for early versus late successional plant species. Because early successional plants escape from herbivores in space and time, Cates and Orians reason that such plants should allocate fewer resources to chemical antiherbivore defenses than the more apparent plants of later stages in succession. Thus, early successional plant species should make better foods for generalized herbivores than later successional and climax plant species. Indeed, experimental studies on slug feeding indicate that early successional annuals were significantly more palatable than later successional species (Cates and Orians, 1975). However, the opposite result was obtained by Otte (1975) in similar experiments with grasshoppers; these generalized herbivores accepted more later successional plant species than early ones. Otte suggests that this difference may arise from the difference in mobility between slugs and grasshoppers. In a survey

TABLE 15.3   **Some of the Suggested Correlates of Plant Apparency**

| Apparent plants | Unapparent plants |
|---|---|
| Common or conspicuous | Rare or ephemeral |
| Woody perennials | Herbaceous annuals |
| Long leaf lifespan | Short-lived leaves |
| Slow growing, competitive species | Faster growing, often fugitive species |
| Late stages of succession, climax | Early stages of succession, second growth |
| Bound to be found by herbivores (cannot escape in time and space) | Protected from herbivores by escape in time and space (but still encountered by wide-ranging generalized herbivores) |
| Produce more expensive qualtitative (broad-based) antiherbivore defenses (tough leaves, tannins) | Produce inexpensive qualitative chemical defenses (poisons or toxins) to discourage generalized herbivores |
| Quantitative defenses constitute effective ecological barriers to herbivores, although perhaps only a weak evolutionary barrier unless supplemented with qualitative defenses | Qualitative defenses may be broken down over evolutionary time by coevolution of appropriate detoxification mechanisms in herbivores (host plant-specific herbivore species result) |

of lepidopteran feeding habits, Futuyma (1976) found greater degrees of specialization to host plant species among insects feeding on herbaceous plants than among those that feed on leaves of shrubs and trees (this pattern neatly fits Feeny's plant apparency dichotomy). Futuyma suggests that plant defense systems are more diverse in floristically rich plant communities than they are in less diverse communities.

Discrepancies among these studies indicate that generalizations are difficult to make and that they will have to allow for exceptions (for further discussion of this interesting area, see Rhoades and Cates, 1976; Gilbert, 1979; and Futuyma and Slatkin, 1983).

Populations of wild ginger *Asarum caudatum* in western Washington are polymorphic for growth rate, seed production, and palatability to a native slug, *Ariolimax columbianus* (Cates, 1975). In habitats where slugs were low in abundance, Cates found that populations of wild ginger were dominated by individuals allocating more energy to growth and seed production and less to production of antiherbivore chemicals. Presumably, less palatable plants have a fitness advantage in habitats with more slugs; even though they grow more slowly, they lose less photosynthetic tissue to slug herbivory.

In some cases plants have actually formed cooperative relationships with animals that result in their protection from certain herbivore species. By means of ant removal experiments, Janzen (1966) showed that some species of *Acacia* deprived of their normal epiphytic ant fauna are highly palatable to herbivorous insects, whereas species that do not normally have ants for protection from herbivores are less palatable. The acacias benefiting from ant protection actually produce nectaries and swollen thorns that attract and in turn benefit the ants! Thus, these plants channel matter and energy into attracting ants that defend their leaves rather than into more direct chemical warfare. This antiherbivore ploy is broad based, since the ants ferociously attack a wide range of herbivores. Ant protection may also reduce competition with other plants, as well as provide a measure of protection from fire, because the ants keep the ground surface clear immediately around their *Acacia* plant.

Many plants protect their seeds either by enclosing them in a toxic matrix or by means of a hard shell. Some seeds are poisonous. Nevertheless, the high nutrient content of seeds has resulted in the evolution of effective seed predators. Predation on seeds may often be heaviest where they occur in greatest concentrations (such as acorns underneath a parent oak) because seed predator populations will generally be largest where the most food is available (Janzen, 1971b). As a result, the probability of an individual seed's surviving to establish itself as a plant may often vary inversely with seed density. In many trees, most seeds fall to the ground near the parent tree, with a continually decreasing number of seeds ending up at distances farther from the parent tree (Figure 15.15). As a result of these opposing processes, Janzen suggests that recruitment is maximized at some distance from the parent tree (Figure 15.15). Janzen's model of seed predation and recruitment may help to explain the high species diversity of tropical trees, which suffer heavy seed losses to specialized seed predators that eat the seeds of particular tree species.

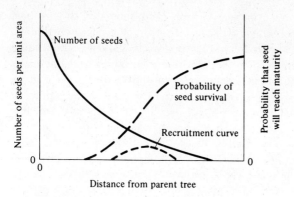

**Figure 15.15** Hypothetical model of seed recruitment versus distance from the parent tree. Near the tree, all seeds are eaten by seed predators; at increasing distances from the tree, the probability of seed survival increases as the density of seeds and their predators decreases. Although the number of seeds drops with distance from the parental tree, recruitment is highest at some distance from the tree. A ring of seedlings should result. [After Janzen (1970).]

(It neglects the question of why there are so many specialized seed predators in the tropics.)

Intricacies of coevolutionary relationships between pine squirrels (*Tamiasciurus*) and their coniferous food trees were studied in the Pacific northwest by C. Smith (1970). Conifer seeds constitute the staple food supply of these small red squirrels; they can effectively strip a tree of most of its cones. Trees reduce the effectiveness of squirrel predation in many different ways: (1) by producing cones that are difficult for the squirrels to reach, open, or carry; (2) by putting fewer seeds into each cone (squirrels eat only the seeds themselves and must "husk" cones to get them); (3) by increasing the thickness of seed coats, requiring that the squirrels spend more time and energy extracting each seed; (4) similarly, by putting less energy into each seed (a drawback is that seedlings from smaller seeds have fewer resources at their disposal and are presumably poorer competitors than seedlings from larger seeds); (5) by shedding seeds from cones early, before the young squirrels of the year begin foraging; and (6) by periodic cone "failures" that decimate the squirrel population, thereby reducing the intensity of predation during the next year. Thus, squirrel predation has had profound evolutionary influences upon various reproductive characteristics of conifers, including details of cone anatomy and location, the number of seeds per cone (and the variability in the number per cone), the time at which the cones shed their seeds, the thickness of seed coats, and annual fluctuations in the size of cone crops. Evolution of these defense mechanisms by the conifers has in turn forced squirrels to adapt in various ways, such as choosing cones carefully and stockpiling them.

Fluctuations in cone crops from year to year are pronounced and are best interpreted as an antisquirrel strategy, because they occur even when climatic conditions are apparently favorable for trees. Apparently, the conifers withhold their products of primary production and store them for later use. Cone crops and failures are often synchronized among different tree species in a given area—a further indication that the phenomenon is directed at the squirrels. Individual trees out of synchrony presumably set fewer seeds and are thus selected against by natural selection. Smith points out that different conifer species have *diverged* from one another in evolution of cone anatomy, size, location, and time of shedding but that these same conifer species have *converged* in their fluctuations in cone crops. Both reduce the squirrel's efficiency.

## SELECTED REFERENCES

### Predation

Errington (1946);   Holt (1977);   Janzen (1971b);   MacArthur (1972);   MacArthur and Connell (1966);   Schall and Pianka (1980);   Wilson and Bossert (1971).

### Predator–Prey Oscillations

Elton (1942);   Errington (1946);   Gause (1934, 1935);   Haigh and Maynard Smith (1972);   Hassell (1980);   Holling (1959a, 1959b, 1961, 1966);   Keith (1963);   Leslie (1948);   Levins (1966);   Lotka (1925);   May (1973, 1981);   Maynard Smith (1974); Pielou (1969, 1974);   Rosenzweig (1971, 1973a, 1973b);   Rosenzweig and MacArthur (1963);   Solomon (1949);   Tanner (1975);   Toft (1986);   Volterra (1926a, 1926b, 1931); Wangersky and Cunningham (1956);   Wilson and Bossert (1971).

### "Prudent" Predation and Optimal Yield

Beverton and Holt (1957);   Gilpin (1975a);   MacArthur (1960b, 1961);   Slobodkin (1968).

### Selected Experiments and Observations

Errington (1946, 1956, 1963);   Force (1972);   Gause (1934, 1935);   Holling (1959a, 1965);   Huffaker (1958);   Luckinbill (1973, 1974);   Maly (1969);   Menge (1972a); Murdoch (1969);   Neill (1972);   Paine (1966);   Salt (1967);   Utida (1957);   Wilbur (1972).

### Evolutionary Consequences of Predation: Escape Tactics

Benson (1933);   Benson (1972);   Benson et al. (1975);   Charnov and Krebs (1975); Cott (1940);   Ehrlich and Raven (1964);   Feeny (1968, 1970);   Fisher (1958b);   Gilbert (1971);   Gordon (1961);   Holt (1977);   Janzen (1966, 1967, 1970, 1971b);   Jeffries and Lawton (1985);   Jones (1962, 1966);   Kettlewell (1956, 1958);   McKey (1974);   Munn (1986);

Otte and Williams (1972);    Rand (1967);    Ricklefs and O'Rourke (1975);    Schall and Pianka (1980);    Sheppard (1959);    Stahl (1888);    Stradling (1976);    Whittaker and Feeny (1971).

## Parasitism

Anderson (1981);    Anderson and May (1979, 1982);    Barbehenn (1969);    Bradley (1972, 1974);    Cornell (1974);    Crofton (1971a, 1971b);    Damian (1964, 1979);    Esch (1977);    Esch et al. (1977);    Ewald (1983, 1987);    Fallis (1971);    Gillespie (1975);    Haldane (1941);    Hirsch (1977);    Holmes (1973, 1983);    Holmes and Bethel (1972);    Jennings and Calow (1975);    Kennedy (1975);    May and Anderson (1979);    Moore (1985);    Park (1948);    Price (1980);    Rohde (1979);    Schad (1963, 1966);    Schall (1982, 1992);    Toft (1986);    Turner (1980);    Warner (1968).

## Epidemiology

Anderson (1976, 1981);    Anderson and May (1979, 1982);    Bailey (1975);    Bartlett (1960);    Haldane (1949);    May and Anderson (1979).

## Coevolution

Brooks and Glen (1982);    Brower (1969);    Brower and Brower (1964);    Caswell et al. (1973);    Cates (1975);    Cates and Orians (1975);    Caughley and Lawton (1981);    Chambers (1970);    Coley et al. (1985);    Dressler (1968);    Ehrlich and Raven (1964);    Faegri and van der Pijl (1971);    Feeny (1970, 1975, 1976);    Feinsinger (1983);    Fraenkel (1959);    Freeland and Janzen (1974);    Futuyma (1976);    Futuyma and Slatkin (1983);    Gilbert (1971, 1972, 1979);    Gilbert and Raven (1975);    Gilbert and Singer (1973, 1975);    Gordon (1961);    Heinrich and Raven (1972);    Howe (1984);    Janzen (1966, 1967, 1971a, 1971b);    Kiester et al. (1984);    Kircher and Heed (1970);    Kircher et al. (1967);    Lawlor and Maynard Smith (1976);    Mitter and Brooks (1983);    Orians (1974);    Otte (1975);    Price (1975);    Rathcke (1976);    Rhoades and Cates (1976);    Rick and Bowman (1961);    C. Smith (1970);    Thompson (1982, 1989);    Willson (1973c).

**Chapter**

*16*

# Community and Ecosystem Ecology

## SYSTEMS AND MACRODESCRIPTORS

Except for a brief treatment of biomes, up until now we have considered the ecology of individuals and populations. We next consider the highest level of biological organization: entire *systems* of interacting populations in a complex and dynamic physical environmental setting. This kind of ecology is known as *synecology*. Synecology is the most abstract and most difficult kind of ecology, but it is also exceedingly tantalizing and vitally important, as well as extremely urgent.

As in any academic endeavor, there are two, almost diametrically opposed, approaches to the ecology of ecosystems: one approach views ecological systems in terms of their component parts, nutrient pools coupled to complex networks of interacting populations. The other approach is more holistic, coming at systems from the top down rather than from the bottom up. These two perspectives each have their own advantages and limitations, but both have proven to be useful.

Community structure concerns all the various ways in which members of communities relate to and interact with one another, as well as any community-level properties that emerge from these interactions. Just as populations have properties that transcend those of the individuals comprising them, communities seem to have both structure and properties that are not possessed by their component populations. Community ecologists are still in the process of developing a vocabulary. Identification of appropriate aggregate variables or *macrodescriptors* (Orians, 1980a) is essential, but constitutes a double-edged sword; macrodescriptors allow progress but simultaneously constrain direction(s) that can be pursued. To be most useful, macrodescriptors must simplify population-level processes while retaining their essence without fatal oversimplification. Examples include trophic structure, connectance, rates of energy fixation and flow, efficiency, diversity, stability, distributions of relative importance among species, guild structure, successional stages, and so on. At this early stage in community ecology, it seems wise not to become overly "locked in" by words and concepts. Even the trophic level concept itself should not be inviolate.

Some system-level properties are simply epiphenomena that arise from pooling components; examples would presumably include trophic levels, subwebs, nutrient cycles, and ecological pyramids. But do communities also possess truly emergent properties that transcend those of mere collections of populations? For example, do patterns of resource utilization among coexisting species become coadjusted so that they mesh together in a meaningful way? If so, truly emergent community-level properties arise as a result of orderly interactions among component populations. This fundamental question needs to be answered. Either way, transcendent phenomena or epiphenomena simply cannot be studied at individual or population levels.

A major problem for community and ecosystem ecologists is that communities are not designed directly by natural selection (as individual organisms are). We must keep clearly in mind that natural selection operates by differential reproductive success of individual organisms. It is tempting but dangerously misleading to view ecosystems as having been "designed" for orderly and efficient function. Antagonistic interactions at the level of individuals and populations (competition, predation, parasitism) must frequently impair certain aspects of ecosystem performance. Effective studies of community organization thus require a plurality of approaches, including all of the following levels: individuals, family groups, populations, trophic levels, community networks, as well as historical and biogeographic studies. All these approaches have something useful to offer. The approach taken must be fitted to the questions asked as well as to the peculiarities of the system under study. Ecosystem-level studies are plagued by difficult problems of scale in both space and time (Figure 1.1). Patch size and dynamics, climatic events, nutrient cycles, disturbance frequency, and dispersal ability are just some of the many factors that vary widely within and among systems, as well as over space from local to geographic areas and through time from the short term to the long term.

Many really interesting questions can be asked about communities: Do they have structure that is transcendent beyond population-level processes? What are the effects of community-level attributes on the component organisms living in a given community? What are the roles of parasitism, predation, mutualism, and interspecific competition in shaping community structure? How important are indirect interactions among species and to what extent do such interactions balance out direct effects? How many, and which, niche dimensions separate species? To what extent are species spread out evenly in niche or resource space? (Such an overdispersion in niche space might be predicted under a competitive null hypothesis, with each species minimizing its interactions with all others.) Or, do clusters of functionally similar species ("guilds") exist? How can such guild structures be detected and measured? What are their components? Are such guilds merely a result of built-in design constraints on consumer species, or do guilds simply reflect natural gaps in resource space? Can guild structure evolve even when resources are continuously distributed as a means of reducing diffuse competition? (A community without guild structure would presumably have greater diffuse competition than one with guild structure.) Do more diverse communities have more guild structure than simpler communities? What factors determine the diversity and stability of communities, and what is their relationship to one another?

Ultimately, we must be able to answer such fundamental questions about *how* natural systems are put together before we will even begin to be able to ask meaningful questions about *why* ecosystems have any particular observed properties, such as "What are the effects of indirect interactions among populations or guild structure on the assembly, structure, stability, and diversity of communities?"

The extreme complexity of most ecosystems makes their study difficult but at the same time quite challenging. The concept of a community is itself an abstraction; communities are seldom clear-cut and distinct but almost always grade into one another. By considering ecological systems as "open" rather than "closed," and by allowing for continual inflow and outflow of materials, energy, and organisms, one can partially overcome this difficulty and the community concept can be quite useful. Thus, communities change both in space and in time, and the picture developed in this chapter is essentially an instantaneous view of a fairly localized portion of a larger community.

## SYSTEMS ECOLOGY

Ecological communities are exceedingly complex, with a myriad of intricate and often quite subtle interactions between and among their component resources, individuals, and populations. Even a process as fundamental as the interaction between a predator and its prey may require elaborate analysis in terms of numerous subcomponents; these include hunger, search, pursuit and capture, functional and numerical response, escape tactics, spatial and temporal distribution of prey, predator learning and interference between predators, and so on (Holling, 1965, 1966). A vigorous branch of ecology, termed *systems ecology*, has been developed to deal with such complexity. Systems ecologists exploit computers to build models of complex ecological systems that allow for various sorts of interactions between and among components, which may themselves have many interacting subcomponents. Using actual data on how each component, at each level, affects others (in practice such data are *very* difficult to obtain), systems ecologists model ecological systems attempting to predict their responses to particular perturbations. Because the systems approach is basically descriptive and deductive, it is limited in that the behavior of a system usually cannot be predicted with accuracy at states outside those used in the original data describing interactions between compartments.

## COMPARTMENTATION

One obvious way to approach a complex community or ecosystem is to attempt to simplify it by recognizing various more or less "natural" subunits. The extent to which systems can be understood in terms of such arbitrary compartments is not clearly evident. Ultimately, we may have to abandon this sort of approach in favor of a more holistic one.

### Trophic Levels

Primary producers, or *autotrophs*, represent the first trophic level. They are the green plants that use solar energy to produce energy-rich chemicals. Primary producers are an essential part of a community in that practically all other organisms in the community are directly or indirectly dependent on them for energy. Organisms other than the primary producers, or *heterotrophs*, include *consumers* and *decomposers*. Herbivores are primary consumers and represent the second trophic level. Carnivores that eat herbivores are *secondary consumers* or *primary carnivores* and are on the third trophic level. Carnivores that eat primary carnivores, in turn, constitute the fourth trophic level and are termed *tertiary consumers* or *secondary carnivores*. Similarly, those that prey on secondary carnivores are *quaternary consumers* or *tertiary carnivores*, and so on. Because many animals, such as omnivores that eat both plant and animal matter, prey on several different trophic levels simultaneously, it is often impossible to assign them to a given trophic level. Such organisms can usually be assigned partial representation in different trophic levels in proportion to the composition of their diet. Another way around this problem is to assign such omnivorous species a position on a "trophic continuum" (Carney et al., 1981; Adams et al., 1983; Cousins, 1985). The trophic level concept has proven to be an extremely useful abstract macrodescriptor for the study of community structure; it facilitates examination of the flow of matter and energy through communities and underscores the fundamental differences between interactions that take place within trophic levels (so-called *horizontal* interactions) as opposed to those that operate between trophic levels (termed *vertical* interactions).

Major components of an ecosystem can be diagrammed conveniently, as shown in Figure 16.1, where each trophic level is treated as a "compartment" and arrows designate the direction of flow of matter and energy. Many materials, including calcium, carbon, nitrogen, and phosphorus, move from compartment to compartment as individual organisms are consumed by others at higher trophic levels, and eventually return to the abiotic "nutrient pool," where they may be reused by primary producers; such movements of matter through ecosystems are termed *biogeochemical cycles* (see following section

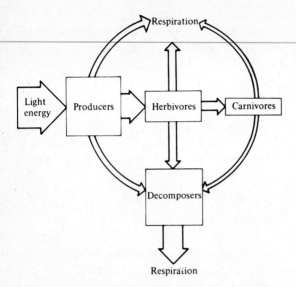

**Figure 16.1** A trophic level "compartment" model of a hypothetical community, with arrows indicating the flow of energy through the system. The width of each arrow reflects the rate of flow of energy between particular parts of the system.

on pp. 351–352). An essential component of any ecosystem is its decomposers, or *reducers*, which return materials to the nutrient pool. Whereas matter continually circulates through the various compartments of an ecosystem and is always being reused, energy can be used only once. All ecosystems thus depend on a continual inflow of energy (see also pp. 352–358).

## Guild Structure

To what extent are species overdispersed in niche or resource space? Do clusters of functionally similar species exist? Root (1967) coined the term "guild" to describe groups of functionally similar species in a community, such as foliage-gleaning insectivorous birds. In competitive communities, guilds would represent arenas with the potential for intense interspecific competition, with strong interactions within guilds but weaker interactions with the remainder of their community. Techniques for objectively defining a guild remain in their infancy, although the "single-linkage" criterion of cluster analysis allows a guild to be defined operationally as follows. *A guild is a group of species separated from all other such clusters by an ecological distance greater than the greatest distance between the two most disparate members of the guild concerned.* This conservative definition allows complex hierarchical patterns in which smaller guilds are nested within larger ones (Figure 16.2).

The numbers of bird species represented in various foraging guilds in tropical forests are shown in Figure 16.3.

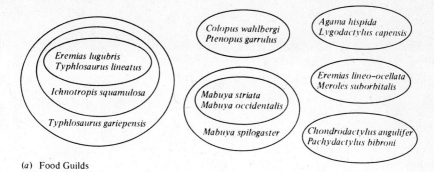

(a) Food Guilds

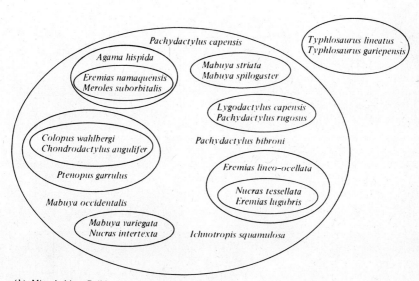

(b) Microhabitat Guilds

**Figure 16.2** The guild structure of the lizard fauna of the Kalahari semidesert, based on stomach contents (food guilds) and microhabitat usage (microhabitat guilds). Of the 21 species, 12 and 14 are members of a two-species food guild or microhabitat guild, respectively (the opposite member of each pair is the other's nearest neighbor in niche space). All but 1 (*Pachydactylus capensis*, a food and microhabitat generalist) of the 21 species is a member of at least one guild. Many two-species guilds are nested inside larger guilds. The microhabitat "superguild" consists of all nonsubterranean species, none of which overlap at all with either of the two subterranean species *Typhlosaurus lineatus* and *T. gariepensis*. (Note that some species in the same food guild, such as *Eremias lineo-ocellata* and *Meroles suborbitalis*, are in different microhabitat guilds.)

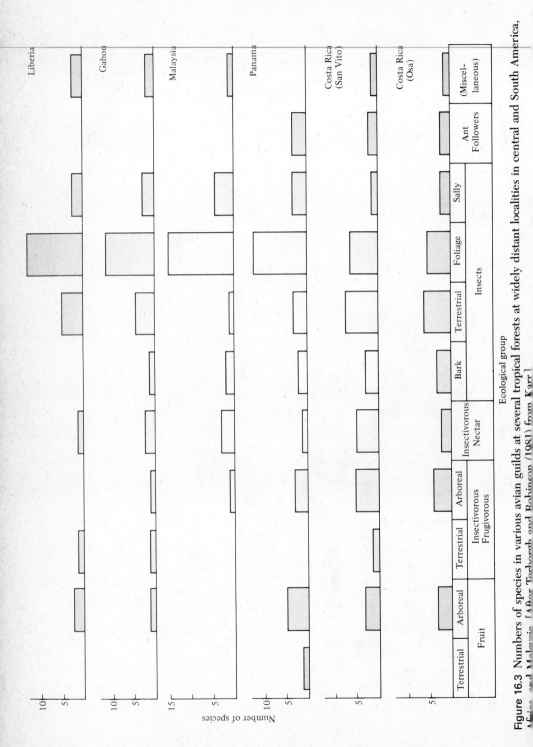

Figure 16.3 Numbers of species in various avian guilds at several tropical forests at widely distant localities in central and South America, Africa, and Malaysia. [After Terborgh and Robinson (1981) from Karr]

342

## Food Webs

Any community can be represented by a *food web*, which is simply a diagram of all the trophic relationships among and between its component species. A food web is generally composed of many *food chains*, each of which represents a single pathway up the food web. Food chains seldom consist of more than five to six links and usually contain only three or four trophic levels (Pimm, 1982; Pimm and Lawton, 1977). A species at the top of a food chain, with no predators of its own, is a *top predator*. Species at the bottom of food chains are termed *basal species*. The direction of flow of matter and energy between species is often shown with arrows, as in Figure 16.4. Horizontal interactions within trophic levels are not usually included. Often such food webs depict only whether or not species in different trophic levels interact, but not the actual intensity of this interaction. A really complete food web would also include the rates of flow of energy and materials among the various populations comprising a community. Gathering accurate data on food webs is not easy and the majority of existing data sets are crude and incomplete. Food webs have recently attracted a lot of interest because they seem to be strongly constrained and exhibit a great deal of interesting pattern. A useful concept is the notion of a *subweb*, or *sink\* food web*, which is a portion of an overall food web that ends up being funneled into a particular top predator (Figure 16.5). Such a subweb may constitute a natural ecological unit of study and may serve as a model network of a larger and more complete food web. Food web complexity increases with the number of cross linkages among component food chains, and this is thought to affect community stability. One means of calculating a simple numerical index of food web complexity is by estimating what is termed *connectance*, defined as the actual number of connections among members of a food web divided by the total possible number of connections (Pimm, 1982). Connectance tends to decrease with increasing numbers of species (more diverse communities have relatively simpler web structures).

## THE COMMUNITY MATRIX

A table of numbers with rows and columns is known as a matrix. Building on the Lotka-Volterra competition equations, Levins (1968) formalized the concept of an alpha matrix, or (more generally) a community matrix. For a community composed of $n$ species, the community matrix is an $n \times n$ matrix that gives the sign and degree of interaction between each pair of species. Figure 16.6 on p. 346 illustrates the community matrix of trophic relationships among ten hypothetical species with a food web as shown. In pairs of competing species, $\alpha_{ij}$ and $\alpha_{ji}$ values both have positive signs, indicating that each inhibits the other's population. The magnitude of the alpha values indicates the intensity of competitive inhibition, which need not be equal and opposite. A prey-predator or

---

\*The opposite concept, probably not as useful, is a *source food web*, which consists of all the consumers that use materials and energy emanating directly or indirectly from a particular entity at a lower trophic level (i.e., a basal species).

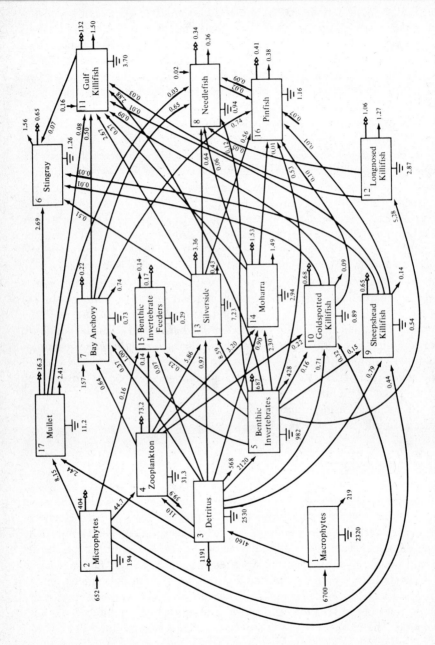

Figure 16.4 A complex network depicting carbon flow in part of a tidal marsh ecosystem in Florida. [From Ulanowicz and Kemp.]

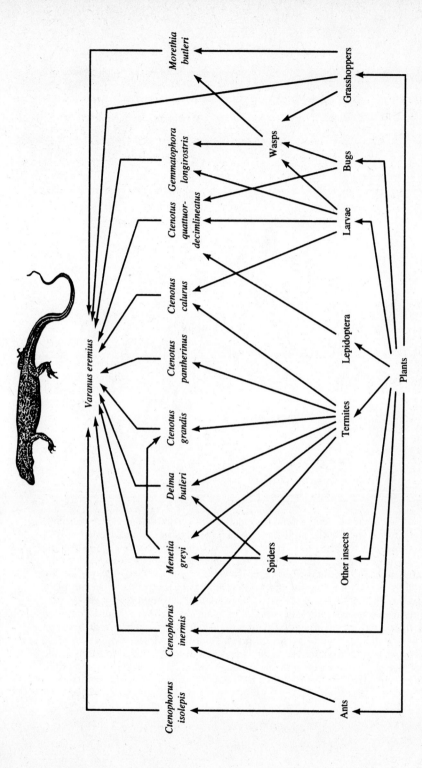

**Figure 16.5** Part of the food web in an Australian sandy desert. The "top" predator, a pygmy monitor lizard, *Varanus eremius*, eats grasshoppers and ten other species of lizards, which in turn have diets dominated by various sorts of arthropods or plants. As in most food webs, taxonomic units are cruder at lower trophic levels than they are at higher ones. A more satisfactory food web would separate all food types into species and would indicate the actual rate of flow of energy up each link, or food chain, in the web.

345

Secondary carnivore:

Primary carnivores:

Herbivores:

Producers:

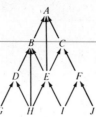

Species having the effect

|  | $A$ | $B$ | $C$ | $D$ | $E$ | $F$ | $G$ | $H$ | $I$ | $J$ |
|---|---|---|---|---|---|---|---|---|---|---|
| $A$ | 1 | − | − | 0 | − | 0 | 0 | 0 | 0 | 0 |
| $B$ | + | 1 | + | − | − | 0 | 0 | − | 0 | 0 |
| $C$ | + | + | 1 | 0 | − | − | 0 | 0 | 0 | 0 |
| $D$ | 0 | + | 0 | 1 | + | + | $\alpha_{DG}$ | − | 0 | 0 |
| $E$ | + | + | + | + | 1 | + | − | − | − | 0 |
| $F$ | 0 | 0 | + | + | + | 1 | 0 | 0 | − | − |
| $G$ | 0 | 0 | 0 | $\alpha_{GD}$ | + | 0 | 1 | + | + | + |
| $H$ | 0 | + | 0 | + | + | 0 | + | 1 | + | + |
| $I$ | 0 | 0 | 0 | 0 | + | + | + | + | 1 | + |
| $J$ | 0 | 0 | 0 | 0 | 0 | + | + | + | + | 1 |

Species affected

Figure 16.6 Food web and interaction matrix for a hypothetical ten-species system. Only the signs of the alphas are given and only direct interactions between and within two trophic levels are considered. All members of a given trophic level are assumed to be in competition. Higher order interactions, such as the effect of a plant's antiherbivore defenses on a carnivore that depends on the plant's herbivores for food, or the indirect effects of the predator on the plant through reducing herbivore densities, are ignored for simplicity's sake. (A community matrix showing the net effect of each species on every other can also be constructed.) One pair of alphas is identified within the matrix, that between Species $D$ (predator) and Species $G$ (prey). Note that $\alpha_{DG}$ is negative, whereas $\alpha_{GD}$ is positive. The community matrix is sometimes formulated in terms of the sensitivity of each species' population density to changes in the density of the others using partial derivatives, in which case the sign structure is reversed.

parasite-host relationship is indicated by $\alpha_{ij}$ and $\alpha_{ji}$ values with opposite signs, with the predator or parasite benefiting from the relationship and the prey or host suffering. Again, the degree of benefit or detriment is represented by the magnitude of the alpha values. Thus, if $\alpha_{ij}$ is negative and $\alpha_{ji}$ is positive, species $i$ eats species $j$. Mutualistic relationships are represented by pairs of negative alphas. Thus, the sign structure of such a matrix reflects the trophic structure of a community.

In competitive communities, a greater number of effective niche dimensions will result in fewer off-diagonal elements of zero and more intense diffuse competition ("connectance" increases). Moreover, the guild structure of a community is implicit in its community matrix, since clusters of species that interact strongly will have large alphas, whereas those that interact weakly will have lower coefficients.

A virtue of the community matrix idea is that it facilitates abstraction and quantification of the interactions among members of a complex community. The concept of the community matrix is actually independent of the Lotka-Volterra equations because it can be formulated equally well in terms of the sensitivities of each species' population density to changes in the densities of other species using partial derivatives (this convention reverses the sign structure of the matrix; see also p. 266). Although one species may affect another in both positive and negative ways simultaneously, the appropriate coefficient in the matrix represents an overall effect. Alpha values in community matrices traditionally represent only the *direct* pairwise effects of each species on every other; however, indirect effects also occur and may either mitigate or enhance direct interactions. An exactly analogous matrix of *net* effects, sometimes termed the *community effect*, between all pairs can also be constructed. Such net community matrices provide useful insights into community structure. For example, a pair of species that competes in direct pairwise interactions may actually have an overall indirect mutualistic relationship provided that each sufficiently reduces the densities of the other species' other competitors. Indirect mutualisms can also arise in several other ways that involve interactions between trophic levels (see pp. 237–239). Indirect effects opposite in sign to direct effects can thus reduce or even actually reverse the overall net effect. Interactions are presumably neither constant nor independent but change both in time and in space and with community composition. Nevertheless, the concept of an ever-changing community matrix is an extremely useful abstraction that helps us to begin to visualize and to model what presumably is actually happening in a complex real community. An emerging idea is that the interactions between any two populations cannot be understood except within the complex network of other interacting populations in which the pair concerned is embedded.

# TYPES OF STABILITY

In ecology, the term *stability* has often been used loosely and left vague and undefined. Many different kinds of stability exist, which can actually vary inversely with one another. Various sorts of mathematical models generate equilibria, such as carrying capacity in single-species models or equilibrium population densities in multispecies models (see Chapters 9, 12, and 16). The notion of such fixed constant equilibria is no doubt an illusion, as equilibria in real systems presumably wander about state space (indeed, some may never be in equilibrium!). The simplest equilibrium structure is the point equilibrium. Point equilibria may be either attractors or repellers. Population trajectories move away from the vicinity of point repellers, but converge on attractors. Associated with point attractors are domains of attraction, bordered regions of state space from within which systems will return to the point attractor. Examples of various stable equilibrium points can be seen in the Lotka-Volterra competition equations (Chapter 12). Local stability is distinguished from global stability (finite versus infinite domains of attraction, respectively). (In the absence of any perturbations, as in a perfectly unchanging world, any system would persist at its equilibrium state.) Obviously, the real world is not unchanging—therefore, a fundamental question of interest is "How do systems respond to various sorts of perturbations?"

Measures of stability are designed to inform us as to the behavior of systems subjected to various sorts of perturbations. Two distinct kinds of perturbations can be recognized: *direct perturbations* to the variables (changes in population densities) versus so-called *structural perturbations* to parameters or species' properties themselves, such as changes in rates of increase or competitive abilities (which will indirectly alter population densities). Most natural perturbations are of the structural sort, which defy analysis because both the equilibria and the domain of attraction are altered. Direct perturbations are understood much better than the structural perturbations.

Numerous different concepts of stability have been applied to populations and communities. *Persistence* through time (the inverse of turnover rate), measured by how long a population lasts before going extinct, is of obvious biological relevance. Various other concepts of stability are represented graphically in Figure 16.7 (after Orians, 1975). Consider the plots in this figure to represent two-dimensional slices through an $n$-dimensional hypervolume that represents the population densities (relative abundances) of all the species in an $n$-species system. All four panels should be considered together, since they represent several fundamental, but different, kinds of stability (two systems can differ in relative degree of stability depending on which particular type of stability is under consideration). In each of the four plots of Figure 16.7, system A is more stable than system B. In part ($a$), raw data on the state of a system at various times is summarized by the frequency of occurrence of states of the system, as indicated by the intensity and extent of stippling. This kind of stability is known as *constancy or variability*. In part ($b$), *inertia* or *resistance* is the degree to which a particular system changes following a given fixed perturbation. In some situations, multiple domains of attraction and multiple alternative stable states may exist, with perturbations causing the system to oscillate between them. In part ($d$), *amplitude*, or *domain of attraction, stability* is illustrated; systems return to initial states following disturbances but only from within a certain delimited region (A is more stable than B because it has a larger domain of attraction). In part ($c$), *resilience* or *elasticity*, the rate at which a system returns to equilibrium following a perturbation, is known as "Lyapunov" stability in

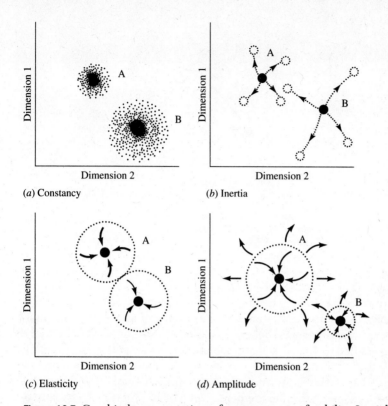

**Figure 16.7** Graphical representations of some concepts of stability. In each graph system A is more stable than system B. (a) Constancy. Frequency of occurrence of states of the systems is indicated by the intensity of stippling. (b) Inertia or resistance. Degree to which the system changes following perturbation. In some situations, multiple alternative states may exist. (c) Resilience or elasticity. Rate of return to equilibrium following perturbation. (d) Amplitude or domains of attraction. Systems return to initial states following disturbances but only from within certain delimited regions. [Adapted from Orians (1975).]

mathematical circles. (Rate of return is measured by the degree of negativity of the dominant eigenvalue of a Jacobian matrix.)

More complex stability structures also exist. The simplest of these is *neutral stability*, wherein the system changes cyclically with the particular trajectory depending solely on initial conditions, as in the Lotka-Volterra predator-prey equations (Chapter 15). Another type of cyclic stability are limit cycles, which exhibit more interesting and constrained behavior: if the system lies within a limit cycle's trajectory, it spirals outward until it reaches the limit which it tracks; if outside, a system spirals in until it reaches its limit cycle attractor (Figure 16.8a). Limit cycles can be represented as ellipsoidal trajectories, indicating the possible states of the system (in such a case, an unstable equilibrium point lies within the ellipse). Under trajectory stability (Figure 16.8b), the system converges to a particular state from a variety of initial states (secondary succession is an example).

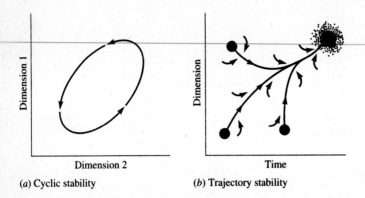

(a) Cyclic stability     (b) Trajectory stability

**Figure 16.8** (a) Cyclic stability. Limit cycles can be represented as ellipsoids, indicating the position of the system at a particular time. (b) Trajectory stability. The system converges to a particular state from a variety of initial states (secondary succession is an example). [Adapted from Orians (1975).]

Still more complex cyclic attractors include toroidal flow, wherein population trajectories follow the three-dimensional surface of a torus (a doughnut-shaped attractor). Quasiperiodic behavior is also approximately cyclic. Chaotic attractors, known too as fractal and strange attractors, are still more complex, in that the system never returns to the same state, but is nevertheless constrained to oscillate within a certain finite hyperspace. Figure 16.9 depicts the Lorenz attractor (named after the meteorologist who discovered chaotic attrac-

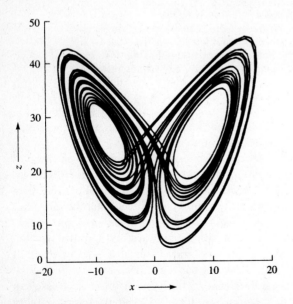

**Figure 16.9** Two dimensions of the Lorenz strange attractor. A third dimension $y$ is not shown.

tors). Note that this system stays on either of two disks that are arranged at angles to each other in three-dimensional space, winding out on one disk until it gets too close to the outside edge, at which point the system then "inserts" to a new trajectory deep within the other disk (such behavior is known as *folding*).

## BIOGEOCHEMICAL CYCLES IN ECOSYSTEMS

Some chemical elements are much more abundant than others; for example, iron, oxygen, and silicon are very common, whereas potassium, calcium, and phosphorus are relatively uncommon, at least on the surface of the planet earth. Organisms expend energy to concentrate and retain certain rare elements in their tissues. A relatively few ions are vitally important to carbon-based living systems, particularly calcium, sodium, potassium, nitrate, phosphate, sulfate, and carbonate. Different nutrients are recycled through ecosystems and nutrient pools at different rates and in different ways. Numerous such cycles exist; among the most important are those involving oxygen, carbon, sodium, calcium, nitrogen, sulfur, and, of course, water. These movements of materials through organisms and various ecosystems are very complex but quite interesting and important. Figure 16.10 depicts one such cycle, that of cal-

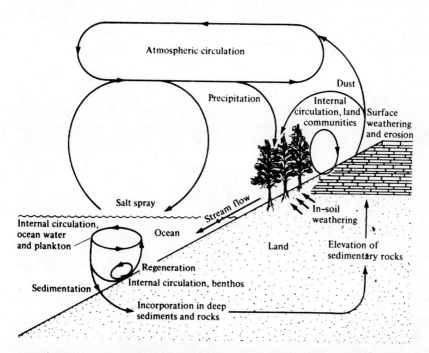

**Figure 16.10** A biogeochemical cycle: major movements of calcium are represented diagrammatically. Many other materials circulate within and between ecosystems in somewhat similar fashions. [From Whittaker (1970). *Communities and Ecosystems.* Reprinted with permission of Macmillan Publishing Company, Inc. Copyright © 1970 by Robert H. Whittaker.]

cium, which circulates in a wide variety of different ways at very different rates, in and between various biotic and abiotic components. Biogeochemical cycles markedly affect the physical environments experienced by organisms. If a biogeochemical cycle is *closed*, nutrients are returned to a nutrient pool and tightly recycled within a system. However, many nutrient cycles are not closed and exhibit losses to other systems. Whereas most gaseous cycles, such as those in the atmosphere, are closed, aqueous and sedimentary cycles are frequently not closed. Rates of turnover of nutrients in various biogeochemical cycles, as well as available reservoirs, determine "residence times" and are of substantial interest in ecosystem ecology because they profoundly affect a system's productivity as well as its resilience. Slow turnover rates reduce stability and productivity, whereas rapid rates of nutrient cycling confer stability on ecosystems and enhance productivity.

The oceans constitute a major reservoir for many materials—rivers carry an unceasing supply of dissolved solutes, as well as various suspended materials, the most important of which are soils (silts). Much of this inexorable flow is essentially one way, but, on a geological time scale, even sea floors are eventually destined to become landmasses. Virtually all materials must make this inevitable journey to the sea. Rainwater dissolves certain rocks, particularly limestones (calcium carbonate); it also leaches all sorts of water-soluble nutrients out of soils. A steady supply of materials such as calcium, potassium, sodium, carbonate, nitrate, and phosphate thus flows steadily into the seas. Except for sodium and chloride, almost all of these ions are removed from seawater either by organisms or by precipitation to wind up on the sea floor in ocean sediments, either as dead organisms or precipitates. Once there, materials remain locked up and out of commission for millennia. Thus, this huge reservoir displays a very long average residence time. Oceanic chemistry is also maintained by a geochemical cycle; its salt concentration, or salinity, of about 35 parts per thousand sodium chloride ($NaCl$), represents an equilibrium, which has probably persisted since before life began.

## PRINCIPLES OF THERMODYNAMICS

An important facet of community ecology is the energy relationships between and among members of a community. Before studying community energetics, however, we must briefly review some fundamentals of thermodynamics.

All organisms require energy to persist and to replace themselves, and the ultimate source of practically all the earth's energy is the sun. One can think of the sun as "feeding" the earth through its radiant energy. But 99 percent or more of this incident solar radiation goes unused by organisms and is lost as heat and heat of evaporation; only about 1 percent is actually captured by plants in photosynthesis and stored as chemical energy. Moreover, energy available from sunlight varies widely over the earth's surface, both in space and in time (see Chapters 3 and 5).

Physics and chemistry have produced two basic laws of thermodynamics that are obeyed by *all* forms of matter and energy, including living organisms. The first law is that of *conservation of matter and energy*, which states that matter and energy cannot be created or destroyed. Matter and energy can be transformed, and energy can be converted from one form into another, but the total of the equivalent amounts of both must always remain constant. Light can be changed into heat, kinetic energy, or potential energy. Whenever energy is converted from one form into another, some of it is given off as heat, which is the most random form of energy. Indeed, the only energy conversion that is 100 percent efficient is conversion to heat, or burning. Burning aliquots of dried organisms in "bomb calorimeters" is a common method of determining how much energy is stored in their tissues (Paine, 1971). Energy can be measured in a variety of different units such as ergs and joules, but the common denominator used in ecology is heat energy or calories.

The second law of thermodynamics states that energy of all sorts, whether it be light, potential, chemical, kinetic, or whatever, tends to change itself spontaneously into a more random, or less organized, form. This law is sometimes stated as *entropy increases*—entropy being random, unavailable energy. Suppose I heat a skillet to cook an egg, and after finishing, I leave it on the stove. At first, heat energy is concentrated near the skillet, which is, relative to the rest of the room, quite nonrandom. But by the next morning the skillet has cooled to air temperature, and the heat energy has radiated throughout the room. That heat energy is now dispersed and unavailable for cooking; the system of the skillet, the room, and the heat has gone toward equilibrium, has become more random, and entropy has increased. Unless an outside source of energy such as a stove, with fuel or electricity, is continually at work to maintain a nonequilibrium state, dispersion of heat results in a random equilibrium state. The same is true for all kinds of energy. According to this law, our solar system and presumably the entire universe should theoretically become a completely random array of molecules and heat in the far distant future.

Life has sometimes been called *reverse entropy* (negentropy) because organisms maintain complex organized states compared to their surroundings. But they obey the second law just as any other system of matter and energy; all organisms must work continually to build and maintain nonrandom assemblages of matter and energy. This process requires energy, and organisms use the energy of the decaying sun (which, of course, also obeys the second law of thermodynamics and tends toward increasing disorder) to "oppose" the second law within their own tissues by producing order out of increasing disorderliness. Wherever there is a live plant or animal, there must be an energy source. Without a continued influx of energy, no organism can survive for very long. Again, this reverse entropy occurs only *within* each organism, and the overall energy relations of the entire solar system are in accord with the second law of thermodynamics, with the overall system continually becoming more and more random.

## PYRAMIDS OF ENERGY, NUMBERS, AND BIOMASS

The rate of flow of energy through a given trophic level decreases with increasing trophic level for several reasons. Because energy transfers are never 100 percent efficient, not all the energy contained in any given prey item is actually available to a predator; some is lost in converting prey tissue into predator tissue, and some is not even assimilated but passes through the predator's intestinal tract unchanged and is then decomposed by reducers. The efficiency of transfer of matter and energy from prey to predator is often greatly reduced by predator-avoidance tactics of prey, such as the chemical defenses of plants. In addition, each organismic unit (and each trophic level) expends some of its available energy on its own activities, further reducing the amount of energy available at higher trophic levels. Ultimately, of course, at equilibrium all the energy captured by primary producers must be expended and dissipated back into space as heat; that is, the amount of energy entering the system must exactly balance that leaving it.

The reduction in the rate of flow of energy from each trophic level to the next higher one determines many of a community's properties, including the total number of trophic levels as well as the proportion of predators to prey. Ecologists estimate that after standardization per unit area and unit time, approximately 10 to 20 percent of the energy at any given trophic level is available to the next higher trophic level. Hence, if a thousand calories are available to primary producers, usually only a few of the thousand are actually available to a secondary carnivore three trophic levels away. A result of this rapid reduction in the availability of energy is that animals at higher trophic levels are generally much rarer than those at lower ones. Moreover, decreasing availability of energy places a distinct upper limit on the number of trophic levels possible, with about five or six being the normal maximum.

A convenient means of expressing the energetic structure of a community is the *pyramid of energy*, which consists simply of the rates of energy flow between various trophic levels (Figure 16.11). The laws of thermodynamics and the preceding considerations dictate that the pyramid of energy can never be inverted; that is, the flow of energy through each trophic level must always decrease with increasing trophic level.

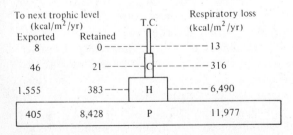

Figure 16.11 Pyramid of energy for Silver Springs, Florida. P = plants, H = herbivores, C = carnivores, TC = top carnivores. [From Phillipson (1966) after Odum.]

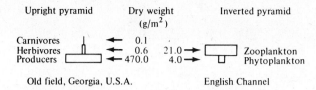

**Figure 16.12** Two pyramids of biomass, one upright and one inverted. [From Phillipson (1966) after Odum.]

Two other types of ecological "pyramids" are the *pyramid of numbers* and the *pyramid of biomass*. These are instantaneous measures rather than rates; they have no time dimension (units of the pyramid of energy are calories per square meter per year, but the units of the pyramid of numbers are numbers per square meter, and the units of the pyramid of biomass are grams per square meter). The pyramid of numbers consists of a set of the densities of individuals in each trophic level; the pyramid of biomass is the biomass (usually measured in grams dry weight) per square or cubic meter in each trophic level. Pyramids of numbers and biomass measure only the *standing crop* (the amount present at an instant) of each trophic level, not their turnover rate. Because they lack a time dimension, these two pyramids may be inverted, with, for example, lower densities or smaller biomasses at lower trophic levels. Thus, one tree may support many insects (an inverted pyramid of numbers); likewise, a rapid rate of turnover allows a small biomass of prey to support a larger biomass of predators with a slower turnover rate. Such an inverted pyramid of biomass often characterizes aquatic ecosystems, where primary producers (phytoplanktonic algae) are small and rapidly dividing, whereas their zooplanktonic predators are larger and longer lived (Figure 16.12).

## ENERGY FLOW AND ECOLOGICAL ENERGETICS

The energy content of a trophic level at any instant (i.e., its standing crop in energy) is usually represented by a capital Greek lambda, ($\Lambda$), subscripted to indicate the appropriate trophic level: $\Lambda_1$ = primary producers, $\Lambda_2$ = herbivores, $\Lambda_3$ = primary carnivores, and so on. Similarly, the rate of flow of energy between trophic levels is designated by lowercase Greek lambdas, $\lambda_{ij}$, where the $i$ and $j$ subscripts indicate the two trophic levels involved, with $i$ representing the level receiving and $j$ the level losing energy. Subscripts of zero denote the world external to the system; subscripts of 1, 2, 3, and so on, indicate trophic level as previously stated.

Using these conventions, one can represent an ecosystem by a compartment model, as in Figure 16.13. At equilibrium, the amount of energy contained in every compartment (trophic level) must be constant, which in turn requires that the rate of flow of energy into each compartment be exactly balanced by the flow of energy out of the compartment concerned. At equilibrium ($d\Lambda_i/dt$ = 0 for all $i$), energy flow in the system portrayed in the figure may

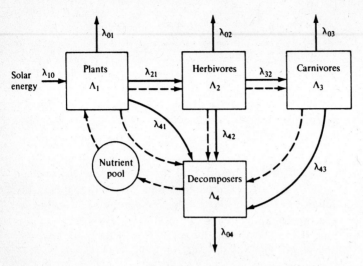

**Figure 16.13** Another compartment model of an ecosystem. Energy flow is shown with solid arrows and the flow of matter with dashed arrows (see text).

thus be represented by a set of simple equations (with inputs on the left and rate of outflow to the right of the equal signs):

$$\lambda_{10} = \lambda_{01} + \lambda_{02} + \lambda_{03} + \lambda_{04}$$

$$\lambda_{10} = \lambda_{21} + \lambda_{01} + \lambda_{41}$$

$$\lambda_{21} = \lambda_{32} + \lambda_{02} + \lambda_{42}$$

$$\lambda_{32} = \lambda_{03} + \lambda_{43}$$

$$\lambda_{41} + \lambda_{42} + \lambda_{43} = \lambda_{04}$$

The rate at which energy is actually captured by plants ($\lambda_{10}$) is estimated at only about 1 percent of the total solar energy hitting the earth's surface. This rate of uptake of solar energy by primary producers, $\lambda_{10}$, is termed the *gross productivity*. It is usually given in *calories per square meter per year*, which represents the *gross annual production* (GAP). Because plants use some of this energy in their own respiration (i.e., $\lambda_{01}$), only a part of the gross annual production is actually available to animals and decomposers; this fraction, $\lambda_{21}$, plus the energy used by decomposers, $\lambda_{41}$, is termed the *net productivity* or (on an annual areal basis) *net annual production* (NAP). Net production may be considerably less than gross production; in some tropical rainforests, plants use as much as 75 to 80 percent of their gross production in respiration. In temperate deciduous forests, respiration is usually 50 to 75 percent of gross primary production, whereas in most other communities, it is about 25 to 50 percent of gross production. Only about 5 to 10 percent of the plant food on land is actually harvested by animal consumers; the remainder of the net primary production is consumed by decomposers. Efficiency of transfer of energy from one trophic level to the next higher trophic level, say, level $i$ to level $j$, may be estimated as $\lambda_{ji}/\lambda_{ih}$, where $j = i + 1$ and $h = i - 1$. Thus, the ratio $\lambda_{21}/\lambda_{10}$ is a measure of the efficiency with which primary producers pass the solar energy they capture on to herbivores and, indirectly, to consumers at higher trophic levels. Such efficiencies of transfer of energy from one trophic level to the next are generally estimated to be between about 5 and 30 percent, and a reasonable average figure is about 10 to 15 percent (Slobodkin, 1960, 1962a, 1962b).

Energy flow diagrams have been constructed for some natural communities (Figures 16.14 and 16.15); these diagrams underscore the relatively minor energetic importance of carnivores in ecological systems. Decomposers often

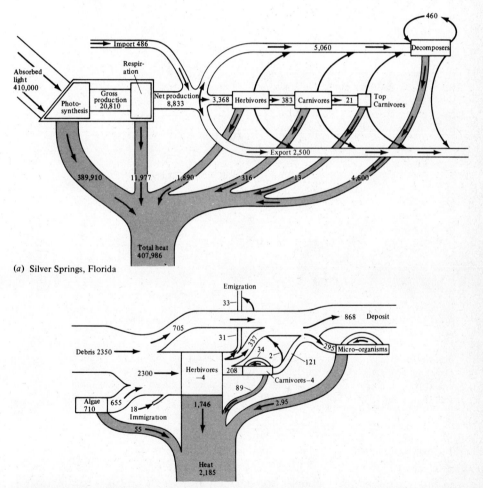

(a) Silver Springs, Florida

(b) Root Spring, Massachusetts

**Figure 16.14** Two energy flow diagrams from actual communities. Figures are in kilocalories per square meter per year; numbers in boxes in (b) represent changes in standing crops. [From Phillipson (1966) after Teal and Odum.]

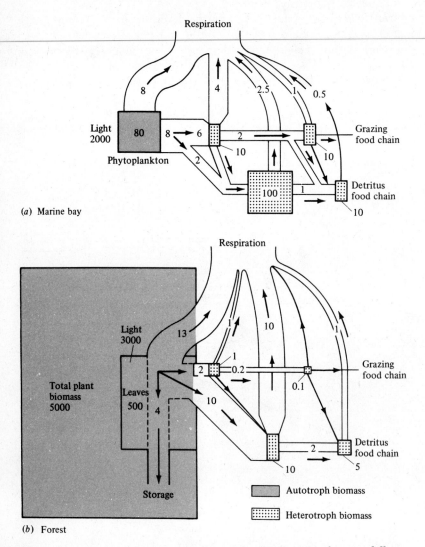

**Figure 16.15** Energy flow diagrams for two ecosystems with very different standing crops. (*a*) A marine bay. (*b*) A forest. Standing crop biomass measured in kilocalories per square meter; energy flow in kilocalories per square meter per day. [From Phillipson (1966) after Odum, Connell, and Davenport.]

play a major energetic role, especially in terrestrial ecosystems where much of the primary production is not consumed by herbivores but instead falls to the ground as dead leaves and other plant material. Indeed, as much as 90 percent of the net annual production in some communities may be consumed by their decomposers. Community ecologists are currently investigating energy flow and efficiency of transfer of energy in natural ecosystems, and much remains to be learned about ecological energetics; such studies have obvious practical significance to human exploitation of ecological systems.

# SECONDARY SUCCESSION

In the southeastern United States, an abandoned field allowed to change naturally is in turn invaded by annual weeds such as crabgrass and asters, then by broomsedge and small perennials, various shrubby species of perennials, pine trees, and finally by oak and hickory trees (Figure 16.16). The composition of the avifauna and bird species diversity also change drastically with these successional changes in the vegetation and are part of the community

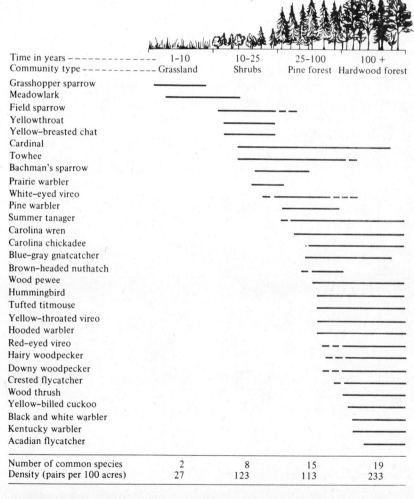

| | | | | |
|---|---|---|---|---|
| Time in years ———————————— | 1-10 | 10-25 | 25-100 | 100 + |
| Community type ——————————— | Grassland | Shrubs | Pine forest | Hardwood forest |
| Grasshopper sparrow | | | | |
| Meadowlark | | | | |
| Field sparrow | | | | |
| Yellowthroat | | | | |
| Yellow–breasted chat | | | | |
| Cardinal | | | | |
| Towhee | | | | |
| Bachman's sparrow | | | | |
| Prairie warbler | | | | |
| White–eyed vireo | | | | |
| Pine warbler | | | | |
| Summer tanager | | | | |
| Carolina wren | | | | |
| Carolina chickadee | | | | |
| Blue–gray gnatcatcher | | | | |
| Brown–headed nuthatch | | | | |
| Wood pewee | | | | |
| Hummingbird | | | | |
| Tufted titmouse | | | | |
| Yellow–throated vireo | | | | |
| Hooded warbler | | | | |
| Red–eyed vireo | | | | |
| Hairy woodpecker | | | | |
| Downy woodpecker | | | | |
| Crested flycatcher | | | | |
| Wood thrush | | | | |
| Yellow–billed cuckoo | | | | |
| Black and white warbler | | | | |
| Kentucky warbler | | | | |
| Acadian flycatcher | | | | |
| Number of common species | 2 | 8 | 15 | 19 |
| Density (pairs per 100 acres) | 27 | 123 | 113 | 233 |

**Figure 16.16** The typical pattern of secondary succession of vegetation and the avifauna on abandoned farmland in the southeastern United States. Number of bird species increases markedly with increased vertical structural complexity of the vegetation. [From MacArthur and Connell (1966) after Odum.]

succession. Plants in each stage modify the environment, presumably making it more suitable for other species in following stages. Typically, shade tolerance and competitive ability increase as succession proceeds. The entire process of secondary succession may take many years (over a hundred years in the preceding example). Only the oak-hickory forest is a stable community in a dynamic equilibrium that replaces itself; such a final stage in succession is termed its *climax*. In deserts, where the open vegetation alters microclimates very little and soil formation is virtually nonexistent, the first plants to invade are usually the climax species, and the succession, if one calls it such, is short. The earth's biomes represent the climax communities that prevail at different localities. Disturbances, both human made and natural (lightning, fires, droughts, landslides, hurricanes, floods), are often frequent enough that extensive areas have not had time enough to reach their own climax state. An equilibrium is reached whereby the proportion of a habitat supporting early successional stages is determined by the frequency of disturbance. Largely undisturbed areas may be primarily in the climax state. During the course of succession, annual production exceeds annual respiration, and organic materials accumulate to form soils and, generally, an increasingly larger biomass of plants and animals. At climax, production equals respiration, and organic materials cease to accumulate (Figure 16.17).

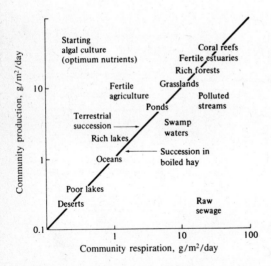

Figure 16.17 Total primary production of various communities plotted against the community's total respiration, in grams per square meter per day (proportional to calories per square meter per day). Communities along the diagonal line are in equilibrium, with production equal to respiration. Production exceeds respiration (autotrophy) in those above the line, whereas respiration exceeds production (heterotrophy) in those below the line. [From Odum (1959) after H. T. Odum.]

TABLE 16.1 **Transition Matrix for Institute Woods in Princeton**
Percent Saplings under Various Species of Trees

| Canopy species | Sapling species (%) | | | | | | | | | | | |
|---|---|---|---|---|---|---|---|---|---|---|---|---|
| | BTA | GB | SF | BG | SG | WO | OK | HI | TU | RM | BE | Total |
| Big-toothed aspen | **3** | 5 | 9 | 6 | 6 | — | 2 | 4 | 2 | 60 | 3 | 104 |
| Gray birch | — | — | 47 | 12 | 8 | 2 | 8 | 0 | 3 | 17 | 3 | 837 |
| Sassafras | 3 | 1 | **10** | 3 | 6 | 3 | 10 | 12 | — | 37 | 15 | 68 |
| Blackgum | 1 | 1 | 3 | **20** | 9 | 1 | 7 | 6 | 10 | 25 | 17 | 80 |
| Sweetgum | — | — | 16 | 0 | **31** | 0 | 7 | 7 | 5 | 27 | 7 | 662 |
| White oak | — | — | 6 | 7 | 4 | **10** | 7 | 3 | 14 | 32 | 17 | 71 |
| Red oak | — | — | 2 | 11 | 7 | 6 | **8** | 8 | 8 | 33 | 17 | 266 |
| Hickory | — | — | 1 | 3 | 1 | 3 | 13 | **4** | 9 | 49 | 17 | 223 |
| Tuliptree | — | — | 2 | 4 | 4 | — | 11 | 7 | **9** | 29 | 34 | 81 |
| Red maple | — | — | 13 | 10 | 9 | 2 | 8 | 19 | 3 | **13** | 23 | 489 |
| Beech | — | — | — | 2 | 1 | 1 | 1 | 1 | 8 | 6 | **80** | 405 |

*Note:* The number of saplings of each species listed in the row at the top, where the abbreviations are self-explanatory, is expressed as a percentage of the total number of saplings (last column) found under individuals of the species listed in the first column. The entries are interpreted as the percentages of individuals of species listed on the left that will be replaced one generation hence by species listed at the top. The percentage of "self-replacements" is shown in boldface.

The total number of recorded saplings is 3286. A dash implies that no saplings of that species were found beneath the canopy; a zero, that the percentage was less than 0.5 percent.

The species are: BTA = *Populus grandidentata*, GB = *Betula populifolia*, SF = *Sassafras albidum*, BG = *Nyssa sylvatica*, SG = *Liquidambar styraciflua*, WO = *Quercus alba*, OK = Section *Erythrobalanus* of *Quercus*, HI = *Carya* spp., TU = *Liriodendron tulipifera*, RM = *Acer rubrum*, BE = *Fagus grandifolia*.

*Source:* From Horn (1975b). Copyright ©1975 by the President and Fellows of Harvard Colleges.

An illuminating analysis of secondary succession was carried out by Horn (1975b) for forests in the northeastern United States. Horn viewed these forests as honeycombs of independent "cells," each occupied by a particular tree. The percentages of saplings of different species observed under various species of canopy trees were used to construct a type of projection matrix, known as a *transition matrix* (Table 16.1). Horn assumed that the proportional representation of various species of saplings under a particular canopy species reflected the probability of its replacement by that sapling species. In Table 16.2, be-

TABLE 16.2 **Observed and Predicted Distributions of Trees**

| Age (years) | BTA | GB | SF | BG | SG | WO | OK | HI | TU | RM | BE |
|---|---|---|---|---|---|---|---|---|---|---|---|
| 25 | 0 | 49 | 2 | 7 | 18 | 0 | 3 | 0 | 0 | 20 | 1 |
| 65 | 26 | 6 | 0 | 45 | 0 | 0 | 12 | 1 | 4 | 6 | 0 |
| 150 | — | — | 0 | 1 | 5 | 0 | 22 | 0 | 0 | 70 | 2 |
| 350[a] | — | — | — | 6 | — | 3 | — | 0 | 14 | 1 | 76 |
| Observed climax (350-year-old forest) | 0 | 0 | 2 | 3 | 4 | 2 | 4 | 6 | 6 | 10 | 63 |

*Source:* From Horn (1975b).

[a] Predicted climax (stationary distribution).

ginning with an observed census of the canopy species in a stand known to be 25 years old, Horn generated the composition of a predicted "climax" forest, which compared reasonably closely to that actually observed in a 350-year-old forest.

## EVOLUTIONARY CONVERGENCE AND ECOLOGICAL EQUIVALENCE

Organisms evolving independently of one another under similar environmental conditions have sometimes responded to similar selective pressures with nearly identical adaptations. Thus, flightless birds such as the emu, ostrich, and rhea fill very similar ecological niches on different continents. Arid regions of South Africa support a wide variety of euphorbeaceous plants, some of which are strikingly close to American cacti phenotypically. A bird of some African prairies and grasslands, the African yellow-throated longclaw (*Macronix croceus*), looks and acts so much like an American meadowlark (*Sturnella magna*) that a competent bird watcher might mistake them for the same species; and yet they belong to different avian families (Figure 16.18g). Such convergent phenotypic responses by different stocks of plants or animals are known as *evolutionary convergence*. Products of convergent evolution—organisms that have evolved independently and yet occupy roughly similar niches in various communities in different parts of the world—are known as *ecological equivalents*. More striking examples of evolutionary convergence (Figure 16.18) usually fall into either or both of two categories. They sometimes occur in relatively simple communities in which biotic interactions are highly predictable and the resulting number of different ways of exploiting the environment are few, or they occur under unusual conditions where selective forces for achievement of a particular mode of existence are particularly strong. Examples of the latter include the independently evolved placental and marsupial "saber-toothed tigers" (Figure 16.18e) and the fusiform shapes of sharks, ichthyosaurs, and dolphins. Evolutionary convergence can easily be read into a situation by placing undue emphasis on superficial similarities but failing to appreciate fully the inevitable dissimilarities between pairs of supposed ecological equivalents.

Often roughly similar ecological systems support relatively few conspicuous ecological equivalents and instead are composed largely of distinctly different plant and animal types. For instance, although the diversities of bird species in the temperate forests of eastern North America and eastern Australia are similar (Recher, 1969, and pp. 385–386), many avian niches appear to be fundamentally different on the two continents. Honeyeaters and parrots are conspicuous in Australia, whereas hummingbirds and woodpeckers are entirely absent. Apparently, different combinations of the various avian ecological activities are possible; thus, an Australian honeyeater might combine aspects

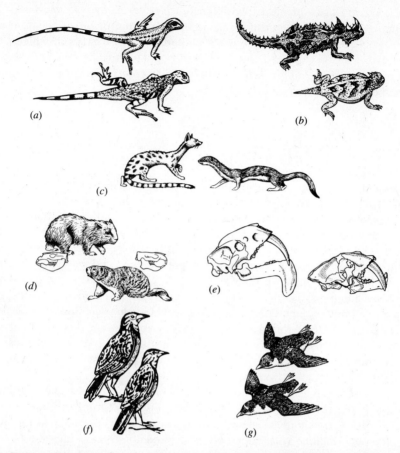

**Figure 16.18** Examples of convergent evolution in animals. Pairs of independently evolved but ecologically similar species that occupy similar niches in different communities are known as "ecological equivalents." (*a*) An Australian agamid lizard, *Amphibolorus cristatus* (left) and a North American iguanid, *Callisaurus draconoides*. (*b*) Another Australian agamid lizard, *Moloch horridus* (left), and an American desert horned lizard, *Phrynosoma coranatum*. (*c*) An African civet (left) and an American weasel. (*d*) An Australian marsupial, a wombat (left), with its skull, and an American placental, a woodchuck, with its skull. (*e*) Skulls of two fossil (but not contemporary) saber-tooth carnivores, the South American marsupial "cat," *Thylacosmilus* (left), and the North and South American placental saber-toothed tiger, *Smilodon*. (*f*) An American icterid, the Eastern Meadowlark, *Sturnella neglecta* (above), and an African motacillid, the yellow-throated longclaw, *Macronix croceus*. (*g*) A North American little auk (above) and a Magellan diving petrel, which belong to two different avian orders. [(*a, b*) After Pianka (1971a, 1986a). (*c, d, e*) After Salthe (1972), *Evolutionary Biology.* Copyright © 1972 by Holt, Rinehart and Winston, Inc. Reprinted by permission of Holt, Rinehart and Winston, Inc. (*f, g*) After Fisher and Peterson (1964).]

of the food and place niches exploited in North America by both warblers and hummingbirds. An analogy can be made by comparing the "total avian niche space" to a deck of cards. There are a limited number of ways in which this niche space can be exploited, and each bird population or species has its own ways of doing things, or its own "hand of cards," determined in part by what other species in the community are doing.

# COMMUNITY EVOLUTION

Many communities change during the lifetimes of the individuals that comprise them. In addition to relatively short-term changes during ecological time, community characteristics are affected by the evolution of and coevolution among the species' populations that are available to form the community over evolutionary time. At the same time, the community itself is a major determinant of the selective milieu of its component populations, and its characteristics presumably dictate many of their adaptations. The so-called *taxon cycle* (pp. 36–37) is thought to be driven by biotic responses to competition and predation, or "counteradaptations" of the other species in a community (Ricklefs and Cox, 1972). Competition within, between, and among species results in the evolution of niche differences, which in turn assures that the resources of a given community, including plants and animals, are utilized more or less in proportion to their effective supply.

As pointed out earlier in this chapter, evolution of the species within a community has still other effects upon community structure. Evolution of prey reduces the efficiency of transfer of energy from one trophic level to the next but increases stability, whereas evolution of predators acts to increase the efficiency of this transfer but reduces stability. The diversity of prey eaten by a predator as well as the predator's ability to alter its diet with changes in prey availability probably influence the stability of prey populations, and therefore of the community.

Can natural selection operate between entire communities? The notions of selection at the levels of communities and ecosystems (Dunbar, 1960, 1968; Lewontin, 1970) constitute apparent extremes of the idea of group selection. Unstable systems would seem to be less likely to persist than stable ones; however, selection is unlikely to occur at these levels, in view of both the limited number of communities and ecosystems and their extremely low rate of turnover. Most important, selection acts only by *differential reproduction*, and it is most difficult to envision reproduction by a community or an ecosystem. Organisms comprising a community are not bound together by obligatory relationships; instead, each evolves in a manner independent of, and often antagonistic to, other members of the community, such as its prey, competitors, and predators. Indeed, community stability may even be incompatible with efficient transfer of energy to higher trophic levels because of the antagonistic interactions between predators and their prey.

# PSEUDOCOMMUNITIES

Communities are so complex that it is exceedingly difficult to study them even when relatively little is known. The situation is exacerbated when a great deal is known—quite simply, communities are unmanageably complicated. Not only is it all but impossible to collect adequate data on such complex systems, but also comparisons between networks are fraught with almost insurmountable difficulties. Manipulative experiments, even if feasible, are of very limited utility due to indirect effects in complex networks. A promising technique for beginning to tease apart complex systems involves construction of various sorts of "pseudocommunities" based on real prototypes, and against which they can be compared. It is no simple matter to construct random pseudocommunities from real ones: inevitably some of the structure of the prototype is mirrored in the randomized replicate. [Colwell and Winkler (1984) term this the *Narcissus effect*.] Such randomly constructed null models of various kinds differ from real systems in interesting ways: similarity among consumers is typically higher and more homogeneous, and guild structure is less evident than in real systems.

Pseudocommunities provide useful insight into the distinction between epiphenomena versus truly emergent community-level properties: to the extent that patterns of resource utilization among coexisting members of a community are "coadjusted" so that they mesh together in a meaningful way, a system has emergent properties. The degree to which consumers actually utilize resources disproportionately to their supply can be quantified with utilizations or "electivities" (Ivlev, 1961; Pianka, 1986a). An assemblage of consumer species can be viewed as being somewhat analogous to a gearbox, with the electivities of various species representing the "cogs" meshing more or less neatly together.* Real assemblages can be compared with "pseudocommunities" in an attempt to ascertain just how good such fits among sympatric consumers actually are.

How well do resource utilization patterns observed among sympatric species in a given system "fit" together? Can evidence be found for ecological adjustments among coexisting species? To address these questions, I undertook a fairly extensive series of artificial "removal-introduction experiments" using data on lizard abundances and diets (Pianka, 1986a). Resource matrices for diets were assembled and analyzed to estimate the "electivities" of each consumer species on each prey resource state over many different study sites. Each "resident" lizard species on every study site was then systematically replaced by the same species, as it was actually observed on each of the other study areas where that species occurred naturally (these are termed *aliens*). A moderately large number of such "transplants" can be made—for a ubiquitous species found on ten study areas, nine alien "introductions" are possible on each site, allowing a total of 90 alien-versus-resident comparisons. Resource utilization spectra of all other resident species are left exactly as observed.

---

*This analogy is dangerous and must not be pursued too far, because communities are not necessarily assembled for orderly and efficient function as a gearbox is, but rather each species of consumer may behave and evolve antagonistically toward the other members of its community.

As one possible measure of the "goodness of fit" among species, assume that the system is approximated by the Lotka-Volterra competition equations. (Consumer species are thus assumed to have reached dynamic equilibria with one another and observed relative abundances are proportional to equilibrium abundances. Moreover, observed dietary similarities are assumed to approximate competition coefficients.) The diffuse competition load on a given target species is estimated by the summation (over all other species) of the products of the alphas times equilibrium densities. Each resident species' observed equilibrium population density is then compared to the theoretical population density that a transplanted alien of the same species would achieve if introduced in its place into its community. By this criterion, residents of most species definitely tend to achieve higher population densities than do aliens. Among all 90-odd species over all 30 study areas, residents outperformed aliens in 1871 out of 3014, or 62 percent, of such "experiments." This trend is much more pronounced when expanded food resource matrices are used: in the Kalahari semidesert, where 46 prey categories (including termite castes) are recognized instead of only 20, residents outperform aliens in 810 out of 1056 cases (a full 76.7 percent!). In a comparable analysis for two Australian study areas using some 300 different prey resource states, residents outperformed aliens in 39 of 52 possible introductions (75 percent of the time). These results strongly suggest that compensatory interactions are occurring among these naturally coexisting lizard species.

Consider now a second, rather more complex, set of pseudocommunities designed to examine whether or not real communities are organized in various ways. Community ecologists struggle with the need to reduce an entire multidimensional system, or a complex network, to a simple graphical state in which they can appreciate the structure and organization of that entire system. As Loehle (1987) aptly said, "the mere attempt to define phenomena operationally can dramatically increase theory maturity." Winemiller (1989, 1990, 1991) worked on neotropical fish in South and Central America. Recently, we developed a promising hybrid approach to compare his aquatic systems with my own desert ones (Winemiller and Pianka, 1990). Our approach adopts a holistic perspective on complex assemblages of interacting species and endeavors to represent the entire assemblage graphically to detect patterns of organization in that system.

The raw data for these analyses are also resource utilization matrices. Some reject the whole approach of resource partitioning, but we maintain that a resource matrix contains vital information about a system. It identifies which species eat which other species as well as which species are potential competitors because they share common foods. Indeed, a resource matrix describes the food web structure of a system.

Considerable tedious work is required to put together a satisfactory resource matrix. Statistical samples of all the species in the system must be collected. If the system is changing in time, this needs to be done quickly; to follow changes in the community through time, adequate samples at different times are necessary. Entries in the resource matrix are used to calculate probabilities, which vary from zero to one, and reflect the probability that a given consumer species will use a particular resource state. Some of these utilization probabilities, $u_{ij}$'s, in the matrix will obviously be zeros because some consumers won't be using certain resources. Without going into all the nitty gritty of the various sorts of probabilistic elements that one can compute to enter into such resource matrices, I want to discuss briefly the concept of *electivity*.

Simple dietary proportions, or $p_i$'s, weight uncommon or very abundant resources disproportionately. Ivlev (1961) suggested resource utilization should be standardized in terms of relative availabilities, however, resource availability is not easily measured in the field. Insects can be sampled with sweep nets, DeVac vacuum cleaners, tanglefoot sticky traps, pit traps, or burliese funnels: each technique yields very different results. Some insects are simply more easily pit trapped than others, whereas others are captured by burliese funnels more than others, and so forth. Pefaur and Duellman (1980, personal communication) studied Andean herps from Colombia to Argentina. They fenced study plots and collected all frogs and squamates (lizards and snakes); inside these plots, all conspicuous insects that a human observer could find were also collected and saved with the intention of using these as standards against the stomach contents of the herps. Humans actually collected only a very few of the insects that were eaten by the herpetofaunas, only about 10 percent (Duellman, personal communication). Incredibly enough, 90 percent of the insects that were in stomachs in fact were not even collected by diligent humans! It is a gross and dangerously misleading oversimplification to accept the idea that a single vector of resource availabilities exists in the real world and that it can adequately describe a real system. Each species experiences its own resource availabilities that depend to some extent on that species' use of space and time, as well as its behavior and foraging mode.

Various solutions to this problem have been proposed. Colwell and Futuyma (1971) suggested a technique that weights resources in proportion to their use in the overall system. We use a variant proposed by Lawlor (1980) exploiting the resource totals in the resource matrix as a measure of resource availability. This constitutes a sort of bioassay. In a system of a hundred species, the diet summed over all the component species represents an estimated resource availability vector. This is used to compute probabilistic analogues of electivity, and an analysis can proceed that is unbiased by resource availability.

In a classic paper, Inger and Colwell (1977) pointed out that there is no consensus as to how to approach community ecology, asserting that there is "no standard protocol for community ecology." That statement is still true today, over 15 years later. Even so, in this paper, Inger and Colwell (1977) made a giant step. They suggested a nearest neighbor approach to look at communities, ranking each species' overlap with every other species from the closest neighbor in niche space to those increasingly more distant. Monotonically declining curves for all the species in the system are generated (Figure 16.19). Some species have high overlap well out into niche space, whereas overlap in others falls off rapidly (such consumers are very distinct, displaying low overlap with most of the other members of the system). The hybrid approach Winemiller and I came

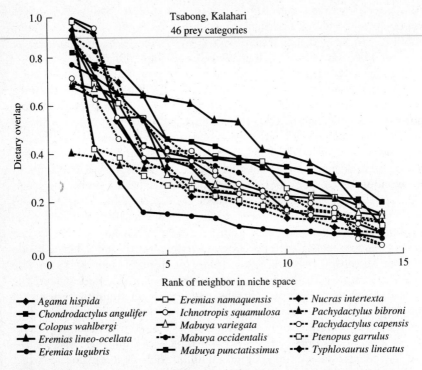

**Figure. 16.19** Plots of dietary overlap with niche neighbors ranked for each lizard species at Tsabong, Botswana. Dietary overlap declines at more distant ranks, but average slopes exhibit large interspecific variation. Overlap at distant ranks is variable in this lizard system. Steep negative slopes indicate relatively unique diets and ecological similarity with very few other species, whereas shallow slopes indicate high or intermediate ecological similarity with many species. [From Winemiller and Pianka (1990).]

up with uses simply the mean overlap at a given rank across all species in the system. Figure 16.19 depicts a system in the Kalahari Desert with 15 species of lizards. This system can be represented with a single curve that is simply the arithmetic average over all 15 species at each rank in niche space.

Another promising technique involves what are called *null models* (Colwell and Winkler, 1984). One of the big challenges in this research is to find something with which to compare a given community. It is extremely difficult to compare a system with someone else's system: that is indeed what provoked us to devise these techniques, to compare fish with lizards. Sale (1974) suggested scrambling the elements of a resource matrix according to some rules to create what I have since come to call pseudocommunities; these are then compared with the original prototype to look for differences to see how the original system is in fact organized. Sale's algorithm involved scrambling all the utilization coefficients, whatever their values are for each consumer in the system (zeros or positive). So one simply takes the first consumer and randomly rearranges all its elements. Rearranged utilization probability $u_{31}$ could fall with equal probability into any slot in that species' utilization vector and $u_{11}$ would be the same. Each species vector of utilization coefficients is scrambled. With a computer, one can easily perform such a rearrangement a hundred times, using a bootstrap approach and exploiting Monte Carlo statistics to generate a distribution against which the observed can be compared (Felsenstein, 1985; Pimm, 1983). Thus one can actually do some statistics and say whether or not any differences are significant.

Lawlor (1980b) suggested a slightly different algorithm which turns out to be equally instructive: Lawlor's algorithm leaves the zero structure of the resource matrix intact. So that if consumer 1 does not eat resource state 3, a zero must remain in cell $u_{31}$; it is frozen and not allowed to change. Elements in the resource matrix are scrambled but only among the resources that are actually used by a given species. (We call Lawlor's method the *conserved-zero* approach, and Sale's method the *scrambled-zero* algorithm, because it destroys the zero structure.)

To exploit these techniques on our real fish and lizard systems (Winemiller and Pianka, 1990), we constructed a test set of hypothetical model systems that had understandable known structure. We built model systems both with and without guilds, and "bench tested" our methodology on these. Figure 16.20 depicts three systems of two guilds of equal size—five species in each, simple little model systems just to see how the randomization algorithms affect them. At the top, there are two guilds with very high, almost total, overlap. At the bottom, there are two guilds with low overlap. In the middle, overlap is intermediate. When the zeros are scrambled, of course, guild structure is destroyed and the result is that the scrambled-zero algorithm results in increased overlap at distant ranks in the niche space. Effectively they float, because the original system (the prototype) had niche segregation in it which was destroyed when the resource matrix elements were scrambled. We also assembled another set of three systems with guilds of different sizes; these behave somewhat the same. We also put together systems like these without any guild structure, but with resource partitioning. It became harder to get pseudocommunities to float, although some conserved-zero pseudocommunities did float, which we interpret as evidence of niche segregation.

We were also interested in the phenomenon of core resources. Both Winemiller's fish and my lizards exploit certain core resources extensively. Among the lizards, these are termites and ants, especially termites. Among the fish, mayflies constitute a core resource. So we created other systems with extensive or total overlap on certain core resources and then unique resources used by each species that were partitioned between species.

Summing up bench tests on the effects of these algorithms: when consumers are piled up on a certain resource state (core resources or guilds with all consumer species within a guild eating the same foods), pseudocommunities generated using the scrambled-zero algorithm display lower overlap than observed real systems (i.e., they tend to "sink" because they fall below the observed system). But, when resources are partitioned, overlap in randomized conserved-zero pseudocommunities is high, and they tend to be above the observed system (i.e., they "float").

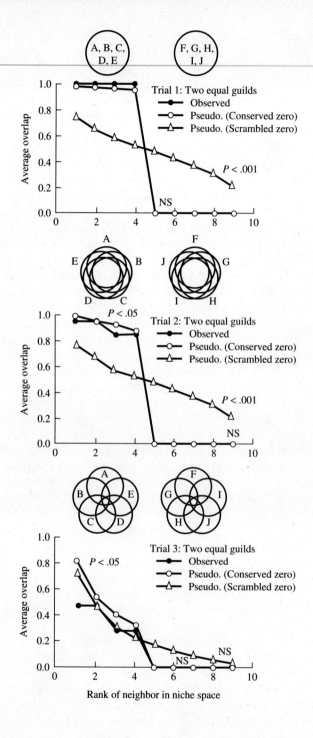

Rank of neighbor in niche space

Winemiller studied ichthyofaunas of aquatic systems in Venezuela and in Costa Rica. One of his study sites has over 80 species of fish in it over the course of an entire annual cycle. Winemiller discovered how to collect virtually an entire freshwater aquatic assemblage. One seine haul during the dry season captured over a thousand fish of dozens of species (plus a "bonus" of a couple of large caiman!). Winemiller's sample sizes are on the order of 300 to 500. While he could not examine the stomachs of all these fish, Winemiller did carry out statistical subsamples, separating his data into wet- versus dry-season resource matrices. Prey content by volume was estimated to the finest discriminatory abilities possible, given our own expertise, usually to insect orders.

We examined 18 different resource matrices, with 2 or 3 from each of eight sites: a wet and dry season for each of four fish sites and microhabitat, plus diet matrices for each of four lizard sites. Numbers of fish species on these sites vary from 19 to 59, and numbers of lizard species vary from 15 to 39. We had between 40 and 217 resource states among the sites analyzed.

An Australian desert site, the L-area near Laverton, Western Australia, has 35 species of lizards. My favorite study area is Red Sands, near Yamarna Homestead in Western Australia. I have collected 47 species of lizards there so far and expect to find several more. The hummock grass tussock plant growth form (spinifex) is very important in the Australian desert. These tussocks, as large as a meter in diameter, house certain lizards that virtually never leave them. Some lizards are highly adapted to spinifex and swim through it with ease. Each lizard collected, some 3000 in Australia and another 2000 from the Kalahari, was weighed and measured in the field, individually tagged, and then permanently preserved by injection with formaldehyde. These specimens are all safely deposited in major museums where they constitute valuable material for systematic research. Many of my lizards have been dissected by people interested in functional anatomy. When the lizards are returned eventually

---

Figure 16.20 Plots of average niche overlap against rank of niche neighbor for three model assemblages with guild structure, and the same plots using means from 100 randomizations based on two algorithms. [Resource matrices for each model system are given in Winemiller and Pianka (1990).] Set theory representations of the systems are depicted above the Colwellian nearest neighbor plots. For systems with high overlap, scrambled-zero pseudocommunities fall below observed systems within guilds but lie above observed systems at more distant between-guild ranks. Trial 1 (two guilds, high overlap) plot shows the observed system exceeding both pseudocommunity overlaps at four out of nine ranks. Trial 2 (two guilds, moderate overlap) plot shows four of nine observed overlaps exceeding pseudocommunity overlaps based on the scrambled-zeros randomization algorithm. Trial 2 conserved-zero pseudocommunities fall above the observed system for three of nine ranks, indicating marginal resource segregation. Scrambled-zero pseudocommunities sink at close-in ranks but float significantly at distant ranks in niche space, as a result of destroying guild structure. In Trial 3 (two guilds, low overlap), conserved-zero pseudocommunities float significantly at the first four ranks in niche space, reflecting segregation, whereas no other differences are significant. Conserved zero probabilities are based on the fraction of randomized means exceeding observed rank means. Scrambled-zero probabilities are based on the fraction of randomized means below observed rank means. [From Winemiller and Pianka (1990).]

to the laboratory, each lizard is measured—ten different body measurements are made on them for anatomical analyses, and then each lizard is dissected and its reproductive state is noted, and relative clutch mass is estimated (testicular cycles can be deduced from serial samples like this), but the most important thing for present purposes is that stomachs are pulled. A competent entomologist went through the stomach contents of the Australian lizards, identifying 100,000 or more prey items to the finest categories possible.

Two Venezuelan fish assemblages are shown in Figure 16.21 during two seasons, wet and dry, based on two different resource matrices (some fish species

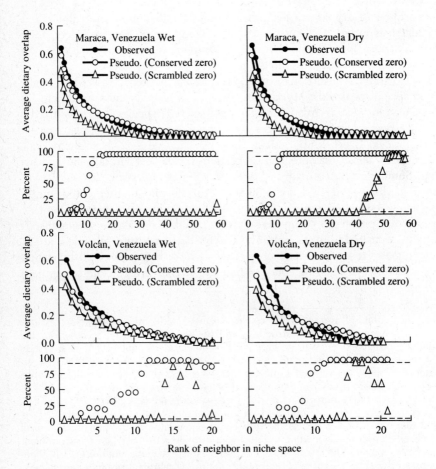

**Figure 16.21** Average observed dietary overlap plotted against rank of niche neighbors for two Venezuelan fish assemblages during wet and dry seasons. Pseudo-community data are based on 100 computer randomizations of the observed prototype. The lower portion of each plot shows the percentage of pseudo-community means greater than the observed mean at each rank in niche space. Conserved-zero pseudocommunity plots lie above observed plots, indicating a high degree of resource segregation. For all assemblages, scrambled-zero pseudocommunity plots fall below observed plots, indicating core resources and guild structure. [From Winemiller and Pianka (1990).]

present in the wet are not there during the dry season). Mean overlap in the observed system is shown in the upper panel of each graph with solid circles. Overlap at each rank in niche space is plotted, this being the average similarity between consumers at the first, second, third, ..., rank. Pseudocommunities are shown with the open symbols: conserved zeros are open circles; scrambled zeros are open triangles. In the lower panels of each figure, the percentage of pseudocommunities that either float or sink are plotted. In this case, sinking of the scrambled-zero pseudocommunities is interesting, as is floating of conserved-zero pseudocommunities, which reflects niche segregation. The dashed lines in the bottom panels are at 5 and 95 percent confidence levels, so when a pseudocommunity lies above the upper dashed line or below the lower dashed line, there is a statistically significant difference between the pseudocommunities and the observed system. At close ranks in niche space, conserved-zero systems don't float, but farther out they clearly differ from observed systems. Scrambled-zero pseudocommunities almost invariably sink at most ranks.

These figures consolidate an enormous amount of information into a single graph. In one case, at Maraca (top panel), 29,000 fish went into preparation of the figure. Winemiller lived in Venezuela for a full year, collected many thousand fish, brought them back, and spent an entire year going through vast numbers of stomachs. All this information can be represented on a single page with a simple graph that one can examine and interpret with a little bit of training.

In Costa Rica, these statistics don't float and sink quite as well as they do in Venezuela, an indication that Central American ichthyofaunas are not as highly organized as the Venezuelan ichthyofaunas.

These aquatic systems are highly organized, with guild structure, core resources, and niche segregation. Consumers are piled up on certain core resources, reflected in the sinking of the scrambled-zero pseudocommunities. But those same consumers are also segregated according to which core resources that they do use, with different species using the same core resources but with different probabilities.

Australian lizards at two study sites are depicted in Figure 16.22. At the very top are microhabitats; in the middle are standard dietary resource matrices (19 prey categories, largely insect orders). At the bottom are expanded dietary resource matrices with 201 different prey categories represented at Red Sands and 217 prey categories recognized at the L-area on the right. Some interesting things emerge from these plots. Scrambled-zero pseudocommunities tend to sink in all cases, which indicates piling up on certain core resources, but this is also indicative of guild structure. Conserved-zero pseudocommunities float somewhat well in microhabitats, which indicates niche segregation: different species use different microhabitats and they float fairly unequivocally, except at the closest ranks at Laverton. In the middle plot, though, pseudocommunities do not show much floating because food resource states are too crudely differentiated, resulting in a piling up of consumers on some resource states. The same data are shown in the bottom panel, but with finer discrimination of prey resource states: note that conserved-zero pseudocommunities float as they did in the fish, indicating segregation.

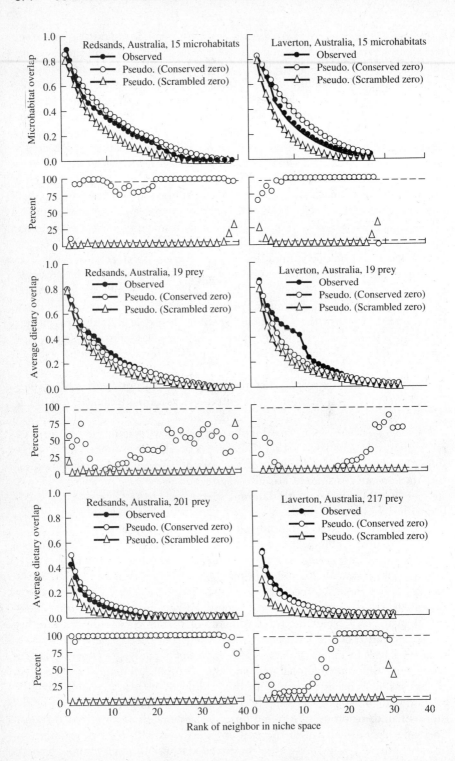

Rank of neighbor in niche space

Kalahari Desert systems are more loosely organized than those in the Australian desert. There are fewer lizard species in the Kalahari and prey could not be distinguished to categories as fine as those in Australia: only 46 different prey resource states were recognized. For microhabitats, there is some hint of floating in the conserved-zero pseudocommunities at Tsabong, Botswana, but not at Bloukrans, South Africa. There is not much niche segregation in diet. There isn't any at all at Tsabong, but there seems to be a little, at least at more distant ranks, at Bloukrans. All in all, the systems we examined tend to be fairly highly organized. This technique should facilitate analyses of other systems and allow comparisons with our own. The pseudocommunity approach is pregnant with potential and would seem to be limited only by our own ingenuity.

## LANDSCAPE ECOLOGY AND MACROECOLOGY

The importance of spatial scale has long been neglected in traditional ecology, although not in the emerging field of landscape ecology. While the implications of the landscape on ecology have long been appreciated, only recently have quantitative methods of study begun to be exploited. In the past, most ecologists, including myself, have focused on local-level processes. Larger scale regional factors, however, also control local phenomena. Local species richness may often be a consequence of regional processes. Relatively little empirical attention has been given to the interaction between these two levels. A recent branch of ecology, known as *landscape ecology*, is concerned with understanding how species and communities persist within a large geographic region. Landscape ecologists adopt a holistic approach and are interested in such things as the origin, size, and shape of habitat patches, habitat fragmentation, and the role corridors play in facilitating dispersal and hence maintaining viable metapopulations. Satellite imagery offers a powerful new tool for such analyses.

To illustrate this approach, I briefly describe some of my own work in progress on the fire succession cycle in inland Western Australia (Pianka, 1992). Few complete closed regions remain unfragmented by human activities in which regional and local phenomena can still be studied simultaneously. The Great Victoria Desert of Western Australia is such an extensive uninhabited area with an extremely high diversity of lizards.

---

Figure 16.22 Plots of average observed overlap in microhabitat (top), and diet (middle) for 19 condensed resource states and (bottom) detailed prey categories against rank of niche neighbors for two Australian lizard assemblages. Pseudocommunity data are based on 100 computer randomizations of the observed prototype, and the lower portion of each plot shows the percentage of pseudocommunity means greater than the observed mean at each rank in niche space. Except for condensed prey resources, conserved-zero pseudocommunioty plots float above observed plots at some but not all ranks, indicating significant resource segregation at these ranks. In each case, scrambled-zero pseudocommunity plots fall below observed plots, indicating significant guild structure. [From Winemiller and Pianka (1990).]

Fires were once a major agent of disturbance in all grassland and semidesert biomes, including the North American tall grass prairies. Most of these ecosystems have now been reduced to mere vestiges, and controlled burning and fire control are practiced by humans almost everywhere. The inland Australian desert is one of the last remaining areas where wildfires remain a regular and dominant feature of an extensive natural area largely undisturbed by humans. An important fire succession cycle, which generates spatial and temporal heterogeneity in microhabitats and habitats, is evident in this region. These regional processes promote local diversity. This system is being studied at the local level in the field in Australia and at the regional level at the University of Texas using aerial photos and multispectral satellite imagery. High-resolution satellite imagery of these areas, which has been collected since 1972, offers a powerful way, heretofore underutilized, to acquire regional-level data on the frequency and phenomenology of wildfires, and thus the systemwide spatial-temporal pattern of disturbance.

Much of the digital satellite data has been acquired by Landsat, but remains archived on magnetic tape. A complete analysis of wildfires for the Great Victoria Desert region will require at least one hundred images. Imagery is being purchased and will be analyzed to detect burned areas. Spectral and spatial statistics will be computed for hundreds of fires through time, and the probability that a given area will be burned will be estimated. Other data to be collated for each fire include: date, location, area, perimeter, compass direction (of burn and prevailing wind), ground-cover characteristics, extent of reticulation, as well as various fractal dimensions. Age and size distributions of burn patches will be estimated. Supporting imagery from other grassland areas, particularly the Kalahari semidesert of southern Africa, will be acquired and used for comparative purposes (fires in the Kalahari do not appear to reticulate to as great an extent as they do in Western Australia).

Fieldwork on the ground will document rates of closure of spinifex, vegetation structure will be mapped, and the presence of various animal species, as well as their abundance, will be assessed at various stages following burns. Low-level aerial photography has been acquired and will be digitized, georeferenced, and analyzed to make detailed maps of vegetation structure for use in computer simulations of fire dynamics. Vertebrate faunas and insects, particularly foods eaten by lizards, will be compared at recently burned sites with those at various stages of postfire recovery to collect data for modeling aspects of the dynamics of the fire succession cycle. Mature spinifex sites have been selected for long-term monitoring. "Burns" will be simulated on the computer to mimic observed fire geometry. Precipitation, rates of accumulation of combustibles, and the insect and vertebrate faunas of these study sites will be monitored roughly every other year for the next decade to collect more precise data on the fire succession cycle.

A major goal of this study is to obtain baseline data on temporal patterns, spatial structure, and distribution of disturbances. These data will form the backdrop for a more detailed study of the population dynamics and dispersal abilities of species. Such data on the component species will be fitted into the overall spatial-temporal mosaic in an effort to explain the persistence of this diverse desert fauna. Ultimately, we plan to model the entire Great Victoria Desert region as a dynamic habitat mosaic so as to understand mechanisms of coexistence of its component species and the effects of fire disturbance in maintaining lizard diversity in this landscape.

An even larger, macroscopic, continent-wide perspective has recently been adopted and deemed "macroecology" by Brown and Maurer (1989) and Brown (1993). Body sizes of entire continental faunas are distributed lognormally— however, size distributions within a given habitat are considerably flatter (Figure 16.23), presumably due to interspecific competition. Brown and Maurer ar-

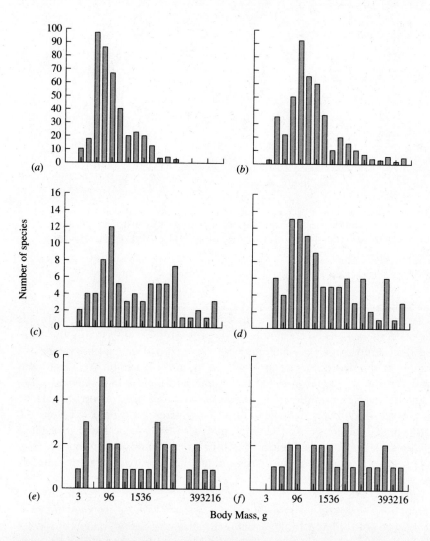

**Figure 16.23** Frequency distributions of body masses among species of North American land birds (*a*) and mammals (*b*) for the entire continent, for land mammals within biomes (*c, d*), and for land mammals within small patches of uniform habitat within biomes (*e, f*). [From Brown and Maurer (1989).]

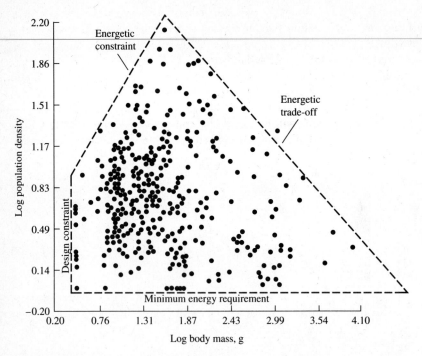

**Figure 16.24** Relationship between average population density and body mass for species of North American land birds. Various hypothesized constraining factors are indicated with dashed lines. [From Wiens (1989) after Brown and Maurer (1989).]

gue that small species are more specialized and have smaller geographic ranges and replace each other more frequently as one moves across the landscape than larger species (which generally have broad geographic ranges). A threshold body size occurs at around 100 grams (Figure 16.24). Up to this critical threshold, species are able to compensate for the higher costs of being smaller by increasing food requirements per unit area of territory. However average population densities fall as body sizes are reduced below 100 grams. Population densities of large animals also tend to be lower than those of moderate-sized animals (Figure 16.24). Large species use more energy than small ones, both on a per-species basis and for all species within a size category (Figure 16.25).

In both North American terrestrial mammals and land birds, long axes of geographic ranges tend to be oriented north-south in small species but east-west in large ones (Figure 16.26, p. 380). However, in Europe, ranges of species of all sizes are aligned east-west. Brown and Maurer offer the following reasonable explanation for these patterns. Species with small ranges (and mostly small body sizes) are limited by habitat types and major topographic features, such

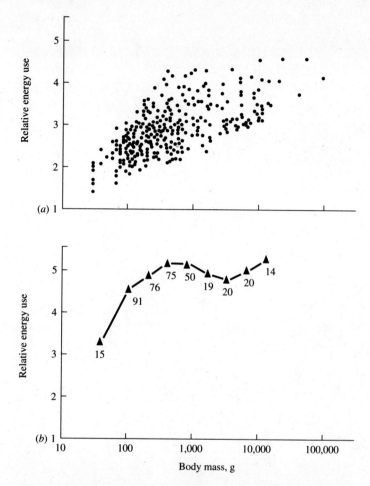

**Figure 16.25** Relative energy usage plotted against body masses for North American land birds. (*a*) Average values for species, and (*b*) values summed for all species in various logarithmic size classes.

as mountain ranges, river valleys, and coastlines. These are oriented predominantly north-south in North America, but east-west in Europe. Species with large ranges are relatively insensitive to such variables and are instead limited by major climatic zones and biome types that are arranged east-west.

In a recent study across a broad range of spatial and temporal scales, Holling (1992) concludes that landscapes are structured hierarchically by a relatively small number of structuring processes into a small number of levels, each characterized by a distinct scale of "architectural" texture and of temporal speed variables. Each of the small number of processes influencing structure

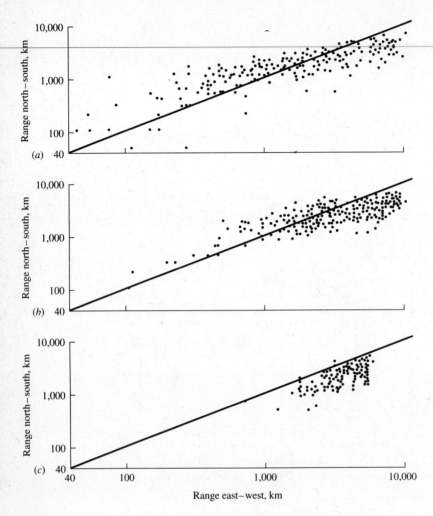

**Figure 16.26** Maximum north-south and east-west extent of geographic ranges of North American terrestrial mammals (*a*) and land birds (*b*). (*c*) The same plot for European land birds.

operates over a limited range of scale (Figure 16.27). In addition, Holling claims to have found evidence that nature is "lumpy," with distinct gaps occurring between such patches of structure. The temporal and architectural structure of ecosystem quanta are determined by three broad groups of processes, each dominating over different ranges of scale. Due to the nonlinear nature of disturbance processes at intermediate ("meso-") scales, fine-scale knowledge of autecology cannot be aggregated to represent behavior at scales beyond the scale of a patch or gap (Holling 1992).

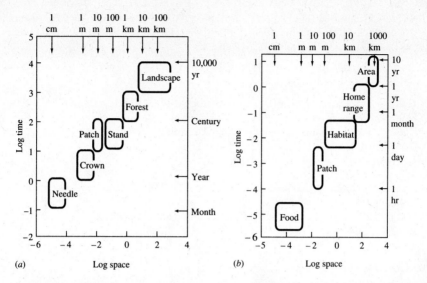

**Figure 16.27** Two examples of how elements of ecosystems scale in time and space. (*a*) Scaling in the boreal forest. (*b*) Scaling in the hierarchy of decisions made by large wading birds. In both plots, time is measured in years and space in kilometers. [From Holling (1992).]

## SELECTED REFERENCES

### Systems and Macrodescriptors

Diamond and Case (1986);   Inger and Colwell (1977);   Kikkawa and Anderson (1986); Levins (1968);   Motomura (1932);   O'Neill et al. (1986);   Orians (1980a).

### Systems Ecology

Berryman (1981);   Bertalanffy (1969);   Bormann and Likens (1967);   Caswell et al. (1972);   Chorley and Kennedy (1971);   Clark et al. (1967);   Dale (1970);   Forrester (1971);   Holling (1959a, 1959b, 1963, 1964, 1965, 1966);   Hubbell (1971, 1973a, 1973b); Huffaker (1971);   Kitching (1983b);   H. Odum (1971);   O'Neill et al. (1986);   Patten (1971, 1972, 1975, 1976);   Reichle (1970); Waterman (1968);   Watt (1968, 1973);   Weinberg (1975).

### Compartmentation

#### Trophic Levels

Adams et al. (1983);   Allee et al. (1949);   Briand and Cohen (1984);   Carney et al. (1981);   Cohen (1978);   Cousins (1985);   Elton (1927, 1949, 1966);   Gallopin

(1972);   Hubbell (1973a, 1973b);   Kozlovsky (1968);   Murdoch (1966a);   Odum (1959, 1963, 1971);   Paine (1966);   Phillipson (1966);   Pimm (1979, 1980, 1982, 1984);   Pimm and Lawton (1977, 1980).

## Guild Structure

Holmes et al. (1979);   Inger and Colwell (1977);   Joern and Lawlor (1981);   MacMahon (1976);   Pianka (1980);   Pianka et al. (1979);   Rathcke (1976);   Root (1967);   Terborgh and Robinson (1986);   Ulfstrand (1977).

## Food Webs

Armstrong (1982);   Auerbach (1984);   Briand (1983);   Briand and Cohen (1984);   Cohen (1978);   Hairston et al. (1960);   Paine (1966, 1977, 1980, 1983, 1988);   Pimm (1979, 1980, 1982, 1984, 1991);   Pimm and Lawton (1977);   Schoener (1989);   Winemiller (1990);   Yodzis (1980, 1981, 1988).

## *The Community Matrix*

Lane (1985);   Lawlor (1980);   Levine (1976);   Levins (1968);   May (1973);   Neill (1974);   Parker and Turner (1961);   Pianka (1987);   Seifert and Seifert (1976);   Vandermeer (1970, 1972a, 1972b, 1980).

## *Types of Stability*

Holling (1973);   May (1975c);   Orians (1975).

## *Biogeochemical Cycles in Ecosystems*

Bormann and Likens (1967);   DeAngelis (1980);   Hutchinson (1950);   Pimm (1982);   Vitousek and Sanford (1986);   Whittaker (1975).

## *Principles of Thermodynamics*

Bertalanffy (1957);   Brody (1945);   Gates (1965);   Odum (1959, 1971);   Paine (1971);   Phillipson (1966);   Wiegert (1968).

## *Pyramids of Energy, Numbers, and Biomass*

Elton (1927);   Leigh (1965);   Odum (1959, 1963, 1971);   Phillipson (1966);   Slobodkin (1962a).

## *Energy Flow and Ecological Energetics*

Bertalanffy (1969);   Borman and Likens (1967);   Engelmann (1966);   Gates (1965);   Golley (1960);   Hairston and Byers (1954);   Hubbell (1971);   Lindemann (1942);   Mann (1969);   Margalef (1963, 1969);   Odum (1959, 1963, 1968, 1969, 1971);   Paine (1966, 1971);   Patten (1959);   Phillipson (1966);   Reichle (1970);   Schultz (1969);   Slobodkin (1960, 1962a);   Teal (1962).

## Secondary Succession

Anderson (1986);   Bazzaz (1975);   Clements (1920);   Drury and Nisbet (1973);   Horn (1971, 1974, 1975a, 1975b, 1976, 1981);   Hutchinson (1941);   Margalef (1968);   Otte (1975);   Pickett (1976);   Usher (1979);   Watt (1947).

## Evolutionary Convergence and Ecological Equivalence

Grinnell (1924);   MacArthur and Connell (1966);   Pianka (1985);   Raunkaier (1934); Recher (1969);   Salthe (1972).

## Community Evolution

Darlington (1971);   Dunbar (1960, 1968, 1972);   Futuyma (1973);   Lewontin (1970); Odum (1969);   Ricklefs and Cox (1972);   Rummell and Roughgarden (1985);   Whittaker (1972);   Whittaker and Woodwell (1971).

## Pseudocommunities

Caswell (1976);   Colwell and Futuyma (1971);   Colwell and Winkler (1984);   Felsenstein (1985);   Inger and Colwell (1977);   Ivlev (1961);   Joern and Lawlor (1980); Lawlor (1980);   Loehle (1987);   Pefaur and Duellman (1980);   Pianka (1986a);   Pimm (1983);   Sale (1974);   Winemiller (1989a, 1990, 1991);   Winemiller and Pianka (1990).

## Landscape Ecology and Macroecology

Brown (1993);   Brown and Maurer (1989);   Forman and Godron (1986);   Holling (1992);   Pianka (1992, 1993);   Ricklefs (1987);   Wiens and Milne (1989).

**Chapter**
**17**

# Biotic Diversity and Community Stability

## SATURATION WITH INDIVIDUALS AND WITH SPECIES

In any closed ecosystem at equilibrium, all the energy of net production must be used up by consumers and decomposers in order for the system to have a balanced energy budget (Figure 16.13 and equations on p. 356). Such an idealized system can be thought of as being saturated with individual organisms because all available energy is used and no more organisms could be supported. However, predators, by reducing densities of organisms at lower trophic levels, can prevent their prey populations from reaching *otherwise maximal stable densities* and thereby effectively preclude true saturation within that lower trophic level. If this is the case, only top predator populations reach a sort of "complete" saturation. Furthermore, antiherbivore defenses of plants render much of the net primary productivity unusable by animal consumers and thus require that many plant tissues be routed directly through a community's decomposers.

Communities, or portions thereof, can be kept from reaching saturation with individuals in other ways. Real ecological systems are almost never truly closed; instead, they usually both receive materials and energy from other systems and lose them to others. A community or a portion of a community that is not closed may be rarefied by continual or sporadic removal of organisms.

As a hypothetical example, consider a lake along a river. Both lake and river contain communities of phytoplankton and zooplankton, but the lake receives an inflow of river water containing no members of the lake community, while losing water containing members of its community. Such a system can never become truly saturated because of the continual removal of organisms from it. Moreover, because the physical environment is changing continually (see Chapters 3 and 5), and because it takes time to respond to these changes, populations and communities probably seldom reach equilibrium, although some very $K$-selected organisms may occasionally approach it.

How much does the degree of saturation with individuals vary within and between communities? And how does efficiency of transfer of energy change with degree of saturation with individuals? It is known (see Chapter 15) that the turnover rate of prey is usually highest at intermediate prey densities; moreover, such prey populations often support higher predator densities than larger prey populations. The immense practical value of such knowledge is readily apparent.

Can communities be saturated with *species*? That is, is there a maximum number of *different* species that can exist within an ecological system? If so, a new species introduced into such a community must either go extinct or cause the extinction of another species (which it then replaces). Conversely, the successful invasion of a new species into a community without the extermination of an existing species would imply that the original community was not saturated with species.

Limited evidence suggests that portions of some communities may indeed be saturated with species, at least within habitats. R. H. MacArthur and his colleagues demonstrated that bird species diversity is strongly correlated with foliage height diversity (Figure 17.1) in a remarkably similar way on three continents: North America, South America, and Australia. Habitats with equal amounts of foliage (measured by leaf surface) in three layers (0 to 0.6, 0.6 to 8, and over 8 meters above ground) are richer in bird species than are habitats with unequal proportions of foliage in the three layers. The diversity of bird species is lowest in habitats with only one of these layers of vegetation, such as a grassland. Interestingly enough, knowledge of plant species diversity does not allow an improvement in the prediction of bird species diversity (MacArthur and MacArthur, 1961), which suggests that birds recognize the structure rather than the type of vegetation. Despite the fact that avian niche space is partitioned in a fundamentally different way in Australia, bird species diversity within a given habitat on that continent is very close to what it is in a habitat of similar structure in North America (Recher, 1969). In addition to illustrating that spatial heterogeneity regulates bird species diversity, the convergence of these data suggests that these avifaunas are saturated with species. However, such neat convergences in species densities of plants, insects, and desert lizards do not occur, which suggests that these groups may not always be saturated with species (Whittaker, 1969, 1970, 1972; Pianka, 1973).

The number of species that can coexist at any point in space may have a distinct upper limit, as previously suggested; however, there is no obvious

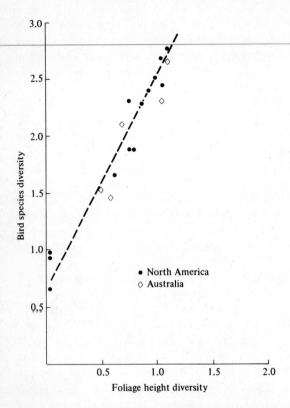

**Figure 17.1** Correlation between foliage height diversity and bird species diversity. North American habitats are shown as solid circles; Australian habitats are represented by diamonds. This is perhaps the best evidence that communities may sometimes be saturated with species. [After Recher (1969).]

limitation on the number of species that can occur in a given *area*, because horizontal replacement of species can allow coexistence of many more species than actually share the use of a common point in space within that area. Indeed, MacArthur (1965) has suggested that the horizontal component of diversity ("between-habitat" diversity) may be increasing continually during evolutionary time, whereas point diversities remain nearly constant. Some upper limit on the horizontal turnover of species also seems likely.

## SPECIES DIVERSITY

Why does one community contain more species than another? Some complex communities, such as tropical rainforests, consist of many thousands of different plant and animal species, whereas other communities, such as tundra communities, support perhaps only a few hundred species. The number of species often varies greatly even at a local level; thus, a grassland habitat typically contains many fewer species of birds than does an adjacent forest. Indeed, different forest communities in the same general region usually vary in numbers as well as in types of plant and animal species. The number of species is referred to as *species richness* or, more frequently, as *species density*.

Communities with similar species densities often differ in yet another way. Some contain a few very common species and many rare ones, whereas others support no very common species but many of intermediate abundance. Abundance is only one way of estimating the *relative importance* of various component species within a community; other measures frequently employed include both the biomass of and the energy flow through various species' populations. The relative importance of species varies within and between communities, and considerable effort has been expended in attempts to document such differences and to understand why they occur. Importances of different species within a community (or a portion of one) can be depicted conveniently by *species importance curves* (Figure 17.2). Several hypothetical distributions

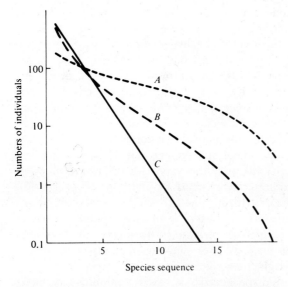

**Figure 17.2** Species importance curves, with species ranked from most important to least important. The line and curves illustrate three different hypothetical distributions based on (A) random niche boundaries [MacArthur (1957, 1960a)], (B) multiple niche dimensions, which generate a lognormal distribution of species importances [Preston (1948, 1960, 1962a, 1962b); Whittaker (1970,1972)], and (C) niche preemption, which leads to a geometric series [Motomura (1932); Whittaker (1970, 1972)]. Data from various real communities fit each hypothetical distribution reasonably well. For a given number of species, diversity would be highest in A and lowest in C. [After Whittaker (1972).]

of the relative importance of species within communities have been suggested that generate different shaped species importance curves (see Figure 17.2 and Whittaker, 1970).

Species density and relative importance have been combined in the concept of *species diversity*, which increases with both increasing species density and increasing equality of importance among the members of a community. Species diversity is high when it is difficult to predict the species of a randomly chosen individual organism and low when an accurate prediction can be made. For example, an organism chosen at random from a cornfield would probably be a stalk of corn, whereas one would not venture to guess the probable identity of a randomly chosen organism from a tropical rainforest. We are currently trying to determine not only why different communities contain different numbers of species with differing relative importances, but how such differences in species richness and importance affect other community properties such as trophic structure and community stability.

Communities can differ in species diversity in several ways. First, more diverse communities may contain a greater range of available resources (i.e., a larger total niche hypervolume space), and second, their component species may, on the average, have smaller niche breadths (i.e., each species might exploit a smaller fraction of the total niche hypervolume). The former corresponds roughly to "more niches" and the latter to "smaller niches." Third, two communities with identical niche space and mean niche breadth can still differ in species diversity if they differ in the average degree of *niche overlap*, because greater niche overlap means that more species can exploit any particular resource (this situation is described as greater resource sharing or "smaller exclusive niches"). Fourth, alternatively, in communities that do not contain all the species they could conceivably support (i.e., those that are "unsaturated" with species), species diversity can vary with the extent to which all available resources are exploited by as many *different* species as possible (i.e., with the degree of saturation with species, or with the number of "empty niches"). Resources are seldom if ever wasted, even in communities that do not contain their full quota of species, because those species that do occur in such communities generally expand their activities and exploit nearly all the available resources, although their efficiency of exploitation may be less than that of some better-adapted species. (Thus, most communities are probably effectively saturated with individuals even if they are not saturated with species.)

Because thorough understanding of community species diversity inevitably requires analysis of niche structure and diversification in component populations, investigations of species diversity have usually gone hand in hand with the study of niches. In practice, one is seldom able to study the species diversity of an entire community, and usually attention is focused on a portion of a community (an "assemblage") such as trees, ants, lizards, seed eaters, or birds. Using the three major niche dimensions (see Chapter 13), we can partition total species diversity of an area into its spatial, temporal, and trophic components. Species replace one another along each of these three niche dimensions, and diversity is generated by separation along each.

I have studied species diversity and niche relations of desert lizards in certain deserts of western North America (the Sonoran and Mojave deserts), southern Africa (the Kalahari Desert), and western Australia (the Great Victoria Desert; Pianka, 1973, 1975, 1986a). Lizard niches in these deserts differ in the three fundamental dimensions (place, time, and food). Moreover, the variety of resources actually used by lizards along each niche dimension, as well as the amount of niche overlap along them, differs markedly among desert systems. Food is a major dimension separating niches of North American lizards; in the Kalahari, food niche separation is slight and differences in the place and time niches are considerable. In the most diverse Australian lizard "communities," all three niche dimensions are important in resource partitioning, and niche overlap is distinctly reduced. Differences between deserts in lizard species diversity are *not* accompanied by conspicuous differences in niche breadths; rather, they stem primarily from differences in the variety of resources used by lizards. Moreover, niche overlap does not increase with diversity but varies inversely with species density and is lowest in the most diverse lizard communities of Australia (Pianka, 1973, 1974, 1975, 1986a).

The spatial component of diversity is due to differential use of space by different populations; for convenience, it can be broken down into horizontal and vertical components. At a gross geographic level, species replace one another horizontally as one moves from one habitat to another (this is the between-habitat component of overall diversity). Similar replacement of species occurs both horizontally and vertically *within* habitats. For instance, birds often tend to partition a given habitat by occupying different vertical strata, such as low bushes, tree trunks, lower foliage, and high canopy. Ground-dwelling mammals and lizards generally partition microhabitats horizontally, with some using open spaces between shrubs and others exploiting the ground beneath or near specific types of vegetation such as grasses, shrubs, and trees. Different populations, by occupying different microhabitats, are thus able to coexist within a given habitat and contribute to within-habitat diversity. The within-habitat component of diversity is most easily distinguished from the between-habitat component in relatively homogeneous communities (heterogeneous communities, such as edge communities and ecoclines, include both components). However, even a homogeneous community has an internal structure in that it consists of a mosaic of repeatable horizontal and vertical patches. Because communities and habitats frequently blend into one another, it is often difficult to distinguish between-habitat from within-habitat diversity. Where does one habitat "stop" and another "begin"? A sand ridge gradually gives way to a sand plain and the intertidal grades into the deeper benthic zone. The problem of defining a habitat can be overcome by the use of "point diversities," which consist of the species diversity occurring at a point in space. Point diversities are difficult to estimate (one might have to wait a *very* long time to see *all* the species that use a particular point!), but point diversity should invariably be lower than any areal estimate of diversity, because the different species in a community have each specialized somewhat as to the microhabitats they use.

Provided that different resources are utilized, temporal separation, both daily and seasonally, of species' populations can allow coexistence of more species and hence may add to community diversity. Many instances of subtle differences in times of activity between populations are known, in addition to such conspicuous distinctions as that between nocturnal and diurnal animals.

Yet another means by which community diversity may be enhanced is by trophic differences. Here again, in addition to the conspicuous differences between trophic levels (such as herbivores, omnivores, and carnivores), there are more subtle but nevertheless important differences between species even within a given trophic level in the prey they eat. Thus, different species of predators living in the same area tend to eat prey of different sizes and types, with the larger species taking larger prey items (this generalization applies to most fish, lizards, carnivorous mammals, and hawks). Moreover, the composition of the diet often varies markedly among potential competitors (see Table 12.3). Finally, diversity of plant defensive chemicals doubtless creates numerous and different potential food niches for herbivores, especially insects (Whittaker, 1969; Whittaker and Feeny, 1971), thereby greatly facilitating trophic diversity at higher trophic levels.

## LATITUDINAL GRADIENTS IN DIVERSITY

A prevalent global pattern of species diversity is of some interest. The diversity of living organisms is usually high near the equator and decreases rather gradually with increasing latitude, both to the north and to the south (Figures 17.3 and 17.4). Such *latitudinal gradients* in diversity are widespread among different plant and animal groups, and it is likely that a general explana-

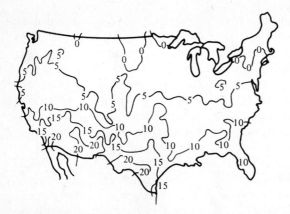

Figure 17.3 Numbers of species of lizards known to occur per square degree of latitude and longitude within the continental United States. [From Schall and Pianka (1978). Copyright © 1978 by the American Association for the Advancement of Science.]

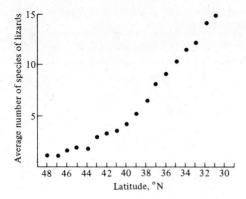

**Figure 17.4**  The latitudinal gradient in the average number of species of lizards per degree square in the continental United States. [From Schall and Pianka (1978).]

tion underlies these ubiquitous patterns. One reason species diversity is higher at lower latitudes than it is in the temperate zones is that often there are more habitats in the tropics. At high altitudes in the tropics, habitats similar to but richer in species than those in temperate zones often occur, whereas true tropical habitats are seldom found in temperate areas. However, the fact that there is a great variety of species where there are more habitats is neither surprising nor theoretically very interesting. Diversity differences *within* a given habitat type are of much greater interest, for they reflect the partitioning of available niche space within habitats.

Considerable speculation about the causes of both local and latitudinal patterns in species diversity has generated many theories and hypotheses for their explanation (Table 17.1), all of which probably operate in some situations (Pianka, 1966a). Various mechanisms for determination of diversity are clearly not independent, and several may often act in concert or in series in any given case. Each hypothesis or theory is briefly outlined subsequently, and some ways in which they could interact are considered.

These mechanisms can be classified as primary, secondary, or tertiary, depending on whether they act mainly through the physical environment alone, through a mixture of both the physical and the biotic environments, or through the biological environment alone, respectively (Poulson and Culver, 1969). Ultimately, a thorough understanding of patterns in diversity requires knowledge of primary-level mechanisms.

**1. Evolutionary Time**  This theory assumes that diversity increases with the age of a community, although the validity of this assumption is still open to question. Temperate habitats are thus considered to be impoverished with species because their component species have not had time enough to adapt to, or to occupy completely, their environment since the recent glaciations and

TABLE 17.1   **Various Hypothetical Mechanisms for the Determination of Species Diversity and Their Proposed Modes of Action**

| Level | Hypothesis or theory | Mode of action |
|---|---|---|
| Primary | 1. Evolutionary time | Degree of unsaturation with species |
| Primary | 2. Ecological time | Degree of unsaturation with species |
| Primary | 3. Climatic stability | Mean niche breadth |
| Primary | 4. Climatic predictability | Mean niche breadth |
| Primary or secondary | 5. Spatial heterogeneity | Range of available resources |
| Secondary | 6. Productivity | Especially mean niche breadth, but also range of available resources |
| Secondary | 7. Stability of primary production | Mean niche breadth and range of available resources |
| Tertiary | 8. Competition | Mean niche breadth |
| Primary, secondary, or tertiary | 9. Disturbance | Degree of allowable niche overlap and level of competition |
| Tertiary | 10. Predation | Degree of allowable niche overlap and level of competition |

other geological disturbances. However, more "mature" tropical communities are more diverse because there has been a longer period without major disturbances for organisms to speciate and diversify within them. The evolutionary time theory does not necessarily imply that temperate communities are unsaturated with individuals; niche expansion may often allow nearly full utilization of available resources even in a habitat that is impoverished in species.

**2. Ecological Time**   This theory is similar to the evolutionary time theory but deals with a shorter, more recent, time span. Here we are concerned primarily with time available for dispersal rather than with time for speciation and evolutionary adaptation. Newly opened or remote areas of suitable habitat, such as a patch of forest burned by lightning, an isolated lake, or a patch of sand dunes, may not have their full complement of species because there has been inadequate time for dispersal into these areas. Dispersal powers of most organisms are good enough that this mechanism may be of relatively minor importance in most communities.

**3. Climatic Stability**   A stable climate is one that does not change much with the seasons. Successful exploitation of environments with unstable climates often requires that organisms have broad tolerance limits to cope with the wide range of environmental conditions they encounter. Thus, by demanding generalization, such variable environments favor organisms with broad niches. Conversely, environments with more constant climates allow finer specialization and narrower niches. For example, plants and animals in the relatively constant tropics are often highly specialized in both the places they forage and the foods they eat. Obviously, in two habitats with the same range of available resources, the one whose component species each use a smaller fraction of these resources will support more species. By these means, the number of species should increase with climatic stability.

**4. Climatic Predictability**   Many aspects of climate, although temporally variable, are nevertheless highly predictable in that they repeat themselves fairly exactly from day to day and year after year. Such cyclical predictability can allow organisms to evolve some degree of dependence on, as well as to specialize on, particular environmental conditions and temporal patterns of re- source availability, thereby enhancing daily or seasonal replacement of species and the temporal component of total diversity. For example, deep freshwater lakes in the temperate zones typically have a consistent annual succession of primary producers; a major causal factor is nutrient availability, which changes markedly during the year, being greatest during the spring and fall turnovers (pp. 78–79). Thus, different species of phytoplankton have adapted to exploit lakes under particular environmental conditions that recur regularly every year. Annual plants in Arizona's Sonoran Desert have adapted to the marked bimodal annual precipitation pattern (Figure 3.12b) with distinct rainfall peaks in winter and summer. There are two distinct sets of species, one whose seeds germinate under wet and cooler conditions (winter annuals) and another set that germi- nate when conditions are wet but warmer (summer annuals that bloom in late summer after the flash floods).

**5. Spatial Heterogeneity**   A forest contains more different species of birds than a grassland, an arboreal desert generally supports more species of lizards than one without trees, and a tidal flat with a great variety of particle sizes and substrate types has more species of mud-dwelling invertebrates than a homogeneous mud flat. Structurally complex habitats obviously offer a greater variety of different microhabitats than simple habitats do. Because there are more different ways of exploiting them, such spatially heterogeneous habitats usually support more species than homogeneous ones do; thus, species replace one another in space with greater frequency and the spatial component of diversity is higher. Correlations between the structural complexity of a habitat and the species diversity of its biota are widespread (one discussed earlier in this chapter is illustrated in Figure 17.1).

**6. Productivity**   In habitats with little food, foraging animals cannot afford to bypass many potential prey items; where food is abundant, however, indi- viduals can be more selective and confine their diets to better prey items (see also pp. 90–92). Thus, more productive habitats, or those where food is dense, offer more prey choice and hence allow greater dietary specialization than do less productive habitats. Because each species uses less of the total range of available foods, the same spectrum of food types will support more species in a more productive environment (Figure 17.5). In addition, productive habitats may support more species than similar, less productive ones do, by virtue of the fact that certain resources, too sparse to support a species in unproductive

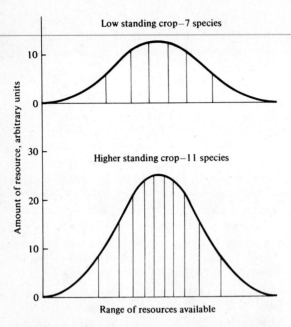

**Figure 17.5** Graphic portrayal of the way in which more abundant resources can support a greater number of species. The horizontal axis represents the different kinds of available resources (ranked in any convenient order). The vertical axis is the amount of each resource type. Both curves cover the same range of resource types, but the lower curve is exactly twice the height of the upper one. All segments under both curves, except the tails, contain about the same area and therefore approximately the same amount of resource. With a low standing crop, component species must have broad niches and only seven species can coexist; when standing crop is doubled, component species can have narrower niches and 11 species may exist in the community. (Resource utilizations by the various species could be represented more realistically with overlapping curves, as in Figure 13.1.) [From Pianka (1971a).]

habitats, are dense enough to be successfully exploited in productive habitats (MacArthur, 1965). An open desert with only one ant nest per hectare might not support a population of specialized ant-eating lizards, whereas another, richer area with several ant nests per hectare would.

**7. Stability of Primary Production** Just as more stable and more predictable climates support more species, areas with temporally stable and predictable patterns of production should often allow coexistence of more species than would be possible in areas with more variable or erratic productivity. This secondary- or tertiary-level mechanism differs from those of climatic stability and climatic predictability in that plants themselves react to climatic conditions and hence can alter temporal variability with their own homeostatic adaptations and storage capacities. Plants both buffer and enhance physical fluctuations, either by releasing the products of primary production (flowers or seeds) gradually and continuously or by expending them in temporally erratic blooms (see also pp. 333–334).

Mechanisms 3, 4, and 7 (climatic stability, climatic predictability, and stability of primary production) could be combined under a more inclusive heading of "temporal heterogeneity" to parallel "spatial heterogeneity" (see also Menge and Sutherland, 1976).

**8. Competition**    In many diverse communities, such as tropical rainforests, populations are thought to be often near their maximal sizes (equilibrium populations of Chapter 9), with the result that intraspecific and interspecific competition are frequently keen. Selection for competitive ability ($K$ selection) is therefore strong, and most successful organisms in these communities have their own zone of competitive superiority. Thus, organisms that have specialized as to foods or habitats are at a competitive advantage (see pp. 281–284), and the resulting narrow niches make high diversity possible. By way of contrast, populations in many less diverse communities, such as temperate and polar ones, are thought to be less stable and often well below their maximal sizes (opportunistic populations of Chapter 8). As a result, portions of the community are often unsaturated with individuals, and intraspecific and interspecific competition are frequently relatively lax. In such communities, it is often the physical world rather than the biotic one that demands adaptation. Selection for competitive ability is weak, whereas selection for rapid reproduction ($r$ selection) is strong. Such $r$-selected populations typically have broad tolerance limits and relatively broad niches.

**9. Disturbance**    Disturbances can result in the continual density-independent removal of organisms from a community. The disturbance hypothesis is essentially an alternative to the competition hypothesis; these two mechanisms would seem to be mutually exclusive. In communities that are not fully saturated with individuals, competition is reduced and coexistence is possible *without competitive exclusion.* Thus, this hypothesis suggests that communities (or portions of communities) can in some sense be oversaturated with species in that more species coexist than would be possible if the system were allowed to become truly saturated with individuals. Disturbance may operate through primary-, secondary-, or tertiary-level mechanisms (see also mechanism 10, Predation). Catastrophic winter cold snaps and subsequent density-independent kills (see Table 9.1) illustrate primary-level disturbance.

An important variant of this hypothesis, termed the *intermediate disturbance hypothesis*, recognizes that, even though infrequent to moderately frequent disturbances enhance diversity, extremely frequent disturbances can be

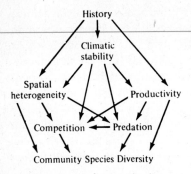

**Figure 17.6** One way in which various mechanisms might interact to determine community diversity. [From Pianka (1971a).]

so decimating that they operate in reverse to reduce diversity. Thus, diversity may actually peak at intermediate levels of disturbance.

**10. Predation** By either selective or random removal of individual prey organisms, predators can act as rarefying agents and effectively reduce the level of competition among their prey. Indeed, as described in Chapter 14 (p. 295), predators can allow the local coexistence of species that are eliminated by competitive exclusion in the absence of a predator. Because many predators prey preferentially upon more abundant prey types, predation is often frequency dependent, which promotes prey diversity.

Clearly, several of these mechanisms may often act together to determine the diversity of a given community, and the relative importance of each mechanism doubtless varies widely from community to community. A multitude of various possible ways in which these mechanisms could interact have been suggested; one is shown in Figure 17.6.

## Tree Species Diversity in Tropical Rainforests

In the lowland tropics, between 50 and 100 different species of trees usually occur together on a single hectare. Although tree species diversity is thus exceedingly high,* many of these trees are phenotypically almost identical, with broad evergreen leaves and smooth bark (indeed, only an expert can distinguish among most species). Many species are quite rare with densities below one tree per hectare. How can so many similar species, apparently all light limited, coexist at such low densities? Explaining why tropical tree diversity is so high is among the most challenging questions facing ecologists. Numerous hypotheses have been proposed but relevant data on this fascinating phenomenon remain distressingly scant.

---

*Of course, such high plant species diversity fosters diversification of animal species, especially insects and birds.

**Seed Predation Hypothesis**    Because seed predation is intense in the tropics, Janzen (1970) argued that seedlings cannot establish themselves in the vicinity of parental trees since the high densities of seeds attract many seed predators (see Figure 15.15, p. 333). This argument predicts that successful recruitment of seeds to seedlings will occur in a ring around (but at some distance from) the parental tree. Inside and outside this ring, other tree species can establish themselves. The variety of seed protection tactics (such as toxic matrices) has forced many seed predators to specialize on the seeds of particular species. Heavy seed predation coupled with species-specific seed predators holds down densities of various tree species and creates a mosaic of conditions for the establishment of seedlings. Hubbell (1980) examines data relevant to this hypothesis and concludes that factors other than seed predation limit the abundance of tropical tree species and prevent single-species dominance.

**Nutrient Mosaic Hypothesis**    The number of ways in which plants can differ is decidedly limited, especially in the wet tropics where variation in soil moisture is relatively slight. One mechanism that could help to maintain high plant species diversity involves differentiation in the use of various materials, such as nitrogen, phosphorus, potassium, calcium, various rare earth metals, and so on. According to this argument, each tree species has its own peculiar set of requirements; the soil underneath each species becomes depleted of those particular resources, making it unsuitable for seedlings of the same species. (Eventually, after the tree falls and is decomposed, these materials reenter the nutrient pool and that species grows again.) Thus, like the seed predation hypothesis, this hypothesis predicts a "shadow" around a parent tree where seedlings of that species will be rare or nonexistent.

**Circular Networks Hypothesis**    In this mechanism, species $A$ is envisioned as being competitively superior to species $B$, while species $B$ in turn excludes species $C$, whereas species $C$ wins in competition with species $A$. Under such a circular hierarchy of competitive ability, the identity of the species occurring at a particular spot will be continually changing from $C$ to $B$ to $A$ and then back to $C$ again, repeating the cycle. Circular networks with many more species could exist. Such nontransitive competitive interactions could help to maintain the high diversity of tropical trees and corals in coral reef communities.

**Disturbance Hypotheses**    Frequent disturbances by fires, floods, and storms might interrupt the process of competitive exclusion locally and allow maintenance of high diversity (Connell, 1978). A variant on this hypothesis involving epiphytes as agents of disturbance (many more epiphytes occur in the tropics than in the temperate zones) has been developed by Strong (1977), who suggests that tree falls (due to epiphyte loads) are frequent in the tropics, continually opening up patches in the forest and fostering local secondary succession.

The preceding hypotheses barely begin to address the question of why hardly any temperate forests support more than a dozen species of trees. If there are more species-specific seed predators in the tropics, why? Why should nutrient differentiation be more pronounced in the tropics? Why doesn't the circular network mechanism foster higher diversity in temperate forests? Are disturbances more frequent in the tropics and, if so, why? Ultimately, latitudinal variation in any of the previously proposed mechanisms will have to be related to underlying variation in primary physical variables such as climate.

A tropical dry forest in Costa Rica was subjected to fairly detailed scrutiny by Hubbell (1979), who mapped 13.44 hectares of forest and analyzed dispersion patterns of the 61 species of trees occurring on this study plot. Prior to Hubbell's work, the traditional generalization had been that tropical trees tend to occur at low densities, more or less uniformly distributed in space (evenly spread out). However, in this Costa Rican dry forest, most tree species, especially rarer ones, were either clumped or randomly dispersed. Moreover, Hubbell found that among the 30 commonest tree species, densities of juvenile trees decreased approximately exponentially with distance away from adults (only 5 of the 30 species displayed sapling "rings" as predicted by Janzen's seed dispersal/predation model, and these rings were very close to adult trees, essentially at the edge of the crown canopy, where a high concentration of seeds might be expected to fall). Hubbell stresses that his results strongly imply that relatively simple physical and biotic factors must govern seed dispersal. Hubbell does not reject the notion of intense seed predation thinning (particularly by bruchid weevil seed predators) among the large-seeded tree species, but he claims that his evidence of clumped or random spacing patterns refutes seed predation by host-specific herbivores as a general mechanism for coexistence of tropical trees. Thus, there seems to be more to high tropical tree species diversity than tree spacing constraints. Hubbell suggests that periodic disturbances are crucial and proposes a random walk-to-extinction model, which generates patterns of relative abundances similar to those observed (along a "disturbance" gradient). In support of this argument, rare species of trees on the study plot had very low reproductive success—they were not replacing themselves locally, although they may have been invading from nearby.

## COMMUNITY STABILITY

Numerous concepts of stability can be and have been applied to communities, including constancy, variability, predictability, persistence, resistance, resilience, and others (Figure 16.7). Too frequently the term "stability" is left vague and undefined. The stability metric of choice is *resilience*, the rate at which a system returns to an original state following a perturbation. A fundamental distinction is made between local or neighborhood stability and global stability.

Traditional ecological "wisdom" holds that more diverse communities are in some sense more stable than simpler ones.* MacArthur (1955) suggested that the stability of populations in a community should increase both with the number of different trophic links between species and with the equitability of energy flow up the various food chains. He postulated that a community with many trophic links provides greater possibilities for checks and balances to operate between and among various species' populations. Should any one population begin to increase markedly, its predators, by changing their diets and feeding selectively on this abundant prey type ("predator switching"), would exert disproportionate negative

**Robert MacArthur**

density-dependent effects on its population growth. Prey refuges result in what Pimm (1982) terms "donor-controlled" population growth, which also enhances stability. Another way in which diversity could confer stability is by means of compensatory indirect interactions that nullify unstable direct interactions (Pianka, 1987).

The "commonsense" generalization that diversity begets stability was challenged by May (1972, 1973), who showed that the probability of finding a stable system actually decreases with complexity in randomly constructed model communities. However, real communities are far from random in construction and must obey various constraints (Lawlor, 1978); there can be no more than five to seven trophic levels, no three-species food-chain "loops" can occur, there must be at least one producer in the system, and so on. Lawlor argues that it is extremely unlikely that *any* of May's random systems even remotely resemble real ecological systems and asserts that the relationship between complexity and stability in large-scale model ecosystems remains unresolved. An effort must be made to perform a comparable analysis on biologically realistic networks to determine whether diversity in such systems enhances or reduces stability.

Empirical studies have produced conflicting results [for reviews, see Goodman (1975), Lawton and Rallison (1979), McNaughton (1977, 1978), and Pimm (1982)]. Watt (1968) found that herbivorous insect species that feed on a wide variety of tree species in Canada actually have *less* stable populations than do similar insects with more restricted diets. Watt does not indicate whether insects with stable populations have a greater variety of potential predators,

---

*This argument is often used to "explain" why our traditional method of agriculture, the monoculture of stands of a single species of plant, tends to lead to unstable ecological systems. Pest populations easily grow at an exponential rate and spread rapidly through the fields. Crop plants are also very vulnerable because they lack a history of adaptation and coevolution. Little wonder we depend on massive applications of pesticides to control such pests!

but population stability did increase with increasing numbers of competing species as would be expected. Clearly, the relationship between diversity and stability remains an important but unresolved problem in community ecology.

## SELECTED REFERENCES

### Saturation with Individuals and with Species

Levins (1968); MacArthur (1965, 1970, 1971, 1972); MacArthur and MacArthur (1961); Pianka (1966a, 1973); Recher (1969); Terborgh and Faaborg (1980); Vandermeer (1972a); Whittaker (1969, 1970, 1972).

### Species Diversity/Latitudinal Gradients in Diversity

Arnold (1972); Baker (1970); Brown (1981); Fischer (1960); Fisher et al. (1943); Futuyma (1973); Gleason (1922); Harper (1969); Huston (1979); Hutchinson (1959); Janzen (1971a); Johnson et al. (1968); Klopfer (1962); Klopfer and MacArthur (1960, 1961); Lack (1945); Leigh (1965); Loucks (1970); MacArthur (1960a, 1965, 1972); MacArthur and MacArthur (1961); MacArthur et al. (1962); MacArthur et al. (1966); Margalef (1958a, 1958b, 1963, 1968); Menge and Sutherland (1976); Murdoch et al. (1972); Odum (1969); Orians (1969a); Paine (1966); Patten (1962); Pianka (1966a, 1973, 1974, 1975, 1986a); Pielou (1975); Poulson and Culver (1969); Prance (1982); Preston (1948, 1960, 1962a, 1962b); Recher (1969); Ricklefs (1966); Schoener (1968a); Schoener and Janzen (1968); Shannon (1948); E. H. Simpson (1949); G. G. Simpson (1969); F. E. Smith (1970a, 1970b, 1972); Sugihara (1980); Tramer (1969); Vandermeer (1970); Watt (1973); Whiteside and Hainsworth (1967); Whittaker (1965, 1969, 1970, 1972); Whittaker and Feeny (1971); Williams (1944, 1953, 1964); Woodwell and Smith (1969).

#### Tree Species Diversity in Tropical Rainforests

Black et al. (1950); Buss and Jackson (1979); Cain (1969); Connell (1978); Dobzhansky (1950); Eggeling (1947); Gilpin (1975b); Hubbell (1979, 1980); Hubbell and Foster (1986); Jackson and Buss (1975); Janzen (1970); Jones (1956); Leigh (1982); Richards (1952); Ricklefs (1977); Strong (1977).

### Community Stability

Armstrong (1982); DeAngelis (1975); Estes and Palmisano (1974); Frank (1968); Futuyma (1973); Goodman (1975); Hairston et al. (1968); Harper (1969); Holling (1973); Hurd et al. (1971); Kikkawa (1986); King and Pimm (1983); Lawlor (1978); Lawton and Rallison (1979); Leigh (1965); Lewontin (1969); Loucks (1970); Lubchenco and Menge (1978); MacArthur (1955, 1965); Margalef (1969); May (1971, 1973, 1975c, 1977); McNaughton (1977, 1978); Milsum (1973); Murdoch (1969); Nunney (1980); Orians (1975); Peterson (1975); Pianka (1987); Pimm (1982, 1984); Rejmanek and Stary (1979); Simenstad et al. (1978); F. E. Smith (1972); Sutherland (1974); Usher and Williamson (1974); Watt (1964, 1965, 1968, 1973); Whittaker (1972); Woodwell and Smith (1969); Yodzis (1981, 1988).

## Chapter
## 18

# Applied Ecology

$T$his book has emphasized basic ecological principles, many of which have obvious and important applications. For example, optimal yields to maximize sustained harvests have long been goals in wildlife management and fisheries biology. An emerging discipline of conservation biology seeks to conserve natural habitats and maintain biotic diversity. Biodiversity constitutes a valuable resource worthy of preservation for many different reasons. Genetic strains of plants with natural resistance to pests are valuable because their genes can be exploited to confer resistance on future crop plants (see next section).

Approximately one drug in four originated in a rainforest: these include analgesics, diuretics, laxatives, tranquilizers, contraceptive pills, and cough drops. Antibiotics were first discovered in fungi, but have now also been found in many species of plants as well. Secondary chemicals of plants have proven to be a vast reservoir for useful pharmaceutical products. Clinically proven drugs derived from higher plants include morphine, codeine, atropine, quinine, digitalis, and many many others. Recently, it was discovered that bark of Pacific yew trees contains taxol, which has proven to be an effective agent in the treatment of certain ovarian cancers. To date, scientists have examined only about 1 percent of existing plant species for useful pharmaceuticals.

# GENETIC ENGINEERING

Modern molecular biotechnological tools,* such as restriction enzymes and gene splicing, now enable geneticists to transfer particular genes from one organism to another. For example, the firefly gene for luciferase has been successfully transferred to tobacco, resulting in transgenic bioluminescent plants. Human insulin and growth hormones are now routinely produced in chemostats of *E. coli* bacteria, which have had human genes spliced into their genomes. Some researchers have even proposed using such transgenic bacteria as live vaccines (the genetically altered bacteria would live within humans and would confer them with resistance to particular diseases, such as hepatitis). Such recombinant DNA technology has also enabled us to produce useful new life forms such as pollutant-eating bacteria that can help us to clean up what's left of our environment. Research is in progress to transfer nitrogen-fixing genes into crop plants. There are legitimate concerns, however, about the safety of research on such synthetic transgenic organisms, particularly the possibility of accidental release of virulent strains that might attack humans. Such concerns have been addressed by implementation of strict containment procedures for recombinant DNA products, as well as by selecting and creating host organisms for foreign DNA that are incapable of surviving outside the laboratory. Obviously, genetically engineered organisms must eventually be designed for release into nature (indeed, genetically engineered tomatoes are now being grown commercially).

Another concern is that genetically engineered organisms could have adverse effects on other species in natural ecosystems. We already have enough natural pests and certainly don't want any humanly engineered ones! Unfortunately, we still know far too little to engineer ecological systems intelligently (obviously, genetic engineers should work hand in hand with ecological engineers). Still another problem is the human tendency to allow short-term financial returns to override long-term prospects.

# CONFLICT BETWEEN HUMAN ACTIVITIES AND PRESERVATION OF WILDERNESS

Most people hold the opinion that the earth exists primarily, or even solely, for human exploitation. Genesis prescribes: "Be fruitful, and multiply, and *replenish the earth*, and subdue it: and have dominion over the fish of the sea, and over the fowl of the air, and over every living thing that moveth upon the earth" (my italics). We have certainly lived up to everything except "replenish the earth."

---

*Other techniques, such as polymerase chain reaction (PCR) amplification of DNA seqments, DNA sequencing, and DNA fingerprinting hold great promise as tools that will allow informative evolutionary studies.

The human population explosion has been fueled by habitat destruction*
—we are usurping resources once exploited by other species. The tall grass
prairies of North America have been replaced with fields of corn and wheat,
native American bison have given way to cattle, and so on. In 1986, humans
consumed an estimated 40 percent of the planet's total production (Vitousek
et al., 1986). Today we consume more than half of the solar energy trapped
by plants. Many species have gone extinct due to human pressures over the
past century, and many more are threatened and endangered. Over a hundred
species of plants and animals now go extinct each and every day due to habitat
destruction by humans.

During the past quarter of a century, world population has increased
from about 3.4 billion people to about 5.6 billion, a 65 percent increase;
5,600,000,000 is a rather large number, a bit like our National debt, and difficult
to comprehend. Each year, the human population increases by approximately
95 million, a daily increase of about one-quarter of a million souls. Nearly
11,000 new people are born each hour—about 3 new babies pop out each
second all day long, every day, day after day.

People everywhere today stand ready to rape and pillage their wilder-
nesses ("wastelands") for whatever they can be forced to yield. Raw materials,
such as ore, lumber, and even sand (used to make glass), are harvested in vast
quantities. Big companies enjoy privileged status, excluding the public from
extensive areas, producing great ugly clear cuts, vast deep open pit mines,
instant, but permanent, manmade mountains, eyesores paying testimony to the
avaricious pursuit of timber, precious metals, and minerals. Deforestation is
nearly complete in many parts of the world. Overgrazing is rampant. Grasses
and the shrub understory have been virtually eliminated over extensive areas.
It is quite instructive to come upon a fenced graveyard, and to see a small
patch of country as it must have been before the land rape by the pastoral industry. Native hardwoods are wasted to make charcoal and burned for firewood.
Lumberjacks will soon be out of work whether or not the remaining timber is
cut. Should forest habitats be saved? Is there enough left to save? This sort of
pillage continues. Virtually everywhere, often with governmental subsidies and
incentives, forests, deserts, and scrublands are being leveled and turned into
fields for crops. Many of these fields are marginal and will soon have to be abandoned, transformed into great vegetationless deserts. More dust bowls are in
the making. In some regions, replacement of the drought-adapted deep-rooted
native vegetation with shallow rooted crop plants has reduced evapotranspiration, thus allowing the water table to rise, bringing deep saline waters to the
surface. Such salinization reduces productivity and seems to be irreversible.
Some deserts have so far been able to resist the tidal wave of advancing
human exploiters, but there are people who dream of the day that technological "advances," such as water plans to move "excess" water or the distillation
of seawater, will make it possible to develop desert regions (i.e., to replace them

---

*Massive consumption of fossil fuels for agriculture has also contributed greatly to overpopulation.

with vast agricultural fields, or even cities). Oxymorons, such as "sustainable development," are strung together by politicians and developers in an attempt to make all this destruction and homogenization seem less offensive.

Most people consider biology, particularly ecology, to be a luxury that they can do without. Even many medical schools no longer require that premedical students major in biology. But basic biology is not a luxury at all; instead, it is an absolute necessity for living creatures such as ourselves. Despite our anthropocentric (human-centered) attitudes, other life forms are not irrelevant to our own existence. As proven products of natural selection that have adapted to natural environments over millennia, they have a right to exist, too. With human populations burgeoning and pressures on space and other limited resources intensifying, we need all the biological knowledge that we can possibly get. For example, in this day and age, a primer on "how to be a successful venereal microbe" has become essential reading for everyone!

Ecological understanding is particularly vital. Basic ecological research is urgent because the worldwide press of humanity is rapidly driving other species extinct and destroying the very systems that ecologists want to understand. No natural community remains pristine. Pathetically, many will disappear without even being adequately described, let alone remotely understood. As existing species go extinct and even entire ecosystems disappear, we lose forever the very opportunity to study them. Knowledge of their evolutionary history and adaptations vanishes with them: we are thus losing access to biological information itself. Just as ecologists are finally beginning to learn to read the "unread" (and rapidly disappearing) book of life, they are encountering governmental and public hostility and having a difficult time attracting support.

Only during the last few generations have biologists been fortunate enough to be able to travel with ease to remote wilderness areas. Panglobal comparisons have broadened our horizons immensely. This is a fleeting and unique opportunity in the history of humanity, for never before could scientists get virtually anywhere. However, all too soon, there won't be any pristine natural habitats left to study.

In a set piece of rational thought, Garrett Hardin (1968) laid out the tragedy of the selfish herdsman on a common grazing land, underscored by the rush to catch the last of the great whales and the ongoing destruction of earth's atmosphere (ozone depletion, acid rain, carbon-dioxide-enhanced greenhouse effect, and so on.). Global weather modification is a very real and an exceedingly serious threat to all of us, as well as to other species of plants and animals.

Earth's atmosphere is unusual in that it has a relatively high oxygen content (about 21 percent). Most other planets have a reducing atmosphere. The free oxygen in today's atmosphere was probably produced largely by the activities of primary producers. The most plausible hypothesis to explain our planet's rather unusual atmosphere is that activities of living organisms, particularly green plants and certain bacteria, play vital roles in the building and maintenance of air. Photosynthetic activities of plants utilize carbon dioxide and water to produce oxygen as a by-product, along with energy-rich reduced-carbon

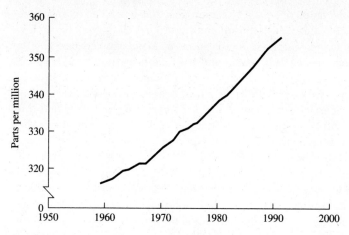

Figure 18.1 Concentration of carbon dioxide in Earth's atmosphere, from 1959 to 1991. [From Brown et al. (1992).]

compounds, such as glucose. Free oxygen is released into the atmosphere by an inanimate process, too. High in the atmosphere above the ozone shield, ionizing solar radiation dissociates water vapor into molecular hydrogen and oxygen. Free oxygen is left behind as the light hydrogen atoms escape into outer space. In a reduced atmosphere, oxidation quickly uses up such free oxygen. Both of these oxygen-generating mechanisms have been important; dissociation was probably much more significant billions of years ago before the ozone layer was formed than it is at present (it will become more important as the ozone layer is further thinned by the release of chlorofluorocarbon gases). Ozone depletion has also increased ultraviolet radiation at the surface, which has almost certainly increased the frequency of skin cancers (though these may not be detectable for another decade).

Our atmosphere is in a complex quasi-equilibrium, although the concentration of carbon dioxide has risen steadily for the last quarter of a century and continues to rise due to deforestation and the burning of fossil fuels (Figure 18.1). Over the past 30 years, consumption of fossil fuels (measured in millions of tons of carbon emissions) increased from 2547 per year to 5600 worldwide (a 120 percent increase).* Carbon emission increased by 55 percent in the United States and by a whopping 170 percent in Australia. Per capita consumption of fossil fuel in both the United States and Australia is 4 to 5 times above the world average. This increase in atmospheric carbon dioxide has enhanced atmospheric heat retention and would have produced global warming sooner except for a fortuitous spinoff of atmospheric pollution—particulate matter increased

---

*To generate this amount of carbon dioxide metabolically, every man, woman, and child on the planet would have to eat approximately 50 kilograms (about 110 pounds) of potatoes each and every day of his or her life!

earth's albedo (reflectance of solar irradiation), so that less solar energy penetrates to the surface (volcanic ash in the atmosphere has the same effect). Until
recently, these two opposing phenomena more or less balanced one another,
but now the balance has clearly shifted and the "greenhouse effect" is leading
to rapid global warming (Figure 18.2). Long-held meterological records the
world over are being broken: 6 of the 7 hottest years on record have occurred
since 1980; a few years ago, the lowest low pressure zone ever recorded (hur-
ricane Gilbert) in late summer was followed in the next winter by the highest
high pressure area ever measured during recorded history.

Desertification has been greatly accelerated during the past century due
to many of the processes already mentioned. Arid areas are in a more precarious and perilous position than wetter areas. As the population burgeons,
the last remaining natural habitats are rapidly being destroyed. Earth's atmosphere is being altered at an ever increasing rate, leading to rapid weather
modification. Global warming is having its impact on virtually all plants and
animals, including humans, and its effects will continue to intensify into the
foreseeable future. Crop failures would seem to be inevitable. Empty shelves
in supermarkets will eventually awaken people to the dire danger of tamper-
ing with earth's atmosphere, but by then it will be much too late to rectify the
situation. People will be appalled that scientists cannot restore the atmosphere
to its former condition. But there can be no quick "technological fix" for earth's
maligned atmosphere. The continuing existence of all the denizens of this poor
beleaguered planet, including ourselves, will ultimately depend more on our
ecological understanding and wisdom than it will on future technological "advances."

Unlimited cheap clean energy, such as that much hoped for in the concept
of cold fusion, would actually be one of the worst things that could possibly
happen to humans. Such energy would enable well-meaning but uninformed

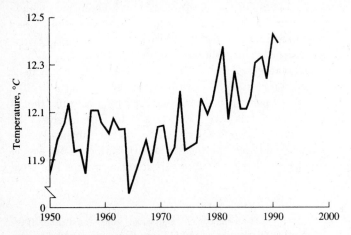

**Figure 18.2** Average global temperatures in degrees Celsius, from
1950 to 1991. [From Brown et al. (1992).]

massive energy consumption (i.e., mountains would be leveled; massive water canals would be dug; ocean water distilled; water would be pumped and deserts turned into green fields of crops). Heat dissipation would of course set limits, for when more heat is produced than can be dissipated, the resulting thermal pollution would quickly warm the atmosphere to the point that all life is threatened, perhaps the ultimate ecocatastrophe. (Diehard technologists will no doubt argue that we will invent ways to shoot our excess heat out into space.)

## TOWARD A SUSTAINABLE FUTURE: AN EQUILIBRIUM ECONOMY

Alas, I am no economist. But, as an interested participant and observer, I cannot resist offering one ecologist's perspective on economic matters that are of great concern to all. I am convinced the very foundations of modern economic thinking can and must be challenged. Economics is based on complementary value differentials, profits for all parties, never-ending growth, and ever-expanding markets (the principle of a chain letter—a Ponzi scheme!). The old adage that "as the rich get richer, the poor get poorer" has some truth in it. While economy is certainly not a zero sum game, the per capita share in all nonrenewable resources such as land and fossil fuels must diminish as the population increases. The same is true even for many renewable resources. For example, in 1990 World Watch reported that, although worldwide grain production has risen steadily in a linear fashion over the past several decades (Figure 18.3), population has increased exponentially, resulting in an actual per capita decrease in grain per person (Figure 18.4) during the past decade. A friend recently calculated the average human's per capita daily share of protein to be about 60 grams (only about one-eighth of a pound). Clearly, in a finite

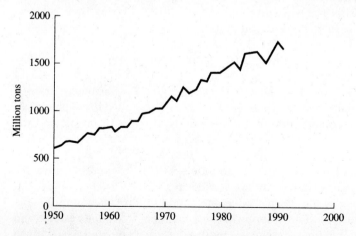

**Figure 18.3** World grain production, from 1950 to 1991. [From Brown et al. (1992).]

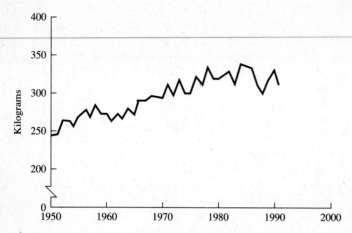

**Figure 18.4** Per capita share of world grain production, from 1950 to 1991. [From Brown et al. (1992).]

world, growth cannot continue indefinitely. As another friend of mine puts it, the "Gulf War" was just the first of the "resource wars." It happened to be fought over oil, but there will soon be other resource wars over food and other much needed resources such as water and land. The recent trend toward increased litigation itself could well be a direct consequence of the declining per capita share. However, litigation has been likened to an "economic black hole," with people standing around trying to get their hands into one another's pockets. Vast amounts of money are poured into litigation, but nothing is ever produced by it!

The economic wars being waged within and between countries depend on the very concept of money as a currency, which itself is seriously flawed. Gold perhaps made some sense, since it is in short supply and nobody could "make" it. Why do governments, who can mint money, insist on taxing their subjects for a percentage of their hard-earned dollars? Is this only to reduce pressures of inflation?

Money leads to runaway greed, which threatens to consume us all. Excessive wealth is not only obscene, it is also absurd. No one needs or should be allowed to own more than he or she can actually use personally. If one's "wealth" was proportional to what one has actually produced, people would have to live within their means. A barter system where both people and nations trade goods and services for other goods and services makes infinitely greater sense than our present monetary systems.

A painfully obvious, self-evident, and timeless truth, known from antiquity, is familiar to everyone, but easily overlooked in the busy hectic competitive heyday of modern-day life: People value one another more when other people are scarce than they do when other people are commonplace. One's esteem and dignity evaporate in cities. Throughout most of heavily populated Asia, human life is of little consequence. People are, quite simply, cheap and dispensable because there are so many of them. Conversely, as wild animals become scarcer and scarcer, they become increasingly more and more valuable and require ever greater protection. As someone put it, once humans were surrounded by wild animals and wilderness but now we surround them. Wilderness and wildlife are integral to human dignity. In southern Africa, endangered rhinos are now under armed guard around the clock and poachers are shot on sight. The bottom line, of course, is that there are quite simply too many of us.

Even now, there is still a great deal of beauty in this world. But those of us who still have the luxury of full bellies and a warm roof over our heads have an obligation to less fortunate souls to use our minds to try to solve the pressing economic and population problems facing everyone on the planet. Continuing population growth and Ponzi scheme economics are quickly leading us all toward total economic collapse. A new, greedless and ecologically responsible, sustainable economy needs to be invented and put into place quickly—a steady-state economy that is energetically balanced in which each of us produces as much as we consume and leaves the planet in roughly the same state as it was in when we entered it. This must include population control. Gains in the standard of living by the few are usually made at the expense of less fortunate brothers and sisters. Improvements in the average standard of living must necessarily be coupled with decreased population in the future. Per capita shares must be equalized. Life as we know and enjoy it simply cannot continue. Cities and their trappings (such as universities), as well as long-distance travel, will have to vanish. Sustainable sources of energy, such as wind and solar energy, must replace fossil fuels. We are all going to have to live like peons, riding bicycles, growing our own vegetables, and trading goods and services.* And the sooner the better. The real problem is how to get from here to there.

## APPLIED BIOGEOGRAPHY: DESIGN OF NATURE PRESERVES

Tall grass prairie covered hundreds of thousands of square kilometers in the midwestern United States just a few hundred years ago; today, this natural community has virtually disappeared. Lowland tropical rainforest is now being destroyed at an alarming rate. Natural communities of all sorts are rapidly being replaced by overgrazed pastures, eroded fields, artificial lakes, golf courses, roads, parking lots, shopping malls, and housing developments. None of the earth's natural communities remain pristine; all have been disturbed either

---

*Of course, this probably won't happen. Instead, the rich will insist on continuing to further enrich themselves at the consequence of the poor becoming even poorer, which must happen inevitably in any case as population builds and resources are further depleted. The question is how much longer can this continue to go on. History flip flops between decades or even centuries of enrichment by the elite few, followed by revolution of the masses, and then back again. An ancient Chinese curse comes to mind: "May you live in interesting times." We are indeed living at one of the most interesting times in human history. We can only wait and see, and experience whatever happens as a consequence of continuing to allow our rapacious greed to have its way.

with pesticides and other pollutants or by way of introductions and extinctions of species. Even the disturbed remnants of the earth's biomes are continually being broken up into smaller and smaller isolated patches or habitat islands (Figure 18.5). As would be expected from the equilibrium theory of island biogeography, faunal and floral diversities are decreasing in such isolates as species go extinct locally (some species, such as the passenger pigeon, have been entirely eradicated). Larger species of animals at higher trophic levels disappear before smaller species and those at lower trophic levels. Habitat fragmentation poses a serious threat to many species, including the spotted owl, the golden-cheeked warbler, and the black-capped vireo. Cowbirds have

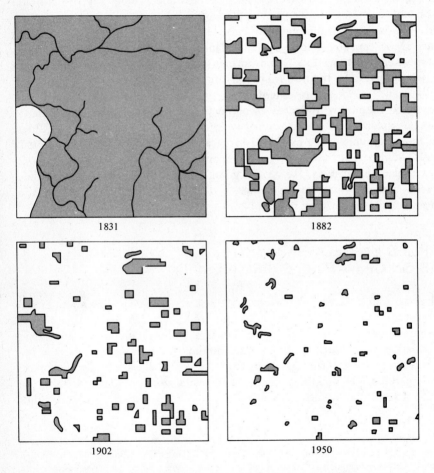

Figure 18.5 Changes in the distribution of forest due to human influence in the past century and a half. The total area depicted, in Wisconsin, is about 10 kilometers on each side. Fragmentation of the forest has created many very small habitat islands. [From J. T. Curtis, The Modification of Mid-latitude Grasslands and Forests by Man, in W. L. Thomas, Jr., ed., *Man's Role in Changing the Face of the Earth.* Copyright © 1956 by The University of Chicago.]

increased greatly in abundance due to human clearing, which has diminished core habitat and increased edge effects. Cowbirds are brood parasites on many small songbirds, which have suffered greatly as a consequence. Unfortunately, much remains to be learned about these vanishing natural communities and their inhabitants.

Biogeographic principles can be used profitably in designing natural preserves to protect endangered habitats and species. Assume that it is desirable to maintain as great a diversity of plants and animals as possible. Clearly, a single large and contiguous reserve will generally be superior to a number of smaller reserves covering an equivalent area. All else being equal, protected areas should be as diverse as possible. Furthermore, the ratio of edge to area should be minimized. The provision of dispersal corridors or "stepping stones" of natural habitat between larger reserves enhances migration and increases diversity (a species can go extinct in one preserve but reinvade from an adjacent one).

In conservation biology, a debate has arisen about whether it is better to have a single large reserve or several small reserves (the so-called SLOSS debate). Because a single large reserve will support only a single population, if this should go extinct that species will be lost; whereas with several smaller reserves, if a population should go extinct in one, it can still reestablish itself from another reserve.

We have now come full circle. In closing, let me remind you of the rapidly disappearing but still unread "book of life" (Rolston, 1985). We must move quickly to preserve as much as possible and to *read* the disappearing pages before they are gone forever.

## SELECTED REFERENCES

### Genetic Engineering

Baba et al. (1992); Cockburn (1991); Simonsen and Levin (1988); Tiedje et al. (1989).

### Conflict Between Human Activities and Preservation of Wilderness

Brown et al. (1989, 1990, 1992); Ehrlich and Ehrlich (1970); Hardin (1968); Soulé (1986, 1987); Vitousek et al. (1986).

### Towards a Sustainable Future: An Equilibrium Economy

Brown et al. (1990, 1991, 1992); Daly (1991); Lubchenco et al. (1991); Wiens (1992).

### Applied Biogeography: Design of Nature Preserves

Boecklen and Bell (1987); Rolston (1985); Terborgh (1974a, 1974b); Wilson and Willis (1975).

# References

Abbott, I., L. K. Abbott, and P. R. Grant. 1977. Comparative ecology of Galápagos ground finches (*Geospiza* Gould): Evaluation of the importance of floristic diversity and interspecific competition. *Ecol. Monogr.* 47: 151–184.

Adams, S. M., B. L. Kimmel, and G. R. Ploskey. 1983. Sources of organic matter for reservoir fish production: A trophic-dynamic analysis. *Can. J. Fish. and Aqu. Sci.* 1480–1495.

Addicott, J. F. 1985. On the population consequences of mutualism. Chapter 25 (pp. 425–436) in J. Diamond and T. J. Case (eds.), *Community ecology*. Harper & Row, New York.

Alcock, J. 1975. *Animal behavior: An evolutionary approach*. Sinauer, Sunderland, Mass.

Alexander, R. D. 1974. The evolution of social behavior. *Ann. Rev. Ecol. Syst.* 5: 325–383.

Alexander, R. D., and G. Borgia. 1978. Group selection, altruism, and the levels of organization of life. *Ann. Rev. Ecol. Syst.* 9: 449–474.

Allee, W. C. 1951. *Cooperation among animals with human implications*. Schuman, New York. (Revised edition of *Social life of animals*, Norton, New York, 1938.)

Allee, W. C., A. E. Emerson, O. Park, T. Park, and K. P. Schmidt. 1949. *Principles of animal ecology*. Saunders, Philadelphia.

Alvarez, L. W., W. Alvarez, F. Asado, and H. V. Michel. 1980. Extraterrestrial cause for the Cretaceous-Tertiary extinction. *Science* 208: 1095–1108.

Anderson, D. J. 1986. Ecological succession. Chapter 12 (pp. 269–285) in J. Kikkawa and D. J. Anderson (eds.), *Community ecology: Pattern and process*. Blackwell Scientific Publications, London.

Anderson, R. M. 1976. Dynamic aspects of parasite population ecology. In C. R. Kennedy (ed.), *Ecological aspects of parasitology*. North Holland Publishing Company, Oxford.

——. 1981. Population ecology of infectious disease agents. Chapter 14 (pp. 318–355) in R. M. May (ed.), *Theoretical ecology* (2nd ed.). Blackwell, Oxford.

——. (ed.). 1982. *The population dynamics of infectious diseases: Theory and applications*. Chapman and Hall, London.

Anderson, R. M., and R. M. May. 1979. Population biology of infectious diseases: Part I. *Nature* 280: 361–367.

——. (eds.). 1982. *The population biology of infectious diseases*. Dahlem Workshop No. 25, Springer-Verlag, New York.

Anderson, W. W. 1971. Genetic equilibrium and population growth under density-regulated selection. *Amer. Natur.* 105: 489–498.

Andrewartha, H. G. 1961. *Introduction to the study of animal populations*. Methuen, London.

——. 1963. Density dependence in the Australian thrips. *Ecology* 44: 218–220.

Andrewartha, H. G., and L. C. Birch. 1953. The Lotka-Volterra theory of interspecific competition. *Aust. J. Zool.* 1: 174–177.

——. 1954. *The distribution and abundance of animals*. Univ. of Chicago Press, Chicago.

Armstrong, R. A. 1982. The effects of connectivity on community stability. *Amer. Natur.* 120: 391–402.

Arnold, S. J. 1972. Species densities of predators and their prey. *Amer. Natur.* 106: 220–236.

Arthur, W. 1987. *The niche in competition and evolution*. Wiley, New York.

Ashmole, N. P. 1963. The regulation of numbers of tropical oceanic birds. *Ibis* 103b: 458–473.

Auerbach, M. J. 1984. Stability, probability, and the topology of food webs. Chapter 24 (pp. 413–438), in D. R. Strong, D. Simberloff, L. G. Abele, and A. B. Thistle (eds.), *Ecological communities: Conceptual issues and the evidence*. Princeton Univ. Press, Princeton, NJ.

Austin, M. P., A. O. Nicholls, and C. R. Margules. 1990. Measurement of the realized qualitative niche: Environmental niches of five *Eucalyptus* species. *Ecol. Monogr.* 60: 161–177.

Avise, J. C. 1989. A role for molecular genetics in the recognition and conservation of endangered species. *Trends Ecol. Evol.* 4: 279–281.

Axelrod, R., and W. D. Hamilton. 1981. The evolution of cooperation. *Science* 211: 1390–1396.

Ayala, F. J. 1965. Genotype, environment, and population numbers. *Science* 162: 1453–1459.

Ayala, F. J., M. E. Gilpin, and J. G. Ehrenfeld. 1973. Competition between species: Theoretical models and experimental tests. *Theoret. Pop. Biol.* 4: 331–355.

Baba, S., O. Akerele, and Y. Kawaguchi. 1992. Natural resources and human health— Plants of medicinal and nutritional value. *Proc. 1st WHO Symposium on Plants and*

*Health for All: Scientific Advancement.* Kobe, Japan, 26–28 August 1991. Elsevier, North Holland.

Bailey, I. W., and E. W. Sinnott. 1916. The climatic distribution of certain types of angiosperm leaves. *Amer. J. Bot.* 3: 24–39.

Bailey, N. T. J. 1975. *The mathematical theory of infectious diseases and its application.* Griffen, London.

Baker, H. G. 1970. Evolution in the tropics. *Biotropica* 2: 101–111.

Baker, J. R. 1938. The evolution of breeding systems. In *Evolution: Essays presented to E. S. Goodrich.* Oxford Univ. Press, London.

Balfour-Lynn, S. 1965. Parthenogenesis in human beings. *Lancet* 1956: 1071–1072.

Barbehenn, K. R. 1969. Host-parasite relationships and species diversity in mammals: An hypothesis. *Biotropica* 1: 29–35.

Bartholomew, G. A. 1972. Body temperature and energy metabolism. Chapter 8 (pp. 298–368) in M. S. Gordon (ed.), *Animal physiology: Principles and adaptations.* Macmillan, New York.

Bartlett, M. S. 1960. *Stochastic population models in ecology and epidemiology.* Methuen, London.

Bartlett, P. N., and D. M. Gates. 1967. The energy budget of a lizard on a tree trunk. *Ecology* 48: 315–322.

Bartz, S. H. 1979. Evolution of eusociality in termites. *Proc. Nat. Acad. Sci., U.S.A.* 76: 5764–5768.

———. 1980. Correction: Evolution of eusociality in termites. *Proc. Nat. Acad. Sci., U.S.A.* 77: 3070.

Bayliss, J. R. 1978. Paternal behavior in fishes: A question of investment, timing, or rate? *Nature* 276: 738.

———. 1981. The evolution of parental care in fishes, with reference to Darwin's rule of male sexual selection. *Env. Biol. Fish.* 6: 223–251.

Bazzaz, F. A. 1975. Plant species diversity in old-field successional ecosystems in southern Illinois. *Ecology* 56: 485–488.

Beard, J. S. 1955. The classification of tropical American vegetation types. *Ecology* 36: 89–100.

Beatty, R. A. 1967. Parthenogenesis in vertebrates. In C. B. Metts and A. Monroy (eds.). *Fertilization: Comparative morphology, biochemistry, and immunology.* Academic Press, New York, pp. 413–440.

Beauchamp, R. S. A., and P. Ullyott. 1932. Competitive relationships between certain species of fresh-water triclads. *J. Ecol.* 20: 200–208.

Begon, M. E., and M. Mortimer. 1981. *Population ecology.* Blackwell Scientific Publications, Oxford.

Bell, G. 1980. The costs of reproduction and their consequences. *Amer. Natur.* 116: 45–76.

———. 1984a. Measuring the cost of reproduction. I. The correlation structure of the life table of a planktonic rotifer. *Evolution* 38: 300–313.

———. 1984b. Measuring the cost of reproduction. II. The correlation structure of the life tables of five freshwater invertebrates. *Evolution* 38: 314–326.

Bender, E. A., T. J. Case, and M. E. Gilpin. 1984. Perturbation experiments in community ecology: Theory and practice. *Ecology* 65: 1–13.

Bengtsson, J. 1989. Interspecific competition increases local extinction rate in a metapopulation system. *Nature* 340: 713–715.

Bennett, A. F., and K. A. Nagy. 1977. Energy expenditure in free-ranging lizards. *Ecology* 58: 697–700.

Benson, S. B. 1933. Concealing coloration among some desert rodents of the southwestern United States. *Univ. Calif. Publ. Zool.* 40: 1–70.

Benson, W. W., K. S. Brown, Jr., and L. E. Gilbert. 1975. Coevolution of plants and herbivores: Passion flower butterflies. *Evolution* 29: 659–680.

Bernal, J. D. 1967. *The origin of life*. World, Cleveland.

Bernays, E., and M. Graham. 1988. On the evolution of host specificity in phytophagous arthropods. *Ecology* 69: 886–892.

Berryman, A. A. 1981. *Population systems: A general introduction*. Plenum Press, New York.

Bertalanffy, L. 1957. Quantitative laws in metabolism and growth. *Quart. Rev. Biol.* 32: 217–231.

——. (ed.). 1969. *General systems theory: Foundations, development, applications*. Braziller, New York.

Beverley, S. M., and A. C. Wilson. 1985. Ancient origin for Hawaiian Drosophilidae inferred from protein comparisons. *Proc. Nat. Acad. Sci., U.S.A.* 82: 4753–4757.

Beverton, R. J. H., and S. J. Holt. 1957. On the dynamics of exploited fish populations. *Great Brit. Min. Agr. Fish, Food, Fish. Invest. Ser. 2*, 19: 1–533.

Billings, W. D. 1964. *Plants and the ecosystem*. Wadsworth, Belmont, Calif.

Birch, L. C. 1948. The intrinsic rate of natural increase of an insect population. *J. Anim. Ecol.* 16: 15–26.

——. 1953. Experimental background to the study of the distribution and abundance of insects. III. The relations between innate capacity for increase and survival of different species of beetles living together on the same food. *Evolution* 7: 136–144.

——. 1957. The meanings of competition. *Amer. Natur.* 91: 5–18.

Birch, L. C., and P. R. Ehrlich. 1967. Evolutionary history and population biology. *Nature* 214: 349–352.

Black, C. A. 1965. *Soil-plant relationships* (2nd ed.). Wiley, New York.

Black, G. A., T. Dobzhansky, and C. Pavan. 1950. Some attempts to estimate the species diversity and population density of trees in Amazonian forests. *Bot. Gaz.* 111: 413–425.

Blair, T. A., and R. C. Fite. 1965. *Weather elements*. Prentice-Hall, Englewood Cliffs, NJ.

Blair, W. F. 1960. *The rusty lizard. A population study*. Univ. of Texas Press, Austin.

Bligh, J. 1973. *Temperature regulation in mammals and other vertebrates*. Elsevier, North Holland.

Blum, H. F. 1968. *Time's arrow and evolution*. Princeton Univ. Press, Princeton, NJ.

Blumenstock, D. I., and C. W. Thornthwaite. 1941. Climate and the world pattern. In *Climate and man* (pp. 98–127). U.S. Department of Agriculture Yearbook. Washington, D.C.

Blumer, L. S. 1979. Male parental care in the bony fishes. *Quart. Rev. Biol.* 54: 149–161.

Boecklen, W. J., and G. W. Bell. 1987. Consequences of faunal collapse and genetic drift for the design of nature reserves. In D. A. Saunders, A. A. Burbidge, and A. J. M. Hopkins (eds.), *Nature conservation: The role of natural remnants.* Surrey Beatty & Sons, Sydney.

Bogue, D. J. 1969. *Principles of demography.* Wiley, New York.

Bonner, J. T. 1965. *Size and cycle: An essay on the structure of biology.* Princeton Univ. Press, Princeton, NJ.

Boorman, S. A., and P. R. Levitt. 1972. Group selection on the boundary of a stable population. *Proc. Nat. Acad. Sci., U.S.A.* 69: 2711–2713.

———. 1973. Group selection at the boundary of a stable population. *Theoret. Pop. Biol.* 4: 85–128.

Borgia, G. 1985. Bower quality, number of decorations and mating success of male satin bowerbirds (*Ptilonorhynchus violaceus*): An experimental analysis. *Animal Behavior* 33: 266–271.

Bormann, F. H., and G. E. Likens. 1967. Nutrient cycling. *Science* 155: 424–429.

Botkin, D. B., and R. S. Miller. 1974. Mortality rates and survival of birds. *Amer. Natur.* 108: 181–192.

Boucher, D. H. (ed.). 1985. *The biology of mutualism.* Oxford Univ. Press, New York.

Boucher, D. H., S. James, and K. H. Keeler. 1982. The ecology of mutualism. *Ann. Rev. Ecol. Syst.* 13: 315–347.

Bovbjerg, R. V. 1970. Ecological isolation and competitive exclusion in two crayfish (*Orconectes virilis* and *Orconectes immunis*). *Ecology* 51: 225–236.

Bowman, R. L. 1961. Morphological differentiation and adaptation in the Galápagos finches. *Univ. Calif. Publ. Zool.* 58: 1–326. Univ. of Calif. Press, Berkeley.

Bradbury, J. W., and G. B. Andersson (eds.). 1987. *Sexual selection: Testing the alternatives.* Wiley, New York.

Bradley, D. J. 1972. Regulation of parasite populations. A general theory of epidemiology and control of parasitic infections. *Trans. Roy. Soc. Trop. Med. Hyg.* 66: 697–708.

———. 1974. Stability in host-parasite systems. In M. B. Usher and M. H. Williamson (eds.), *Ecological stability.* Chapman & Hall, London.

Bradshaw, A. D. 1965. Evolutionary significance of phenotypic plasticity in plants. *Adv. Gen.* 13: 115–155.

Braun-Blanquet, J. 1932. *Plant sociology: The study of plant communities* (translated and edited by G. D. Fuller and H. C. Conard). McGraw-Hill, New York.

Brian, M. V. 1956. Exploitation and interference in interspecies competition. *J. Anim. Ecol.* 25: 339–347.

Briand, F. 1983. Environmental control of food web structure. *Ecology* 64: 253–263.

Briand, F., and J. E. Cohen. 1984. Community food webs have scale-invariant structure. *Nature* 307: 264–266.

Briand, F., and P. Yodzis. 1982. The phylogenetic distribution of obligate mutualism. *Oikos* 39: 273–275.

Brockelman, W. Y. 1975. Competition, the fitness of offspring, and optimal clutch size. *Amer. Natur.* 109: 677–699.

Brockelman, W. Y., and R. M. Fagen. 1972. On modeling density-independent population change. *Ecology* 53: 944–948.

Brody, S. 1945. *Bioenergetics and growth*. Van Nostrand Reinhold, New York.

Brooks, D. R. 1988. Scaling effects in historical biogeography: A new view of space, time, and form. *Syst. Zool.* 38: 237–244.

Brooks, D. R., and D. R. Glen. 1982. Pinworms and primates: A case study in coevolution. *Proc. Helm. Soc. Wash.* 49: 76–85.

Brooks, D. R., and D. A. McLennan. 1991. *Phylogeny, ecology, and behavior.* Univ. Chicago Press, Chicago.

Brooks, G. R., Jr. 1967. Population ecology of the ground skink, *Lygosoma laterale* (Say). *Ecol. Monogr.* 37: 71–87.

Brower, L. P. 1969. Ecological chemistry. *Sci. Amer.* 220 (March): 22–29.

Brower, L. P., and J. Brower. 1964. Birds, butterflies, and plant poisons: A study in ecological chemistry. *Zoologica* 49: 137–159.

Brown, J. H. 1971. Mammals on mountaintops: Non-equilibrium insular biogeography. *Amer. Natur.* 105: 467–478.

——. 1981. Two decades of homage to Santa Rosalia: Towards a general theory of diversity. *Amer. Zoologist* 21: 877–888.

——. 1993. *Macroecology*. In press.

Brown, J. H., and C. R. Feldmeth. 1971. Evolution in constant and fluctuating environments: Thermal tolerances of desert pupfish (*Cyprinodon*). *Evolution* 25: 390–398.

Brown, J. H., and R. C. Lasiewski. 1972. Metabolism of weasels: The cost of being long and thin. *Ecology* 53: 939–943.

Brown, J. H., and B. A. Maurer. 1987. Evolution of species assemblages: Effects of energetic constraints and species dynamics on the diversification of the North American avifauna. *Amer. Natur.* 130: 1–17.

——. 1989. Macroecology: The division of food and space among species on continents. *Science* 243: 1145–1150.

Brown, J. H., and R. A. Ojeda. 1987. Granivory: Patterns, proceses, and consequences of seed consumption on two continents. *Revista Chilena de Historia Natural* 60: 337–349.

Brown, J. H., D. W. Davidson, J. C. Munger, and R. S. Inouye. 1986. Experimental community ecology: The desert granivore system. Chapter 3 (pp. 41–61) in J. Diamond and T. J. Case (eds.), *Community ecology.* Harper & Row, New York.

Brown, J. L. 1964. The evolution of diversity in avian territorial systems. *Wilson Bull.* 76: 160–169.

——. 1966. Types of group selection. *Nature* 211: 870.

——. 1969. Territorial behavior and population regulation in birds. *Wilson Bull.* 81: 293–329.

——. 1975. *The evolution of behavior.* Norton, New York.

Brown, J. L., and G. H. Orians. 1970. Spacing patterns in mobile animals. *Ann. Rev. Ecol. Syst.* 1: 239–262.

Brown, K. S. 1982. Paleoecology and regional patterns of evolution in neotropical forest butterflies. Chapter 16 (pp. 255–308) in G. T. Prance (ed.), *Biological diversification in the tropics*. Columbia Univ. Press, New York.

Brown, L. R., A. Durning, C. Flavin, H. French, J. Jacobson, N. Lenssen, M. Lowe, S. Postel, M. Renner, J. Ryan, L. Starke, and J. Young. 1991. *State of the world 1991*. Norton, New York.

Brown, L. R., A. Durning, C. Flavin, H. French, J. Jacobson, N. Lenssen, M. Lowe, S. Postel, M. Renner, J. Ryan, L. Starke, and J. Young. 1992. *State of the world 1992*. Norton, New York.

Brown, L. R., A. Durning, C. Flavin, H. French, J. Jacobson, M. Lowe, S. Postel, M. Renner, L. Starke, and J. Young. 1990. *State of the world 1990*. Norton, New York.

Brown, L. R., C. Flavin, and H. Hane. 1992. *Vital signs, 1992. The trends that shape our future*. Norton, New York.

Brown, W. L., and E. O. Wilson. 1956. Character displacement. *Syst. Zool.* 5: 49–64.

Brues, C. T. 1920. The selection of food-plants by insects with special reference to lepidopterous larvae. *Amer. Natur.* 54: 313–332.

——. 1924. The specificity of food-plants in the evolution of phytophagous insects. *Amer. Natur.* 58: 127–144.

Bryant, J. H., F. S. Chapin, and D. R. Klein. 1983. Carbon/nutrient balance of boreal plants in relation to vertebrate herbivory. *Oikos* 40: 357–368.

Bryant, J. P. 1980a. Selection of winter forage by subarctic browsing vertebrates: The role of plant chemistry. *Ann. Rev. Ecol. Syst.* 11: 261–285.

——. 1980b. The regulation of snowshoe hare feeding behavior during winter by plant antiherbivore chemistry. In K. Meyers (ed.), *Proc. First Int. Lagomorph Conference* (pp. 69–98). Guelph Univ. Press, Guelph.

Bull, J. J. 1983. *Evolution of sex determining mechanisms*. Benjamin/Cummings, Menlo Park, California.

Bull, J. J., and P. H. Harvey. 1989. A new reason for having sex. *Nature* 339: 260–261.

Burbidge, A. A., K. A. Johnson, P. J. Fuller, and R. I. Southgate. 1988. Aboriginal knowledge of the mammals of the central deserts of Australia. *Aust. Wildlife Research* 15: 9–39.

Burges, A., and F. Raw (eds.). 1967. *Soil biology*. Academic Press, New York.

Burley, N. 1977. Parental investment, mate choice, and mate quality. *Proc. Nat. Acad. Sci., U.S.A.* 74: 3476–3479.

——. 1981a. Sex ratio manipulation and selection for attractiveness. *Science* 211: 721–722.

——. 1981b. Mate choice by multiple criteria in a monogamous species. *Amer. Natur.* 117: 515–528.

——. 1986. Sex ratio manipulation in color-banded populations of zebra finches. *Evolution* 40: 1191–1206.

Burley, N., and N. Moran. 1979. The significance of age and reproductive experience in the mate preferences of feral pigeons, *Columba livia. Anim. Behav.* 27: 686–695.

Bush, G. L. 1974. The mechanism of sympatric host race formation in the true fruit flies. In M. J. D. White (ed.), *Genetic mechanisms of speciation in insects* (pp. 3–23). Australian and New Zealand Book Company, Sydney.

———. 1975. Modes of animal speciation. *Ann. Rev. Ecol. Syst.* 6: 339–364.

Buss, L. W., and J. B. C. Jackson. 1979. Competitive networks: Nontransitive competitive relationships in cryptic coral reef environments. *Amer. Natur.* 113: 223–234.

Buss, L. W. 1986. Competition and community organization on hard surfaces in the sea. Chapter 31 (pp. 517–536) in J. Diamond and T. J. Case (eds.), *Community ecology.* Harper & Row, New York.

Byers, H. G. 1954. The atmosphere up to 30 kilometers. In G. P. Kuiper (ed.), *The earth as a planet.* Univ. Chicago Press, Chicago.

Caccone, A., and J. R. Powell. 1989. DNA divergence among hominoids. *Evolution* 43: 925–942.

Cain, A. J. 1969. Speciation in tropical environments: Summing up. *Biol. J. Linn. Soc.* 1: 233–236.

Cain, S. A. 1944. *Foundations of plant geography.* Harper, New York.

———. 1950. Life-forms and phytoclimate. *Bot. Rev.* 16: 1–32.

Calder, W. A. 1984. *Size, function and life history.* Harvard Univ. Press, Cambridge, Mass.

Caldwell, M. M. 1979. Root structure: The considerable cost of below ground function. In O. T. Solbrig, S. Jain, G. B. Johnson, and P. H. Raven (eds.), *Topics in plant population biology* (pp. 408–427). Columbia Univ. Press, New York.

Caldwell, M. M., and O. A. Fernandez. 1975. Dynamics of great basin shrub root systems. In N. F. Hadley (ed.), *Environmental physiology of desert organisms* (pp. 38–51). Halstead, New York.

Calvin, M. 1969. *Chemical evolution.* Oxford Univ. Press, New York.

Carlquist, S. 1965. *Island life: A natural history of the islands of the world.* Natural History Press, Garden City, NY.

Carney, J. H., D. L. DeAngelis, R. H. Gardner, J. B. Mankin, and W. M. Post. 1981. *Calculation of probabilities of transfer, recurrence intervals, and positional indices for linear compartment models.* Oak Ridge National Laboratory, Technical Manual 7379, Oak Ridge, Tenn.

Carpenter, C. R. 1958. Territoriality: A review of concepts and problems. In A. Roe and G. G. Simpson (eds.), *Behavior and evolution* (pp. 224–250). Yale Univ. Press, New Haven.

Carpenter, F. L. 1987. The study of territoriality: Complexities and future directions. *Amer. Zool.* 27: 401–409.

Carpenter, S. R. 1989. Replication and treatment strength in whole-lake experiments. *Ecology* 70: 453–463.

Carpenter, S. R., T. M. Frost, J. F. Kitchell, and T. K. Kratz. 1993. Species dynamics and global environmental change: A perspective from ecosystem experiments. Chapter 16 (pp. 267–279) in P. M. Kareiva, J. G. Kingsolver, and R. B. Huey (eds.), *Biotic interactions and global change.* Sinauer, Sunderland, Mass.

Carson, H. L. 1973. Ancient chromosomal polymorphism in Hawaiian *Drosophila. Nature* 241: 200–202.

———. 1983. Chromosomal sequences and interisland colonization in Hawaiian *Drosophila. Genetics* 103: 465–482.

Carson, H. L., and K. Y. Kaneshiro. 1976. *Drosophila* of Hawaii: Systematics and ecological genetics. *Ann. Rev. Ecol. Syst.* 7: 311–345.

Case, T. J., and E. A. Bender. 1981. Testing for higher order interactions. *Amer. Natur.* 118: 920–929.

Case, T. J., and M. E. Gilpin. 1974. Interference competition and niche theory. *Proc. Nat. Acad. Sci., U.S.A.* 71: 3073–3077.

Case, T. J., J. Faaborg, and R. Sidell. 1983. The role of body size in the assembly of West Indian bird communities. *Evolution* 37: 1062–1074.

Caswell, H. 1976. Community structure: A neutral model analysis. *Ecol. Monogr.* 46: 327–354.

——. 1989. *Matrix population models*. Sinauer, Sunderland, Mass.

Caswell, H., and J. E. Cohen. 1991. Disturbance, interspecific interaction and diversity in metapopulations. Pages 193–218 In M. E. Gilpin and I. Hanski (eds.), *Metapopulation dynamics: Empirical and theoretical investigations. Biol. J. Linn. Soc.* 42:1–336.

Caswell, H., H. E. Koenig, J. A. Resh, and Q. E. Ross. 1972. An introduction to systems science for ecologists. In B. Patten (ed.), *Systems analysis and simulation in ecology* (pp. 4–78). Vol. II. Academic Press, New York.

Caswell, H., F. Reed, S. N. Stephenson, and P. A. Werner. 1973. Photosynthetic pathways and selective herbivory: A hypothesis. *Amer. Natur.* 107: 465–480.

Cates, R. G. 1975. The interface between slugs and wild ginger: Some evolutionary aspects. *Ecology* 56: 391–400.

Cates, R. G., and G. H. Orians. 1975. Successional status and the palatability of plants to generalized herbivores. *Ecology* 56: 410–418.

Caughley, G. 1966. Mortality patterns in mammals. *Ecology* 47: 906–918.

Caughley, G., and J. H. Lawton. 1981. Plant-herbivore systems. Chapter 7 (pp. 132–166) in R. M. May (ed.), *Theoretical ecology: Principles and applications* (2nd ed.). Blackwell, Oxford.

Chambers, K. L. (ed.). 1970. Biochemical coevolution. *29th Biology Colloquium*. Oregon State Univ. Press, Eugene.

Charlesworth, B. 1971. Selection in density-regulated populations. *Ecology* 52: 469–474.

Charnov, E. L. 1973. Optimal foraging: Some theoretical explorations. Ph.D. dissertation. Univ. of Washington, Seattle.

——. 1976a. Optimal foraging: Attack strategy of a mantid. *Amer. Natur.* 110: 141–151.

——. 1976b. Optimal foraging: The marginal value theorem. *Theoret. Pop. Biol.* 9: 129–136.

——. 1982. *The theory of sex allocation*. Princeton Univ. Press, Princeton, NJ.

Charnov, E. L., and J. R. Krebs. 1973. On clutch size and fitness. *Ibis* 116: 217–219.

——. 1975. The evolution of alarm calls: Altruism or manipulation? *Amer. Natur.* 109: 107–112.

Charnov, E. L., G. H. Orians, and K. Hyatt. 1976. Ecological implications of resource depression. *Amer. Natur.* 110: 247–259.

Chesson, P. L. 1989. A general model of the role of environmental variability in communities of competing species. In A. Hastings (ed.), *Community ecology* (pp. 68–83). Springer-Verlag, New York.

Chesson, P. L., and N. Huntly. 1989. Short-term instabilities and long-term community dynamics. *Trends Ecol. Evol.* 4: 293–298.

Chitty, D. 1960. Population processes in the vole and their relevance to general theory. *Canad. J. Zool.* 38: 99–113.

———. 1967a. The natural selection of self-regulatory behavior in animal populations. *Proc. Ecol. Soc.* Australia 2: 51–78.

———. 1967b. What regulates bird populations? *Ecology* 48: 698–701.

Chorley, R. J., and B. A. Kennedy. 1971. *Physical geography: A systems approach.* Prentice-Hall, London.

Christian, J. J., and D. E. Davis. 1964. Endocrines, behavior, and population. *Science* 146: 1550–1560.

Christianson, F. B., and T. M. Fenchel. 1977. *Theories of populations in biological communities. Ecological studies.* Vol. 20. Springer-Verlag, Heidelberg.

Clapham, W. B. 1973. *Natural ecosystems.* Macmillan, New York.

Clark, L. R., P. W. Geier, R. D. Hughes, and R. F. Morris. 1967. *The ecology of insect populations in theory and practice.* Methuen, London.

Clark, W. C. 1985. Scales of climate impacts. *Climatic Change* 7: 5–27.

Clarke, B. C. 1972. Density-dependent selection. *Amer. Natur.* 106: 1–13.

Clarke, G. L. 1954. *Elements of ecology.* Wiley, New York.

Clausen, J., D. D. Keck, and W. M. Hiesey. 1948. Experimental studies on the nature of species. III. Environmental responses of climatic races of *Achillea.* Publication No. 581 (pp. 1–129). Carnegie Institute, Washington, D.C.

Clements, F. E. 1920. Plant succession: An analysis of the development of vegetation. Publication No. 290. Carnegie Institute, Washington, D.C.

———. 1949. *Dynamics of vegetation.* Hafner, New York.

Cloudsley-Thompson, J. L. 1971. *The temperature and water relations of reptiles.* Merrow, Watford Herts, England.

Cockburn, A. 1991. *An introduction to evolutionary ecology.* Blackwell, Oxford.

Cody, M. L. 1966. A general theory of clutch size. *Evolution* 20: 174–184.

———. 1968. On the methods of resource division in grassland bird communities. *Amer. Natur.* 102: 107–147.

———. 1971. Ecological aspects of reproduction. Chapter 10 (pp. 461–512) in D. S. Farner and J. R. King (eds.), *Avian biology.* Vol. I. Academic Press, New York.

———. 1974. *Competition and the structure of bird communities.* Princeton Univ. Press, Princeton, NJ.

Cohen, J. 1978. *Food webs and niche space.* Princeton Univ. Press, Princeton, NJ.

Cole, B. J. 1983. Multiple mating and the evolution of social behavior in the hymenoptera. *Behav. Ecol. Sociobiol.* 12: 191–201.

Cole, G. A. 1975. *Textbook of limnology.* Mosby, St. Louis.

Cole, J. J., G. M. Lovett, and S. E. G. Findlay (eds.). 1991. *Comparative analyses of ecosystems: Patterns, mechanisms and theories.* Springer-Verlag, New York.

Cole, L. C. 1951. Population cycles and random oscillations. *J. Wildl. Management.* 15: 233–251.

———. 1954a. Some features of random cycles. *J. Wildl. Management* 18: 2–24.

——. 1954b. The population consequences of life history phenomena. *Quart. Rev. Biol.* 29: 103–137.

——. 1958. Sketches of general and comparative demography. *Cold Spring Harbor Symposia Quantitative Biology* 22: 1–15.

——. 1960. Competitive exclusion. *Science* 132: 348–349.

——. 1965. Dynamics of animal population growth. In M. C. Sheps and J. C. Ridley (eds.), *Public health and population change* (pp. 221–241). Univ. Pittsburgh Press, Pittsburgh.

Coley, P. D., J. P. Bryant, and F. S. Chapin. 1985. Resource availability and plant anti-herbivore defense. *Science* 230: 895–899.

Colinvaux, P. A. 1973. *Introduction to ecology.* Wiley, New York.

Collier, B., G. W. Cox, A. W. Johnson, and P. C. Miller. 1973. *Dynamic ecology.* Prentice-Hall, Englewood Cliffs, NJ.

Colwell, R. K. 1973. Competition and coexistence in a simple tropical community. *Amer. Natur.* 107: 737–760.

Colwell, R. K., and E. R. Fuentes. 1975. Experimental studies of the niche. *Ann. Rev. Ecol. Syst.* 6: 281–310.

Colwell, R. K., and D. J. Futuyma. 1971. On the measurement of niche breadth and overlap. *Ecology* 52: 567–576.

Colwell, R. K., and D. W. Winkler. 1984. A null model for null models in biogeography. Chapter 20 (pp. 344–359) in D. R. Strong, D. Simberloff, L. G. Abele, and A. B. Thistle (eds.), *Ecological communities: Conceptual issues and the evidence.* Princeton Univ. Press, Princeton, NJ.

Connell, J. H. 1961a. The effects of competition, predation by *Thais lapillus*, and other factors on natural populations of the barnacle *Balanus balanoides. Ecol. Monogr.* 31: 61–104.

——. 1961b. The influence of interspecific competition and other factors on the distribution of the barnacle, *Chthamalus stellatus. Ecology* 42: 710–723.

——. 1978. Diversity in tropical rain forests and coral reefs. *Science* 199: 1302–1310.

Connor, E. F., and E. D. McCoy. 1979. The statistics and biology of the species-area relationship. *Amer. Natur.* 113: 791–833.

Connor, E. F., and D. Simberloff. 1979. The assembly of species communities: chance or competition? *Ecology* 60: 1132–1140.

Cornell, H. 1974. Parasitism and distributional gaps between allopatric species. *Amer. Natur.* 108: 880–883.

Cott, H. B. 1940. *Adaptive coloration in animals.* Oxford Univ. Press, London.

Cousins, S. H. 1985. The trophic continuum in marine ecosystems: Structure and equations for a predictive model. In R. E. Ulanowicz and T. Platt (eds.), *Ecosystem theory for biological oceanography* (pp. 76–93). Dept. Fisheries and Oceans, Canadian Bureau of Fisheries and Aquatic Science, Ottawa.

Cowles, R. B., and C. M. Bogert. 1944. A preliminary study of the thermal requirements of desert reptiles. *Bull. Amer. Mus. Natur. Hist.* 83: 261–296.

Cowles, R. B., and C. E. Brambel. 1936. A study of the environmental conditions in a bog pond with special reference to the diurnal vertical distribution of *Gonyostomum semen. Biol. Bull. Mar. Biol. Lab. Woods Hole* 71: 286–298.

Cracraft, J. 1974. Continental drift and vertebrate distribution. *Ann. Rev. Ecol. Syst.* 5: 215–261.

Crocker, R. L. 1952. Soil genesis and the pedogenic factors. *Quart. Rev. Biol.* 27: 139–168.

Crocker, R. L., and J. Major. 1955. Soil development in relation to vegetation and surface age at Glacier Bay, Alaska. *J. Ecol.* 43: 427–448.

Crofton, H. D. 1971a. A quantitative approach to parasitism. *Parasitology* 62: 179–193.

———. 1971b. A model of host-parasite relationships. *Parasitology* 63: 343–364.

Crombie, A. C. 1947. Interspecific competition. *J. Anim. Ecol.* 16: 44–73.

Crook, J. H. 1962. The adaptive significance of pair formation types in weaver birds. *Symp. Zool. Soc. London* 8: 57–70.

———. 1963. Monogamy, polygamy and food supply. *Discovery* (Jan.): 35–41.

———. 1964. The evolution of social organization and visual communication in the weaver birds (Ploceinae). *Behaviour* 10: 1–178.

———. 1965. The adaptive significance of avian social organization. In P. E. Ellis (ed.), *Social organization of animal communities* (pp. 181–218). *Symp. Zool. Soc. London*, Vol. 14. Zoological Society of London.

———. 1972. Sexual selection, dimorphism, and social organization in the primates. In B. G. Campbell (ed.), *Sexual selection and the descent of man (1871–1971)* (pp. 231–281). Aldine-Atherton, Chicago.

Crow, J. F. 1986. *Basic concepts in population, quantitative, and evolutionary genetics.* Freeman, New York.

Crow, J. F., and M. Kimura. 1970. *An introduction to population genetics.* Harper & Row, New York.

Crowell, K. L. 1962. Reduced interspecific competition among the birds of Bermuda. *Ecology* 43: 75–88.

Curtis, J. T. 1956. The modification of mid-latitude grasslands and forests by man. In W. L. Thomas, Jr. (ed.), *Man's role in changing the face of the earth.* Univ. Chicago Press, Chicago.

Dale, M. B. 1970. Systems analysis and ecology. *Ecology* 51: 2–16.

Daly, H. E. 1991. *Steady-state economics.* Island Press, Washington, D. C.

Daly, M. 1979. Why don't male mammals lactate? *J. Theoret. Biol.* 78: 323–346.

Damian, R. T. 1964. Molecular mimicry: Antigen sharing by parasite and host and its consequences. *Amer. Natur.* 98: 129–149.

———. 1979. Molecular mimicry in biological adaptation. In B. B. Nichol (ed.), *Host-parasite interfaces* (pp. 103–126). Academic Press, New York.

Dammerman, K. W. 1948. The fauna of Krakatau 1883–1933. *Verhandel. Kontnkl. Ned. Akad. Wetenschap. Afdel. Natuurk.* 44: 1–594.

Dansereau, P. 1957. *Biogeography: An ecological perspective.* Ronald, New York.

Dapson, R. W. 1979. Phenologic influences on cohort-specific reproductive strategies in mice (*Peromyscus polionotus*). *Ecology* 60: 1125–1131.

Darlington, C. D., and K. Mather. 1949. *The elements of genetics.* Allen & Unwin, London.

Darlington, P. J. 1957. *Zoogeography: The geographical distribution of animals.* Wiley, New York.

——. 1959. Area, climate, and evolution. *Evolution* 13: 488–510.

——. 1965. *Biogeography of the southern end of the world.* Harvard Univ. Press, Cambridge, Mass.

——. 1971. Non-mathematical models for evolution of altruism, and for group selection. *Proc. Nat. Acad. Sci., U.S.A.* 69: 293–297.

Darnell, R. M. 1970. Evolution and the ecosystem. *Amer. Zool.* 10: 9–15.

Darwin, C. 1859. *The origin of species by means of natural selection* (numerous editions). Murray, London.

——. 1871. *The descent of man, and selection in relation to sex* (numerous editions). Murray, London.

Daubenmire, R. F. 1947. *Plants and environment.* Wiley, New York.

——. 1956. Climate as a determinant of vegetation distribution in eastern Washington and northern Idaho. *Ecol. Monogr.* 26: 131–154.

——. 1968. *Plant communities.* Harper & Row, New York.

Davidson, J., and H. G. Andrewartha. 1948. Annual trends in a natural population of *Thrips imaginis* (Thysanoptera). *J. Anim. Ecol.* 17: 193–222.

Dawkins, R. 1976. *The selfish gene.* Oxford Univ. Press, Oxford.

——. 1982. *The extended phenotype.* Oxford Univ. Press, Oxford.

Dawkins, R., and T. R. Carlisle. 1976. Parental investment, mate desertion and a fallacy. *Nature* 262: 131–133.

Dawson, P. S., and C. E. King (eds.). 1971. *Readings in population biology.* Prentice-Hall, Englewood Cliffs, NJ.

Dawson, W. R. 1975. On the physiological significance of the preferred body temperatures of reptiles. In D. M. Gates and R. B. Schmerl (eds.), Perspectives of biophysical ecology, *Ecological studies,* Vol. 12. (pp. 443–473). Springer-Verlag, New York.

Dayton, P. K. 1971. Competition, disturbance, and community organization: The provision and subsequent utilization of space in a rocky intertidal community. *Ecol. Monogr.* 41: 351–389.

Dean, A. M. 1983. A simple model of mutualism. *Amer. Natur.* 121: 409–417.

DeAngelis, D. L. 1975. Stability and connectance in food web models. *Ecology* 56: 238–243.

——. 1980. Energy flow, nutrient cycling and ecosystem resilience. *Ecology* 61: 764–771.

DeAngelis, D. L., and J. C. Waterhouse. 1987. Equilibrium and non-equilibrium concepts in ecological models. *Ecol. Monogr.* 57: 1–21.

DeAngelis, D. L., W. M. Post, and G. Sugihara. 1983. *Current trends in food web theory: Report on a food web workshop.* Oak Ridge National Laboratory Technical Manual 5983, Oak Ridge, Tenn.

DeBach, P. 1966. The competitive displacement and coexistence principles. *Ann. Rev. Entomol.* 11: 183–212.

Deevey, E. S., Jr. 1947. Life tables for natural populations of animals. *Quart. Rev. Biol.* 22: 283–314.

——. (ed.). 1972. Growth by intussusception. *Trans. Conn. Acad. Arts. Sci.* 44: 1–443.

Denny, M. 1980. Locomotion: The cost of gastropod crawling. *Science* 208: 1288–1290.

DeVries, P. J. 1991a. The mutualism between *Thisbe irenea* and ants, and the role of ant ecology in the evolution of larval-ant associations. *Biol. J. Linn. Soc.* 43: 179–195.

——. 1991b. Call production by myrmecophilous Riodinid and Lycaenid butterfly caterpillars (Lepidoptera): Morphological, acoustical, functional, and evolutionary patterns. *Amer. Mus. Novitates* 3025: 1–23.

——. 1992. Singing caterpillars, ants and symbiosis. *Sci. Amer.* 267: 76–82.

Diamond, J., and T. Case (eds.). 1986. *Community ecology.* Harper & Row, New York.

Dice, L. R. 1952. *Natural communities.* Univ. of Michigan Press, Ann Arbor.

Dietz, R. S., and J. C. Holden. 1970. The breakup of pangaea. *Sci. Amer.* 223 (Oct.): 30–41.

Dobzhansky, T. 1950. Evolution in the tropics. *Amer. Sci.* 38: 208–221.

——. 1970. *Genetics of the evolutionary process.* Columbia Univ. Press, New York.

Docters van Leeuwen, W. M. 1936. Krakatau, 1833 to 1933. *Ann. Jard. Botan. Buitetizorg* 56–57: 1–506.

Doeksen, J., and J. van der Drift. 1963. *Soil organisms.* North-Holland, Amsterdam.

Downhower, J. F., and K. B. Armitage. 1971. The yellow-bellied marmot and the evolution of polygamy. *Amer. Natur.* 105: 355–370.

Drake, E. T. (ed.). 1968. *Evolution and environment.* Yale Univ. Press, New Haven, Conn.

Drake, J. A. 1990. Communities as assembled structures: Do rules govern pattern? *Trends Ecol. Evol.* 5: 159–164.

Drake, J. A., H. A. Mooney, F. diCastri, R. H. Groves, F. J. Kruger, M. Rejamanek, and M. Williamson (eds.). 1989. *Biological invasions: A global perspective.* SCOPE Report No. 37. Wiley, New York.

Dressler, R. L. 1968. Pollination by euglossine bees. *Evolution* 22: 202–210.

Drury, W. H., and I. C. T. Nisbet. 1973. Succession. *J. Arnold Arboretum* 54: 331–368.

Duellman, W. E., and L. Trueb. 1986. *Biology of amphibians.* McGraw-Hill, New York.

Duggins, D. O. 1978. Kelp beds and sea otters: An experimental approach. *Ecology* 61: 447–453.

Dunbar, M. J. 1960. The evolution of stability in marine environments: Natural selection at the level of the ecosystem. *Amer. Natur.* 94: 129–136.

——. 1968. *Ecological development in polar regions.* Prentice-Hall, Englewood Cliffs, NJ.

——. 1972. The ecosystem as unit of natural selection. In E. S. Deevey (ed.), Growth by intussusception. *Trans. Conn. Acad. Arts Sci.* 44: 113–130.

Dunham, A. 1980. An experimental study of interspecific competition between the iguanid lizards *Sceloporus merriami* and *Urosaurus ornatus*. *Ecol. Monogr.* 50: 309–330.

Dunson, W. A., and J. Travis. 1991. The role of abiotic factors in community organization. *Amer. Natur.* 138: 1067–1091.

Du Toit, A. L. 1937. *Our wandering continents. An hypothesis of continental drifting.* Oliver and Boyd, Edinburgh, London.

Dykhuizen, D. E., and A. M. Dean. 1990. Enzyme activity and fitness: Evolution in solution. *Trends Ecol. Evol.* 5: 257–263.

Ebeling, A. W., and D. R. Laur. 1988. Fish populations in kelp forests without sea otters: Effects of severe storm damage and destructive sea urchin grazing. In E. R. Van Blaricom and J. Al. Estes (eds.), *The community ecology of sea otters* (pp. 169–191). Springer-Verlag, Berlin.

Ebenhard, T. 1991. Colonization in metapopulations: A review of theory and observations. *Biol. J. Linn. Soc.* 42: 105–121.

Eberhard, M. J. W. 1975. The evolution of social behavior by kin selection. *Quart. Rev. Biol.* 50: 1–33.

Eggeling, W. J. 1947. Observations on the ecology of the Budongo rain forest, Uganda. *J. Ecol.* 34: 20–87.

Ehrlich, P. R., and L. C. Birch. 1967. "The balance of nature" and "population control." *Amer. Natur.* 101: 97–107.

Ehrlich, P. R., D. E. Breedlove, P. F. Brussard, and M. A. Sharp. 1972. Weather and the "regulation" of subalpine populations. *Ecology* 53: 243–247.

Ehrlich, P. R., and A. H. Ehrlich. 1970. *Population resources environment.* Freeman, New York.

———. 1981. *Extinction.* Random House, New York.

Ehrlich, P. R., and R. W. Holm. 1963. *The process of evolution.* McGraw-Hill, New York.

Ehrlich, P. R., and P. H. Raven. 1964. Butterflies and plants: A study in coevolution. *Evolution* 18: 586–608.

———. 1969. Differentiation of populations. *Science* 165: 1228–1231.

Ehrman, L. 1983. Endosymbiosis. Chapter 6 (pp. 128–136) in D. Futuyma and M. Slatkin (eds.), *Coevolution.* Sinauer, Sunderland, Mass.

Eldredge, N., and J. Cracraft. 1980. *Phylogenetic patterns and the evolutionary process.* Columbia Univ. Press, New York.

Elton, C. S. 1927. *Animal ecology.* Sidgwick and Jackson, London.

———. 1942. *Voles, mice and lemmings. Problems in population dynamics.* Oxford Univ. Press, London.

———. 1946. Competition and the structure of ecological communities. *J. Anim. Ecol.* 15: 54–68.

———. 1949. Population interspersion: An essay on animal community patterns. *J. Ecol.* 37: 1–23.

———. 1958. *The ecology of invasions by animals and plants.* Methuen, London.

———. 1966. *The pattern of animal communities.* Methuen, London.

Emerson, A. E. 1960. The evolution of adaptation in population systems. In S. Tax (ed.), *Evolution after Darwin* (pp. 307–348). Vol. I. Univ. Chicago Press, Chicago.

Emlen, J. M. 1966. The role of time and energy in food preference. *Amer. Natur.* 100: 611–617.

——. 1968a. Optimal choice in animals. *Amer. Natur.* 102: 385–390.

——. 1968b. A note on natural selection and the sex ratio. *Amer. Natur.* 102: 94–95.

——. 1970. Age specificity and ecological theory. *Ecology* 51: 588–601.

——. 1973. *Ecology: An evolutionary approach.* Addison-Wesley, Reading, Mass.

Emlen, S. T., and L. W. Oring. 1977. Ecology, sexual selection, and the evolution of mating systems. *Science* 197: 215–223.

Emlen, S. T. and P. H. Wrege. 1988. The role of kinship in helping decisions among white-fronted bee-eaters. *Behav. Ecol. Sociobiol.* 23: 305–315.

——. 1989. A test of alternative hypotheses for helping behavior in white-fronted bee-eaters of Kenya. *Behav. Ecol. Sociobiol.* 25: 303–319.

Endler, J. A., and A. M. Lyles. 1989. Bright ideas about parasites. *Trends Ecol. Evol.* 4: 246–248.

Engelmann, M. D. 1966. Energetics, terrestrial field studies, and animal productivity. *Adv. Ecol. Res.* 3: 73–115.

Erckmann, W. J. 1983. The evolution of polyandry in shorebirds: An evaluation of hypotheses. In S. K. Wasser (ed.), *Social behavior of female vertebrates* (pp. 113–168). Academic Press, New York.

Errington, P. L. 1946. Predation and vertebrate populations. *Quart. Rev. Biol.* 21: 144–177.

——. 1956. Factors limiting higher vertebrate populations. *Science* 124: 304–307.

——. 1963. *Muskrat populations.* Iowa State Univ. Press, Ames.

Erwin, D. H. 1989. The end-Permian mass extinction: What really happened and did it matter? *Trends. Ecol. Evol.* 4: 225–229.

Erwin, D. H., J. W. Valentine, and J. J. Sepkoski. 1987. A comparative study of diversification events: The early Paleozoic versus the Mesozoic. *Evolution* 41: 1177–1186.

Erwin, T. L. 1988. The tropical forest canopy: The heart of biotic diversity. In E. O. Wilson (ed.), *Biodiversity* (pp. 123–129). Nat. Acad. Press, Washington, D. C.

Esch, G. W. (ed.). 1977. *Regulation of parasite populations.* Academic Press, New York.

Esch, G. W., T. C. Hazen, and J. M. Aho. 1977. Parasitism and *r* and *K* selection. In G. W. Esch (ed.), *Regulation of parasite populations* (pp. 9–62). Academic Press, New York.

Eshel, I. 1972. On the neighbor effect and the evolution of altruistic traits. *Theoret. Pop. Biol.* 3: 258–277.

Esser, M. H. M. 1946a. Tree trunks and branches as optimal mechanical supports of the crown. I. The trunk. *Bull. Math. Biophys.* 8: 65–74.

——. 1946b. Tree trunks and branches as optimal mechanical supports of the crown. II. The branches. *Bull. Math. Biophys.* 8: 95–100.

Estes, J.A., and J.F. Palamisano. 1974. Sea otters: their role structuring nearshore communities. *Science* 185: 1058–1060.

Evans, F. C., and F. E. Smith. 1952. The intrinsic rate of natural increase for the human louse, *Pediculus humanus L. Amer. Natur.* 86: 299–310.

Evans, H. E. 1977. Extrinsic versus intrinsic factors in the evolution of insect sociality. *BioScience* 27: 613–617.

Evans, L. T. 1971. Evolutionary, adaptive and environmental aspects of the photosynthetic pathway: Assessment. In M. D. Hatch, C. B. Osmond, and R. O. Slayter (eds.), *Photosynthesis and photorespiration*. Wiley, New York.

Ewald, P. W. 1983. Host-parasite relations, vectors, and the evolution of disease severity. *Ann. Rev. Ecol. Syst.* 14: 465–485.

——. 1987. Transmission modes and evolution of the parasitism-mutualism controversy. *Ann. New York Acad. Sci.* 503: 295–306.

Eyre, S. R. 1963. *Vegetation and soils: A world picture*. Aldine, Chicago.

Faegri, K., and J. van der Pijl. 1971. *The principles of pollination ecology*. Pergamon Press, London.

Falconer, D. S. 1981. *Quantitative genetics* (2nd ed.). Second edition. Longman, New York.

Fallis, A. M. (ed.). 1971. *Ecology and physiology of parasites*. Adam Hilger, London.

Falls, J. B. 1969. Functions of territorial songs in the white-throated sparrow. In R. A. Hinde (ed.), *Bird vocalizations* (pp. 207–232). Cambridge Univ. Press, Cambridge, England.

Feder, M. E., and G. V. Lauder (eds.). 1986. *Predator-prey relationships*. Univ. Chicago Press, Chicago.

Feeny, P. P. 1968. Effects of oak leaf tannins on larval growth of the winter moth *Operophtera brumata. J. Insect Physiol.* 14: 805–817.

——1970. Seasonal changes in oak leaf tannins and nutrients as a cause of spring feeding by winter moth caterpillars. *Ecology* 51: 565–581.

——. 1975. Biochemical coevolution between plants and their insect herbivores. In L. E. Gilbert and P. H. Raven (eds.), *Coevolution of animals and plants* (pp. 3–19). Univ. of Texas Press, Austin.

——. 1976. Plant apparency and chemical defense. *Rec. Adv. Phytochemistry* 10: 1–40.

Feinsinger, P. 1983. Coevolution and pollination. Chapter 13 (pp. 282–310) in D. Futuyma and M. Slatkin (eds.), *Coevolution*. Sinauer, Sunderland, Mass.

Felsenstein, J. 1985. Phylogenies and the comparative method. *Amer. Natur.* 125: 1–15.

——1988. Phylogenies and quantitative characters. *Ann. Rev. Eco. Sys.* 19: 445–471.

Fenchel, T. 1974. Intrinsic rate of natural increase: The relationship with body size. *Oecologia* 14: 317–326.

——. 1975. Character displacement and coexistence in mud snails (Hydrobiidae). *Oecologia* 20: 19–32.

Fenchel, T., and L. H. Kofoed. 1976. Evidence for exploitative interspecific competition in mud snails (Hydrobiidae). *Oikos* 27: 367–376.

Fielder, P. L. and S. K. Jain. 1992. *Conservation biology*. Chapman and Hall, London.

Finch, V. C., and G. T. Trewartha. 1949. *Physical elements of geography*. McGraw-Hill, New York.

Fischer, A. G. 1960. Latitudinal variation in organic diversity. *Evolution* 14: 64–81.

Fisher, J., and R. T. Peterson. 1964. *The world of birds*. Doubleday, New York.

Fisher, R. A. 1930. *The genetical theory of natural selection*. Clarendon Press, Oxford.

——. 1935. *The design of experiments*. Oliver and Boyd, Edinburgh, Great Britain.

——. 1958a. *The genetical theory of natural selection* (2nd ed.). Dover, New York.

——. 1958b. Polymorphism and natural selection. *J. Anim. Ecol.* 46: 289–293.

Fisher, R. A., A. S. Corbet, and C. B. Williams. 1943. The relation between the number of species and the number of individuals in a random sample of an animal population. *J. Anim. Ecol.* 12: 42–58.

Fitzpatrick, L. C. 1973. Energy allocation in the Allegheny Mountain salamander. *Desmognathus ochrophaeus*. *Ecol. Monogr.* 43: 43–58.

Flohn, H. 1969. *Climate and weather*. McGraw-Hill, New York.

Florey, E. 1966. *An introduction to general and comparative physiology*. Saunders, Philadelphia.

Folk, G. E., Jr. 1974. *Textbook of environmental physiology* (2nd ed.). Lea & Febiger, Philadelphia.

Fons, W. L. 1940. Influence of forest cover on wind velocity. *J. Forestry* 38: 481–486.

Force, D. C. 1972. *r*- and *K*-strategists in endemic host-parasitoid communities. *Bull. Entomol. Soc. Amer.* 18: 135–137.

Ford, E. B. 1931. *Mendelism and evolution*. Methuen, London.

——. 1964. *Ecological genetics*. Methuen, London.

Ford, R. F., and W. E. Hazen. 1972. *Readings in aquatic ecology*. Saunders, Philadelphia.

Forman, R. T. and M. Godron. 1986. *Landscape ecology*. Wiley, New York.

Forrester, J. W. 1971. *World dynamics*. Wright-Allen Press, Cambridge, Mass.

Fox, L. R., and P. A. Morrow. 1981. Specialization: Species property or local phenomenon? *Science* 211: 887–893.

Fox, S. W., and K. Dose. 1972. *Molecular evolution and the origin of life*. Freeman, San Francisco.

Fraenkel, G. S. 1959. The raison d'être of secondary plant substances. *Science* 129: 1466–1470.

Frank, P. W. 1968. Life histories and community stability. *Ecology* 49: 355–357.

Frazzetta, T. H. 1975. *Complex adaptations in evolving populations*. Sinauer, Sunderland, Mass.

Freeland, W. J. 1974. Vole cycles: Another hypothesis. *Amer. Natur.* 108: 238–245.

——. 1986. Arms races and covenants: The evolution of parasite communities. Chapter 13 (pp. 289–308) in J. Kikkawa and D. J. Anderson (eds.), *Community ecology: Pattern and process*. Blackwell, London.

Freeland, W. J., and D. H. Janzen. 1974. Strategies in herbivory by mammals: The role of plant secondary compounds. *Amer. Natur.* 108: 269–289.

Fretwell, S. D. 1972. *Populations in a seasonal environment*. Princeton Univ. Press, Princeton, NJ.

Fretwell, S. D., and H. L. Lucas, Jr. 1969. On territorial behavior and other factors influencing habitat distribution in birds. I. Theoretical development. *Acta Biotheoretica* 19: 16–36.

Frey, D. G. 1963. *Limnology in North America*. Univ. of Wisconsin Press, Madison.

Fricke, H., and S. Fricke. 1977. Monogamy and sex change by aggressive dominance in coral reef fish. *Nature* 266: 830–832.

Fried, M., and H. Broeshart. 1967. *The soil-plant system in relation to inorganic nutrition*. Academic Press, New York.

Furness, R. W., and J. J. D. Greenwood. 1992. *Birds as monitors of environmental change*. Chapman and Hall, London.

Futuyma, D. J. 1973. Community structure and stability in constant environments. *Amer. Natur.* 107: 443–446.

——. 1976. Food plant specialization and environmental predictability in lepidoptera. *Amer. Natur.* 110: 285–292.

——. 1986. *Evolutionary biology*. (2nd Ed.). Sinauer, Sunderland, Mass.

Futuyma, D. J., and G. Moreno. 1988. The evolution of ecological specialization. *Ann. Rev. Ecol. Syst.* 19: 207–233.

Futuyma, D. J., and M. Slatkin (eds.). 1983. *Coevolution*. Sinauer, Sunderland, Mass.

Gadgil, M., and W. H. Bossert. 1970. Life historical consequences of natural selection. *Amer. Natur.* 104: 1–24.

Gadgil, M., and O. T. Solbrig. 1972. The concept of *r* and *K* selection: Evidence from wild flowers and some theoretical considerations. *Amer. Natur.* 106: 14–31.

Gaffney, P. M. 1975. Roots of the niche concept. *Amer. Natur.* 109: 490.

Gallopin, G. C. 1972. Structural properties of food webs. In B. Patten (ed.), *Systems analysis and simulation in ecology* (pp. 241–282). Vol. II. Academic Press, New York.

Garland, T. R. 1992. Rate tests for phenotypic evolution using phylogenetically independent contrasts. *Amer. Natur.* 140: 509–519.

Garland, T. R., P. H. Harvey, and A. R. Ives. 1992. Procedures for the analysis of comparative data using phylogenetically independent contrasts. *Syst. Biol.* 41: 18–32.

Garland, T. R., B. Huey, and A. F. Bennett. 1991. Phylogeny and coadaptation of thermal physiology in lizards: A reanalysis. *Evolution* 45: 1969–1975.

Gaston, K. and J. H. Lawton. 1989. Insect herbivores on bracken do not support the core-satellite hypothesis. *Amer. Natur.* 134: 761–777.

Gates, D. M. 1962. *Energy exchange in the biosphere*. Harper & Row, New York.

——. 1965. Energy, plants and ecology. *Ecology* 46: 1–13.

——. 1972. *Man and his environment: Climate*. Harper & Row, New York.

Gates, D. M., and R. B. Schmerl. 1975. *Perspectives of biophysical ecology. Ecological studies*. Vol. 12. Springer-Verlag, New York.

Gause, G. F. 1934. *The struggle for existence*. Hafner, New York. (Reprinted in 1964 by Williams & Wilkins, Baltimore, Md.)

——. 1935. Experimental demonstration of Volterra's periodic oscillations in the numbers of animals. *J. Exp. Biol.* 12: 44–48.

Geiger, R. 1966. *The climate near the ground*. Harvard Univ. Press, Cambridge, Mass.

Gentry, A. H. 1969. A comparison of some leaf characteristics of tropical dry forest and tropical wet forest in Costa Rica. *Turrialba* 19: 419–428.

———.1988. Tree species richness of upper Amazonian forests. *Proc. Nat. Acad. Sci.* 85: 156–159.

Gibb, J. A. 1956. Food, feeding habits, and territory of the Rock Pipit, *Anthus spinoletta. Ibis* 98: 506–530.

———. 1960. Populations of tits and goldcrests and their food supply in pine plantations. *Ibis* 102: 163–208.

Gilbert, F. S. 1980. The equilibrium theory of island biogeography: Fact or fiction? *J. Biogeography* 7: 209–235.

Gilbert, L. E. 1971. Butterfly-plant coevolution: *Has Passiora adenopoda* won the selectional race with Heliconiine butterflies? *Science* 172: 585–586.

———. 1972. Pollen feeding and reproductive biology of *Heliconius* butterflies. *Proc. Nat. Acad. Sci., U.S.A.* 69: 1403–1407.

———. 1979. Development of theory in the analysis of insect-plant interactions. Chapter 5 (pp. 117–154) in D. J. Horn, R. Mitchell, and G. R. Stairs (eds.), *Analysis of ecological systems.* Ohio State Univ. Press, Columbus.

Gilbert, L. E., and P. H. Raven (eds.). 1975. *Coevolution of animals and plants.* Univ. of Texas Press, Austin.

Gilbert, L. E., and M. C. Singer. 1973. Dispersal and gene flow in a butterfly species. *Amer. Natur.* 107: 58–72.

———. 1975. Butterfly ecology. *Ann. Rev. Ecol. Syst.* 6: 365–397.

Gill, D. E. 1972. Intrinsic rates of increase, saturation densities, and competitive ability. I. An experiment with *Paramecium. Amer. Natur.* 106: 461–471.

Gillespie, J. H. 1975. Natural selection for resistance to epidemics. *Ecology* 56: 493–495.

Gilpin, M. E. 1973. Do hares eat lynx? *Amer. Natur.* 107: 727–730.

———. 1975a. *Group selection in predator-prey communities.* Princeton Univ. Press, Princeton, NJ.

———. 1975b. Limit cycles in competition communities. *Amer. Natur.* 109: 51–60.

Gilpin, M. E., and I. Hanski (eds.). 1991. *Metapopulation dynamics: Empirical and theoretical investigations. Biol. J. Linn. Soc.* 42:1–336.

Gindell, I. 1973. *A new ecophysiological approach to forest-water relationships in arid climates.* Junk, the Hague.

Ginzburg, L. R., and E. M. Golenberg. 1985. *Lectures in theoretical population biology.* Prentice-Hall, Englewood Cliffs, NJ.

Gisborne, H. T. 1941. How the wind blows in the forest of northern Idaho. *Northern Rocky Mountain Forest Range Experimental Station.*

Givnish, T. J. 1979. On the adaptive significance of leaf form. In O. T. Solbrig, S. Jain, G. B. Johnson, and P. H. Raven (eds.), *Topics in plant population biology* (pp. 375–409). Columbia Univ. Press, New York.

———.(ed.). 1986. *On the economy of plant form and function.* Cambridge Univ. Press.

Givnish, T. J., and G. J. Vermeij. 1976. Sizes and shapes of liane leaves. *Amer. Natur.* 110: 743–778.

Glasser, J. W. 1983. Variation in niche breadth with trophic position: On the disparity between expected and observed species packing. *Amer. Natur.* 122: 542–548.

Gleason, H. A. 1922. On the relation between species and area. *Ecology* 3: 158–162.

———. 1929. The significance of Raunkiaer's law of frequency. *Ecology* 10: 406–408.

Gleason, H. A., and A. Cronquist. 1964. *The natural geography of plants*. Columbia Univ. Press, New York.

Golley, F. B. 1960. Energy dynamics of a food chain of an old-field community. *Ecol. Monogr.* 30: 187–206.

Goodman, D. 1974. Natural selection and a cost ceiling on reproductive effort. *Amer. Natur.* 108: 247–268.

———. 1975. The theory of diversity-stability relationships in ecology. *Quart. Rev. Biol.* 50: 237–266.

Goodman, L. A. 1971. On the sensitivity of the intrinsic growth rate to changes in the age-specific birth and death rates. *Theoret. Pop. Biol.* 2: 339–354.

Gordon, H. T. 1961. Nutritional factors in insect resistance to chemicals. *Ann. Rev. Entomol.* 6: 27–54.

Gordon, M. S. (ed.). 1972. *Animal physiology: Principles and adaptations*. Macmillan, New York.

Gotelli, N. J. 1991. Metapopulation models: The rescue effect, the propagule rain, and the core-satellite hypothesis. *Amer. Natur.* 138: 768–776.

Gould, S. J., and R. C. Lewontin. 1979. The spandrels of San Marco and the Panglossian paradigm: A critique of the adaptationist programme. *Proc. Roy. Soc. London* B205: 581–598.

Graham, R. W. 1986. Response of mammalian communities to environmental changes during the late quaternary. Chapter 18 (pp. 300–313) in J. Diamond and T. Case (eds.), *Community ecology*. Harper & Row, New York.

Grant, P. R. 1967. Bill length variability in birds of the Tres Marias Islands, Mexico. *Canad. J. Zool.* 45: 805–815.

———. 1971. Variation in the tarsus length of birds in island and mainland regions. *Evolution* 25: 599–614.

———. 1972. Convergent and divergent character displacement. *Biol. J. Linnean Soc.* 4: 39–68.

———. 1981. Speciation and adaptive radiation of Darwin's finches. *Amer. Sci.* 69: 653–663.

———. 1986. *Ecology and evolution of Darwin's finches*. Princeton Univ. Press, Princeton, NJ.

Grant, P. R., and I. Abbott. 1980. Interspecific competition, null hypotheses, and island biogeography. *Evolution* 34: 332–341.

Grassle, J. F., and J. P. Grassle. 1974. Opportunistic life histories and genetic systems in marine benthic polychaetes. *J. Marine Res.* 32: 253–284.

Green, R. H. 1969. Population dynamics and environmental variability. *Amer. Zool.* 9: 393–398.

———. 1971. A multivariate statistical approach to the Hutchinsonian niche: Bivalve mollusks of central Canada. *Ecology* 52: 543–556.

Greenberg, J. 1979. Genetic component of bee odor in kin recognition. *Science* 206: 1095–1097.

Greene, E. 1989. A diet-induced developmental polymorphism in a caterpillar. *Science* 243: 643–646.

Greene, H. W. 1982. Dietary and phenotypic diversity in lizards: Why are some organisms specialized? In D. Mossakowski and G. Roths (eds.), *Environmental adaptation and evolution* (pp. 107–128). Gustav Fischer, Stuttgart.

Greenslade, P. J. M. 1968. Island patterns in the Solomon islands bird fauna. *Evolution* 22: 751–761.

Greig-Smith, P. 1964. *Quantitative plant ecology* (2nd ed.). Butterworth, London.

Grice, G. D., and A. D. Hart. 1962. The abundance, seasonal occurrence and distribution of the epizooplankton between New York and Bermuda. *Ecol. Monogr.* 32: 287–307.

Griffin, D. R. 1958. *Listening in the dark*. Yale Univ. Press, New Haven, Conn.

Grime, J. P. 1977. Evidence for the existence of three primary strategies in plants and its relevance to ecological and evolutionary theory. *Amer. Natur.* 111: 1169–1194.

———. 1979. *Plant strategies and vegetation processes*. Wiley, New York.

Grinnell, J. 1917. The niche relationships of the California thrasher. *Auk* 21: 364–382.

———. 1924. Geography and evolution. *Ecology* 5: 225–229.

———. 1928. The presence and absence of animals. *Univ. Calif. Chronicle* 30: 429–450. (Reprinted in 1943 as Joseph Grinnell's *Philosophy of nature*. Univ. of California Press, Berkeley, pp. 187–208.)

Grodzinski, W., and A. Gorecki. 1967. Daily energy budgets of small rodents. In K. Petrusewicz (ed.), *Secondary productivity of terrestrial ecosystems* (pp. 295–314). Vol. I. Warsaw.

Gross, M. R., and R. Shine. 1981. Parental care and mode of fertilization in ectothermic vertebrates. *Evolution* 35: 775–793.

Guilday, J. E. 1967. Differential extinction during late Pleistocene and recent times. In P. S. Martin and H. E. Wright (eds.), *Pleistocene extinctions: The search for a cause* (pp. 121–140). Yale Univ. Press, New Haven.

Gunter, G. 1941. Death of fishes due to cold on the Texas coast, January 1940. *Ecology* 22: 203–208.

Gur, C. R., and H. A. Sackeim. 1979. Self-deception: a concept in search of a phenomenon. *J. Personality and Social Psychology* 37: 147–169.

Guyton, A. C., and D. Horrobin (eds.). 1974. *Environmental physiology*. Physiology, Vol. 7, Series One. Butterworth, London.

Gwynne, D. T. 1991. Sexual competition among females: What causes courtship-role reversal? *Trends Ecol. Evol.* 6: 118–121.


Haartman, L. V. 1969. Nest-site and evolution of polygamy in European passerine birds. *Ornis Fenn.* 46: 1–12.

Hadley, N. F. (ed.). 1975. *Environmental physiology of desert organisms*. Dowden, Hutchinson & Ross, Stroudsburg, Penn.

Haigh, J., and J. Maynard Smith. 1972. Can there be more predators than prey? *Theoret. Pop. Biol.* 3: 290–299.

Hairston, N. G. 1951. Interspecies competition and its probable influence upon the vertical distribution of Appalachian salamanders of the genus *Plethodon. Ecology* 32: 266–274.

———. 1989. *Ecological experiments. Purpose, design, and execution.* Cambridge Univ. Press, Cambridge, U.K.

Hairston, N. G., J. D. Allan, R. K. Colwell, D. J. Futuyma, J. Howell, M. D. Lubin, J. Mathias, and J. H. Vandermeer. 1968. The relationship between species diversity and stability: An experimental approach with protozoa and bacteria. *Ecology* 49: 1091–1101.

Hairston, N. G., and G. W. Byers. 1954. The soil arthropods of a field in southern Michigan: A study in community ecology. *Contrib. Lab. Vert. Biol., Univ. of Michigan* 64: 1–37.

Hairston, N. G., F. E. Smith, and L. B. Slobodkin. 1960. Community structure, population control, and competition. *Amer. Natur.* 94: 421–425.

Haldane, J. B. S. 1932. *The causes of evolution.* Cornell Univ. Press, Ithaca, NY. (Reprinted in 1966).

———. 1941. *New paths in genetics.* Harper, London.

———. 1949. Disease and evolution. *Riorca Sci.* (Suppl.) 19: 3–10.

———. 1964. A defense of bean bag genetics. *Perspectives Biol. Med.* 7: 343–359.

Hallam, A. 1973. *A revolution in the earth sciences: From continental drift to plate tectonics.* Clarendon Press, Oxford.

Hallet, J. G., and S. L. Pimm. 1979. Direct estimation of competition. *Amer. Natur.* 113: 593–600.

Hamilton, T. H. 1961. On the functions and causes of sexual dimorphism in breeding plumage of North American species of warblers and orioles. *Amer. Natur.* 45: 121–123.

Hamilton, T. H., and I. Rubinoff. 1963. Isolation, endemism, and multiplication of species in the Darwin finches. *Evolution* 17: 388–403.

———. 1967. On predicting insular variation in endemism and sympatry for the Darwin finches in the Galápagos archipelago. *Amer. Natur.* 101: 161–172.

Hamilton, W. D. 1964. The genetical evolution of social behavior (two parts). *J. Theoret. Biol.* 7: 1–52.

———. 1966. The moulding of senescence by natural selection. *J. Theoret. Biol.* 12: 12–45.

———. 1967. Extraordinary sex ratios. *Science* 156: 477–488.

———. 1970. Selfish and spiteful behaviour in an evolutionary model. *Nature* 228: 1218–1220.

———. 1971. Geometry for the selfish herd. *J. Theoret. Biol.* 31: 295–311.

———. 1972. Altruism and related phenomena, mainly in insects. *Ann. Rev. Ecol. Syst.* 3: 193–232.

Hamilton, W. D., and M. Zuk. 1982. Heritable true fitness and bright birds: A role for parasites? *Science* 218: 384–387.

Hamilton, W. J., III. 1973. *Life's color code.* McGraw-Hill, New York.

Handford, P., and M. A. Mares. 1985. The mating systems of ratites and tinamous: An evolutionary perspective. *Biol. J. Linn. Soc.* 25: 77–104.

Hanski, I. 1982. Dynamics of regional distribution: The core and satellite species hypothesis. *Oikos* 38: 210–221.

——. 1983. Coexistence of competitors in patchy environment. *Ecology* 64: 493–500.

Hanski, I., and E. Ranta. 1983. Coexistence in a patchy environment: Three species of *Daphnia* in rock pools. *J. Anim. Ecol.* 52: 263–279.

Hanski, I., L. Hansson, and H. Henttonen. 1991. Specialist predators, generalist predators, and the microtine rodent cycle. *J. Anim. Ecol.* 60: 353–367.

Hanski, I., and M. Gilpin. 1991. Metapopulation dynamics: Brief history and conceptual domain. In M. E. Gilpin and I. Hanski (eds)., *Metapopulation dynamics: Empirical and theoretical investigations.* (pp. 3–16). *Biol. J. Linn. Soc.* 42:1–336.

Hardin, G. 1960. The competitive exclusion principle. *Science* 131: 1292–1297.

——. 1968. The tragedy of the commons. *Science* 162: 1243–1248.

——. 1982. Discriminating altruisms. *Zygon* 17: 163–186.

Harper, J. L. 1961a. Approaches to the study of plant competition. *Soc. Exp. Biol. Symp.* 15: 1–39.

——. 1961b. The evolution and ecology of closely related species living in the same area. *Evolution* 15: 209–227.

——. 1966. The reproductive biology of the British poppies. In J. G. Hawkes (ed.), *Reproductive biology and taxonomy of vascular plants* (pp. 26–39). Pergamon & Botanical Society of the British Isles, Oxford.

——. 1967. A Darwinian approach to plant ecology. *J. Ecol.* 55: 247–270.

——. 1969. The role of predation in vegetational diversity. *Brookhaven Symp. Biol.* 22: 48–62.

Harper, J. L., and J. Ogden. 1970. The reproductive strategy of higher plants. I. The concept of strategy with special reference to *Senecio vulgaris* L. *J. Ecol.* 58: 681–698.

Harper, J. L., and J. White. 1974. The demography of plants. *Ann. Rev. Ecol. Syst.* 5: 419–463.

Harper, J. L., P. H. Lovell, and K. G. Moore. 1970. The shapes and sizes of seeds. *Ann. Rev. Ecol. Syst.* 1: 327–356.

Harrison, S. 1991. Local extinction in a metapopulation context: An empirical evaluation. In M. E. Gilpin and I. Hanski (ed.), *Metapopulation dynamics: empirical and theoretical investigations* (pp. 73–88). *Biol. J. Linn. Soc.* 42:1–336.

Harrison, S., D. Murphy and P. R. Ehrlich. 1988. Distribution of the bay checkerspot butterfly, *Euphydryas editha bayensis*: Evidence for a metapopulation model. *Amer. Natur.* 132: 360–382.

Harvey, P. H., and M. D. Pagel. 1991. *The comparative method in evolutionary biology.* Oxford Univ. Press, Oxford.

Harvey, P. H., and M. D. Purvis. 1991. Comparative methods for explaining adaptations. *Nature* 351: 619–624.

Harvey, P. H., N. Birley, and T. H. Blackstock. 1975. The effect of experience on the selective behaviour of song thrushes feeding on artificial populations of *Cepaea* (Held.). *Genetica* 45: 211–216.

Haskell, E. F. 1947. A natural classification of societies. *N.Y. Acad. Sci., Trans. Series* 2, 9: 186–196.

———. 1949. A clarification of social science. *Main Currents in Modern Thought* 7: 45–51.

Haslett, J. R. 1990. Geographic information systems: A new approach to habitat definition and the study of distributions. *Trends Ecol. Evol.* 5: 214–218.

Hassell, M. P. 1980. *The dynamics of arthropod predator-prey systems.* Princeton Univ. Press, Princeton, NJ.

Hassell, M. P., and R. M. May. 1989. The population biology of host-parasite and host-parasitoid associations. Chapter 22 (pp. 319–347) in J. Roughgarden, R. M. May, and S. A. Levin, *Perspectives in ecological theory.* Princeton Univ. Press, Princeton, NJ.

———. 1990. Population regulation and dynamics. *Proceedings of a Royal Society discussion meeting held on 23 and 24 May 1990.* The Royal Society, London.

Hastings, A. 1986. Interacting age-structured populations. In G. Hallam and S. Levin (eds.), *Mathematical ecology* (pp. 287–294). Springer-Verlag, New York.

———. 1991. Structured models of metapopulation dynamics. In M. E. Gilpin and I. Hanski (eds.), *Metapopulation dynamics: Empirical and theoretical investigations. Biol. J. Linn. Soc.* 42: 57–71.

———. (ed.), 1988. *Community ecology* (pp. 68–83.) Springer-Verlag, New York.

Hastings, A., and C. L. Wolin. 1989. Within-patch dynamics in a metapopulation. *Ecology* 70: 1261–1266.

Haurwitz, B., and J. M. Austin. 1944. *Climatology.* McGraw-Hill, New York.

Hazen, W. E. 1964. *Readings in population and community ecology* (1st ed.). Saunders, Philadelphia.

———. 1970. *Readings in population and community ecology* (2nd ed.). Saunders, Philadelphia.

Heatwole, H. 1965. Some aspects of the association of cattle egrets with cattle. *Anim. Behavior* 13: 79–83.

Heatwole, H., and R. Levins. 1972. Trophic structure stability and faunal changes during recolonization. *Ecology* 53: 531–534.

Hedrick, P. W. 1983. *Genetics of populations.* Van Nostrand, New York.

Heinrich, B., and P. H. Raven. 1972. Energetics and pollination ecology. *Science* 176: 597–602.

Henderson, L. J. 1913. *The fitness of the environment.* Macmillan, New York.

Hennig, W. 1966. *Phylogenetic systematics.* Univ. Illinois Press, Urbana.

Hensley, M. M., and J. B. Cope. 1951. Further data on removal and repopulation of the breeding birds in a spruce-fir forest community. *Auk* 68: 483–493.

Hespenhide, H. 1971. Food preference and the extent of overlap in some insectivorous birds, with special reference to Tyrannidae. *Ibis* 113: 59–72.

Hesse, R., W. C. Allee, and K. P. Schmidt. 1951. *Ecological animal geography* (2nd ed.). Wiley, New York.

Hickman, J. C. 1975. Environmental unpredictability and plastic energy allocation strategies in the annual *Polygonum cascadeense* (Polygonaceae). *J. Ecol.* 63: 689–701.

Hilborn, R., and S. C. Stearns. 1982. On inference in ecology and evolutionary biology: The problem of multiple causes. *Acta Biotheoretica* 31: 145–164.

Hirsch, R. P. 1977. Use of mathematical models in parasitology. In G. W. Esch (ed.), *Regulation of parasite populations* (pp. 169–207). Academic Press, New York.

Hochachka, P. W., and G. N. Somero. 1973. *Strategies of biochemical adaptation*. Saunders, Philadelphia.

Holdridge, L. R. 1947. Determination of world plant formations from simple climatic data. *Science* 105: 367–368.

———. 1959. Simple method for determining potential evapotranspiration from temperature data. *Science* 130: 572.

———. 1967. *Life zone ecology*. Tropical Science Center, San José, Costa Rica.

Holling, C. S. 1959a. The components of predation as revealed by a study of small mammal predation of the European pine sawfly. *Canad. Entomol.* 91: 293–320.

———. 1959b. Some characteristics of simple types of predation and parasitism. *Canad. Entomol.* 91: 385–398.

———. 1961. Principles of insect predation. *Ann. Rev. Entomol.* 6: 163–182.

———. 1963. An experimental component analysis of population processes. *Mem. Entomol. Soc. Canada* 32: 22–32.

———. 1964. The analysis of complex population processes. *Canad. Entomol.* 96: 335–347.

———. 1965. The functional response of predators to prey density and its role in mimicry and population regulation. *Mem. Entomol. Soc. Canada* 45: 1–60.

———. 1966. The functional response of invertebrate predators to prey density. *Mem. Entomol. Soc. Canada* 48: 1–87.

———. 1973. Resilience and stability of ecological systems. *Ann. Rev. Ecol. Syst.* 4: 1–23.

———. 1992. Cross-scale morphology, geometry, and dynamics of ecosystems. *Ecol. Monogr.* 62: 447–502.

Holm, C. H. 1973. Breeding sex ratios, territoriality, and reproductive success in the red-winged blackbird (*Agelaius phoeniceus*). *Ecology* 54: 356–365.

Holmes, J. C. 1973. Site selection of parasitic helminthes: Interspecific interactions, site segregation, and their importance to the development of helminth communities. *Can. J. Zool.* 51: 333–347.

———. 1983. Evolutionary relationships between parasitic helminthes and their hosts. Chapter 8 (pp. 161–185) in D. Futuyma and M. Slatkin (eds.), *Coevolution*. Sinauer, Sunderland, Mass.

Holmes, J. C., and W. M. Bethel. 1972. Modification of intermediate host behaviour by parasites. In E. U. Canning and C. A. Wright (eds.), *Behavioural aspects of parasite transmission. J. Linn. Soc.* 51(Suppl. 1): 123–149.

Holmes, R. T., R. E. Bonney, and S. W. Pacala. 1979. Guild structure of the Hubbard Brook bird community: A multivariate approach. *Ecology* 60: 512–520.

Holt, R. D. 1977. Predation, apparent competition, and the structure of prey communities. *Theor. Pop. Biol.* 12: 197–229.

Horn, H. S. 1966. Measurement of overlap in comparative ecological studies. *Amer. Natur.* 100: 419–424.

———. 1968a. Regulation of animal numbers: A model counterexample. *Ecology* 49: 776–778.

———. 1968b. The adaptive significance of colonial nesting in the Brewer's blackbird (*Euphagus cyanocephalus*). *Ecology* 49: 682–694.

———. 1971. *The adaptive geometry of trees.* Princeton Univ. Press, Princeton, NJ.

———. 1974. The ecology of secondary sucession. *Ann. Rev. Ecol. Syst.* 5: 25–37.

———. 1975a. Forest succession. *Sci. Amer.* 232 (May): 90–98.

———. 1975b. Markovian properties of forest succession. In M. L. Cody and J. M. Diamond (eds.), *Ecology and evolution of communities* (pp. 196–211). Harvard Univ. Press, Cambridge, Mass.

———. 1976. Succession. Chapter 10 (pp. 187–204) in R. M. May (ed.), *Theoretical ecology: Principles and applications.* Blackwell, Oxford, U.K.

———. 1979. Adaptation from perspective of optimality. In O. T. Solbrig, S. Jain, G. B. Johnson, and P. H. Raven (eds.), *Topics in plant population biology* (pp. 48–61). Columbia Univ. Press, New York.

Horn, H. S. 1981. Succession. Chapter 11 (pp. 253–271) in R. M. May (ed.), *Theoretical ecology: principles and applications* (2nd ed.). Blackwell Scientific Publications, Oxford, U.K.

Horn, H. S., and R. H. MacArthur. 1972. Competition among fugitive species in a harlequin environment. *Ecology* 53: 749–752.

Horn, H. S., and R. M. May. 1977. Limits to similarity among coexisting competitors. *Nature* 270: 660–661.

Horn, H. S., H. H. Shugart, and D. L. Urban. 1989. Simulators as models of forest dynamics. Chapter 17 (pp. 256–267) in J. Roughgarden, R. M. May, and S. A. Levin (eds.). *Perspectives in ecological theory.* Princeton Univ. Press, Princeton, NJ.

Houde, A. E., and J. A. Endler. 1990. Correlated evolution of female mating preferences and male color patterns in the guppy *Poecilia reticulata. Science* 248: 1405–1408.

Howard, H. E. 1920. *Territory in bird life.* Murray, London. (Reprinted in 1964 by Atheneum, New York.)

Howard, R. D. 1974. The influence of sexual selection and interspecific competition on mockingbird song (*Mimulus polyglottos*). *Evolution* 28: 428–438.

Howe, H. F. 1984. Constraints on the evolution of mutualisms. *Amer. Natur.* 123: 764–777.

Howland, H. C. 1962. Structural, hydraulic, and "economic" aspects of leaf venation and shape. In E. E. Bernard and M. R. Kare (eds.), *Biological prototypes and synthetic systems.* Vol. 1. Cornell Univ. Press, Ithaca, NY.

Hrdy, S. B. 1981. *The woman that never evolved.* Harvard Univ. Press, Cambridge, Mass.

Hubbell, S. P. 1971. Of sowbugs and systems: The ecological bioenergetics of a terrestrial isopod. In B. Patten (ed.), *Systems analysis and simulation in ecology* (pp. 269–324). Vol. I. Academic Press, New York.

———. 1973a. Populations and simple food webs as energy filters. I. One-species systems. *Amer. Natur.* 107: 94–121.

———. 1973b. Populations and simple food webs as energy filters. II. Two-species systems. *Amer. Natur.* 107: 122–151.

———. 1979. Tree dispersion, abundance, and diversity in a tropical dry forest. *Science* 203: 1299–1309.

——. 1980. Seed predation and the coexistence of tree species in tropical forests. *Oikos* 35: 214–229.

Hubbell, S. P., and R. B. Foster. 1986. Biology, chance and history and the structure of tropical rainforest tree communities. In J. Diamond and T. J. Case (eds.), *Community ecology* (pp. 314–329). Harper & Row, New York.

Huey, R. B., and A. F. Bennett. 1987. Phylogenetic studies of coadaptation: Preferred temperatures versus optimal performance temperatures of lizards. *Evolution* 41: 1098–1115.

Huey, R. B., and E. R. Pianka. 1977. Seasonal variation in thermoregulatory behavior and body temperature of diurnal Kalahari lizards. *Ecology* 58: 1066–1075. (With an Appendix by J. A. Hoffman.)

——. 1981. Ecological consequences of foraging mode. *Ecology* 62: 991–999.

Huey, R. B., and M. Slatkin. 1976. Costs and benefits of lizard thermoregulation. *Quart. Rev. Biol.* 51: 363–384.

Huey, R. B., and R. D. Stevenson. 1979. Integrating thermal physiology and ecology of ectotherms: A discussion of approaches. *Amer. Zool.* 19: 357–366.

Huey, R. B., E. R. Pianka, M. E. Egan, and L. W. Coons. 1974. Ecological shifts in sympatry: Kalahari fossorial lizards (*Typhlosaurus*). *Ecology* 55: 304–316.

Huffaker, C. B. 1958. Experimental studies on predation: Dispersion factors and predator-prey oscillations. *Hilgardia* 27: 343–383.

——. 1971. *Biological control.* Plenum, New York.

Hughes, A. L. 1988. *Evolution and human kinship.* Oxford Univ. Press, Oxford.

Hurd, J. E., M. V. Mellinger, L. L. Wolf, and S. J. McNaughton. 1971. Stability and diversity at three trophic levels in terrestrial successional ecosystems. *Science* 173: 1134–1136.

Hurlbert, S. H. 1984. Pseudoreplication and the design of ecological field experiments. *Ecol. Monog.* 54: 187–211.

Huston, M. 1979. A general hypothesis of species diversity. *Amer. Natur.* 113: 81–101.

Hutchinson, G. E. 1941. Ecological aspects of succession in natural populations. *Amer. Natur.* 75: 406–418.

——. 1949. Circular causal systems in ecology. *Ann. N.Y. Acad. Sci.* 51: 221–246.

——. 1950. The biogeochemistry of vertebrate excretion. *Bull. Amer. Mus. Nat. Hist.* 96: 1–554.

——. 1951. Copepodology for the ornithologist. *Ecology* 32: 571–577.

——. 1953. The concept of pattern in ecology. *Proc. Nat. Acad. Sci., U.S.A.* 105: 1–12.

——. 1957a. Concluding remarks. *Cold Spring Harbor Symp. Quant. Biol.* 22: 415–427.

——. 1957b. *A treatise on limnology. Vol. I. Geography, physics and chemistry.* Wiley, New York.

——. 1959. Homage to Santa Rosalia, or why are there so many kinds of animals? *Amer. Natur.* 93: 145–159.

——. 1961. The paradox of the plankton. *Amer. Natur.* 95: 137–145.

——. 1965. *The ecological theater and the evolutionary play.* Yale Univ. Press, New Haven, Conn.

——. 1967. *A treatise on limnology. Vol. II. Introduction to lake biology and the limnoplankton.* Wiley, New York.

——. 1978. *An introduction to population ecology.* Yale Univ. Press, New Haven, Conn.

Hutchinson, G. E., and R. H. MacArthur. 1959. A theoretical ecological model of size distributions among species of animals. *Amer. Natur.* 93: 117–125.

Imbrie, J., et al. 1984. The orbital theory of Pleistocene climate: Support for a revised chronology of the marine $\delta^{18}O$ record. In A. L. Berger et al. (eds.), *Milankovitch and climate* (pp. 269–306). Reidel, Dordrecht, Holland.

Inger, R., and R. K. Colwell. 1977. Organization of contiguous communities of amphibians and reptiles in Thailand. *Ecol. Monogr.* 47: 229–253.

Istock, C. A. 1967. The evolution of complex life cycle phenomena: An ecological perspective. *Evolution* 21: 592–605.

Itô, Y. 1980. *Comparative ecology.* (Translated from the Japanese.) Cambridge Univ. Press, New York.

Ivlev, V. S. 1961. *Experimental feeding ecology of fishes.* Yale Univ. Press, New Haven, Conn.

Jackson, J. B. C., and L. Buss. 1975. Allelopathy and spatial competition among coral reef invertebrates. *Proc. Nat. Acad. Sci., U.S.A.* 72: 5160–5163.

Jaeger, R. G. 1971. Competitive exclusion as a factor influencing the distributions of two species of terrestrial salamanders. *Ecology* 52: 632–637.

Janzen, D. H. 1966. Coevolution of mutualism between ants and acacias in Central America. *Evolution* 20: 249–275.

——. 1967. Fire, vegetation structure, and the ant-acacia interaction in Central America. *Ecology* 48: 26–35.

——. 1970. Herbivores and the number of tree species in tropical forests. *Amer. Natur.* 104: 501–528.

——. 1971a. Euglossine bees as long-distance pollinators of tropical plants. *Science* 171: 203–205.

——. 1971b. Seed predation by animals. *Ann. Rev. Ecol. Syst.* 2: 465–492.

——. 1976. *Ecology of plants in the tropics.* Arnold, London.

——. 1987. Insect diversity of a Costa Rican dry forest: Why keep it and how? *Biol. J. Linn. Soc.* 30: 343–356.

Jeffries, M. J., and J. H. Lawton. 1985. Enemy free space and the structure of ecological communities. *Biol. J. Linn. Soc.* 23: 269–286.

Jelgersma, S. 1966. Sea-level changes during the last 10,000 years. In J. S. Sawyer (ed.), *World climate from 8,000 to 0 B.C. Proc. Int. Symp. on World Climate 8,000 to 0 B. C.* (pp. 54–71). Imperial College, London 1966. Royal Meteorological Society, London.

Jennings, J. B., and P. Calow. 1975. The relationship between high fecundity and the evolution of entoparasitism. *Oecologia* 21: 109–115.

Jenny, H. 1941. *Factors of soil formation.* McGraw-Hill, New York.

Jermy, T. 1987. Can predation lead to narrow food specialization in phytophagous insects? *Ecology* 69: 902–904.

Joern, A., and L. R. Lawlor. 1980. Food and microhabitat utilization by grasshoppers: Comparison with neutral models. *Ecology* 61: 591–599.

——. 1981. Guild structure in grasshopper assemblages based on food and microhabitat resources. *Oikos* 37: 93–104.

Joffe, J. S. 1949. *Pedology.* Pedology, New Brunswick, NJ.

Johnson, M. P., and S. A. Cook. l968. "Clutch size" in buttercups. *Amer. Natur.* 102: 405–411.

Johnson, M. P., L. G. Mason, and P. H. Raven. 1968. Ecological parameters and plant species diversity. *Amer. Natur.* 102: 297–306.

Johnston, R. F. 1954. Variation in breeding season and clutch size in song sparrows of the Pacific coast. *Condor* 56: 265–273.

Jones, D. A. 1962. Selective eating of the acyanogenic form of the plant *Lotus corniculatus* by various animals. *Nature* 193: 1109–1110.

——. 1966. On the polymorphism of cyanogenesis in *Lotus corniculatus*. Selection by animals. *Canad. J. Genet. Cytol.* 8: 556–567.

Jones, E. W. 1956. Ecological studies on the rain forest of southern Nigeria. *J. Ecol.* 44: 83–117.

Kamil, A. C., and T. D. Sargent (eds.). 1982. *Foraging behavior.* Garland Press, NY.

Kamil, A. C., J. R. Krebs, and H. R. Pulliam (eds.). 1986. *Foraging behavior.* Plenum Press, New York.

Kaneshiro, K. Y., and C. R. B. Boake. 1987. Sexual selection and speciation: Issues raised by Hawaiian *Drosophila*. *Trends Ecol. Evol.* 2: 207–212.

Kareiva, P. M. 1987. Habitat fragmentation and the stability of predator-prey interactions. *Nature* 321: 388–391.

——. 1989. Renewing the dialogue between theory and experiments in population ecology. Chapter 5 (pp. 68–88) in J. Roughgarden, R. M. May, and S. A. Levin, *Perspectives in ecological theory.* Princeton Univ. Press, Princeton, NJ.

——. 1990. Population dynamics in spatially complex environments: Theory and data. *Phil. Trans. R. Soc. Lond. B* 53–68.

Kareiva, P. M., and M. Andersen. 1988. Spatial aspects of species interactions: The wedding of models and experiments. In A. Hastings (ed.), *Community ecology: Proceedings of a workshop held at Davis, California, April 1986* (pp. 35–50). Springer-Verlag, New York.

Kareiva, P. M., J. G. Kingsolver, and R. B. Huey (eds.). 1993. *Biotic interactions and global change.* Sinauer, Sunderland, Mass.

Karr, J. R. 1980. Geographical variation in the avifaunas of tropical forest undergrowth. *Auk* 97: 283–298.

Keith, L. B. 1963. *Wildlife's ten-year cycle.* Univ. of Wisconsin Press, Madison.

——. 1974. Some features of population dynamics in mammals. *Proc. Int. Congr. Game Biol.* Stockholm 11: 17–58.

Kendeigh, S. C. 1961. *Animal ecology.* Prentice-Hall, Englewood Cliffs, NJ.

Kennedy, C. R. 1975. *Ecological animal parasitology.* Wiley, New York.

Kerfoot, W. C., and A. Sih (eds.). 1987. *Predation: Direct and indirect impacts on aquatic communities.* Univ. Press of New England, Hanover, NH.

Kershaw, K. A. 1964. *Quantitative and dynamic ecology.* Arnold, London.

Kettlewell, H. B. D. 1956. Further selection experiments on industrial melanism in the Lepidoptera. *Heredity* 10: 287–301.

——. 1958. Industrial melanism in the Lepidoptera and its contribution to our knowledge of evolution. *Proc. 10th Int. Congr. Entomol.* 2: 831–841.

Keyfitz, N. 1968. *Introduction to the mathematics of population*. Addison-Wesley, Reading, Mass.

Keyfitz, N., and W. Flieger. 1971. *Populations: Facts and methods of demography*. Freeman, San Francisco.

Kiester, A. R., R. Lande, and D. W. Schemske. 1984. Models of coevolution and speciation in plants and their pollinators. *Amer. Natur.* 124: 220–243.

Kikkawa, J. 1986. Complexity, diversity and stability. Chapter 4 (pp. 41–62) in J. Kikkawa and D. J. Anderson (eds.), *Community ecology: Pattern and process*. Blackwell, London.

Kikkawa, J. and D. J. Anderson (eds.). 1986. *Community ecology: Pattern and process*. Blackwell, London.

King, A. W., and S. L. Pimm. 1983. Complexity, diversity, and stability: A reconciliation of theoretical and empirical results. *Amer. Natur.* 122: 229–239.

King, C. E. 1971. Resource specialization and equilibrium population size in patchy environments. *Proc. Nat. Acad. Sci., U.S.A.* 68: 2634–2637.

King, C. E., and W. W. Anderson. 1971. Age-specific selection. II. The interaction between $r$ and $K$ during population growth. *Amer. Natur.* 105: 137–156.

King, M. C., and A. C. Wilson. 1975. Evolution at two levels: Molecular similarities and biological differences between humans and chimpanzees. *Science* 188: 107–116.

Kircher, H. W., and W. B. Heed. 1970. Phytochemistry and host plant specificity in *Drosophila*. In C. Steelink and V. C. Runeckles (eds.), *Recent advances in phytochemistry* (pp. 191–209). Vol. 3. Appleton, New York.

Kircher, H. W., W. B. Heed, J. S. Russell, and J. Grove. 1967. Senita cactus alkaloids: Their significance to Sonoran desert *Drosophila* ecology. *J. Insect Physiol.* 13: 1869–1874.

Kirkpatrick, M. 1982. Sexual selection and the evolution of female choice. *Evolution* 36: 1–12.

Kirkpatrick, M., and M. J. Ryan. 1991. The evolution of mating preferences and the paradox of the lek. *Nature* 350: 33–38.

Kitching, R. L. 1983a. Community structure in water-filled treeholes in Europe and Australia—some comparisons and speculations. In P. Lounibos and H. Frank (eds.), *Phytotelmata: Terrestrial plants as hosts for aquatic insect communities* (pp. 205–222). Plexus, Medford, NJ.

——. 1983b. Systems ecology: *An introduction to ecological modeling*. Univ. Queensland Press, Brisbane.

Klomp, H. 1970. The determination of clutch size in birds. *Ardea* 58: 1–124.

Klopfer, P. H. 1962. *Behavioral aspects of ecology*. Prentice-Hall, Englewood Cliffs, NJ.

Klopfer, P. H., and R. H. MacArthur. 1960. Niche size and faunal diversity. *Amer. Natur.* 94: 293–300.

——. 1961. On the causes of tropical species diversity: Niche overlap. *Amer. Natur.* 95: 223–226.

Knight, C. B. 1965. *Basic concepts of ecology*. Macmillan, New York.

Kohn, A. J. 1959. The ecology of *Conus* in Hawaii. *Ecol. Monogr.* 29: 47–90.

——. 1968. Microhabitats, abundance and food of *Conus* on atoll reefs in the Maldive and Chagos Islands. *Ecology* 49: 1046–1062.

Kolman, W. A. 1960. The mechanism of natural selection for the sex ratio. *Amer. Natur.* 94: 373–377.

Kozlovsky, D. G. 1968. A critical evaluation of the trophic level concept. I. Ecological efficiencies. *Ecology* 49: 48–60.

Krebs, C. J. 1964. The lemming cycle at Baker Lake, Northwest Territories, during 1959–62. *Arctic Inst. of North America Tech. Paper No.* 1–5. 104 pp.

——. 1966. Demographic changes in fluctuating populations of *Microtus californicus*. *Ecol. Monogr.* 36: 239–273.

——. 1970. *Microtus* population biology: Behavioral changes associated with the population cycle in *M. ochrogaster* and *M. pennsylvanicus*. *Ecology* 51: 34–52.

——. 1972. *Ecology: The experimental analysis of distribution and abundance*. Harper & Row, New York.

——. 1978. A review of the Chitty hypothesis of population regulation. *Can. J. Zool.* 56: 2463–2480.

——. 1989. *Ecological methodology*. Harper & Row, New York.

Krebs, C. J., and K. T. DeLong. 1965. A *Microtus* population with supplemental food. *J. Mammal.* 46: 566–573.

Krebs, C. J., and J. H. Myers. 1974. Population cycles in small mammals. *Adv. Ecol. Res.* 8: 268–399.

Krebs, C. J., B. L. Keller, and J. H. Myers. 1971. *Microtus* population densities and soil nutrients in southern Indiana grasslands. *Ecology* 52: 660–663.

Krebs, C. J., B. L. Keller, and R. H. Tamarin. 1969. *Microtus* population biology: I. Demographic changes in fluctuating populations of *M. ochrogaster* and *M. pennsylvanicus* in southern Indiana, 1965–1967. *Ecology* 50: 587–607.

Krebs, J. R. 1978. Optimal foraging: Decision rules for predators. Chapter 2 (pp. 23–63) in J. R. Krebs and N. B. Davies (eds.), *Behavioral ecology: An evolutionary approach*. Blackwell, Oxford.

Kurtén, B. 1969. Continental drift and evolution. *Sci. Amer.* 220 (March): 54–64.

LaBarbera, M. 1983. Why the wheels won't go. *Amer. Natur.* 121: 395–408.

Lack, D. 1945. The ecology of closely related species with special reference to cormorant (*Phalacrocorax carbo*) and shag (*P. aristotelis*). *J. Anim. Ecol.* 14: 12–16.

——. 1947. *Darwin's finches*. Cambridge Univ. Press, Cambridge, England. (Reprinted in 1961 by Harper & Row, New York.)

——. 1948. Natural selection and family size in the starling. *Evolution* 2: 95–110.

——. 1954. *The natural regulation of animal numbers*. Oxford Univ. Press, New York.

——. 1956. Further notes on the breeding biology of the swift, *Apus apus*. *Ibis* 98: 606–619.

——. 1966. *Population studies of birds*. Oxford Univ. Press, New York.

——. 1968. *Ecological adaptations for breeding in birds*. Methuen, London.

——. 1971. *Ecological isolation in birds*. Blackwell, Oxford.

Lack, D., and E. Lack. 1951. The breeding biology of the swift, *Apus apus*. *Ibis* 93: 501–546.

Lande, R. 1982. A quantitative genetic theory of life history evolution. *Ecology* 63: 607–615.

Lande, R., and S. J. Arnold. 1983. The measurement of selection on correlated characters. *Evolution* 37: 1210–1226.

Lane, P. A. 1985. A food web approach to mutualism in lake communities. Chapter 14 (pp. 344–374) in D. H. Boucher (ed.), *The biology of mutualism*. Oxford Univ. Press, New York.

Lawlor, L. R. 1978. A comment on randomly constructed model ecosystems. *Amer. Natur.* 112: 445–447.

——. 1979. Direct and indirect effects of *n*-species competition. *Oecologia* 43: 355–364.

——. 1980. Structure and stability in natural and randomly constructed competitive communities. *Amer. Natur.* 116: 394–408.

Lawlor, L. R., and J. Maynard Smith. 1976. The coevolution and stability of competing species. *Amer. Natur.* 110: 79–99.

Lawton, J. H. 1984. Noncompetitive populations, non-convergent communities, and vacant niches: The herbivores of bracken. Chapter 6 (pp. 67–100) in D. R. Strong, D. Simberloff, L. G. Abele, and A. B. Thistle, (eds.), *Ecological communities: Conceptual issues and the evidence*. (pp. 67–100). Princeton Univ. Press, Princeton, NJ.

Lawton, J. H. and S. P. Rallison. 1979. Stability and diversity in grassland communities. *Nature* 279: 351.

Leigh, E. G. 1965. On the relation between the productivity, biomass, diversity, and stability of a community. *Proc. Nat. Acad. Sci., U.S.A.* 53: 777–783.

——. 1977. How does selection reconcile individual advantage with the good of the group? *Proc. Nat. Acad. Sci.* 74: 4542–4546.

——. 1981. Average lifetime of a population in a varying environment. *J. Theor. Biol.* 90: 213–239.

——. 1982. Introduction: Why are there so many kinds of tropical trees? In E. G. Leigh, A. S. Rand, and D. M. Windsor (eds.), *The ecology of a tropical rainforest* (pp. 63–66). Smithsonian Institution Press, Washington, D. C.

——. 1990. Community diversity and environmental stability: A re-examination. *Trends Ecol. Evol.* 5: 340–344.

Lerner, I. M., and F. K. Ho. 1961. Genotype and competitive ability of *Tribolium* species. *Amer. Natur.* 95: 329–343.

Leslie, P. H. 1945. On the use of matrices in certain population mathematics. *Biometrika* 33: 183–212.

——. 1948. Some further notes on the use of matrices in population mathematics. *Biometrika.* 35: 213–245.

Leslie, P. H., and T. Park. 1949. The intrinsic rate of natural increase of *Tribolim castaneum* Herbst. *Ecology* 30: 469–477.

Levin, S. 1974. Dispersion and population interactions. *Amer. Natur.* 108: 207–228.

——. 1992. The problem of pattern and scale in ecology. *Ecology* 73: 1943–1967.

Levin, S. A., and R. T. Paine. 1974. Disturbance, patch formation, and community structure. *Proc. Nat. Acad. Sci.* 71: 2744–2747.

Levine, S. H. 1976. Competitive interactions in ecosystems. *Amer. Natur.* 110: 903–910.

Levins, R. 1966. The strategy of model building in population biology. *Amer. Sci.* 54: 421–431.

——. 1968. *Evolution in changing environments.* Princeton Univ. Press, Princeton, N. J.

——. 1969. Some demographic and genetic consequences of environmental heterogeneity for biological control. *Bull. Ent. Soc. Amer.* 15: 237–240.

——. 1970. Extinction. In M. Gerstenhaber (ed.), *Some mathematical questions in biology* (pp. 75–108). American Mathematical Society, Providence, R.I.

——. 1975. Evolution in communities near equilibrium. In M. L. Cody and J. M. Diamond (eds.), *Ecology and evolution of communities* (pp. 16–50). Harvard Univ. Press, Cambridge, Mass.

Levins, R., and D. Culver. 1971. Regional coexistence of species and competition between rare species. *Proc. Nat. Acad. Sci., U.S.A.* 68: 246–248.

Levitt, J. 1972. *Responses of plants to environmental stresses.* Academic Press, New York.

Lewontin, R. C. 1965. Selection for colonizing ability. In H. G. Baker and G. L. Stebbins (eds.), *The genetics of colonizing species* (pp. 77–94). Academic Press, New York.

——. 1969. The meaning of stability. *Brookhaven Symp. Biol.* 22: 13–24.

——. 1970. The units of selection. *Ann. Rev. Ecol. Syst.* 1: 1–18.

——. 1974. *The genetic basis of evolutionary change.* Columbia Univ. Press, New York.

Lidicker, W. Z. 1988. Solving the enigma of microtine "cycles." *J. Mammalogy* 69: 225–235.

Liebig, J. 1840. *Chemistry in its application to agriculture and physiology.* Taylor and Walton, London.

Liem, K. F., and D. B. Wake. 1985. Morphology: Current approaches and concepts. Chapter 18 (pp. 366–377) in M. Hildebrand, D. M. Bramble, K. F. Liem, and D. B. Wake (eds.), *Functional vertebrate morphology.* Harvard Univ. Press, Cambridge, Mass.

Lightbody, J. P. and P. J. Weatherhead. 1987. Polygyny in yellow-headed blackbirds: female choice versus male competition. *Anim. Behav.* 35: 1670–1684.

——. 1988. Female settling patterns and polygyny: Tests of a neutral-mate-choice hypothesis. *Amer. Natur.* 132: 20–33.

Lindemann, R. I. 1942. The trophic-dynamic aspect of ecology. *Ecology* 23: 399–418.

Lloyd, J. E. 1965. Aggressive mimicry in *Photuris:* Firefly femmes fatales. *Science* 149: 653–654.

——. 1966. Studies on the flash communication system in *Photinus* fireflies. *Misc. Publ. Mus. Zool., Univ. Mich.* 130: 1–95.

——. 1971. Bioluminescent communication in insects. *Ann. Rev. Entomology* 16: 97–122.

——. 1975. Aggressive mimicry in *Photuris* fireflies: Signal repertoires by femmes fatales. *Science* 187: 452–453.

Loehle, C. 1987. Hypothesis testing in ecology: Psychological aspects and the importance of theory manipulation. *Quart. Rev. Biol.* 62: 397–409.

Lotka, A. J. 1922. The stability of the normal age distribution. *Proc. Nat. Acad. Sci., U.S.A.* 8: 339–345.

——. 1925. *Elements of physical biology.* Williams and Wilkins, Baltimore, Md. (Reprinted in 1956 as *Elements of mathematical biology* by Dover, New York.)

——. 1956. *Elements of mathematical biology.* Dover, New York.

Loucks, O. L. 1970. Evolution of diversity, efficiency, and community stability. *Amer. Zool.* 10: 17–25.

Lowry, W. P. 1969. *Weather and life: An introduction to biometeorology.* Academic Press, New York.

Lubchenco, J. 1978. Plant species diversity in a marine intertidal community: Importance of herbivore food preference and algal competitive abilities. *Amer. Natur.* 112: 23–39.

——. 1980. Algal zonation in the New England rocky intertidal community: An experimental analysis. *Ecology* 61: 333–344.

——, and fifteen coauthors. 1991. The sustainable biosphere initiative: an ecological research agenda. *Ecology* 72: 371–412.

Lubchenco, J., and B. A. Menge. 1978. Community development and persistence in a low rocky interidal zone. *Ecol. Monogr.* 48: 67–94.

Luckinbill, L. S. 1973. Coexistence in laboratory populations of *Paramecium aurelia* and its predator *Didinium nastutum. Ecology* 54: 1320–1327.

——. 1974. The effects of space and enrichment on a predator-prey system. *Ecology* 55: 1142–1147.

——. 1979. Selection and the *r-K* continuum in experimental populations of Protozoa. *Amer. Natur.* 113: 427–437.

Luke, C. 1986. Convergent evolution of lizard toe fringes. *Biol. J. Linn. Soc.* 27: 1–16.

Mabberley, D. J. 1991. *Tropical rain forest ecology* (2nd ed.). Chapman and Hall, London.

MacArthur, R. H. 1955. Fluctuations of animal populations, and a measure of community stability. *Ecology* 36: 533–536.

——. 1957. On the relative abundance of bird species. *Proc. Nat. Acad. Sci., U.S.A.* 43: 293–295.

——. 1958. Population ecology of some warblers of northeastern coniferous forests. *Ecology* 39: 599–619.

——. 1959. On the breeding distribution pattern of North American migrant birds. *Auk* 76: 318–325.

——. 1960a. On the relative abundance of species. *Amer. Natur.* 94: 25–36.

——. 1960b. On the relation between reproductive value and optimal predation. *Proc. Nat. Acad. Sci., U.S.A.* 46: 143–145.

——. 1961. Population effects of natural selection. *Amer. Natur.* 95: 195–199.

——. 1962. Some generalized theorems of natural selection. *Proc. Nat. Acad. Sci. U.S.A.* 48: 1893–1897.

——. 1964. Environmental factors affecting bird species diversity. *Amer. Natur.* 98: 387–397.

——. 1965. Patterns of species diversity. *Biol. Reviews* 40: 510–533.

——. 1968. The theory of the niche. In R. C. Lewontin (ed.), *Population biology and evolution* (pp. 159–176). Syracuse Univ. Press, Syracuse, NY.

——. 1970. Species packing and competitive equilibrium for many species. *Theoret. Pop. Biol.* 1: 1–11.

——. 1971. Patterns of terrestrial bird communities. Chapter 5 (pp. 189–221) in D. S. Farner and J. R. King (eds.), *Avian Biology.* Vol. I. Academic Press, New York.

——. 1972. *Geographical ecology: Patterns in the distribution of species.* Harper & Row, New York.

MacArthur, R. H., and J. H. Connell. 1966. *The biology of populations.* Wiley, New York.

MacArthur, R. H., and R. Levins. 1964. Competition, habitat selection, and character displacement in a patchy environment. *Proc. Nat. Acad. Sci., U.S.A.* 51: 1207–1210.

——. 1967. The limiting similarity, convergence, and divergence of coexisting species. *Amer. Natur.* 101: 377–385.

MacArthur, R. H., and J. W. MacArthur. 1961. On bird species diversity. *Ecology* 42: 594–598.

MacArthur R. H., and E. R. Pianka. 1966. On optimal use of a patchy environment. *Amer. Natur.* 100: 603–609.

MacArthur, R. H., and E. O. Wilson. 1963. An equilibrium theory of insular zoogeography. *Evolution* 17: 373–387.

——. 1967. *The theory of island biogeography.* Princeton Univ. Press, Princeton, NJ.

MacArthur, R. H., J. M. Diamond, and J. R. Karr. 1972. Density compensation in island faunas. *Ecology* 53: 330–342.

MacArthur, R. H., J. W. MacArthur, and J. Preer. 1962. On bird species diversity. II. Prediction of bird census from habitat measurements. *Amer. Natur.* 96: 167–174.

MacArthur, R. H., H. Recher, and M. Cody. 1966. On the relation between habitat selection and species diversity. *Amer. Natur.* 100: 319–332.

Macfadyen, A. 1963. *Animal ecology.* Pitman, London.

Machin, K. E., and H. W. Lissman. 1960. The mode of operation of the electric receptors in *Gymnarchus niloticus. J. Exp. Biol.* 37: 801–811.

MacMahon, J. A. 1976. Species and guild similarity of North American desert mammal faunas: A functional analysis of communities. In D. W. Goodall (ed.), *Evolution of desert biota* (pp. 133–148). Univ. of Texas Press, Austin.

Maguire, B. 1963. The passive dispersal of small aquatic organisms and their colonization of isolated bodies of water. *Ecol. Monogr.* 33: 161–185.

——. 1967. A partial analysis of the niche. *Amer. Natur.* 101: 515– 523.

——. 1971. Phytotelmata: Biota and community structure determination in plant-held waters. *Ann. Rev. Ecol. Syst.* 2: 439–464.

——. 1973. Niche response structure and the analytic potentials of its relationship to the habitat. *Amer. Natur.* 107: 213–246.

Main, A. R. 1976. Adaptation of Australian vertebrates to desert conditions. Chapter 5 (pp. 101–131) in D. W. Goodall (ed.), *Evolution of Desert Biota.* Univ. of Texas Press, Austin.

Main, A. R. 1982. Rare species: Precious or dross? pp. 163–174 in R. H. Graves and W. D. L. Ride (eds.), *Species at Risk: Research in Australia.* Australian Academy of Science, Canberra.

Maly, E. J. 1969. A laboratory study of the interaction between the predatory rotifer *Asplanchna* and *Paramecium*. *Ecology* 50: 59–73.

Mann, K. H. 1969. The dynamics of aquatic ecosystems. *Adv. Ecol. Res.* 6: 1–81.

Margalef, R. 1958a. Information theory in ecology. *Gen. Syst.* 3: 36–71.

——. 1958b. Temporal succession and spatial heterogeneity in phytoplankton. In Buzzati-Traverso (ed.), *Perspectives in marine biology*. Univ. California Press, Berkeley.

——. 1963. On certain unifying principles in ecology. *Amer. Natur.* 97: 357–374.

——. 1968. *Perspectives in ecological theory*. Univ. of Chicago Press, Chicago.

——. 1969. Diversity and stability: A practical proposal and a model of interdependence. *Brookhaven Symp. Biol.* 22: 25–37.

Margulis, L. 1970. *Origin of eukaryotic cells*. Yale Univ. Press, New Haven, Conn.

——. 1974. Introduction to origin and evolution of the eukaryotic cell. *Taxon* 23: 225–226.

——. 1976. Genetic and evolutionary consequences of symbiosis. *Exp. Parasitology* 39: 277–349.

Margulis, L., and D. Sagan. 1986. *Origins of sex*. Yale Univ. Press, New Haven, Conn.

Martin, P. S. 1967. Prehistoric overkill. In P. S. Martin and H. E. Wright (eds.), *Pleistocene extinctions: The search for a cause* (pp. 75–120). Yale Univ. Press, New Haven, Conn.

Martin, P. S., and R. G. Klein (eds.). 1984. *Quaternary extinctions. A prehistoric revolution*. Univ. Arizona Press, Tucson.

Martin, P. S., and P. J. Mehringer, Jr. 1965. Pleistocene pollen analysis and biogeography of the southwest. In H. E. Wright and Z. Fry (eds.), *The Quaternary of the U.S.* (pp. 433–451). Princeton, Univ. Press, Princeton, NJ.

Martin, P. S., and H. E. Wright (eds.). 1967. *Pleistocene extinctions: The search for a cause*. Yale Univ. Press, New Haven, Conn.

Marvin, U. B. 1973. *Continental drift: The evolution of a concept*. Smithsonian Institution Press, Washington, D. C.

May, R. M. 1971. Stability in multispecies community models. *Math. Biosci.* 12: 59–79.

——. 1972. Will a large complex system be stable? *Nature* 238: 413–414.

——. 1973. *Stability and complexity in model ecosystems*. Princeton Univ. Press, Princeton, NJ.

——. 1974. On the theory of niche overlap. *Theoret. Pop. Biol.* 5: 297–332.

——. 1975a. Patterns of species abundance and diversity. Chapter 4 (pp. 81–120) in M. L. Cody and J. M. Diamond (eds.), *Ecology and evolution of communities*. Harvard Univ. Press, Cambridge, Mass.

——. 1975b. Some notes on estimating the competition matrix, $\alpha$. *Ecology* 56: 737–741.

——. 1975c. Stability in ecosystems: Some comments. In W. H. van Dobben and R. H. Lowe-McConnell (eds.), *Unifying concepts in ecology* (pp. 161–168). Junk, The Hague.

——. (ed.). 1976a. *Theoretical ecology: Principles and applications*. Blackwell, Oxford.

——. 1976b. Estimating r: A pedagogical note. *Amer. Natur.* 110: 496–499.

——. 1977. Thresholds and breakpoints in ecosystems with a multiplicity of stable states. *Nature* 269: 471–477.

——. 1981. Patterns in multi-species communities. Chapter 9 (pp. 197–227) in R. M. May (ed.), *Theoretical ecology: Principles and applications* (2nd ed.). Blackwell, Oxford.

——. 1982. Mutualistic interactions among species. *Nature* 296: 803–804.

——. 1986a. How many species are there? *Nature* 324: 514–515.

——. 1986b. The search for patterns in the balance of nature: Advances and retreats. *Ecology* 67: 1115–1126.

——. 1987. Chaos and the dynamics of biological populations. *Proc. Roy. Soc.* A413: 27–44.

——. 1988. How many species are there on earth? *Science* 241: 1441–1449.

May, R. M., and R. M. Anderson. 1979. Population biology of infectious diseases. Part II. *Nature* 280: 455–461.

——. 1990. Parasite-host coevolution. *Parasitology* 100: 89–101.

May, R. M., and R. H. MacArthur. 1972. Niche overlap as a function of environmental variability. *Proc. Nat. Acad. Sci., U.S.A.* 69: 1109–1113.

Maynard Smith, J. 1956. Fertility, mating behavior, and sexual selection in *Drosophila subobscura*. *J. Genet.* 54: 261–279.

——. 1958. *The theory of evolution.* Penguin, Baltimore, Md.

——. 1964. Group selection and kin selection: A rejoinder. *Nature* 201: 1145–1147.

——. 1965. *Mathematical ideas in biology.* Cambridge Univ. Press, Cambridge, England.

——. 1971. The origin and maintenance of sex. In G. C. Williams (ed.), *Group selection* (pp. 163–175). Aldine, Chicago.

——. 1974. *Models in ecology.* Cambridge Univ. Press, Cambridge, England.

——. 1976. A comment on the Red Queen. *Amer. Natur.* 110: 325–330.

Mayr, E. 1959. Where are we? *Cold Spring Harbor Symp. Quant. Biol.* 24: 1–14.

——. 1961. Cause and effect in biology. *Science* 134: 1501–1506.

——. 1963. *Animal species and evolution.* Harvard Univ. Press, Cambridge.

McIntosh, R. P. 1967. The continuum concept of vegetation. *Bot. Rev.* 33: 130–187.

McKelvey, K., B. R. Noon, and R. H. Lamberson. 1993. Conservation planning for species occupying fragmented landscapes: The case of the northern spotted owl. Chapter 26 (pp. 424–450) in P. M. Kareiva, J. G. Kingsolver, and R. B. Huey (eds.), *Biotic interactions and global change.* Sinauer, Sunderland, Mass.

McKey, D. 1974. Adaptive patterns in alkaloid physiology. *Amer. Natur.* 108: 305–320.

McLain, D. K. 1991. The r-K continuum and the relative effectiveness of sexual selection. *Oikos* 60: 263–265.

McLaren, I. A. 1971. *Natural regulation of animal populations.* Atherton, New York.

McNab, B. K. 1963. Bioenergetics and the determination of home range size. *Amer. Natur.* 97: 133–140.

McNaughton, S. J. 1977. Diversity and stability of ecological communities: A comment on the role of empiricism in ecology. *Amer. Natur.* 111: 515–525.

——. 1978. Stability and diversity of ecological communities. *Nature* 274: 251–253.

McNaughton, S. J., and L. L. Wolf. 1970. Dominance and the niche in ecological systems. *Science* 167: 131–139.

Medawar, P. B. 1957. *The uniqueness of the individual.* Methuen, London.

Mendel, G. 1865. Versuche uber Pflanzenhybriden. *Verh. naturforsch. Verein Brunn* 4: 3–17. (Translated and reprinted in W. Bateson. 1909. *Mendel's principles of heredity.* Cambridge Univ. Press, Cambridge, England.)

Menge, B. A. 1972a. Foraging strategy of a starfish in relation to actual prey availability and environmental predictability. *Ecol. Monogr.* 42: 25–50.

———. 1972b. Competition for food between two intertidal starfish species and its effect on body size and feeding. *Ecology* 53: 635–644.

———. 1974. Effect of wave action and competition on brooding and reproductive effort in the seastar, *Leptasterias hexactis. Ecology* 55: 84–93.

Menge, B. A., and T. M. Farrell. 1989. Community structure and interaction webs in shallow marine hard-bottom communities: Tests of an environmental stress model. *Adv. Ecol. Res.* 19: 189–262.

Menge, J. L., and B. A. Menge. 1974. Role of resource allocation, aggression and spatial heterogeneity in coexistence of two competing intertidal starfish. *Ecol. Monogr.* 44: 189–209.

Menge, B. A., and A. M. Olson. 1990. Role of scale and environmental factors in regulation of community structure. *Trends. Ecol. Evol.* 5: 52–57.

Menge, B. A., and J. P. Sutherland. 1976. Species diversity gradients: Synthesis of the roles of predation, competition, and temporal heterogeneity. *Amer. Natur.* 110: 351–369.

Merriam, C. H. 1890. Results of a biological survey of the San Francisco mountain region and the desert of the Little Colorado, Arizona. *North American Fauna* 3: 1–113.

Mertz, D. B. 1970. Notes on methods used in life-history studies. In J. H. Connell, D. B. Mertz, and W. W. Murdoch (eds.), *Readings in ecology and ecological genetics* (pp. 4–17). Harper & Row, New York.

———. 1971a. Life history phenomena in increasing and decreasing populations. In E. C. Pielou and W. E. Waters (eds.), *Statistical ecology. Vol. II. Sampling and modeling biological populations and population dynamics* (pp. 361–399). Pennsylvania State Univ. Press, University Park.

———. 1971b. The mathematical demography of the California condor population. *Amer. Natur.* 105: 437–453.

———. 1975. Senescent decline in flour beetle strains selected for early adult fitness. *Physiol. Zool.* 48: 1–23.

Mettler, J. E., and T. G. Gregg. 1969. *Population genetics and evolution.* Prentice-Hall, Englewood Cliffs, NJ.

Meyer, A., T. D. Kocher, P. Basasibwaki, and A. C. Wilson. 1990. Monophyletic origin of Lake Victoria cichlid fishes suggested by mitochondrial DNA sequences. *Nature* 347: 550–553.

Meyer, B. S., D. B. Anderson, and R. H. Bohning. 1960. *Introduction to plant physiology.* Van Nostrand, New York.

Michod, R. E. 1982. The theory of kin selection. *Ann. Rev. Ecol. Syst.* 13: 23–55.

Millar, J. S. 1973. Evolution of litter size in the pika, *Ochotona princeps*. *Evolution* 27: 134–143.

Miller, P. L. 1979. Quantitative plant ecology. In D. J. Horn, R. Mitchell, and G. R. Stairs (eds.), *Analysis of ecological systems* (pp. 179–231). Ohio State Univ. Press, Columbus.

Miller, R. S. 1964. Ecology and distribution of pocket gophers *(Geomyidae)* in Colorado. *Ecology* 45: 256–272.

——. 1967. Pattern and process in competition. *Adv. Ecol. Res.* 4: 1–74.

Milne, A. 1961. Definition of competition among animals. In F. L. Milthorpe (ed.), *Mechanisms in biological competition* (pp. 40–61). Symp. Soc. Exp. Biol. No. 15. Cambridge Univ. Press, London.

Milsum, J. H. 1973. A short note on "stability in multispecies community models." *Math. Biosci.* 17: 189–190.

Milthorpe, F. L. (ed.). 1961. *Mechanisms in biological competition.* Symp. Soc. Exp. Biol. No. 15. Cambridge Univ. Press, London.

Mitter, C., and D. R. Brooks. 1983. Phylogenetic aspects of coevolution. Chapter 4 (pp. 65–98) in D. Futuyma and M. Slatkin (eds.), *Coevolution.* Sinauer, Sunderland, Mass.

Mitter, C., B. Farrell, and B. Wiegmann. 1988. The phylogenetic study of adaptive zones: Has phytophagy promoted insect diversification? *Amer. Natur.* 132: 107–128.

Moen, J. 1989. Diffuse competition—a diffuse concept. *Oikos* 54: 260–263.

Mohr, C. O. 1940. Comparative populations of game, fur and other mammals. *Amer. Midl. Natur.* 24: 581–584.

——. 1943. Cattle droppings as ecological units. *Ecol. Monogr.* 13: 215–298.

Mooney, H. A., O. Bjorkman, and J. Berry. 1975. Photosynthetic adaptations to high temperature. In N. Hadley (ed.), *Environmental physiology of desert organisms* (pp. 138–151). Dowden, Hutchinson & Ross. Stroudsburg, Penn.

Moore, J. 1985. Science as a way of knowing—human ecology. *Amer. Zool.* 25: 483–637.

Morse, D. H. 1971. The insectivorous bird as an adaptive strategy. *Ann. Rev. Ecol. Syst.* 2: 177–200.

Morton, S. R. and C. D. James. 1988. The diversity and abundance of lizards in arid Australia: A new hypothesis. *Amer. Natur.* 132: 237–256.

Motomura, I. 1932. A statistical treatment of associations (in Japanese). *Japan. J. Zool.* 44: 379–383.

Moulton, M. P., and S. L. Pimm. 1986. The extent of competition in shaping an introduced avifaun. In J. Diamond and T. J. Case (eds.), *Community ecology* (pp. 80–97). Harper & Row, New York.

Mulroy, T. W., and P. W. Rundel. 1977. Annual plants: Adaptations to desert environments. *BioScience* 27: 109–114.

Munger, J. C., and J. H. Brown, 1981. Competition in desert rodents: An experiment with semi-permeable enclosures. *Science* 211: 510–512.

Munn, C. A. 1986. Birds that 'cry wolf.' *Nature* 319: 143–145.

Munroe, E. G. 1948. The geographical distribution of butterflies in the West Indies. Ph.D. dissertation, Cornell Univ., Ithaca, NY.

Murdoch, W. W. 1966a. Community structure, population control, and competition: A critique. *Amer. Natur.* 100: 219–226.

——. 1966b. Population stabiity and life history phenomena. *Amer. Natur.* 100: 5–11.

——. 1969. Switching in general predators: Experiments on predator specificity and stability of prey populations. *Ecol. Monogr.* 39: 335–354.

——. 1970. Population regulation and population inertia. *Ecology* 51: 497–502.

Murdoch, W. W., F. C. Evans, and C. H. Peterson. 1972. Diversity and pattern in plants and insects. *Ecology* 53: 819–829

Murphy, G. I. 1968. Pattern in life history and the environment. *Amer. Natur.* 102: 391–403.

Murray, J. D. 1989. *Mathematical biology.* Springer-Verlag, New York.

Myles, T. G. and W. L. Nutting, 1988. Termite eusocial evolution: A re-examination of Bartz's hypothesis and assumptions. *Quart. Rev. Biol.* 63: 1-23.

Nagy, K. A. 1987. Field metabolic rate and food requirement scaling in mammals and birds. *Ecol. Monogr.* 57: 111–128.

National Academy of Science. 1969. *Eutrophication: Causes, consequences and correctives.* Int. Symp. Eutrophication, Washington, D. C.

Neill, W. E. 1972. Effects of size-selective predation on community structure in laboratory aquatic microcosms. Ph.D. dissertation, Univ. of Texas, Austin.

——. 1974. The community matrix and interdependence of the competition coefficients. *Amer. Natur.* 108: 399–408.

——. 1975. Experimental studies of microcrustacean competition, community composition and efficiency of resource utilization. *Ecology* 56: 809–826.

Nelson, G., and N. Platnick. 1981. *Systematics and biogeography: Cladistics and vicariance.* Columbia Univ. Press, New York.

Newbigin, M. I. 1936. *Plant and animal geography.* Methuen, London.

Newell, N. D. 1949. Phyletic size increase: An important trend illustrated by fossil invertebrates. *Evolution* 3: 103–124.

Neyman, J., T. Park, and E. J. Scott. 1956. Struggle for existence. The *Tribolium* model: Biological and statistical aspects. In *Proc. Third Berkeley Symp. on Mathematical Statistics and Probability* (pp. 41–79). Vol. IV. Univ. Calif. Press, Berkeley.

Nicholson, A. J. 1933. The balance of animal populations. *J. Anim. Ecol.* 2: 132–178.

——. 1954. An outline of the dynamics of animal populations. *Aust. J. Zool.* 2: 9–65.

——. 1957. The self-adjustment of populations to change. *Cold Spring Harbor Symp. Quant. Biol.* 22: 153–173.

Nilsson, T. 1983. *The pleistocene. Geology and life in the quaternary ice age.* Reidel, Dordrecht, Holland.

Nunney, L. 1980. The stability of complex model ecosystems. *Amer. Natur.* 115: 639–649.

Odum, E. P. 1959. *Fundamentals of ecology* (2nd ed.). Saunders, Philadelphia.

——. 1963. *Ecology.* Holt, Rinehart and Winston, New York.

——. 1968. Energy flow in ecosystems: A historical review. *Amer. Zool.* 8: 11–18.

——. 1969. The strategy of ecosystem development. *Science* 164: 262–270.

——. 1971. *Fundamentals of ecology* (3rd ed.). Saunders, Philadelphia.

Odum, H. T. 1971. *Environment, power, and society.* Wiley, New York.

Ødum, S. 1965. Germination of ancient seeds. *Dansk Botanisk Arkiv* 24: 1–70.

Olsen, P. E. 1986. A 40-million-year lake record of early mesozoic orbital climatic forcing. *Science* 234: 842–848.

O'Neill, R. V., D. L. DeAngelis, J. B. Waide, and T. F. H. Allen. 1986. *A hierarchical concept of ecosystems*. Princeton Univ. Press, Princeton, NJ.

Oosting, H. J. 1958. *The study of plant communities*. (2nd ed.). Freeman, San Francisco.

Oparin, A. I. 1957. *The origin of life on the earth* (3rd ed.). Oliver and Boyd, London.

Orgel, L. E., and F. H. C. Crick. 1980. Selfish DNA: The ultimate parasite. *Nature* 284: 604–607.

Orians, G. H. 1962. Natural selection and ecological theory. *Amer. Natur.* 96: 257–263.

——. 1969a. The number of bird species in some tropical forests. *Ecology* 50: 783–797.

——. 1969b. On the evolution of mating systems in birds and mammals. *Amer. Natur.* 103: 589–603.

——. 1971. Ecological aspects of behavior. Chapter 11 (pp. 513–546) in D. S. Farner and J. R. King (eds.), *Avian biology*. Vol. I. Academic Press, New York.

——. 1972. The adaptive significance of mating systems in the Icteridae. *Proc. XV Int. Ornith. Congr.* 389–398.

——. 1974. An evolutionary approach to the study of ecosystems. In *Structure, functioning and management of ecosystems* (pp. 198–200). Proc. First Int. Congr. Ecol., The Hague, Netherlands.

——. 1975. Diversity, stability and maturity in natural ecosystems. In W. H. van Dobben and R. H. Lowe-McConnell (eds.), *Unifying concepts in ecology* (pp. 139–150). Junk, The Hague.

——. 1980a. Micro and macro in ecological theory. *BioScience* 30: 79.

——. 1980b. *Some adaptations of marsh-nesting blackbirds*. Princeton Univ. Press, Princeton, NJ.

——. 1982. The influence of tree falls in tropical forests on tree species richness. *Trop. Ecol.* 23: 255–279.

——. 1993. Policy implications of global climate change. Chapter 28 (pp. 467–479) in P. M. Kareiva, J. G. Kingsolver, and R. B. Huey (eds.), 1993. *Biotic interactions and global change*. Sinauer, Sunderland, Mass.

Orians, G. H., and H. S. Horn. 1969. Overlap in foods and foraging of four species of blackbirds in the potholes of central Washington. *Ecology* 50: 930–938.

Orians, G. H., and N. Pearson. 1979. On the theory of central place foraging. In D. J. Horn, R. Mitchell, and G. R. Stairs (eds.), *Analysis of ecological systems* (pp. 155–177). Ohio State Univ. Press, Columbus.

Orians, G. H. and R. T. Paine. 1983. Convergent evolution at the community level. Chapter 19 (pp. 431–458) in D. Futuyma and M. Slatkin (eds.), *Coevolution*. Sinauer, Sunderland, Mass.

Orians, G. H., and O. T. Solbrig. 1977. A cost-income model of leaves and roots with special reference to arid and semiarid areas. *Amer. Natur.* 111: 677–690.

Orians, G. H. and O. T. Solbrig (eds.), 1977. *Convergent evolution in warm deserts: An examination of strategies and patterns in deserts of Argentina and the United States*. U. S./I. B. P. Synthesis Series, vol. 3. Dowden, Hutchinson and Ross, Inc., Stroudsburg, Pennsylvania.

Orians, G. H., and M. F. Willson. 1964. Interspecific territories of birds. *Ecology* 45: 136–145.

Orians, G. H., G. M. Brown, W. E. Kunin, and J. E. Swierzbinski (eds.). 1990. *The preservation and valuation of biological resources.* Univ. Washington Press, Seattle.

Osmond, C. B., O. Björkman, and D. J. Anderson. 1981. *Physiological processes in plant ecology.* Springer-Verlag, Berlin.

Oster, G., and E. O. Wilson. 1979. *Caste and ecology in the social insects.* Princeton Univ. Press, Princeton, NJ.

Otte, D. 1975. Plant preference and plant succession. A consideration of evolution of plant preference in *Schistocerca. Oecologia* 18: 129–144.

Otte, D., and J. A. Endler (eds.). 1989. *Speciation and its consequences.* Sinauer, Sunderland, Mass.

Otte, D., and K. Williams. 1972. Environmentally induced color dimorphisms in grasshoppers, *Syrbula admirabilis, Dicromorpha viridis,* and *Chortophaga viridifasciata. Ann. Entomol. Soc. Amer.* 65: 1154–1161.

Owen, D. F. 1977. Latitudinal gradients in clutch size: An extension of David Lack's theory. In B. Stonehouse and C. M. Perrins (eds.), *Evolutionary ecology* (pp. 171–179). Macmillan, London.

Pacala, S. W., M. P. Hassell, and R. M. May. 1990. Host-parasitoid associations in patchy environments. *Nature* 344: 150–153.

Paine, R. T. 1966. Food web complexity and species diversity. *Amer. Natur.* 100: 65–76.

———. 1971. The measurement and application of the calorie to ecological problems. *Ann. Rev. Ecol. Syst.* 2: 145–164.

———. 1974. Intertidal community structure: Experimental studies on the relationship between a dominant competitor and its principal predator. *Oecologia* (Berlin) 15: 93–120.

———. 1977. Controlled manipulations in the marine intertidal zone, and their contributions to ecological theory. In C. E. Goulden (ed.), *The changing scenes in natural sciences, 1776–1976* (pp. 245–270). *Acad. Nat. Sci. Phil. Spec. Publ.* 12.

———. 1980. Food webs, interaction strength, linkage and community infrastructure. *J. Anim. Ecol.* 49: 667–685.

———. 1983. Intertidal food webs: Does connectance describe their essence? In *Current trends in food web theory. Report on a food web workshop* (pp. 11–15). Oak Ridge National Laboratory Technical Manual 5983, Oak Ridge, Tenn.

———. 1988. Food webs: Road maps of interactions or grist for theoretical development? *Ecology* 69: 1648–1654.

Park, T. 1948. Experimental studies of interspecific competition. I. Competition between populations of flour beetles *Tribolium confusum* Duval and *T. castaneum.* Herbst. *Ecol. Monogr.* 18: 265–307.

———. 1954. Experimental studies of interspecific competition. II. Temperature, humidity, and competition in two species of *Tribolium. Physiol. Zool.* 27: 177–238.

———. 1962. Beetles, competition, and populations. *Science* 138: 1369–1375.

Park, T., P. H. Leslie, and D. B. Mertz. 1964. Genetic strains and competition in populations of *Tribolium. Physiol. Zool.* 37: 97–162.

Parker, B. C., and B. J. Turner. 1961. "Operational niche" and "community-interaction values" as determined from *in vitro* studies of some soil algae. *Evolution* 15: 228–238.

Parkhurst, D. F., and O. L. Loucks. 1971. Optimal leaf size in relation to environment. *J. Ecol.* 60: 505–537.

Partridge, L., and P. H. Harvey. 1988. The ecological context of life history evolution. *Science* 241: 1449–1455.

Patten, B. C. 1959. An introduction to the cybernetics of the ecosystem: The trophic-dynamic aspect. *Ecology* 40: 221–231.

———. 1961. Competitive exclusion. *Science* 134: 1599–1601.

———. 1962. Species diversity in net phytoplankton of Raritan Bay. *J. Marine Res.* 20: 57–75.

———(ed.). 1971. *Systems analysis and simulation in ecology.* Vol. I. Academic Press, New York.

———(ed.). 1972. *Systems analysis and simulation in ecology.* Vol. II. Academic Press, New York.

———(ed.). 1975. *Systems analysis and simulation in ecology.* Vol. III. Academic Press, New York.

———(ed.). 1976. *Systems analysis and simulation in ecology.* Vol. IV. Academic Press, New York.

———. 1983. On the quantitative dominance of indirect effects in ecosystems. In W. K. Lauenroth, G. V. Skogerboe, and M. Elug (eds.), *Analysis of ecological systems: State of the art in ecological modelling* (pp. 27–37). Elsevier, North Holland.

Patten, D. T., and E. M. Smith. 1975. Heat flux and the thermal regime of desert plants. In N. F. Hadley (ed.), *Environmental physiology of desert organisms* (pp. 1–19). Dowden, Hutchinson & Ross, Stroudsburg, Penn.

Patterson, H. E. H. 1982. Perspective on speciation by reinforcement. *South African J. Sci.* 78: 53–57.

Pauling, L. 1970. *Vitamin C and the common cold.* Freeman, San Francisco.

Payne, R. B. 1979. Sexual selection and intersexual differences in variance of breeding success. *Amer. Natur.* 114: 447–466.

Pearl, R. 1922. *The biology of death.* Lippincott, Philadelphia.

———. 1927. The growth of populations. *Quart. Rev. Biol.* 2: 532–548.

———. 1928. *The rate of living.* Knopf, New York.

———. 1930. *The biology of population growth.* Knopf, New York.

Pearson, D. L. 1977. A pantropical comparison of bird community structure on six lowland tropical forest sites. *Condor* 79: 232–244.

Pearson, O. P. 1948. Metabolism and energetics. *Sci. Monthly* 66: 131–134.

Pease, J. L., R. H. Vowles, and L. B. Keith. 1979. Interaction of snowshoe hares and woody vegetation. *J. Wildl. Management* 43: 43–60.

Pefaur, J. E., and W. E. Duellman. 1980. Community structure in high Andean herpetofaunas. *Trans. Kansas Acad. Science* 83: 45–65.

Perkins, E. J. 1974. *The biology of estuaries and coastal waters*. Academic Press, New York.

Perrins, C. M. 1964. Survival of young swifts in relation to brood-size. *Nature* 201: 1147–1149.

——. 1965. Population fluctuations and clutch size in the great tit, *Parus major* L. *J. Anim. Ecol.* 34: 601–647.

Perrone, M., and T. M. Zaret. 1979. Parental care patterns of fishes. *Amer. Natur.* 113: 351–361.

Peterson, C. H. 1975. Stability of species and of community for the benthos of two lagoons. *Ecology* 56: 958–965.

Phillipson, J. 1966. *Ecological energetics*. Arnold, London.

Pianka, E. R. 1966a. Latitudinal gradients in species diversity: A review of concepts. *Amer. Natur.* 100: 33–46.

——. 1966b. Convexity, desert lizards, and spatial heterogeneity. *Ecology* 47: 1055–1059.

——. 1969. Sympatry of desert lizards (*Ctenotus*) in western Australia. *Ecology* 50: 1012–1030.

——. 1970. On *r* and *K* selection. *Amer. Natur.* 102: 592–597.

——. 1971a. Species diversity. In *Topics in the study of life* (pp. 401–406). Harper & Row, New York.

——. 1971b. Ecology of the agamid lizard *Amphibolurus isolepis* in Western Australia. *Copeia* 1971: 527–536.

——. 1972. *r* and *K* selection or *b* and *d* selection? *Amer. Natur.* 106: 581–588.

——. 1973. The structure of lizard communities. *Ann. Rev. Ecol. Syst.* 4: 53–74.

——. 1974. Niche overlap and diffuse competition. *Proc. Nat. Acad. Sci., U.S.A.* 71: 2141–2145.

——. 1975. Niche relations of desert lizards. Chapter 12 (pp. 292–314) in M. Cody and J. M. Diamond (eds.), *Ecology and evolution of communities*. Harvard Univ. Press, Cambridge, Mass.

——. 1976a. Competition and niche theory. Chapter 7 (pp. 114–141) in R. M. May (ed.), *Theoretical ecology: Principles and applications*. Blackwell, Oxford.

——. 1976b. Natural selection of optimal reproductive tactics. *Amer. Zool.* 16: 775–784.

——. 1980. Guild structure in desert lizards. *Oikos* 35: 194–201.

——. 1981a. Resource acquisition and allocation among animals. In C. Townsend and P. Calow (eds.), *Physiological ecology: An evolutionary approach to resource use* (pp. 300–314). Blackwell, Oxford.

——. 1981b. Competition and niche theory. Chapter 8 (pp. 167–196) in R. M. May (ed.), *Theoretical ecology: Principles and applications*. Blackwell, Oxford.

——. 1985. Some intercontinental comparisons of desert lizards. *Nat. Geogr. Res.* 1: 490–504.

——. 1986a. *Ecology and natural history of desert lizards. Analyses of the ecological niche and community structure*. Princeton Univ. Press, Princeton, NJ.

——. 1986b. Ecological phenomena in evolutionary perspective. Chapter 16 (pp. 325–336) in N. Polunin (ed.), *Ecosystem theory and application*. Wiley, New York.

——. 1987. The subtlety, complexity and importance of population interactions when more than two species are involved. *Revista Chilena de Historia Natural* 60: 351–362.

——. 1989a. Desert lizard diversity: Additional comments and some data. *Amer. Natur.* 134: 344–364.

——. 1989b. The role of plants in evolutionary ecology. *Evol. Trends Plants* 3: 75–80.

——. 1992. Fire ecology. Disturbance, spatial heterogeneity, and biotic diversity: Fire succession in arid Australia. *Res. Expl.* 8:352–371.

——. 1993. The many dimensions of a lizard's ecological niche. Pages 1–34. In E. D. Valakos (ed.), *First International Congress on Lacertids of the Mediterranean Basin.* University of Athens, Greece.

Pianka, E. R., and W. S. Parker. 1975a. Ecology of horned lizards: A review with special reference to *Phrynosoma platyrhinos. Copeia* 1975: 141–162.

——. 1975b. Age-specific reproductive tactics. *Amer. Natur.* 109: 453–464.

Pianka, E. R., R. B. Huey, and L. R. Lawlor. 1979. Niche segregation in desert lizards. In D. J. Horn, R. Mitchell, and G. R. Stairs (eds.), *Analysis of ecological systems* (pp. 67–115). Ohio State Univ. Press, Columbus.

Pickett, S. T. A. 1976. Succession: An evolutionary interpretation. *Amer. Natur.* 110: 107–119.

Pickett, S. T. A., and M. J. McDonnell. 1989. Changing perspectives in community dynamics: A theory of successional forces. *Trends Ecol. Evol.* 4: 241–245.

Pickett, S. T. A., and P. S. White (eds.). 1986. *The ecology of natural disturbance and patch dynamics.* Academic Press, New York.

Pielou, E. C. 1969. *An introduction to mathematical ecology.* Wiley-Interscience, New York.

——. 1972. Niche width and niche overlap: A method for measuring them. *Ecology* 53: 687–692.

——. 1974. *Population and community ecology: Principles and methods.* Gordon and Breach, New York.

——. 1975. *Ecological diversity.* Wiley, New York.

Pierce, N. E. 1985. Lycaenid butterflies and ants: Selection for nitrogen-fixing and other protein-rich food plants. *Amer. Natur.* 125: 888–895.

Pimentel, D. 1968. Population regulation and genetic feedback. *Science* 159: 1432–1437.

Pimm, S. L. 1979. The structure of food webs. *Theoret. Pop. Biol.* 16: 144–158.

——. 1980. Properties of food webs. *Ecology* 61: 219–225.

——. 1982. *Food webs.* Chapman and Hall, London.

——. 1983. Appendix: Monte Carlo analyses in ecology. Pages 290–296. In R. B. Huey, T. W. Schoener, and E. R. Pianka (eds.), *Lizard ecology: Studies on a model organism.* Harvard Univ. Press, Cambridge, Mass.

——. 1984. The complexity and stability of ecosystems. *Nature* 307: 321–326.

——. 1989. The geometry of niches. In A. Hastings (ed.), *Community ecology.* Springer-Verlag, New York.

——. 1991. *The balance of nature?* Univ. Chicago Press, Chicago.

Pimm, S. L., and J. H. Lawton. 1977. Number of trophic levels in ecological communities. *Nature* 268: 329–331.

——. 1980. Are food webs divided into compartments? *J. Animal Ecol.* 49: 879–898.

Pitelka, F. A. 1964. The nutrient-recovery hypothesis for arctic microtine cycles. I. Introduction. In D. J. Crisp (ed.), *Grazing in terrestrial and marine environments* (pp. 55–56). Brit. Ecol. Soc. Symposium, Blackwell Scientific Publications, Oxford.

Pittendrigh, C. S. 1961. Temporal organization in living systems. *Harvey Lecture Series* 56: 93–125. Academic Press, New York.

Platnick, N. I., and G. Nelson. 1978. A method of analysis for historical biogeography. *Syst. Zool.* 27: 1–16.

Platt, J. R. 1964. Strong inference. *Science* 146: 347–353.

Pleszczynska, W. K. 1978. Microgeographic prediction of polygyny in the lark bunting. *Science* 201: 935–937.

Pomerantz, M. J. 1980. Do "higher" order interactions in competition systems really exist? *Amer. Natur.* 117: 583–591.

Ponnamperuma, C. 1972. *The origins of life.* Dutton, New York.

Poole, R. W. 1974. *An introduction to quantitative ecology.* McGraw-Hill, New York.

Popper, K. R. 1959. *The logic of scientific discovery.* Basic Books, New York.

Porter, W. P., and D. M. Gates. 1969. Thermodynamic equilibria of animals with environment. *Ecol. Monogr.* 39: 227–244.

Porter, W. P., J. W. Mitchell, W. A. Beckman, and C. B. DeWitt. 1973. Behavioral implications of mechanistic ecology—thermal and behavioral modeling of desert ectotherms and their microenvironment. *Oecologia* 13: 1–54.

Post, W. M., C. C. Travis, and D. L. DeAngelis. 1985. Mutualism, limited competition and positive feedback. In D. H. Boucher (ed.), *The biology of mutualism* (pp. 305–325). Oxford Univ. Press, New York.

Poulson, T. L., and D. D. Culver. 1969. Diversity in terrestrial cave communities. *Ecology* 50: 153–158.

Power, M. E., W. J. Matthews, and A. J. Stewart. 1985. Grazing minnows, piscivorous bass, and stream algae: Dynamics of a strong interaction. *Ecology* 66: 1448–1456.

Prance, G. T. (ed.). 1982. *Biological diversification in the tropics.* Columbia Univ. Press, New York.

Preston, F. W. 1948. The commonness and rarity of species. *Ecology* 29: 254–283.

——. 1960. Time and space and the variation of species. *Ecology* 41: 611–627.

——. 1962a. The canonical distribution of commonness and rarity. I. *Ecology* 43: 185–215.

——. 1962b. The canonical distribution of commonness and rarity. II. *Ecology* 43: 410–432.

Price, G., and J. Maynard Smith. 1973. The logic of animal conflict. *Nature* 246: 15–18.

Price, P. W. 1975. *Insect ecology.* Wiley, New York.

——. 1980. *Evolutionary biology of parasites.* Princeton Univ. Press, Princeton, NJ.

Price, P. W., M. Westoby, and B. Rice. 1988. Parasite-mediated competition: Some predictions and tests. *Amer. Natur.* 131: 544–555.

Prokopy, R. J., A. L. Averill, S. S. Cooley, and B. D. Roitberg. 1982. Associative learning in egg-laying site selection by apple maggot flies. *Science* 218: 76–77.

Prosser, C. L. (ed.). 1973. *Comparative animal physiology.* Saunders, Philadelphia.

Pulliam, H. R. 1974. On the theory of optimal diets. *Amer. Natur.* 108: 50–65.

——. 1988. Sources, sinks, and population regulation. *Amer. Natur.* 132: 652–661.

Pulliam, H. R., and B. J. Danielson. 1991. Sources, sinks, and habitat selection: A landscape perspective on population dynamics. *Amer. Natur.* 137 (symposium supplement): S50–S66.

Putman, R. J., and S. D. Wratten. 1984. *Principles of ecology.* Univ. California Press, Berkeley.

Quinn, J. A. 1987. *Complex patterns of genetic differentiation and phenotypic plasticity versus an outmoded ecotype terminology. Differentiation patterns in higher plants.* Academic Press, New York.

Quinn, J. F., and A. E. Dunham. 1983. On hypothesis testing in ecology and evolution. *Amer. Natur.* 122: 602–617.

Quinn, J. F., and S. Harrison. 1988. Effects of habitat fragmentation and isolation on species richness: Evidence from biogeographic patterns. *Oecologia* 75: 132–140.

Quinn, J. F., and P. W. Signor. 1989. Death stars, ecology, and mass extinctions. *Ecology* 70: 824–834.

Rabenold, P., et al. 1991. Using DNA fingerprinting to assess kinship and genetic structure in avian populations. In E. C. Dudley (ed.), *The unity of evolutionary biology* (pp. 611–620). Proc. Fourth Int. Congress of Systematic and Evolutionary Biology. Discorides Press, Portland, Oregon.

Ralls, K., and P. H. Harvey. 1985. Geographic variation in size and sexual dimorphism of North American weasels. *Biol. J. Linn. Soc.* 25: 119–167.

Rand, A. S. 1967. Predator-prey interactions and the evolution of aspect diversity. *Atas do Simposio sobre a Biota Amazonica* 5: 73–83.

Randolph, P. A., J. C. Randolph, and C. A. Barlow. 1975. Age-specific energetics of the pea aphid, *Acyrothosiphon pisum. Ecology* 56: 359–369.

Rapport, D. J. 1971. An optimization model of food selection. *Amer. Natur.* 105: 575–587.

Rathcke, B. J. 1976. Insect-plant patterns and relationships in the stem-boring guild. *Amer. Midl. Natur.* 96: 98–117.

Raunkaier, C. 1934. *The life form of plants and statistical plant geography.* Clarendon Press, Oxford.

Recher, H. F. 1969. Bird species diversity and habitat diversity in Australia and North America. *Amer. Natur.* 103: 75–80.

Reichle, D. (ed.). 1970. *Analysis of temperate forest ecosystems.* Springer-Verlag, Heidelberg.

Rejmanek, M., and P. Stary. 1979. Connectance in real biotic communities and critical values for stability of model ecosystems. *Nature* 280: 311–313.

Reynoldson, T. B. 1964. Evidence for intraspecific competition in field populations of triclads. *J. Anim. Ecol.* 33: 187–207.

Rhoades, D. F., and R. G. Cates. 1976. Toward a general theory of plant antiherbivore chemistry. In J. Wallace and R. Mansell (eds.), *Biochemical interactions between plants and insects* (pp. 168–213). Recent Advances in Phytochemistry, Vol. 10. Plenum Press, New York.

Richards, B. N. 1974. *Introduction to the soil ecosystem.* Longman, New York.

Richards, P. W. 1952. *The tropical rain forest.* Cambridge Univ. Press, New York.

Rick, I. M., and R. I. Bowman. 1961. Galápagos tomatoes and tortoises. *Evolution* 15: 407–417.

Ricklefs, R. E. 1966. The temporal component of diversity among species of birds. *Evolution* 20: 235–242.

———. 1973. *Ecology.* Chiron Press, Portland, Ore.

———. 1977. Environmental heterogeneity and plant species diversity: A hypothesis. *Amer. Natur.* 111: 376–381.

———. 1987. Community diversity: Relative roles of local and regional processes. *Science* 235: 167–171.

———. 1989. Speciation and diversity: The integration of local and regional processes. Chapter 24 (pp. 599–622) in D. Otte and J. Endler (eds.), *Speciation and its consequences.* Sinauer Sunderland, Mass.

Ricklefs, R. E., and G. W. Cox. 1972. The taxon cycle in the land bird fauna of the West Indies. *Amer. Natur.* 106: 195–219.

Ricklefs, R. E., and K. O'Rourke. 1975. Aspect diversity in moths: A temperate-tropical comparison. *Evolution* 29: 313–324.

Ridley, M. 1978. Paternal care. *Animal Behavior* 26: 904–932.

———. 1983. *The explanation of organic diversity: The comparative method and adaptations for mating.* Oxford Univ. Press, Oxford.

Robbins, C.S., J. R. Sauer, R. S. Greenberg, and S. Droege. 1989. Population declines in North American birds that migrate to the tropics. *Proc. Nat. Acad. Sci.* 86: 7658–7662.

Rohde, K. 1979. A critical evaluation of intrinsic and extrinsic factors responsible for niche restriction in parasites. *Amer. Natur.* 114: 648–671.

Rolston, H. 1985. Duties to endangered species. *BioScience* 35: 718–726.

Romanoff, A. L., and A. J. Romanoff. 1949. *The avian egg.* Wiley, New York.

Root, R. B. 1967. The niche exploitation pattern of the blue-gray gnatcatcher. *Ecol. Monogr.* 37: 317–350.

Rosen, D. E. 1978. Vicariant patterns and historical explanation in biogeography. *Syst. Zool.* 27: 159–188.

Rosen, R. 1967. *Optimality principles in biology.* Plenum, New York.

Rosenzweig, M. L. 1968. Net primary productivity of terrestrial communities: Prediction from climatological data. *Amer. Natur.* 102: 67–74.

———. 1971. The paradox of enrichment: Destabilization of exploitation ecosystems in ecological time. *Science* 171: 385–387.

———. 1973a. Exploitation in three trophic levels. *Amer. Natur.* 107: 275–294.

———. 1973b. Evolution of the predator isocline. *Evolution* 27: 84–94.

———. 1979. Optimal habitat selection in two-species competitive systems. *Fortschr. Zool.* 25: 283–293.

Rosenzweig, M. L., and R. H. MacArthur. 1963. Graphical representation and stability conditions of predator-prey interactions. *Amer. Natur.* 97: 209–223.

Rosenzweig, M. L., J. S. Brown, and T. L. Vincent. 1987. Red queens and ESS: The coevolution of evolutionary rates. *Evol. Ecol.* 1: 59–94.

Ross, H. H. 1957. Principles of natural coexistence indicated by leafhopper populations. *Evolution* 11: 113–129.

——. 1958. Further comments on niches and natural coexistence. *Evolution* 12: 112–113.

Rothstein, S. I. 1973. The niche-variation model—is it valid? *Amer. Natur.* 107: 598–620.

Roughgarden, J. 1971. Density-dependent natural selection. *Ecology* 52: 453–468.

——. 1972. Evolution of niche width. *Amer. Natur.* 106: 683–718.

——. 1974a. Species packing and the competition function with illustrations from coral reef fish. *Theoret. Pop. Biol.* 5: 163–186.

——. 1974b. Niche width: Biogeographic patterns among *Anolis* lizard populations. *Amer. Natur.* 108: 429–442.

——. 1974c. The fundamental and realized niche of a solitary population. *Amer. Natur.* 108: 232–235.

——. 1975. Evolution of marine symbiosis—A simple cost-benefit model. *Ecology* 56: 1201–1208.

——. 1976. Resource partitioning among competing species: A coevolutionary approach. *Theoret. Pop. Biol.* 9: 388–424.

——. 1983. Competition and theory in community ecology. *Amer. Natur.* 122:583–601.

Roughgarden, J., and M. Feldman. 1975. Species packing and predation pressure. *Ecology* 56: 489–492.

Royama, T. 1969. A model for the global variation of clutch size in birds. *Oikos* 20: 562–567.

——. 1970. Factors governing the hunting behaviour and selection of food by the great tit (*Parus major* L.). *J. Anim. Ecol.* 39: 619–668.

Ruibal, R. 1961. Thermal relations of five species of tropical lizards. *Evolution* 15: 98–111.

Ruibal, R., and R. Philibosian. 1970. Eurythermy and niche expansion in lizards. *Copeia* 1970: 645–653.

Rummell, J. D., and J. Roughgarden. 1985. A theory of faunal buildup for competition communities. *Evolution* 39: 1009–1033.

Russell-Hunter, W. D. 1970. *Aquatic productivity: An introduction to some basic aspects of biological oceanography and limnology.* Macmillan, New York.

Ruttner, F. 1953. *Fundamentals of limnology.* Univ. of Toronto Press, Toronto, Canada.

Ryan, M. J. 1990. Signals, species and sexual selection. *Amer. Sci.* 78: 46–62.

Ryder, V. 1954. On the morphology of leaves. *Bot. Rev.* 20: 263–276.

Sadlier, R. M. 1973. *Reproduction of vertebrates.* Academic Press, New York.

Safriel, U. N. 1975. On the significance of clutch size in nidifugous birds. *Ecology* 56: 703–708.

Sale, P. 1974. Overlap in resource use and interspecific competition. *Oecologia* 17: 245–256.

Salisbury, E. J. 1942. *The reproductive capacity of plants: Studies in quantitative biology.* Bell, London.

Salt, G. W. 1967. Predation in an experimental protozoan population (*Woodruffia-Paramecium*). *Ecol. Monogr.* 37: 113–144.

——. 1983. Roles: Their limits and responsibilities in ecological and evolutionary research. *Amer. Natur.* 122: 697–706.

Salthe, S. N. 1972. *Evolutionary biology*. Holt, Rinehart and Winston, New York.

Salzburg, M. A. 1984. *Anolis sagrei* and *Anolis cristatellus* in southern Florida: A case study in interspecific competition. *Ecology* 65: 14–19.

Savage, J. M. 1958. The concept of ecologic niche with reference to the theory of natural coexistence. *Evolution* 12: 111–121.

Sawyer, J. S. (ed.). 1966. *World climate from 8,000 to 0 B.C.* Proc. Int. Symp. on World Climate 8,000 to 0 B.C. Imperial College, London. Royal Meteorological Society, London.

Schad, G. A. 1963. Niche diversification in a parasitic species flock. *Nature* 198: 404–407.

——. 1966. Immunity, competition, and natural regulation of Helminth populations. *Amer. Natur.* 100: 359–364.

Schaffer, W. M. 1974. Selection for optimal life histories: Effects of age structure. *Ecology* 55: 291–303.

Schaffer, W. M., and R. H. Tamarin. 1973. Changing reproductive rates and population cycles in lemmings and voles. *Evolution* 27: 111–124.

Schall, J. J. 1982. Lizard malaria: Parasite-host ecology. Chapter 5 (pp. 84–100) in R. B. Huey, E. R. Pianka, and T. W. Schoener (eds.), *Lizard ecology: Studies of a model organism*. Harvard Univ. Press, Cambridge, Mass.

——. 1992. Parasite-mediated competition in *Anolis* lizards. *Oecologia* 92: 58–64.

Schall, J. J., and E. R. Pianka. 1978. Geographical trends in numbers of species. *Science* 201: 679–686.

——. 1980. Evolution of escape behavior diversity. *Amer. Natur.* 115: 551–566.

Schaller, F. 1968. *Soil animals*. Univ. of Michigan Press, Ann Arbor.

Schindler, D. W. 1990. Experimental perturbations of whole lakes as tests of hypotheses concerning ecosystem structure and function. *Oikos* 57: 25–41.

Schluter, D. 1988. Character displacement and the adaptive divergence of finches on islands and continents. *Amer. Natur.* 131: 799–824.

Schluter, D., T. D. Price, and P. R. Grant. 1985. Ecological character displacement in Darwin's finches. *Science* 227: 1056–1059.

Schmidt-Nielsen, K. 1964. *Desert animals: Physiological problems of heat and water.* Oxford Univ. Press, London.

——. 1972. Locomotion: Energy cost of swimming, flying, and running. *Science* 177: 222–228.

——. 1975. *Animal physiology: Adaptation and environment*. Cambridge Univ. Press, London.

Schmidt-Nielsen, K., and W. R. Dawson. 1964. Terrestrial animals in dry heat: Desert reptiles. In D. B. Dill (ed.), *Handbook of physiology, Section 4: Adaptation to the environment* (pp. 467–480). American Physiological Society, Washington, D.C.

Schoener, A. 1974. Experimental zoogeography: Colonization of marine mini-islands. *Amer. Natur.* 108: 715–738.

Schoener, T. W. 1965. The evolution of bill size differences among sympatric congeneric species of birds. *Evolution* 19: 189–213.

———. 1967. The ecological significance of sexual dimorphism in size in the lizard *Anolis conspersus. Science* 155: 474–477.

———. 1968a. The *Anolis* lizards of Bimini: Resource partitioning in a complex fauna. *Ecology* 49: 704–726.

———. 1968b. Sizes of feeding territories among birds. *Ecology* 49: 123–141.

———. 1969a. Models of optimal size for solitary predators. *Amer. Natur.* 103: 277–313.

———. 1969b. Optimal size and specialization in constant and fluctuating environments: An energy-time approach. *Brookhaven Symp. Biol.* 22: 103–114.

———. 1970. Nonsynchronous spatial overlap of lizards in patchy habitats. *Ecology* 51: 408–418.

———. 1971. Theory of feeding strategies. *Ann. Rev. Ecol. Syst.* 2: 369–404.

———. 1973. Population growth regulated by intraspecific competition for energy or time: Some simple representations. *Theoret. Pop. Biol.* 4: 56–84.

———. 1974a. Resource partitioning in ecological communities. *Science* 185: 27–39.

———. 1974b. The compression hypothesis and temporal resource partitioning. *Proc. Nat. Acad. Sci., U.S.A.* 71: 4169–4172.

———. 1975a. Competition and the form of habitat shift. *Theoret. Pop. Biol.* 5: 265–307.

———. 1975b. Presence and absence of habitat shift in some widespread lizard species. *Ecol. Monogr.* 45: 232–258.

———. 1976a. The species-area relation within archipelagos: Models and evidence from island land birds. *Proc. 16th Int. Ornith. Congr.*

———. 1976b. Alternatives to Lotka-Volterra competition: Models of intermediate complexity. *Theoret. Pop. Biol.* 10: 309–333.

———. 1977. Competition and the niche. In D. W. Tinkle and C. Gans (eds.), *Biology of the Reptilia* (pp. 35–136). Academic Press, New York.

———. 1982. The controversy over interspecific competition. *Amer. Sci.* 70: 586–595.

———. 1983. Field experiments on interspecific competition. *Amer. Natur.* 122: 240–285.

———. 1984. Size differences among sympatric, bird-eating hawks: A worldwide survey. In D. R. Strong, D. Simberloff, L. G. Abele, and A. B. Thistle (eds.), *Ecological communities: Conceptual issues and the evidence* (pp. 254–281). Princeton Univ. Press, Princeton, NJ.

———. 1986. Resource partitioning. Chapter 6 (pp. 91–126) in J. Kikkawa and D. J. Anderson (eds.), *Community ecology: Pattern and process.* Blackwell, London.

———. 1988. Ecological interactions. pp. 257–297 in A. A. Myers and P. S. Giller (eds.), Analytical Biogeography. Chapman and Hall.

———. 1989. Food webs from the small to the large. *Ecology* 70: 1559–1589.

Schoener, T. W., and G. C. Gorman. 1968. Some niche differences in three Lesser Antillean lizards of the genus *Anolis. Ecology* 49: 819–830.

Schoener, T. W., and D. H. Janzen. 1968. Notes on environmental determinants of tropical versus temperate insect size patterns. *Amer. Natur.* 102: 207–224.

Schultz, A. M. 1964. The nutrient-recovery hypothesis for arctic microtine cycles. In D. J. Crisp (ed.), *Grazing in terrestrial and marine environments* (pp. 57–68). Brit. Ecol. Soc. Symposium. Blackwell Scientific Publications, Oxford.

———. 1969. A study of an ecosystem: The arctic tundra. In G. Van Dyne (ed.), *The ecosystem concept in natural resource management* (pp. 77–93). Academic Press, New York.

Seger, J., and H. J. Brockmann. 1987. What is bet-hedging? *Oxford Survey Evol. Biol.* 4: 182–211.

Seger, J., and W. D. Hamilton. 1988. Parasites and sex. In R. E. Michod and B. R. Levin (eds.), *The Evolution of Sex* (pp. 176–193). Sinauer, Sunderland, Mass.

Seifert, R. P., and F. H. Seifert. 1976. A community matrix analysis of *Heliconia* insect communities. *Amer. Natur.* 110: 461–483.

Selander, R. K. 1965. On mating systems and sexual selection. *Amer. Natur.* 99: 129–141.

———. 1966. Sexual dimorphism and differential niche utilization in birds. *Condor* 68: 113–151.

———. 1972. Sexual selection and dimorphism in birds. In B. G. Campbell (ed.), *Sexual selection and the descent of man (1871–1971)* (pp. 180–230). Aldine-Atherton, Chicago.

Shannon, C. E. 1948. The mathematical theory of communication. In C. E. Shannon and W. Weaver (eds.), *The mathematical theory of communication* (pp. 3–91). Univ. of Illinois Press, Urbana.

Shapiro, D. Y. 1980. Serial female sex changes after simultaneous removal of males from social groups of a coral reef fish. *Science* 209: 1136–1137.

Sheldon, A. L. 1972. Comparative ecology of *Arcynopteryx* and *Diura* (Plecoptera) in a California stream. *Arch. Hydrobiol.* 69: 521–546.

Shelford, V. E. 1913a. *Animal communities in temperate America.* Univ. Chicago Press, Chicago.

———. 1913b. The reactions of certain animals to gradients of evaporating power and air. A study in experimental ecology. *Biol. Bull.* 25: 79–120.

———. 1963. *The ecology of North America.* Univ. of Illinois Press, Urbana.

Sheppard, P. M. 1951. Fluctuations in the selective value of certain phenotypes in the polymorphic land snail *Cepaea nemoralis. Heredity* 5: 125–134.

———. 1959. *Natural selection and heredity.* Hutchinson Univ. Library, London.

Sherman, P. W. 1977. Nepotism and the evolution of alarm calls. *Science* 197: 1246–1253.

Sherman, P. W., J. U. M. Jarvis, and R. D. Alexander (eds.). 1991. *The biology of the naked mole-rat.* Princeton Univ. Press, Princeton, NJ.

Shimwell, D. W. 1971. *Description and classification of vegetation.* Univ. of Washington Press, Seattle.

Shine, R. 1989. Ecological causes for the evolution of sexual dimorphism: A review of the evidence. *Quart. Rev. Biol.* 64: 419–461.

Shmida, A., and M. V. Wilson. 1985. Biological determinants of species diversity. *J. Biogeography* 12: 1–20.

Shugart, H. H. 1984. *A theory of forest dynamics.* Springer-Verlag, New York.

Shugart, H. H., and B. C. Patten. 1972. Niche quantification and the concept of niche pattern. In B. Patten (ed.), *Systems analysis and simulation in ecology* (pp. 284–327). Vol. II. Academic Press, New York.

Shukla, J., C. Nobre, and P. Sellers. 1990. Amazon deforestation and climate change. *Science* 240: 1322–1325.

Sibley, C. G., J. E. Ahlquist, and B. L. Monroe. 1988. A classification of the living birds of the world based on DNA-DNA hybridization studies. *Auk* 105: 409–423.

Sibley, C. G., J. A. Comstock, and J. E. Ahlquist. 1990. DNA hyubridization evidence of hominoid phylogeny: A reanalysis of the data. *J. Molecular Evolution* 30: 202–236.

Sih, A. 1987. Predators and prey lifestyles: An evolutionary and ecological overview. In W. C. Kerfoot, and A. Sih (eds.), *Predation: Direct and indirect impacts on aquatic communities.* Univ. Press of New England, Hanover, NH.

Sih, A. P., P. Crowley, M. McPeek, J. Petranka, and K. Strohmeier. 1985. Predation, competition, and prey communities: A review of field experiments. *Ann. Rev. Ecol. Syst.* 16: 269–311.

Simberloff, D. S. 1976. Trophic structure determination and equilibrium in an arthropod community. *Ecology* 57: 395–398.

———. 1988. The contribution of population and community biology to conservation science. *Ann. Rev. Ecol. Syst.* 19: 473–511.

Simberloff, D. S., and E. O. Wilson. 1970. Experimental zoogeography of islands. A two-year record of colonization. *Ecology* 51: 934–937.

Simenstad, C. A., J. A. Estes, and K. W. Kenyon. 1978. Aleuts, sea otters, and alternate stable-state communities. *Science* 200: 403–441.

Simon, C. 1987. Hawaiin evolutionary biology: An introduction. *Trends Ecol. Evol.* 2: 175–178.

Simonsen, L., and B. R. Levin. 1988. Evaluating the risk of releasing genetically engineered organisms. *Trends Ecol. Evol.* 2: 175–178.

Simpson, E. H. 1949. Measurement of diversity. *Nature* 163: 688.

Simpson, G. G. 1969. Species density of North American recent mammals. *Syst. Zool.* 13: 57–73.

Sinclair, A. R. E., J. M. Gosline, G. Holdsworth, C. J. Krebs, S. Boutin, J. N. M. Smith, R. Boonstra, and M. Dale. 1993. Can the solar cycle and climate synchronize the snowshoe hare cycle in Canada? Evidence from tree rings and ice cores. *Amer. Natur.* 141: 173–198.

Sjogren, P. 1991. Extinction and isolation gradients in metapopulations: The case of the pool frog (*Rana lessonae*). *Biol. J. Linn. Soc.* 42: 135–147.

Skellam, J. G. 1951. Random dispersal in theoretical populations. *Biometrika* 38: 196–218.

Skutch, A. F. 1949. Do tropical birds rear as many young as they can nourish? *Ibis* 91: 430–455.

———. 1967. Adaptive limitation of the reproductive rate of birds. *Ibis* 109: 579–599.

Slatkin, M. 1974. Competition and regional coexistence. *Ecology* 55: 128–134.

Slobodkin, L. B. 1960. Ecological energy relationships at the population level. *Amer. Natur.* 94: 213–236.

———. 1962a. Energy in animal ecology. *Adv. Ecol. Res.* 1: 69–101.

———. 1962b. *Growth and regulation of animal populations.* Holt, Rinehart and Winston, New York.

——. 1968. How to be a predator. *Amer. Zool.* 8: 43–51.

Smith, A. D. 1940. A discussion of the application of a climatological diagram, the hythergraph, to the distribution of natural vegetation types. *Ecology* 21: 184–191.

Smith, C. C. 1968. The adaptive nature of social organization in the genus of tree squirrels *Tamiasciurus*. *Ecol. Monogr.* 38: 31–63.

——. 1970. The coevolution of pine squirrels (*Tamiasciurus*) and conifers. *Ecol. Monogr.* 40: 349–371.

Smith, C. C., and S. D. Fretwell. 1974. The optimal balance between size and number of offspring. *Amer. Natur.* 108: 499–506.

Smith, F. E. 1952. Experimental methods in population dynamics. A critique. *Ecology* 33: 441–450.

——. 1954. Quantitative aspects of population growth. In E. Boell (ed.), *Dynamics of growth processes* (pp. 277–294). Princeton Univ. Press, Princeton, NJ.

——. 1961. Density dependence in the Australian thrips. *Ecology* 42: 403–407.

——. 1963a. Population dynamics in *Daphnia magna* and a new model for population growth. *Ecology* 44: 651–663.

——. 1963b. Density dependence. *Ecology* 44: 220.

——. 1970a. Analysis of ecosystems. In D. Reichle (ed.), *Analysis of temperate forest ecosystems* (pp. 7–18). Springer-Verlag, Berlin.

——. 1970b. Effects of enrichment in mathematical models. In *Eutrophication: Causes, consequences, correctives* (pp. 631–645). National Academy of Sciences, Washington, D. C.

——. 1972. Spatial heterogeneity, stability, and diversity in ecosystems. In E. S. Deevey (ed.), *Growth by intussusception: Ecological essays in honor of G. Evelyn Hutchinson. Trans. Conn. Acad. Arts. Sci.* 44: 309–335.

Smith, J. N. M., and M. P. A. Sweatman. 1976. Feeding habits and morphological variation in Cocos finches. *Condor* 78: 244–248.

Smith, R. 1966. *Ecology and field biology.* Harper & Row, New York.

Smith, T. B. 1990. Resource use by bill morphs of an African finch: Evidence for intraspecific competition. *Ecology* 71: 1246–1257.

Smouse, P. E. 1971. The evolutionary advantages of sexual dimorphism. *Theoret. Pop. Biol.* 2: 469–481.

Snell, T. W., and D. G. Burch. 1975. The effects of density on resource partitioning in *Chamaesyce hirta* (Euphorbiacae). *Ecology* 56: 742–746.

Snell, T. W., and C. E. King. 1977. Lifespan and fecundity patterns in rotifers: The cost of reproduction. *Evolution* 31: 882–890.

Sokal, R. R. 1970. Senescence and genetic load: Evidence from *Tribolium. Science* 167: 1733–1734.

Solomon, M. E. 1949. The natural control of animal populations. *J. Animal Ecol.* 18: 1–32.

——. 1972. *Population dynamics.* Arnold, London.

Somero, G. N. 1969. Enzymic mechanisms of temperature compensation. *Amer. Natur.* 103: 517–530.

Soulé, M. (ed.). 1986. *Conservation biology: The science of scarcity and diversity.* Sinauer, Sunderland, Mass.

―― (ed.). 1987. *Viable populations for conservation.* Cambridge Univ. Press, Cambridge, England.

Soulé, M., and B. R. Stewart. 1970. The "niche-variation" hypothesis: A test and alternatives. *Amer. Natur.* 104: 85–97.

Southwood, T. R. E. 1966. *Ecological methods with particular reference to the study of insect populations.* Methuen, London.

Spinage, C. A. 1972. African ungulate life tables. *Ecology* 53: 645–652.

St. Amant, J. L. S. 1970. The detection of regulation in animal populations. *Ecology* 51: 823–828.

Stahl, E. 1888. Pflanzen und Schnecken. Biolosche Studie uber die Schutzmittel der Pflanzen gegen Schneckenfrass. *Jena Z. Med. Natururw.* 22: 557–684.

Stearns, S. C. 1976. Life-history tactics: A review of the ideas. *Quart. Rev. Biol.* 51: 3–47.

――(ed.). 1987. *The evolution of sex and its consequences.* Birkhauser Verlag, Basel.

――. 1989. Trade- offs in life history evolution. *Funct. Ecol.* 3: 259–268.

Stehli, F. G., and S. D. Webb (eds.). 1985. *The great American biotic interchange.* Plenum, New York.

Stephens, D. W., and J. R. Krebs. 1986. *Foraging theory.* Princeton Univ. Press, Princeton, NJ.

Stevans, G. C. 1989. The latitudinal gradient in geographical range: How so many species coexist in the tropics. *Amer. Natur.* 133: 240–256.

Stewart, R. E., and J. W. Aldrich. 1951. Removal and repopulation of breeding birds in a spruce-fir community. *Auk* 68: 471–482.

Stowe, L. G., and J. L. Brown. 1981. A geographic perspective on the ecology of compound leaves. *Evolution* 35: 818–821.

Stradling, D. J. 1976. The nature of the mimetic patterns of the brassolid genera, *Caligo* and *Eryphanis. Ecol. Entomol.* 1: 135–138.

Strauss, S. Y. 1991. Indirect effects in community ecology: Their definition, study, and importance. *Trends Ecol. Evol.* 6: 206–210.

Straussmann, J. E. 1989. Altruism and relatedness at colony foundation in social insects. *Trends Ecol. Evol.* 4: 371–374.

Strobeck, C. 1973. *n*-species competition. *Ecology* 54: 650–654.

Strong, D. R. 1977. Epiphyte loads, tree falls, and perennial forest disruption: A mechanism for maintaining higher tree species richness in the tropics without animals. *J. Biogeography* 4: 215–218.

Strong, D. R., D. Simberloff, L. G. Abele, and A. B. Thistle (eds.). 1984. *Ecological communities: Conceptual issues and the evidence.* Princeton Univ. Press, Princeton, NJ.

Sugihara, G. 1980. Minimal community structure: An explanation of species abundance patterns. *Amer. Natur.* 116: 770–787.

Sugihara, G., and R. M. May. 1990. Applications of fractals in ecology. *Trends Ecol. Evol.* 5:79–86.

Sultan, S. E. 1987. Evolutionary implications of phenotypic plasticity in plants. *Evol. Biol.* 21: 127–178.

Sutherland, J. P. 1974. Multiple stable points in natural communities. *Amer. Natur.* 108: 859–873.

Sverdrup, H. U., M. W. Johnson, and R. H. Fleming. 1942. *The oceans: Their physical chemistry and general biology.* Prentice-Hall, Englewood Cliffs, N. J.

Tamarin, R. H., and C. J. Krebs. 1969. *Microtus* population biology. II. Genetic changes at the transferrin locus in fluctuating populations of two vole species. *Evolution* 23: 183–211.

Tanner, J. T. 1966. Effects of population density on the growth rates of animal populations. *Ecology* 47: 733–745.

——. 1975. The stability and the intrinsic growth rates of prey and predator populations. *Ecology* 56: 855–869.

Tarling, D. H., and S. K. Runcorn (eds.). 1973. Implications of continental drift to the earth sciences (2 volumes). Academic Press, London.

Taylor, F. B. 1910. Bearing of the Tertiary mountain belt on the origin of the earth's plan. *Bull. Geol. Soc. Amer.* 21: 179–226.

Taylor, G. 1920. *Australian meteorology.* Clarendon Press, Oxford.

Taylor, H. M., R. S. Gourley, C. E. Lawrence, and R. S. Kaplan. 1974. Natural selection of life history attributes: An analytical approach. *Theoret. Pop. Biol.* 5: 104–122.

Teal, J. M. 1962. Energy flow in the salt marsh ecosystem of Georgia. *Ecology* 43: 614–624.

Terborgh, J. 1971. Distribution on environmental gradients: Theory and a preliminary interpretation of distributional patterns in the avifauna of the Cordillera Vilcabamba, Peru. *Ecology* 52: 23–40.

——. 1974a. Faunal equilibria and the design of wildlife preserves. In F. Golley and E. Medina (eds.), *Tropical ecological systems: Trends in terrestrial and aquatic research.* Springer-Verlag, New York.

——. 1974b. Preservation of natural diversity: The problem of extinction prone species. *BioScience* 24: 715–722.

Terborgh, J., and J. M. Diamond. 1970. Niche overlap in feeding assemblages of New Guinea birds. *Wilson Bull.* 82: 29–52.

Terborgh, J., and J. Faaborg. 1980. Saturation of bird communities in the West Indies. *Amer. Natur.* 116: 178–195.

Terborgh, J. and S. Robinson. 1986. Guilds and their utility in ecology. Chapter 5 (pp. 65–90) in J. Kikkawa and D. J. Anderson (eds.), *Community ecology: Pattern and process.* Blackwell, London.

Terborgh, J., and J. S. Weske. 1969. Colonization of secondary habitats by Peruvian birds. *Ecology* 50: 765–782.

Thompson, J. N. 1982. *Interaction and coevolution.* Wiley, New York.

——. 1989. Concepts of coevolution. *Trends Ecol. Evol.* 4: 179–183.

Thornes, J. 1992. *Deforestation.* Chapman and Hall, London.

Thornhill, A. R., 1976. Sexual selection and nuptial feeding behavior in *Bittacus apicalis* (Insecta: Mecoptera). *Amer. Natur.* 110: 529–548.

——. 1976. Sexual selection and parental investment in insect. *Amer. Natur.* 110: 153–163.

Thornhill, A. R., and J. Alcock. 1983. *The evolution of insect mating systems.* Harvard Univ. Press, Cambridge, Mass.

Thornthwaite, C. W. 1948. An approach toward a rational classification of climate. *Geogr. Rev.* 38: 55–94.

Tiedje, J. M., R. K. Colwell, Y. L. Grossman, R. E. Hodson, R. E. Lenski, R. N. Mack, and P. J. Regal. 1989. The planned introduction of genetically engineered organisms: Ecological considerations and recommendations. *Ecology* 70: 298–351.

Tinbergen, N. 1957. The functions of territory. *Bird Study* 4: 14–27.

Tinkle, D. W. 1967. The life and demography of the side-blotched lizard, *Uta stansburiana. Misc. Publ. Mus. Zool.,* Univ. Mich. No. 132: 1–182.

——. 1969. The concept of reproductive effort and its relation to the evolution of life histories of lizards. *Amer. Natur.* 103: 501–516.

Tinkle, D. W., H. M. Wilbur, and S. G. Tilley. 1970. Evolutionary strategies in lizard reproduction. *Evolution* 24: 55–74.

Toft, C. A. 1985. Resource partitioning in amphibians and reptiles. *Copeia* 1985: 1–21.

——. 1986. Communities of species with parasitic life styles. In J. Diamond and T. J. Case (eds.), *Community ecology* (pp. 445–463). Harper & Row, New York.

Tosi, J. A. 1964. Climatic control of terrestrial ecosystems: A report on the Holdridge model. *Econ. Geogr.* 40: 173–181.

Townsend, C. R., and P. Calow (eds.). 1981. *Physiological ecology: An evolutionary approach to resource use.* Blackwell, Oxford.

Tramer, E. J. 1969. Bird species diversity: Components of Shannon's formula. *Ecology* 501: 927–929.

Trewartha, G. T. 1943. *An introduction to weather and climate.* McGraw-Hill, New York.

Trivers, R. L. 1971. The evolution of reciprocal altruism. *Quart. Rev. Biol.* 46: 35–57.

——. 1972. Parental investment and sexual selection. In B. G. Campbell (ed.), *Sexual selection and the descent of man (1871-1971)* (pp. 136–179). Aldine-Atherton, Chicago.

——. Parent-offspring conflict. *Amer. Zool.* 14: 249–264.

——. 1985. *Social evolution.* Benjamin/Cummings, Menlo Park, Calif.

Trivers, R. L., and H. Hare. 1976. Haplodiploidy and the evolution of social insects. *Science* 191: 249–263.

Trivers, R. L., and D. E. Willard. 1973. Natural selection of parental ability to vary the sex ratio of offspring. *Science* 179: 90–92.

Tucker, V. A. 1975. The energetic cost of moving about. *Amer. Sci.* 63: 413–419.

Tullock, G. 1970a. The coal tit as a careful shopper. *Amer. Natur.* 104: 77–80.

——. 1970b. Switching in general predators: A comment. *Bull. Ecol. Soc. Amer.* 51: 21–23.

Turelli, M. 1981. Niche overlap and invasion of competitors in random environments. I. Models without demographic stochasticity. *Theoret. Pop. Biol.* 20: 1–56.

Turner, F. B., G. A. Hoddenbach, P. A. Medica, and R. Lannom. 1970. The demography of the lizard *Uta stansburiana* (Baird and Girard), in southern Nevada. *J. Anim. Ecol.* 39: 505–519.

Turner, F. B., R. I. Jennrich, and J. D. Weintraub. 1969. Home ranges and body size of lizards. *Ecology* 50: 1076–1081.

Turner, M. 1980. How trypanosomes change coats. *Nature* 284: 13–14.

Udvardy, M. D. F. 1959. Notes on the ecological concepts of habitat, biotope, and niche. *Ecology* 40: 725–728.

——. 1969. Dynamic zoogeography, with special reference to land animals. Van Nostrand Reinhold, New York.

Ulanowicz, R. E., and T. Platt (eds.). 1985. *Ecosystem theory for biological oceanography.* Dept. Fisheries and Oceans, Canadian Bureau of Fisheries and Aquatic Science, Ottawa.

Ulfstrand, S. 1977. Foraging niche dynamics and overlap in a guild of passerine birds in a south Swedish coniferous woodland. *Oecologia* 27: 23–45.

Underwood, A. J. 1986. The analysis of competition by field experiments. Chapter 11 (pp. 240–268) in J. Kikkawa and D. J. Anderson (eds.), *Community ecology: Pattern and process.* Blackwell, Oxford.

United Nations. 1968. *Demographic year book.* United Nations, New York.

United States Department of Agriculture. 1941. *Climate and man.* Washington, D.C.

Usher, M. B. 1979. Markovian approaches to ecological succession. *J. Animal Ecol.* 48: 413–426.

Usher, M. B., and M. H. Williamson (eds.). 1974. *Ecological stability.* Halstead Press, New York.

Utida, S. 1957. Population fluctuation, an experimental and theoretical approach. *Cold Spring Harbor Symp. Quant. Biol.* 22: 139–151.

Uyenoyama, M. K. 1979. Evolution of altruism under group selection in large and small populations in fluctuating environments. *Theoret. Pop. Biol.* 15: 58–85.

Van Valen, L. 1965. Morphological variation and width of the ecological niche. *Amer. Natur.* 94: 377–390.

——. 1971. Group selection and the evolution of dispersal. *Evolution* 25: 591–598.

——. 1973. A new evolutionary law. *Evol. Theory* 1: 1–30.

——. 1975. Life, death, and energy of a tree. *Biotropica* 7: 260–269.

Van Valen, L., and P. R. Grant. 1970. Variation and niche width reexamined. *Amer. Natur.* 104: 589–590.

Vandermeer, J. H. 1968. Reproductive value in a population of arbitrary age distribution. *Amer. Natur.* 102: 586–589.

——. 1969. The competitive structure of communities: An experimental approach with protozoa. *Ecology* 50: 362–371.

——. 1970. The community matrix and the number of species in a community. *Amer. Natur.* 104: 73–83.

——. 1972a. On the covariance of the community matrix. *Ecology* 53: 187–189.

——. 1972b. Niche theory. *Ann. Rev. Ecol. Syst.* 3: 107–132.

——. 1973. Generalized models of two species interactions: A graphical analysis. *Ecology* 54: 809–818.

——. 1975. Interspecific competition: A new approach to the classical theory. *Science* 188: 253–255.

——. 1980. Indirect mutualism: Variations on a theme by Stephen Levine. *Amer. Natur.* 116: 441–448.

Vandermeer, J. H., and D. H. Boucher. 1978. Varieties of mutualistic interactions in population models. *J. Theoret. Biol.* 74: 549–558.

Vandermeer, J., B. Hazlett, and B. Rathcke. 1985. Indirect facilitation and mutualism. Chapter 13 (pp. 326–343) in D. H. Boucher (ed.), *The biology of mutualism*. Oxford University Press, New York.

Vaurie, C. 1951. Adaptive differences between two sympatric species of nuthatches (*Sitta*). *Proc. Int. Ornithol. Congr.* 19: 163–166.

Verboom, J., K. Lankester, and J. A. J. Mertz. 1991. Linking local and regional population dynamics in stochastic metapopulation models. *Biol. J. Linn. Soc.* 42: 39–55.

Vernberg, F. J. 1975. *Physiological adaptation to the environment*. Intext Educational Publishers, New York.

Vernberg, F. J., and W. B. Vernberg (eds.). 1974. *Pollution and the physiological ecology of estuarine and coastal water organisms*. Academic Press, New York.

Verner, J. 1964. Evolution of polygamy in the long-billed marsh wren. *Evolution* 18: 252–261.

——. 1965. Breeding biology of the long-billed marsh wren. *Condor* 67: 6–30.

Verner, J., and G. H. Engelsen. 1970. Territories, multiple nest building, and polygyny in the long-billed marsh wren. *Auk* 87: 557–567.

Verner, J., and M. F. Willson. 1966. The influence of habitats on mating systems of North American passerine birds. *Ecology* 47: 143–147.

Vitousek, P. M., and R. L. Sanford. 1986. Nutrient cycling in moist tropical forest. *Ann. Rev. Ecol. Syst.* 17: 137–167.

Vitousek, P. M., P. Ehrlich, and A. Ehrlich. 1986. Human appropriation of the products of photosynthesis. *BioScience* 36: 368–373.

Vitt, L. J., and J. D. Congdon. 1978. Body shape, reproductive effort, and relative clutch mass in lizards: Resolution of a paradox. *Amer. Natur.* 112: 595–608.

Vitt, L. J., J. D. Congdon, and N. Dickson. 1977. Adaptive strategies and energetics of tail autonomy in lizards. *Ecology* 58: 326–337.

Vogel, S. 1970. Convective cooling at low air speeds and the shape of broad leaves. *J. Exp. Bot.* 21: 91–101.

Volterra, V. 1926a. Fluctuations in the abundance of a species considered mathematically. *Nature* 188: 558–560.

——. 1926b. Variazioni e fluttuazioni del numero d'individui in specie animali conviventi. *Mem. Acad. Lincei* 2: 31–113.

——. 1931. Variation and fluctuations of the number of individuals in animal species living together. English translation published in an Appendix (pp. 409–448) to R. N. Chapman (1939), *Animal ecology*. McGraw-Hill, New York.

Wade, M. J. 1976. Group selection among laboratory populations of *Tribolium*. *Proc. Nat. Acad. Sci., U.S.A.* 73: 4604–4607.

——. 1977. An experimental study of group selection. *Evolution* 31: 134–153.

——. 1978. A critical review of the models of group selection. *Quart. Rev. Biol.* 53: 101–114.

Waksman, S. A. 1952. *Soil microbiology.* Wiley, New York.

Wald, G. 1964. The origins of life. *Proc. Nat. Acad. Sci., U.S.A.* 52: 595–611.

Wallace, A. R. 1876. *The geographical distribution of animals* (2 volumes). Harper, New York. (Reprinted in 1962 by Hafner, New York.)

Wallace, B. 1973. Misinformation, fitness, and kin selection. *Amer. Natur.* 107: 1–7.

Walter, H. 1939. Grassland, savanne und busch der arideren teile Afrikas. *Jahrbucher für Wissenschaftliche Botanik* 87: 750–860.

Wangersky, P. J., and W. J. Cunningham. 1956. On time lags in equations of growth. *Proc. Nat. Acad. Sci., U.S.A.* 42: 699–702.

Warburg, M. 1965. The evolutionary significance of the ecological niche. *Oikos* 16: 205–213.

Warner, R. E. 1968. The role of introduced diseases in the extinction of endemic Hawaiian avifauna. *Condor* 70: 101–120.

Waterman, T. H. 1968. Systems theory and biology—View of a biologist. In M. D. Mesarovic (ed.), *Systems theory and biology.* Proc. Third Syst. Symp. Case Inst. Tech. Springer-Verlag, New York.

Watt, A. S. 1947. Pattern and process in the plant community. *J. Ecol.* 35: 1–22.

Watt, K. E. F. 1964. Comments on fluctuations of animal populations and measures of community stability. *Canad. Entomol.* 96: 1434–1442.

——. 1965. Community stability and the strategy of biological control. *Canad. Entomol.* 97: 887–895.

——. (ed.). 1966. *Systems analysis in ecology.* Academic Press, New York.

——. 1968. *Ecology, and resource management.* McGraw-Hill, New York.

——. 1973. *Principles of environmental science.* McGraw-Hill, New York.

Watts, D. 1971. *Principles of biogeography.* McGraw-Hill, New York.

Weatherley, A. H. 1963. Notions of niche and competition among animals, with special reference to freshwater fish. *Nature* 197: 14–17.

Weaver, J., and F. E. Clements. 1929. *Plant ecology.* McGraw-Hill, New York.

Weaver, J., and F. E. Clements. 1938. *Plant ecology* (2nd ed.). McGraw-Hill, New York.

Weedon, J. S., and J. B. Falls. 1959. Differential responses of male ovenbirds to recorded songs of neighboring and more distant individuals. *Auk* 76: 343–351.

Wegener, A. 1924. *The origin of continents and oceans* (English translation). W. A. Skerl, London.

Weinberg, G. M. 1975. *An introduction to general systems thinking.* Wiley, New York.

Wiens, J. A. 1989. Spatial scaling in ecology. *Functional Ecology* 3: 385–397.

Welch, P. S. 1952. *Limnology* (2nd ed.). McGraw-Hill, New York.

Wellington, W. G. 1957. Individual differences as a factor in population dynamics: The development of a problem. *Canad. J. Zool.* 35: 293–323.

——. 1960. Qualitative changes in natural populations during changes in abundance. *Canad. J. Zool.* 38: 289–314.

Werner, E. E. 1977. Species packing and niche complementarity in three sunfishes. *Amer. Natur.* 111: 553–578.

——. 1986. Amphibian metamorphosis: Growth rate, predation risk, and optimal size at transformation. *Amer. Natur.* 128: 319–341.

Werner, E. E., and D. J. Hall. 1974. Optimal foraging and size selection of prey by the bluegill sunfish. *Ecology* 55: 1042–1052.

——. 1976. Niche shifts in sunfishes: Experimental evidence and significance. *Science* 191: 404–406.

Werren, J. H., M. R. Gross, and R. Shine. 1980. Paternity and the evolution of male parental care. *J. Theoret. Biol.* 82: 619–631.

West-Eberhard, M. J. 1986. Alternative adaptations, speciation, and phylogeny. *Proc. Nat. Acad. Sci.* 83: 1388–1392.

West-Eberhard, M. J. 1989. Phenotypic plasticity and the origins of diversity. *Ann. Rev. Ecol. Syst.* 20: 249–278.

Wetzel, R. G. 1975. *Limnology.* Saunders, Philadelphia.

Weyl, P. K. 1970. *Oceanography: An introduction to the marine environment.* Wiley, New York.

Whiteside, M. C., and R. B. Hainsworth. 1967. Species diversity in chydorid (*Cladocera*) communities. *Ecology* 48: 664–667.

Whitmore, T. C., and J. A. Sayer. 1992. *Tropical deforestation and species extinction.* Chapman and Hall, London.

Whittaker, R. H. 1953. A consideration of climax theory: The climax as a population and pattern. *Ecol. Monogr.* 23: 41–78.

——. 1962. Classification of natural communities. *Bot. Rev.* 28: 1–239.

——. 1965. Dominance and diversity in land plant communities. *Science* 147: 250–260.

——. 1967. Gradient analysis of vegetation. *Biol. Rev.* 42: 207–264.

——. 1969. Evolution of diversity in plant communities. *Brookhaven Symp. Biol.* 22: 178–196.

——. 1970. *Communities and ecosystems.* Macmillan, New York.

——. 1972. Evolution and measurement of species diversity. *Taxon* 21: 213–251.

——. 1975. *Communities and ecosystems.* Macmillan, New York.

Whittaker, R. H., and P. P. Feeny. 1971. Allelochemics: Chemical interactions between species. *Science* 171: 757–770.

Whittaker, R. H., and S. A. Levin (eds.). 1975. *Niche: Theory and application.* Dowden, Hutchinson & Ross, New York.

Whittaker, R. H., and G. M. Woodwell. 1971. Evolution of natural communities. In J. A. Wiens (ed.), *Ecosystem structure and function* (pp. 137–159). Proc. 3lst Ann. Biol. Coll., Oregon State Univ. Press, Corvallis.

Whittaker, R. J., M. B. Bush, and K. Richards. 1989. Plant recolonization and vegetation succession on the Krakatau Islands, Indonesia. *Ecol. Monogr.* 59: 59–123.

Whittaker, R. H., S. A. Levin, and R. B. Root. 1973. Niche, habitat, and ecotope. *Amer. Natur.* 107: 321–338.

Whittaker, R. H., R. B. Walker, and A. R. Kruckeberg. 1954. The ecology of serpentine soils. *Ecology* 35: 258–288.

Whittingham, L. A., P. D. Taylor, and R. J. Robertson. 1992. Confidence of paternity and male parental care. *Amer. Natur.* 139: 1115–1125.

Whittow, G. C. (ed.). 1970. *Comparative physiology of thermoregulation.* Academic Press, New York.

Wiegert, R. C. 1968. Thermodynamic considerations in animal nutrition. *Amer. Zool.* 8: 71–81.

Wiens, J. A. 1966. Group selection and Wynne-Edward's hypothesis. *Amer. Sci.* 54: 273–287.

———. 1977. On competition and variable environments. *Amer. Sci.* 65: 590–597.

———. 1989. The ecology of bird communities. 1. Foundations and patterns. Cambridge Univ. Press, Cambridge, England.

———. 1992. Ecology 2000: An essay on future directions in ecology. *Bull. Ecol. Soc. Amer.* 73: 165–170.

Wiens, J. A., and B. T. Milne. 1989. Scaling of 'landscapes' in landscape ecology, or, landscape ecology from a beetle's perspective. *Landscape Ecol.* 3: 87–96.

Wieser, W. (ed.). 1973. *Effects of temperature on ectothermic organisms.* Springer-Verlag, Berlin.

Wilbur, H. M. 1972. Competition, predation, and the structure of the *Ambystoma-Rana sylvatica* community. *Ecology* 53: 3–21.

———. 1977. Propagule size, number, and dispersion pattern in *Ambystoma* and *Asclepias.* *Amer. Natur.* 111: 43–68.

Wilbur, H. M., D. W. Tinkle, and J. P. Collins. 1974. Environmental certainty, trophic level and resource availability in life history evolution. *Amer. Natur.* 108: 805–817.

Wiley, E. O. 1981. *Phylogenetics.* Wiley, New York.

Wiley, R. H. 1974. Evolution of social organization and life-history patterns among grouse. *Quart. Rev. Biol.* 49: 201–227.

Williams, C. B. 1944. Some applications of the logarithmic series and the index of diversity to ecological problems. *J. Ecol.* 32: 1–44.

———. 1953. The relative abundance of different species in a wild animal population. *J. Animal Ecol.* 22: 14–31.

———. 1964. *Patterns in the balance of nature.* Academic Press, New York.

Williams, G. C. 1957. Pleiotropy, natural selection, and the evolution of senescence. *Evolution* 11: 398–411.

———. 1966a. *Adaptation and natural selection.* Princeton Univ. Press, Princeton, NJ.

———. 1966b. Natural selection, the costs of reproduction, and a refinement of Lack's principle. *Amer. Natur.* 100: 687–690.

———. 1971. *Group selection.* Aldine-Atherton, Chicago.

———. 1975. *Sex and evolution.* Princeton Univ. Press, Princeton, NJ.

Williamson, M. H. 1967. Introducing students to the concepts of population dynamics. In J. M. Lambert (ed.), *The teaching of ecology* (pp. 169–176). Symp. Brit. Ecol. Soc. No. 7, London.

———. 1971. *The analysis of biological populations.* Arnold, London.

———. 1981. *Island populations.* Oxford Univ. Press, Oxford.

Williamson, P. 1971. Feeding ecology of the red-eyed vireo (*Vireo olivaceous*) and associated foliage gleaning birds. *Ecol. Monogr.* 41: 129–152.

Willson, M. F. 1969. Avian niche size and morphological variation. *Amer. Natur.* 103: 531–542.

——. 1971. Life history consequences of death rates. *The Biologist* 53: 49–56.

——. 1972a. Evolutionary ecology of plants: A review. I. Introduction and energy budgets. *The Biologist* 54: 140–147.

——. 1972b. Evolutionary ecology of plants: A review. II. Ecological life histories. *The Biologist* 54: 148–162.

——. 1973a. Evolutionary ecology of plants: A review. III. Ecological genetics and life history. *The Biologist* 55: 1–12.

——. 1973b. Evolutionary ecology of plants: A review. IV. Niches and competition. *The Biologist* 55: 74–82.

——. 1973c. Evolutionary ecology of plants: A review. V. Plant/animal interactions. *The Biologist* 55: 89–105.

——. 1990. Sexual selection in plants and animals. *Trends Ecol. Evol.* 5: 210–214.

Willson, M. F., and N. Burley. 1983. Mate choice in plants. Princeton Univ. Press, Princeton, NJ.

Willson, M. F., and E. R. Pianka. 1963. Sexual selection, sex ratio, and mating system. *Amer. Natur.* 97: 405–407.

Wilson, D. S. 1975. A theory of group selection. *Proc. Nat. Acad. Sci. U.S.A.* 72: 143–146.

——. 1980. *The natural selection of populations and communities*. Benjamin/Cummings, Menlo Park, Calif.

——. 1983. The group selection controversy: History and current status. *Ann. Rev. Ecol. Syst.* 14: 159–187.

——. 1986. Adaptive indirect effects. Chapter 26 (pp. 437–444) in J. Diamond and T. J. Case (eds.), *Community ecology*. Harper & Row, New York.

Wilson, D. S., and P. A. Keddy. 1986a. Measuring diffuse competition along an environmental gradient: Results from a shoreline plant community. *Amer. Natur.* 127: 862–869.

——. 1986b. Species competitive ability and position along a natural environmental gradient. *Ecology* 67: 1236–1242.

Wilson, E. O. 1961. The nature of the taxon cycle in the Melanesian ant fauna. *Amer. Natur.* 95: 169–193.

——. 1969. The species equilibrium. *Brookhaven Symp. Biol.* 22: 38–47.

——. 1971. *The insect societies*. Belknap Press, Cambridge, Mass.

——. 1973. Group selection and its significance for ecology. *BioScience* 23: 631–638.

——. 1975. *Sociobiology: The new synthesis*. Harvard Univ. Press, Cambridge, Mass.

——. 1976. The central problems of sociobiology. In R. M. May (ed.), *Theoretical ecology: Principles and applications* (pp. 205–217). Blackwell, Oxford.

——. 1985. The biological diversity crisis. *BioScience* 35: 700–706.

——. 1987. Causes of ecological success: The case of the ants. *J. Anim. Ecol.* 56: 1–9.

—— (ed.). 1988. *Biodiversity*. National Academy Press, Washington, D. C.

Wilson, E. O., and W. H. Bossert. 1971. *A primer of population biology.* Sinauer, Stamford, Conn.

Wilson, E. O., and E. O. Willis. 1975. Applied biogeography. Chapter 18 (pp. 522–534) in M. L. Cody and J. M. Diamond (eds.), *Ecology and evolution of communities.* Harvard Univ. Press, Cambridge, Mass.

Wilson, J. T. 1971. Continental drift. In *Topics in animal behavior, ecology, and evolution* (pp. 88–92). Harper & Row, New York.

——(ed.). 1973. *Continents adrift. A collection of articles from Scientific American.* Freeman, San Francisco.

Winemiller. K. O. 1989a. Must connectance decrease with species richness? *Amer. Natur.* 134: 960–968.

——. 1989b. Patterns of variation in life history among South American fishes in seasonal environments. *Oecologia* 81: 225–241.

——. 1990. Spatial and temporal variation in tropical fish trophic networks. *Ecol. Monogr.* 60: 331–367.

——. 1991. Ecomorphological diversification in lowland freshwater fish assemblages from five biotic regions. *Ecol. Monogr.* 61: 343–365.

——. 1992. Life-history strategies and the effectiveness of sexual selection. *Oikos* 63: 318–327.

Winemiller, K. O., and E. R. Pianka. 1990. Organization in natural assemblages of desert lizards and tropical fishes. *Ecol. Monogr.* 60: 27–55.

Wiseman, J. D. H. 1966. Evidence for recent climatic changes in cores from the ocean bed. In J. S. Sawyer (ed.), *World climates from 8,000 to 0 B.C.* Imperial College, London, 1966. Royal Meteorological Society, London.

Wittenberger, J. F. 1976. The ecological factors selecting for polygyny in altricial birds. *Amer. Natur.* 110: 779–799.

Wittow, G. C. (ed.). 1970. *Comparative physiology of thermoregulation.* Academic Press, New York.

Wolf, L. L., and F. R. Hainsworth. 1971. Time and energy budgets of territorial hummingbirds. *Ecology* 52: 980–988.

Wolf, L. L., F. R. Hainsworth, and F. G. Stiles. 1972. Energetics of foraging: Rate and efficiency and nectar extraction by hummingbirds. *Science* 176: 1351–1352.

Wolin, C. L. 1985. The population dynamics of mutualistic systems. Chapter 10 (pp. 248–269) in D. H. Boucher (ed.), *The biology of mutualism.* Oxford Univ. Press, New York.

Wolin, C. L., and L. R. Lawlor. 1984. Models of facultative mutualism: Density effects. *Amer. Natur.* 124: 843–862.

Woodwell, G. M., and H. Smith (eds.). 1969. *Diversity and stability in ecological systems.* Brookhaven Symp. Biol. No. 22. Upton, N.Y.

Woodwell, G. M., and R. H. Whittaker. 1968. Primary production in terrestrial communities. *Amer. Zool.* 98: 19–30.

Wootton, R. J. 1979. Energy costs of egg production and environmental determinants of fecundity in Teleost fishes. *Symp. Zool. Soc. London.* 44: 133–159.

Wright, H. E., and D. Frey (eds.). 1965. *The quaternary of the United States.* Princeton Univ. Press, Princeton, NJ.

Wright, S. 1931. Evolution in Mendelian populations. *Genetics* 16: 97–159.

———. 1968. *Evolution and the genetics of populations.* Vol. I. Univ. Chicago Press, Chicago.

———. 1969. *Evolution and the genetics of populations.* Vol. II. Univ. Chicago Press, Chicago.

———. 1977. *Evolution and the genetics of populations.* Vol. III. Univ. Chicago Press, Chicago.

———. 1978. *Evolution and the genetics of populations.* Vol. IV. Univ. Chicago Press, Chicago.

Wynne-Edwards, V. C. 1955. Low reproductive rates in birds, especially seabirds. *Acta XI Int. Orn. Congr., Basel* 1954: 540–547.

———. 1962. *Animal dispersion in relation to social behaviour.* Oliver and Boyd, Edinburgh.

———. 1964. Group selection and kin selection. *Nature* 201: 1145–1147.

———. 1965a. Self-regulating systems in populations of animals. *Science* 147: 1543–1548.

———. 1965b. Social organization as a population regulator. *Symp. Zool. Soc. London* 14: 173–178.

Xia, X. 1992. Uncertainty of paternity can select against paternal care. *Amer. Natur.* 139: 1126–1129.

Yodzis, P. 1980. The connectance of real ecosystems. *Nature* 284: 544–545.

———. 1981. The stability of real ecosystems. *Nature* 289: 674–676.

———. 1988. The indeterminacy of ecological interactions as perceived through perturbation experiments. *Ecology* 69: 508–515.

———. 1989. *Introduction to theoretical ecology.* Harper & Row, New York.

Yousef, M. K., S. M. Horvath, and R. W. Bullard (eds.). 1972. *Physiological adaptations: Desert and mountain.* Academic Press, New York.

Zahavi, A. 1975. Mate selection—A selection for a handicap. *J. Theor. Biol.* 53: 205–214.

———. 1977. The cost of honesty (further remarks on the handicap principle). *J. Theor. Biol.* 67: 603–605.

Zeuthen, E. 1953. Oxygen uptake as related to body size in organisms. *Quart. Rev. Biol.* 28: 1–12.

Zuk, M. 1991. Sexual ornaments as animal signals. *Trends Ecol. Evol.* 6: 228–231.

Zweifel, R. G., and C. H. Lowe. 1966. The ecology of a population of *Xantusia vigilis*, the desert night lizard. *Amer. Mus. Novitates* 2247: 1–57.

# Index